38.50
100h

Positron–Electron Pairs in Astrophysics

(Goddard Space Flight Center, 1983)

AIP Conference Proceedings
Series Editor: Hugh C. Wolfe
Number 101

Positron–Electron Pairs in Astrophysics
(Goddard Space Flight Center, 1983)

Edited by
**Michael L. Burns, Alice K. Harding
and Reuven Ramaty**
NASA Goddard Space Flight Center

American Institute of Physics
New York   1983

Copying fees: The code at the bottom of the first page of each article in this volume gives the fee for each copy of the article made beyond the free copying permitted under the 1978 US Copyright Law. (See also the statement following "Copyright" below). This fee can be paid to the American Institute of Physics through the Copyright Clearance Center, Inc., Box 765, Schenectady, N.Y. 12301.

Copyright © 1983 American Institute of Physics

Individual readers of this volume and non-profit libraries, acting for them, are permitted to make fair use of the material in it, such as copying an article for use in teaching or research. Permission is granted to quote from this volume in scientific work with the customary acknowledgment of the source. To reprint a figure, table or other excerpt requires the consent of one of the original authors and notification to AIP. Republication or systematic or multiple reproduction of any material in this volume is permitted only under license from AIP. Address inquiries to Series Editor, AIP Conference Proceedings, AIP, 335 E. 45th St., New York, N. Y. 10017

L.C. Catalog Card No. 83-71926
ISBN 0-88318-200-9
DOE CONF- 830136

PREFACE

A workshop on Positron-Electron Pairs in Astrophysics was held at the Goddard Space Flight Center in January 1983. This workshop brought together observers and theorists actively engaged in the study of astrophysical sites, as well as the physical processes therein, where positron-electron pairs have a profound influence on both the overall dynamics of the source region and on the properties of the emitted radiation. This was the first meeting of its kind to be devoted exclusively to positron-electron pairs in astrophysical sources. We hope that the present volume, which constitutes the workshop proceedings, will be a valuable reference on the subject.

The most obvious signature of positron-electron pairs is their annihilation radiation, generally observed as a line at 0.511 MeV. This line has been seen from the Galactic Center, where observations performed with detectors flown on balloons and on the Third High Energy Astronomical Observatory (HEAO-3) have revealed a very narrow, very intense and time variable line at precisely 0.511 MeV. These observations were summarized at the workshop, together with the theories that were suggested to account for this unusual source at the nucleus of our galaxy. It has been proposed that the energy source for the positrons responsible for the observed radiation is a massive black hole. The positrons are produced either by multibillion degree photon distributions or in nonthermal processes associated with a beam of high energy photons and particles accelerated by induced electric fields. Other proposed possibilities were a recent supernova or a pulsar along the line of sight to the Galactic Center.

Positron-electron annihilation radiation is observed also at energies other than 0.511 MeV. The best known example is the emission line at about 0.4 MeV observed from gamma-ray bursts by detectors on the Venera and ISEE-3 spacecraft. The shift from energies greater than 0.5 MeV is generally believed to be due to the gravitational redshift at surfaces of neutron stars, and this result provides strong evidence that gamma-ray bursts originate from these objects.

Exciting new observations of gamma-ray bursts were the recurrent events from the source direction of the March 5, 1979 burst presented by E. P. Mazets. This last-minute contribution has not been submitted for publication in the present volume. Other interesting gamma-ray burst observations concern the optical identification of bursts and high time resolution measurements of burst spectra, which indicate variations of the spectral shape during the course of a single event. A confirmation of cyclotron line observations, using data from HEAO-1, was also presented.

Even though annihilation radiation has not yet been directly observed from pulsars, the role of positron-electron pairs in pulsar magnetospheres has long been recognized. In particular, the radio and gamma-ray emission is generally believed to be closely associated with an electromagnetic cascade. These cascades are thought to produce the large pair densities in the magnetosphere necessary, in most models, for the observed coherent radio emission.

Among new observational results on pulsars presented at the workshop was the recent discovery, at Arecibo, of a millisecond pulsar. This object, because of its very low magnetic field, may be able to test our current ideas about the relation of positron-electron pairs to pulsar radio emission.

Another class of astrophysical sites where positron-electron pairs may play an important role are active galaxies. Even though the annihilation line has not yet been seen from such objects, the large observed photon luminosities above pair production threshold and the compact sizes of these objects lead to strong expectations for positive detections of annihilation radiation from active galaxies. Some proposed sites of high energy radiation and subsequent positron-electron production in active galaxies are electromagnetic cascades in relativistic jets, accretion shocks around black holes and hot accretion disks.

The large number of papers on physical processes involving positron-electron pairs indicates that this is currently a very active field of research. It also suggests that the understanding of the role of positron-electron pairs in astrophysical sources requires a deeper study of the basic processes. Results were presented on equilibria in relativistic pair plasmas, both with and without magnetic fields, pair production and annihilation in superstrong magnetic fields, and stimulated annihilation processes.

Two papers on gamma-ray lines from solar flares were also presented. These lines were observed by an instrument that is currently flying on the Solar Maximum Mission. Although positrons do not play an important dynamical role in flares, the Sun, because of its proximity, is the only site from which a full spectrum of gamma-ray lines has so far been seen. As such, solar flares can be used to test mechanisms of gamma-ray production and particle acceleration, both of which are important for the compact astrophysical sites that constituted the main topic of the workshop.

In summary, we believe that an important new astrophysical research topic is opening up at the present time, one that involves relativistic and highly magnetized plasmas with large concentrations of positron-electron pairs. As attested by the papers presented in this volume, a growing body of observations provides evidence for the existence of such plasmas and a considerable amount of theoretical work reveals the exciting new physics that govern them. We hope that this volume will serve as a useful summary of these observations and theories.

The success of the workshop depended on many individuals. In addition to ourselves, the scientific organizing committee consisted of Jonathan Arons, University of California at Berkeley, Marvin Leventhal, Bell Laboratories, Alan P. Lightman, Harvard-Smithsonian Center for Astrophysics and Richard E. Lingenfelter, University of California at San Diego. W.A. Hilley provided outstanding secretarial help prior to and during the workshop and M.E. Schronce provided valuable assistance in the preparation of this volume. We especially thank J.M. McKinley for his help in editing the papers. We are also indebted to the various support elements of the Goddard Space Flight Center for providing transportation, projection equipment and meeting rooms. Financial support for the publication of this volume was provided by NASA.

Michael L. Burns
Alice K. Harding
Reuven Ramaty

Laboratory for High Energy Astrophysics
Goddard Space Flight Center
May 1983

TABLE OF CONTENTS

PREFACE
    M. L. Burns, A. K. Harding and R. Ramaty......................v

WELCOMING REMARKS
    G. F. Pieper..1

I.   GAMMA-RAY LINES FROM SOLAR FLARES

    SOLAR γ-RAY LINES
        D. J. Forrest...3

    POSITRON ANNIHILATION RADIATION FROM SOLAR FLARES
        G. H. Share, E. L. Chupp, D. J. Forrest and
        E. Rieger..15

II. GAMMA-RAY BURSTS

    GAMMA-RAY BURSTS
        K. Hurley..21

    ENERGY SPECTRA OF THE COSMIC GAMMA RAY BURSTS
        E. P. Mazets, S. V. Golenetskii, Yu. A. Guryan,
        R. L. Aptekar, V. N. Ilyinskii and V. N. Panov...........36

    FINE TIME RESOLUTION SPECTRAL ANALYSIS OF THE 1978
    NOVEMBER 4 AND 19 GAMMA RAY BURSTS
        C. Barat...54

    UPPER LIMITS ON NARROW ANNIHILATION LINES IN GAMMA RAY
    BURSTS
        P. L. Nolan, G. H. Share, D. J. Forrest,
        E. L. Chupp, S. Matz and E. Rieger.......................59

    CANDIDATES FOR A GAMMA RAY BURSTER OPTICAL COUNTERPART
        B. E. Schaefer, P. Seitzer and H. V. Bradt...............64

    THEORIES OF GAMMA RAY BURSTS
        J. I. Katz...65

    EMISSION MODEL OF GAMMA-RAY BURSTS
        E. P. Liang..76

    A FIREBALL MODEL FOR THE MARCH 25, 1978 GAMMA RAY BURST
        G. J. Hueter and R. E. Lingenfelter......................89

    POSITRONS FROM GAMMA BURSTS
        S. A. Colgate and A. G. Petschek.........................94

III. PULSARS

    RADIO PULSARS: INTENSITY, POLARIZATION AND ROTATION
    FLUCTUATIONS
        J. M. Cordes... 98

    THE EFFECT OF NULLS ON THE DRIFTING SUBPULSES IN
    PSR 0809+74
        A. V. Filippenko, A. C. S. Readhead and M. S. Ewing...... 113

    DISCOVERY OF A MILLISECOND PULSAR
        S. Kulkarni... 118

    HIGH ENERGY OBSERVATIONS OF PULSARS
        G. Kanbach.. 129

    PROPERTIES OF THE CRAB PULSAR INFERRED FROM THE PHASE-
    AVERAGED SPECTRUM
        F. K. Knight.. 141

    COS-B UPPER LIMITS ON GAMMA RAY EMISSION FROM RADIO
    PULSARS
        R. Buccheri... 147

    THE QUEST FOR ELUSIVE GEMINGA: A UNIQUE OBJECT
    PROPOSED AS THE COUNTERPART OF 2CG 195+04
        P. A. Caraveo, G. F. Bignami and R. C. Lamb.............. 152

    ENERGY SPECTRA OF VERY HIGH ENERGY γ-RAYS FROM THE
    CRAB AND VELA PULSARS
        S. K. Gupta, P. V. Ramana Murthy, B. V. Sreekantan,
        S. C. Tonwar and P. R. Viswanath......................... 157

    ELECTRON POSITRON PAIRS IN RADIO PULSARS
        J. Arons.. 163

    PAIR PRODUCTION NEAR THRESHOLD IN PULSAR MAGNETIC
    FIELDS
        A. K. Harding and J. K. Daugherty........................ 194

    POLARIZATION MODE COUPLING IN RADIO PULSARS
        D. R. Stinebring and J. M. Cordes........................ 199

    PULSAR GAMMA-RAY LINES FROM QUARKONIUM SYSTEMS
        G. Kanbach and R. Schlickeiser........................... 204

IV. THE GALACTIC CENTER

    OBSERVATIONS OF ANNIHILATION RADIATION FROM THE
    GALACTIC CENTER REGION
        C. J. MacCallum and M. Leventhal......................... 211

GAMMA-RAY SPECTRUM OF THE GALACTIC CENTER REGION
G. R. Riegler, J. C. Ling, W. A. Mahoney,
W. A. Wheaton and A. S. Jacobson......................... 230

EFFECTS OF LINE WIDTH AND SPATIAL EXTENT ON
MEASUREMENTS OF THE 0.51 MeV GALACTIC CENTER LINE
P. P. Dunphy, E. L. Chupp and D. J. Forrest.............. 237

LOW-TEMPERATURE POSITRON ANNIHILATION
R. J. Drachman... 242

GAMMA-GUN IN THE CENTER OF THE GALAXY
N. S. Kardashev, I. D. Novikov, A. G. Polnarev
and B. E. Stern.. 253

THE ORIGIN OF THE GALACTIC CENTER ANNIHILATION
RADIATION
R. E. Lingenfelter and R. Ramaty......................... 267

POSITRONS FROM SUPERNOVA AND THE ORIGIN OF THE GALACTIC
CENTER POSITRON ANNIHILATION RADIATION
S. A. Colgate.. 273

AN ELECTRON-POSITRON JET MODEL FOR THE GALACTIC CENTER
M. L. Burns.. 281

POSITRON PRODUCTION BY A HOT, YOUNG PULSAR
K. Brecher and A. Mastichiadis........................... 287

PHENOMENA IN THE CENTER OF THE GALAXY AS A
CONSEQUENCE OF A RECENT SUPERNOVA OUTBURST
I. Shklovskii.. 291

V. ACTIVE GALAXIES

X-RAYS AND GAMMA-RAYS FROM ACTIVE GALAXIES
J. L. Matteson... 292

ARE MILDLY ACTIVE GALAXIES SOURCES OF $e^{\pm}$
ANNIHILATION RADIATION?
A. P. Marscher, K. Brecher, W. A. Wheaton, J. C. Ling,
W. A. Mahoney and A. S. Jacobson......................... 303

UPPER LIMITS ON THE ANNIHILATION RADIATION LUMINOSITY
OF CENTAURUS A
N. Gehrels, T. L. Cline, W. S. Paciesas,
B. J. Teegarden, J. Tueller, P. Durouchoux and
J. M. Hameury.. 309

ELECTRON/POSITRON/GAMMA RAY BEAMS IN COSMIC RADIO SOURCES
R. V. E. Lovelace and C. B. Ruchti....................... 314

COMPACTNESS AND PAIR PRODUCTION IN ACTIVE GALACTIC NUCLEI
M. Salvati, A. Cavaliere, E. Costa and E. Massaro........332

ELECTRON-POSITRON PROCESSES AND SPECTRAL EVOLUTION IN
BLACK HOLE ACCRETION DISK DYNAMO MODELS FOR AGN
SOURCES OF THE COSMIC X-RAY AND GAMMA RAY BACKGROUNDS
D. Leiter...337

e^+-e^- ANNIHILATION AND THE COSMIC X-RAY BACKGROUND
D. Kazanas and R. A. Shafer.............................343

HIGH ENERGY SPECTRUM OF SPHERICALLY ACCRETING BLACK HOLES
P. Mészáros and J. P. Ostriker..........................348

THE HIGH ENERGY SPECTRUM OF HOT ACCRETION DISKS
J. A. Eilek and M. Kafatos..............................353

HOT ACCRETION DISKS AND γ-RAY COSMIC SOURCES
M. Kafatos and J. A. Eilek..............................354

VI. PHYSICAL PROCESSES IN RELATIVISTIC AND MAGNETIZED PLASMAS

FUNDAMENTAL PROCESSES IN PAIR PLASMAS
A. P. Lightman..359

TEMPORAL EVOLUTION OF ELECTRON-POSITRON PLASMAS
W. Brinkmann..368

PAIR PRODUCTION IN THERMAL PLASMAS: A COMPUTER MODEL
S. Stepney..373

MONTE CARLO CALCULATIONS OF PAIR ANNIHILATION AND ITS
INVERSE
P. D. Noerdlinger.......................................377

THE PARTICLE AND PHOTON SPECTRUM OF AN OPTICALLY THICK
RELATIVISTIC WIND
L. J. Caroff, J. A. Eilek and P. D. Noerdlinger.........382

PAIR PRODUCTION AND ANNIHILATION IN STRONG MAGNETIC
FIELDS
J. K. Daugherty and A. K. Harding.......................387

EQUILIBRIUM PAIR DENSITY IN A RELATIVISTIC PLASMA WITH
MAGNETIC FIELDS
F. Takahara and M. Kusunose.............................400

ANNIHILATION LINES FROM CONFINED PLASMAS
P. W. Guilbert..405

ELECTRON-POSITRON PAIR ANNIHILATION AND CREATION IN
SUPERSTRONG MAGNETIC FIELDS
   G. Wunner, H. Herold and H. Ruder......................411

COMPARISON OF PHOTON-PHOTON AND PHOTON-MAGNETIC FIELD
PAIR PRODUCTION RATES
   M. L. Burns and A. K. Harding..........................416

CONDITIONS FOR STIMULATED ANNIHILATION IN A DEGENERATE
e^--e^+ FLUID AT THE SURFACE OF PULSARS
   C. M. Varma..421

CRITERIA FOR GRASAR ACTION IN ASTROPHYSICAL SOURCES
   J. M. McKinley and R. Ramaty...........................428

PAIR PRODUCTION AND NON-THERMAL RADIO STARS
   W. T. Vestrand...433

RESONANT FREE-FREE EMISSION FROM ELECTRONS IN MAGNETIC
POLAR REGIONS OF ACCRETING NEUTRON STARS
   R. Lieu..438

THE PRODUCTION OF SPINLESS HADRON PAIRS VIA VIRTUAL
PHOTON EXCHANGE IN UNIFORM MAGNETIC FIELDS
   D. White...444

WELCOMING REMARKS

George F. Pieper
Director of Sciences
Goddard Space Flight Center
Greenbelt, MD 20771

It is a pleasure for me to welcome you here today to the Goddard Space Flight Center for what promise to be 3 busy and exciting days devoted to electron positron pairs in astrophysics.

It is particularly appropriate for this workshop to be taking place at this time just after the 50th anniversary of the discovery of the positron and the 10th anniversary of the detection of annihilation radiation from the Sun.

The positron was discovered just over 50 years ago on August 2, 1932, by Carl Anderson, a young post-doctoral research fellow at Cal Tech, using a Wilson Cloud Chamber in a strong magnetic field. The existence of the positron had been predicted by Paul Andre Maurice Dirac in his relativistic solutions to the Schroedinger Equation, although he originally thought that the unfilled holes in the distribution of negative energy electrons would be protons. The existence of the annihilation radiation that occurs when a positron and electron come together was implied in Dirac's theory. It was first explicitly stated by Blackett and Occhialini in a paper published in the proceedings of the Royal Society on February 7, 1933, in which they said,

". . . it is necessary to come to the same remarkable conclusion that has already been drawn by Anderson from similar photographs. This is that some of the tracks must be due to particles with a positive charge but whose mass is much less than that of a proton."

"The existence of positive electrons in these showers raises immediately the question of why they have hitherto eluded observation. It is clear that they can have only a limited life as free particles since they do not appear to be associated with matter under normal conditions. It is conceivable that they can enter into combination with other elementary particles to form stable nuclei and so cease to be free, but it seems more likely that they disappear by reacting with a negative electron to form two more quanta. This latter mechanism is given immediately by Dirac's theory of electrons."

Later on June 9, 1933, in a paper in the Physical Review, Oppenheimer and Plesset wrote,

"This is what we should expect from the pairs, which should lose practically all of their kinetic energy in passing through matter, and in which the anti-electron near the end of its range should combine with an electron with the radiation of two quanta of about a half-million volts."

That of course was all 50 years ago. Ten years ago, almost precisely on the 40th anniversary of Anderson's discovery, Ed Chupp and his colleagues from the University of New Hampshire first saw the solar 511-keV line from the August 4, 1972 solar flare in their OSO-7 gamma-ray detector. Later Jack Trombka saw annihilation radiation from the moon on Apollo flights. Marvin Leventhal observed it from the galactic center from a balloon and Professor E. P. Mazets from Leningrad first saw annihilation radiation from a gamma-ray burst. All of these people are here today, in fact practically everyone who has ever seen positron electron annihilation radiation of astrophysical origin is here. What you are finding out about the Universe with this tool is just as interesting as the original studies which led to the discovery of the positron. You have a large and enthusiastic group concerned with an exciting subject. We're very pleased you are here. I know you will have a good time. Welcome to Goddard.

SOLAR γ-RAY LINES

D. J. Forrest*
University of New Hampshire, Durham, NH 03824

ABSTRACT

The Gamma Ray Spectrometer on the Solar Maximum Mission (SMM) satellite has observed emissions produced by nuclear reactions in over 20 separate solar flares. The observed intensity from different flares of some of these emissions ranges over a factor of 100 and the time scale for their production ranges from 10 s pulses, to complete events lasting over 1000 s. The emissions include narrow and broadened prompt γ-ray lines from numerous isotopes from ^{7}Li to ^{56}Fe and cover the energy range 0.431 MeV (^{7}Be) to 7.12 MeV (^{16}O). The instrument has also observed emissions at energies greater than 10 MeV from the decay of $\pi 0$ mesons, from electron bremsstrahlung, and from the direct observation of $> 10^2$ MeV solar neutrons. The intensity, temporal and spectral properties of these emissions will be reviewed from the point of view that solar flares represent an astrophysical particle acceleration site.

INTRODUCTION

The discovery of cosmic rays in 1908 marks the beginning of the general field of high energy astrophysics[1]. This discovery immediately led to questions concerning the location and the mechanism responsible for the production of these energetic particles. These questions are still unresolved and studies of cosmic rays have shed little light on the process, other than to show that the process must still be going on today.

Recently, the location of a nearby acceleration process was identified with the discovery of solar cosmic rays associated with solar flares[2,3]. Again, studies of solar cosmic rays have not led directly to a full understanding of either the exact location or the mechanism of the acceleration process, beyond reconfirmation of its close association with the multifaceted phenomenon known as a solar flare. Lingenfelter and Ramaty[4] pointed out that another way of investigating any acceleration process was to study the properties of the energetic X-rays, γ-rays, and neutrons which must always be produced as the accelerated particles interact with matter in and near the acceleration region.

The first successful observations of nuclear γ-rays from solar flares were made in August 1972[5]. These observations have been extensively studied and have led to a model with two separate steps of acceleration or heating[6,7]. The first step is relatively impulsive in time and produces energetic electrons up to > 100 keV. The second step follows the first by minutes and produces a

*Presented in behalf of the SMM Gamma Ray Astronomy Team.

nonthermal spectrum of energetic nucleons and relativistic electrons. It is thought that some of the particles from this second process stay in the solar atmosphere to produce the observed nuclear γ-ray emission and some escape from the Sun to be observed near 1 AU as solar cosmic rays.

One of the advantages of this model is that the time-scale of the second acceleration process is consistent with a Fermi-type process, a process which has been considered for the acceleration of cosmic rays in other astrophysical sources. Hence, the theoretical pictures were complementary and it was felt that the acceleration process in solar flares, if not completely understood, at least was on firm theoretical ground [8].

In this review I will summarize some of the observations made with the Gamma Ray Spectrometer (GRS) on the SMM spacecraft, which was launched on 1980 February 14. For the most part, the results presented in this review come from analyses of all of the 70 flares observed at energies > 300 keV during the first 2 years of operation. These results include electron bremsstrahlung X-rays and prompt γ-ray lines from excited nuclei produced in a variety of nuclear reactions. They also include secondary γ-ray lines from the annihilation of positrons, the capture of neutrons, and the first observation of solar neutrons. I will use these results to discuss the temporal evolution of the acceleration process; the range of observed photon yields from this sample of 70 flares; and in one flare I will set limits on the time-constant of the spectral shape evolution of the accelerated particles. Finally, I will discuss the current status of our efforts to determine the relative isotopic abundance in both the accelerated particles and the target region.

TEMPORAL EVOLUTION

During a solar flare observation the (4.1-6.4) MeV energy band in the GRS is dominated by prompt ($< 10^{-12}$ s), broad and narrow nuclear line emissions from carbon, nitrogen, and oxygen (CNO)[10]. The spectral and temporal properties of these nuclear emissions have been shown to be most consistent with ions interacting in a high density target[11]. Because of the rapid rate of particle energy loss in the high density target, it is believed that these emission time-profiles also closely represent the yield of energetic ions from the acceleration process.

Fig. 1 shows the counting rate time-profile in this energy band for 2 flares which represent extremes in temporal profiles among the 70 flares observed by GRS. The flare on 1980 June 21 is one of the "fastest" and the flare on 1981 April 27 is one of the "slowest". As can be seen, the majority of the emission in both flares is made up of semi-symmetric pulses or injections of accelerated particles. A characteristic time for these pulses is ~ 10 s (FWHM) for the fast flare and ~ 200 s (FWHM) for the slower one. A study of the slow 1981 April 27 flare, at higher time resolution, shows that its emission is not made up of a large number of ~ 10 s (FWHM) pulses. In general, we do not observe a mixture of fast and slow pulses in any given flare. We conclude that the bulk of energetic particles produced in a flare are accelerated in multiple separate pulses, each one maintaining a memory of the time-scale of those preceding it.

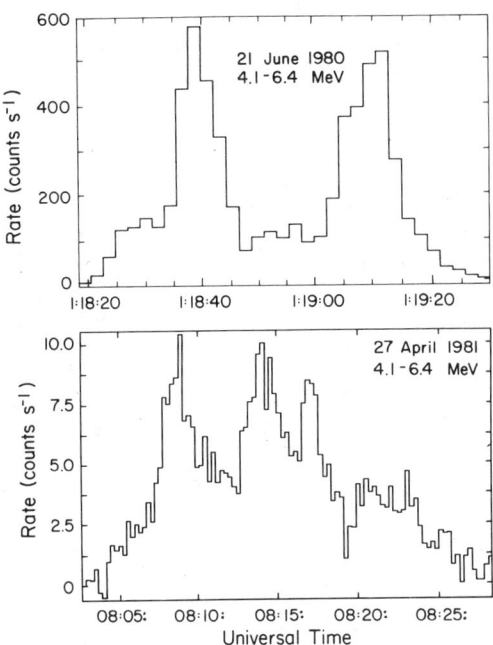

Fig. 1. The observed time-histories in an energy band dominated by γ-ray lines from a fast γ-ray flare (top panel) and a slow γ-ray flare (bottom panel).

In addressing the question of lower and upper limits for the time-scales in solar flares we must recognize possible experimental selection effects. We can be confident that the GRS can detect very fast events due to the large number of impulsive cosmic γ-ray burst events detected in the same time period[12,13]. However, in order for a detected event to be called a solar flare, it must be in coincidence with other independent solar observations, usually observations in Hα. Since we do not know the detection efficiency of small and fast Hα events, it is possible that some of the fast γ-ray events, presently labeled cosmic γ-ray events in the GRS archives, may be solar flares.

Selection effects for long, slow events can be caused by the 90 minute orbit of the SMM spacecraft and also due to the fact that the statisical detection limit increases as the square root of the event time-scale. Hence, longer events will not only be occulted but they must of necessity be the larger, less frequent events. In spite of these uncertainties there is nothing in the GRS data suggesting high energy emission from flares either significantly faster or slower than the two events shown in Fig. 1. Hence, our tentative conclusion is that the time-scales of acceleration in the large majority of solar flares fall within the scales represented in Fig. 1. As a final point I note that the fast 1980 June 21 event produced twice the integrated counts in the (4.1-6.4) MeV band than the slower 1981 April 27 event. Hence, there is not a direct relationship between the acceleration time-scale and the

RANGE OF PHOTON YIELDS

In order to discuss the photon yields from solar flares we must first separate the electron bremsstrahlung from the nuclear line component observed in a flare spectrum. Fig. 2 shows the results of this separation for 65 of the 70 flares identified during the first 2 years of operation. In this figure the bremsstrahlung yield is an integration over energy of a power-law photon spectrum, determined by the GRS spectra between (0.3-1) MeV. The nuclear line yield is the difference between the observed flux and the extension of the above continuum in the (4-8) MeV band[14,15]. This separation of the total emission into electron bremsstrahlung X-rays and prompt nuclear lines is entirely consistent with the nuclear emission based on the 2.223 MeV line emission from the delayed reaction $n(^1H,\gamma)^2H$[16].

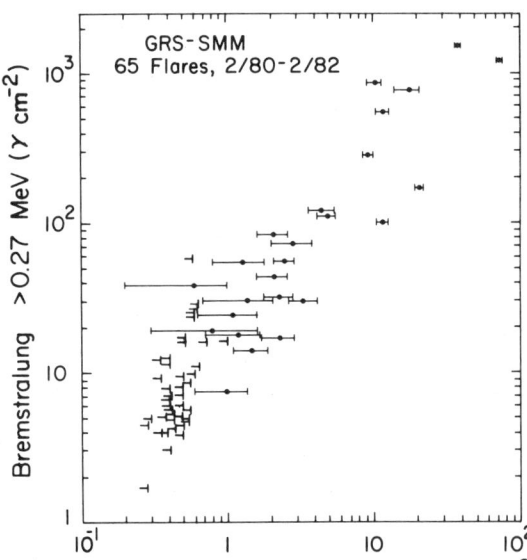

Fig. 2. *A correlation plot of the electron bremsstrahlung and nuclear line emissions from 65 flares.*

Fig. 2 shows that the detection of flares is dominated by the low photon energy bremsstrahlung yield and that GRS has a detector threshold of $F(>0.27$ MeV$) \sim 4$ γcm^{-2}. These data also show that above the GRS nuclear line threshold of $F(4-8$ MeV$) \sim 1$ γ cm^{-2}, these 2 emissions are well correlated over 2 orders of magnitude and that there is a direct proportionality between the number of energetic electrons and ions produced by the acceleration process. Finally, the data in Fig. 2 are not inconsistent with ion acceleration in all of the flares.

Although the photon fluence from both electron bremsstrahlung and nuclear lines is measured directly, the information of most importance is the numbers and spectra of the electrons and nucleons required to produce these observations. A full discussion of the model-dependent interpretation of the photon spectrum is beyond the scope of this review. Let it suffice to point out that the flare of 0312 UT on 1980 June 7, located in Fig. 2 at coordinates

F(>0.27 MeV) = 285 γ cm^{-2} and F(4-8) = 9.4 γ cm^{-2}, has been relatively well studied. Kane et al.[17] have shown that the hard X-ray emission from this event is consistent with thick target bremsstrahlung from a distribution of $\sim 10^{36}$ electrons >100 keV. These electrons have a differential spectral shape of E^{-4} and contain $\sim 3 \times 10^{29}$ ergs. Ramaty[11] interpreted the nuclear line emission from this same flare and found that this emission is consistent with that produced by a Bessel function spectral shape of nucleons with a spectral parameter of $\alpha T = 0.015$ and containing 2×10^{29} ergs. The largest flare observed by GRS to date occured on 1982 June 3. This extremely intense flare had a duration of ~ 60 s and produced ~ 400 γ cm^{-2} in the (4-8) MeV range[18] as well as a large neutron flux extending beyond 10^3 MeV in neutron energy[19]. A simple scaling of the observed fluxes suggests that this flare accelerated ions containing $\sim 10^{31}$ ergs in a time-scale of ~ 60 s. Within this range of yields, we find no evidence for an upper limit cutoff in the total number of particles accelerated, nor is it clear that we have found an upper limit cutoff to the energy that can be given to a single nucleon.

SPECTRAL EVOLUTION

An intensity time-profile, such as shown in Fig. 1, must always represent a convolution of the time-dependence of the number of accelerated particles (i.e. beam current) and the temporal evolution of the spectrum of the particles (i.e. beam energy). To separate these two effects it is most useful to start with the emission requiring the highest energy particle. Fig. 3 shows the high energy solar neutron flux observed within minutes after the photon flux from the 21 June 1980 flare[20]. This data is plotted as a neutron flux versus t_n, the neutron travel time from the Sun to 1 AU. In this figure t_n = 8.33 minute was assumed to occur at 0118:36 UT. Note that this implies that the photons produced simultaneously with the neutrons on the Sun, will be observed at the instrument at 0118:36 UT. With this assumption the neutron energy can be calculated from the arrival time and the observed counting rate converted to a neutron flux. As an aid, the point 8.33 minute is marked with a vertical arrow and the two solid lines show the predicted distributions for neutrons with a differential spectrum at the Sun of $E^{-\alpha}$ for α = 3.5 and 4.0. Fig. 4 shows the photon flux time-history on an expanded time-scale over the energy range 40 keV to >25 MeV for this same flare. As can be seen, the time 0118:36 UT occurs near the center of the first large impulsive peak. The data show that the assumption of t_n = 8.33 minute occuring at 0118:36 UT is, in a sense, a worst-case assumption. Any delay in the neutron production interval will require even higher neutron energies since the travel time must be smaller. As an example, if we assumed that t_n = 8.33 minute occurred at 0119:37 UT, a delay of 1 minute, the time axis point marked 10 minute for the neutron spectrum in Fig. 3 would become 9 minute and the corresponding neutron energy must change from 758 MeV to 1541 MeV.

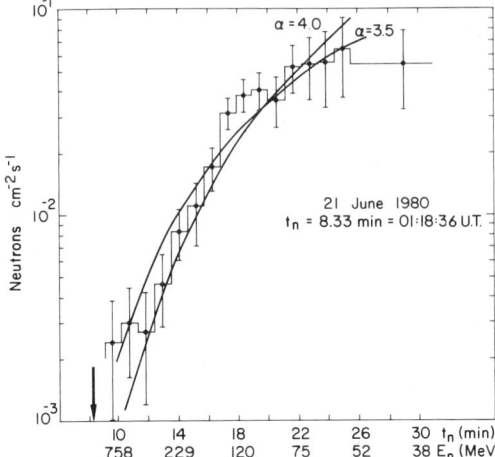

Fig. 3. The high energy neutron flux observed after the 1980 June 21 flare plotted as a function of the travel time, t_n. The vertical arrow indicates 0118:36 UT and the solid lines show the calculated response for differential power-law neutron emission spectral shapes at the Sun.

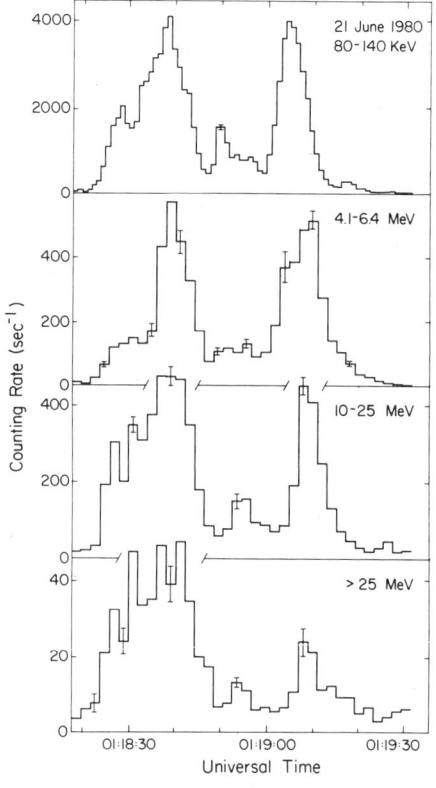

Fig. 4. An expanded time-history for the 1980 June 21 flare showing the photon count rates from 80 keV to 25 MeV.

Combining the γ-ray and neutron histories with the knowledge that the energetic particles needed to produce this spectrum of neutrons must also produce a large flux of nuclear lines in the (4-8) MeV band, we arrive at the conclusion that the emission profiles shown in Fig. 4 represent the profiles of the very energetic nuclear reactions responsible for the high energy neutrons. In this case, the photon flux > 25 MeV shown in Fig. 4 must be a mixture of π^0 decay photons from very energetic nuclear reactions and bremsstrahlung from electrons > 25 MeV. On the other hand, Ramaty et al.[21] have interpreted the overall γ-ray and neutron spectra from this event to show that the π^0 meson flux could be very small. If his interpretation is correct, then the observed flux greater than 25 MeV must be all electron bremsstrahlung. In either case, the observation shows the spectral evolution time-constant for the production of highest energy ions and/or electrons is very small ($\lesssim 5$ s).

COMPOSITION

In a previous section we showed briefly how the observed γ-ray spectrum between (0.3-9) MeV can be separated into its electron bremsstrahlung and nuclear emission components. This separation provides information on the relative number of electrons and ions which are accelerated. In a similar way, the separation of the nuclear emission into individual nuclear γ-ray line components provides information on the relative isotopic abundances. Furthermore, the relative abundance of a given isotope both in the target and in the accelerated particle population can be determined from the spectral properties of the associated nuclear γ-ray lines.

The width of a γ-ray line is controlled by the kinematics of the nuclear reaction which produces the line[22]. Proton and α induced reactions on ambient high Z isotopes result in γ-ray lines which are broadened by 1 to 2%. However, the inverse reaction from energetic high Z isotopes interacting with ambient H and He result in lines which are broadened by ∼ 25%. Since the energy resolution of the SMM GRS ranges from 7% at 835 keV to 4% at 4.4 MeV, the broad and narrow line components can in theory be easily separated over the entire measured energy range. However, experimentally the total broad line spectrum from all of the reactions produces an energy loss spectrum which is not easily separated from the true electron bremsstrahlung spectrum. This is illustrated in Fig. 5. The photon spectrum shown on the bottom of Fig. 5 is the result of a Monte-Carlo calculation for thick-target interaction of a Bessel function spectrum of energetic particles having solar abundances and $\alpha T = 0.02$[28]. The intensity is scaled such that there is 1 photon cm^{-2} in the narrow 4.439 MeV line from ^{12}C. We list in Table I some of the isotopes producing well resolved γ-ray lines along with the abundances assumed in calculating the spectra shown in Fig. 5. The spectrum on the top of the figure is the "observed" energy loss spectrum from this same photon spectrum after being folded through the response function of the GRS. Both of these spectra are binned in 20 keV intervals. As can be seen, the signature of numerous strong narrow lines is clearly observed, while the response of the weaker narrow lines and all of the broad lines appears as a very hard, continuum-like spectrum.

	ENERGY (MeV)	ISOTOPE	ASSUMED ABUNDANCE
1	0.431	^7Be	—
2	0.478	^7Li	—
3	0.847	^{56}Fe	0.075
4	1.238	^{56}Fe	0.075
5	1.369	^{24}Mg	0.075
6	1.634	^{20}Ne	0.22
7	1.779	^{28}Si	0.083
8	2.313	^{14}N	0.22
9	4.439	^{12}C	1
10	6.129	^{16}O	1.67

Table I Prominent Gamma Ray Lines Ordered by Energy (MeV) with Their Assumed Isotopic Abundance
($^{12}C:^1H = 4.15 \times 10^{-4}$)

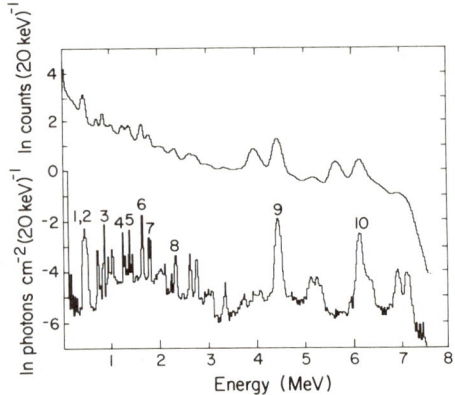

Fig. 5. The bottom curve shows a theoretical solar γ-ray line spectrum produced assuming solar abundances in both target and energetic particles. The top curve shows the GRS count spectrum from this input spectrum. The numbered features are keyed to Table I.

The nuclear line spectrum shown in Fig. 5 along with a power-law photon continuum have been used in an attempt to model the observed energy loss spectrum from the flare of 1981 April 27. Note that this model has only 3 free parameters: The continuum intensity and its power-law index, and one common intensity for the entire nuclear line spectrum. Fig. 6 shows an example of this effort. In this figure the solid line is the observed energy loss spectrum after removing the energy loss spectrum produced by a continuum photon spectrum given by $175 \ (E/1 \ MeV)^{-2.1} \gamma \ cm^{-2} \ MeV^{-1}$. The dotted line is the energy loss spectrum shown in Fig. 5 scaled to fit the ^{20}Ne line at 1.63 MeV. This scaling corresponds to a photon flux of 55 $\gamma \ cm^{-2}$ at 1.63 MeV and 90 $\gamma \ cm^{-2}$ in the line at 4.439 MeV. A comparison of the structure in the measured spectrum with that in the predicted spectrum clearly shows that most of the intense nuclear lines are resolved. However, Fig. 6 also clearly shows a large and significant disagreement in the relative intensities of the lines from ^{20}Ne, ^{24}Mg, ^{28}Si and ^{56}Fe in the (0.8-2) MeV band as compared to those in the (4-8) MeV band from CNO. As an example, the observed line flux at 4.439 MeV is 25 $\gamma \ cm^{-2}$. We have not been able to reduce this disagreement with searches over numerous continuum spectral shapes and intensities. We conclude that assuming the model-spectrum shown in Fig. 5 is correct, the observed spectrum from the 1981 April 27 flare requires a relative enhancement in the target of Ne, Mg, Si and Fe over CNO by a factor of $\lesssim 3$. This result, using a model which includes both narrow and broad lines, is consistent with an earlier result which includes only narrow lines [23]. In this latter case, the required enhancement was a factor > 5.

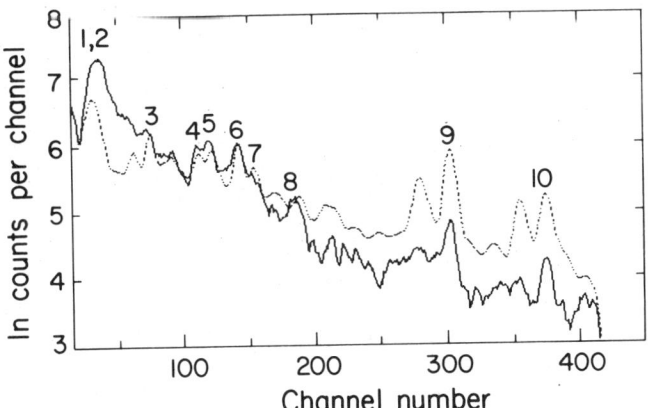

Fig. 6. A comparison of the observed nuclear emission from the 1981 April 27 flare with the calculated spectrum shown in Fig. 5.

The interpretation of the disagreement in intensity for ^{7}Li and ^{7}Be in the (400-550) keV band is complicated by the fact that the measured spectrum contains a delayed line at 511 keV from β^+ annihilation while the model spectrum does not. This is also true near 2.313 MeV(^{14}N) due to the delayed 2.223 MeV line from neutron capture.

Specific results concerning the isotopic abundance in the energetic particles are not as clear. However, the agreement in spectral shape shown in Fig. 6, particularly at lower energies would seem to argue for a broad energy enhancement in the photon flux which is consistent with that produced by broad lines from high Z isotopes in the accelerated particles.

SUMMARY

In this section we summarize some of the properties of the particle acceleration in solar flares. In doing this it is necessary to keep in mind that the GRS detects X-ray emission from electrons > 300 keV and nuclear reaction secondaries from ions >10 MeV. Hence, these observations represent the higher energy portion of the total accelerated particle population. Beyond this, the wealth of new, and in many cases unexpected results from the GRS have not been fully analyzed in every detail. When this is done, and further, when the observations from the more recent flares are analyzed, some of the views presented here may change.

One of the first conclusions we can make is that the majority of energetic particles in a flare are produced in individual acceleration pulses. A solar flare seems to always produce at least several of these pulses. Although it is not clear if these multiple pulses are produced in the same exact physical location, it is observationally required that each of the successive pulses in a given flare maintains a memory of the time-profile of those preceding it.

The emission time-profiles have been clearly measured over the range of $\sim$10 s (FWHM) to $\sim$200 s (FWHM) from flares of total duration $\sim$50 s to $\sim$1400 s respectively. There is evidence that the 10 s pulses may be near the fastest limit of the flare acceleration process since most of the faster events appear to be cosmic γ-ray bursts. The slower limit is not as clear, however, mainly because of instrumental selection effects which make the identification and analysis of low intensity, long duration events difficult. However, at present there is no compelling evidence for flare emissions with time-constants significantly longer than those observed for the 1981 April 27 event.

In spite of the observed variation in the acceleration time-profile, studies of the spectral properties show that the ratio of electron bremsstrahlung to ion nuclear line emission remains constant to about a factor of 2. Hence, this property of the acceleration process does not appear to change with the time-scale of the acceleration process. Finally, we note that there is no evidence that longer and slower flares produce more accelerated particles, nor is there any evidence for a cutoff in either the total number of particles accelerated or in the maximum energy per particle.

One of the most unexpected results from the GRS observations is the rapidity with which the highest energy particles are produced. The separation of the observed flux into its spectral shape and its intensity dependence is most easily done with rapid and intense flares. The flare of 1980 June 21 requires no spectral shape evolution at all, down to a statistical time limits of ~ 1 s. If this interpretation is correct it implies that the final spectral shape was fully developed, up to ion energies of > 500 MeV, in a time-scale of < 5 s.

Finally, we have presented some of our preliminary efforts to determine isotopic abundances in both the target region and in the accelerated particles. Although the analysis is not complete, the results to date seem to require a relative enhancememt by factors of 2-4 in the abundances of Ne, Mg, Si and Fe relative to CNO in the flare target region. This result comes from tests with a model which assumes the acceleration process accelerates all isotopes with equal efficiency. Earlier tests with a model that assumed only energetic protons and α particles required an even larger enhancement of the Ne, Mg, Si and Fe isotopes.

In the INTRODUCTION we outlined a flare acceleration process which seemed to describe the observations made before the launch of SMM. The results from SMM are clearly in conflict with the earlier picture, in particular with respect to the rapidity with which electrons and ions are accelerated to very high energies and any observational requirements for two separate acceleration processes[24]. Theoretical efforts and calculations are currently in progress to determine if refinements in a Fermi acceleration process can explain these new observations[25,26,27]. It seems entirely possible that a new acceleration process will be required, possibly one that has not yet been thought of. In either case, this is an exciting prospect since the acceleration process in solar flares can be relatively easily measured compared to most astrophysical sources.

It is difficult to determine the significance the acceleration process in solar flares has for our understanding of particle acceleration in other astrophysical sources until the process in solar flares is better understood.

ACKNOWLEDGEMENTS

I would like to thank R. Ramaty and R. Murphy for permission to use their calculated γ-ray line spectrum. I would like to acknowledge important discussions with E. Chupp, S. Matz and G. Share. Finally, I thank M. Chupp for editing and manuscript preparation. This work was partially supported at the University of New Hampshire by NASA and the Air Force under contracts NAS 5-23761; at the Naval Research Laboratory by S.70926A; at the Max-Planck-Institut, FRG, by 010K017ZA/WS/WRK 0275:4.

REFERENCES

1. G. W. Clark, Phys. Today 13, No. 11, 26 (1982).
2. S. E. Forbush, Phys. Rev. 70, 771 (1946).
3. A. Ehmert, Z. Naturforsch. 3a, 264 (1948).
4. R. E. Lingenfelter and R. Ramaty, in High Energy Nuclear Reactions in Astrophysics, (W. A. Benjamin N. Y. 1967) p. 99.
5. E. L. Chupp et al., Nature 241, 333 (1973).
6. R. Ramaty et al. Space Science Rev., 18, 341 (1975).
7. R. Ramaty et al., in P. Sturrock (ed.) Solar Flares, (Colorado Associated University Press, Boulder 1980) p. 117.
8. J. C. Brown and D. F. Smith, Rep. Prog. Phys. 43, 125 (1980).
9. D. J. Forrest et al., Solar Phys. 65, 15 (1980).
10. D. J. Forrest et al., in Proceedings of the 17th Intl. Cosmic Ray Conf. (Paris) 10, 5 (1981).
11. R. Ramaty, in P. A. Sturrock (ed.) The Physics of the Sun, (Stanford University Press, Stanford 1983) in press.
12. G. H. Share et al., in Gamma Ray Transients and Related Astrophysical Phenomena, (American Institute of Physics, N. Y. 1982) p. 45.
13. P. L. Nolan, et al., this volume (1983).
14. D. J. Forrest et al., Bull. Am. Astron. Soc. 14, 875 (1982).
15. D. J. Forrest et al., in preparation.
16. G. H. Share et al., Bull. Am. Astron. Soc. 14, 875 (1982).
17. S. R. Kane et al., Astrophys. J. in press.
18. T. A. Prince et al., Bull. Am. Astron. Soc. 14, 875 (1982).
19. E. L. Chupp et al., Bull. Am. Astron. Soc. 14, 875 (1982).
20. E. L. Chupp et al., Astrophys. J. (Letters) 263, L95 (1982).
21. R. Ramaty et al., in Recent Advances in the Understanding of Solar Flares, Solar Phys. in press.
22. R. Ramaty et al., Astrophys. J. Supp. 40, 487 (1979).
23. D. J. Forrest et al., Bull. Am. Astron. Soc. 13, 903 (1981).
24. D. J. Forrest and E. L. Chupp to be submitted to Nature (1983).
25. T. Bai et al., Submitted to Astrophys. J. (1982).
26. M. E. Pesses and R. B. Decker, Bull. Am. Astron. Soc. 14, 608 (1982).
27. M. A. Forman et al., in P. A. Sturrock (ed.) The Physics of the Sun, (Stanford University Press, Stanford 1983) in press.
28. R. Ramaty, Personal communication. (1983).

POSITRON ANNIHILATION RADIATION FROM SOLAR FLARES*

G. H. Share
E.O. Hulburt Center for Space Research
Naval Research Laboratory, Washington, D.C. 20375

E. L. Chupp, D. J. Forrest
University of New Hampshire, Durham, N. H. 03824

E. Rieger
Max Planck Institute for Physics and Astrophysics
Garching bei Munchen, West Germany

ABSTRACT

Positron annihilation radiation has been observed from the 21 June 1980 and 3 June 1982 flares by the gamma-ray spectrometer on the Solar Maximum Mission satellite. The observed 0.511 MeV line fluences from the flares were 14.6 ± 3.3 γ/cm^2 and 103 ± 8 γ/cm^2, respectively. Measurement of the line width allows us to set an upper limit to the temperature in the annihilation region of 3×10^6 °K. The time dependence of the 0.511 MeV line during the 21 June 1980 flare is consistent with Ramaty et al.'s calculations[1] for positrons created in the decay of radioactive nuclei. The time dependence of 0.511 MeV line for the 3 June 1982 flare is more complex and requires more detailed study.

INTRODUCTION

Positrons are primarily produced during solar flares from the decay of radioactive nuclei and π^+ mesons resulting from nuclear interactions of accelerated ions on ambient solar material.[2,3] Radioactive nuclei are the principal source of positrons for accelerated ion spectra which are relatively steep. Following production the positrons can either escape, annihilate in flight, or slow-down prior to annihilating.[4] The ambient temperature and density determine the relative contributions from free annihilation and decay of positronium, as well as the width of the annihilation line.[5] The relative fluxes of the 0.511 MeV line and the continuum emission from decay of positronium are dependent on the ambient density.

The first detection of positron annihilation radiation from an extraterrestrial source was made by the OSO-7 gamma-ray spectrometer during the intense solar flares of 4 and 7 August 1972.[6] Neither of the flares was observed in its entirety and lower limits to the 0.511 MeV fluences were ~35 γ/cm^2 and ~20 γ/cm^2, respectively. The partial observations and limited statistics prevented detailed time profiles of the annihilation radiation to be deduced.

*Supported by NASA contracts S.70926A at NRL and NAS 5-23761 at UNH, and 010K017-ZA/WS/WRK 0275:4 at MPI.

0094-243X/83/1010015-06 $3.00 Copyright 1983 American Institute of Physics

In this paper we present detailed observations of positron annihilation radiation from two flares observed by the Solar Maximum Mission (SMM) gamma-ray spectrometer[7], which was launched in February 1980. During its first 2 1/2 years in operation this instrument has observed over twenty flares for which there is clear evidence for the acceleration and interaction of ions.[8] A summary of some of the important results obtained to date can be found elsewhere in these proceedings.[9]

OBSERVATIONS

The flares of 21 June 1980 and 3 June 1982 were the two most intense observed by SMM in its first 2 1/2 years in operation. Some of their characteristics are summarized in Table I.

Table I Characteristics of the flares

Date	21 June 1980	3 June 1982
Time (UT)	01h18m	11h43m
Solar Coordinates	20°N, 90°W	09°S, 72°E
Importance, brightness	1B	2B
40-140 keV fluence	~8×10^4 γ/cm^2	~3×10^5 γ/cm^2
Peak 40-140 kev intensity	~2.5×10^3 γ/cm^2-s	~8×10^3 γ/cm^2-s

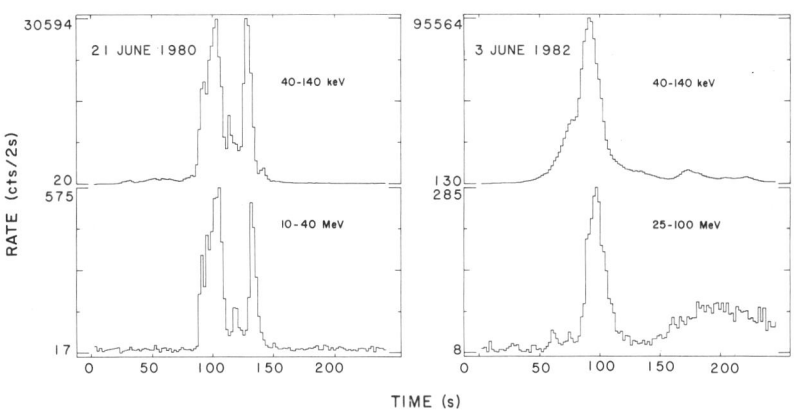

Fig. 1 Time profile of the flares in two energy bands.

These flares were also exceptional because they produced the first detectable fluxes of neutrons at the earth[10] and they were detectable up to ≳50 MeV in gamma rays. The time histories of the impulsive phase of the flares are shown in Figure 1 for the 40-140 keV and high-energy gamma-ray bands. The flares were so intense at the peak of the impulsive phase that significant dead time and gain shifts occurred. This was especially true for the 3 June flare. The impulsive phase of the 21 June flare lasted for about 1 minute and exhibited at least 3 peaks. There was no evidence for significant nuclear interactions following this phase. The primary peak of the 3 June flare had a duration of about 30 sec. A broad peak evident primarily at the higher energies rose just after the main peak and lasted for at least two minutes. There is evidence for the continued production of prompt gamma-rays following this feature and extending for a period of $\gtrsim 10$ minutes. This indicates that nuclear interactions occurred up to several minutes after the impulsive phase of the 3 June flare, in contrast to the 21 June flare.

In order to study the 0.511 MeV line from these flares, it is necessary to subtract contributions to this line which come from the earth's atmosphere and from radioactive nuclei produced in the detector and satellite. These background features vary significantly with time and are dependent on the aspect of the satellite, geomagnetic cutoff rigidity, and interval from the last traversal through the intense radiation of the South Atlantic Anomaly (SAA). This correction can be made by using data from previous and subsequent days when the background conditions are similar. Data from 15 orbits earlier on June 20 and 15 orbits later on June 22 were averaged to estimate the background 0.511 MeV intensity profile during the 21 June flare. The dominant source of background for the 3 June flare came from radioactivity created by passage through an intense portion of the SAA 20 minutes earlier. SMM passed through a similar intense region of the SAA 15 orbits earlier on June 2, but the SAA intensity 15 orbits later on June 4 was significantly weaker. Therefore only data from June 2 were used to estimate the background during the 3 June flare.

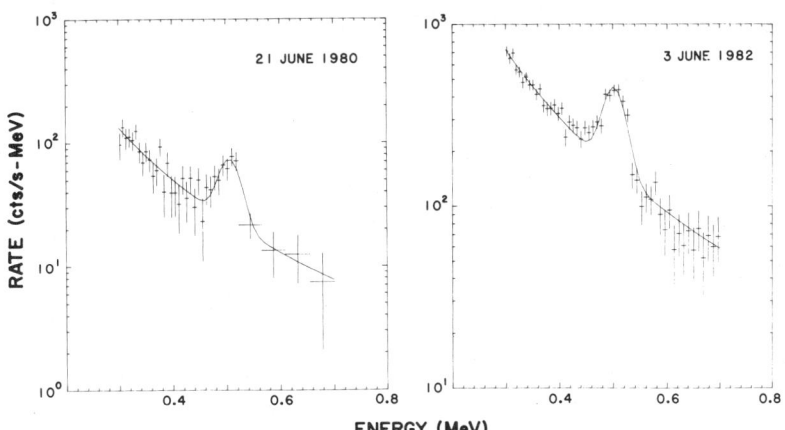

Fig. 2 Fits to the annihilation line following the impulsive phase of the flares.

Least square fits to the 0.511 MeV solar line emission following the impulsive phase of the two flares are shown in Figure 2. Comparison of these fits with those made on the background 0.511 MeV line during the same orbits enables us to establish an upper limit to the intrinsic line width of the solar positron annihilation line. The 95% confidence limit is 20 keV FWHM.

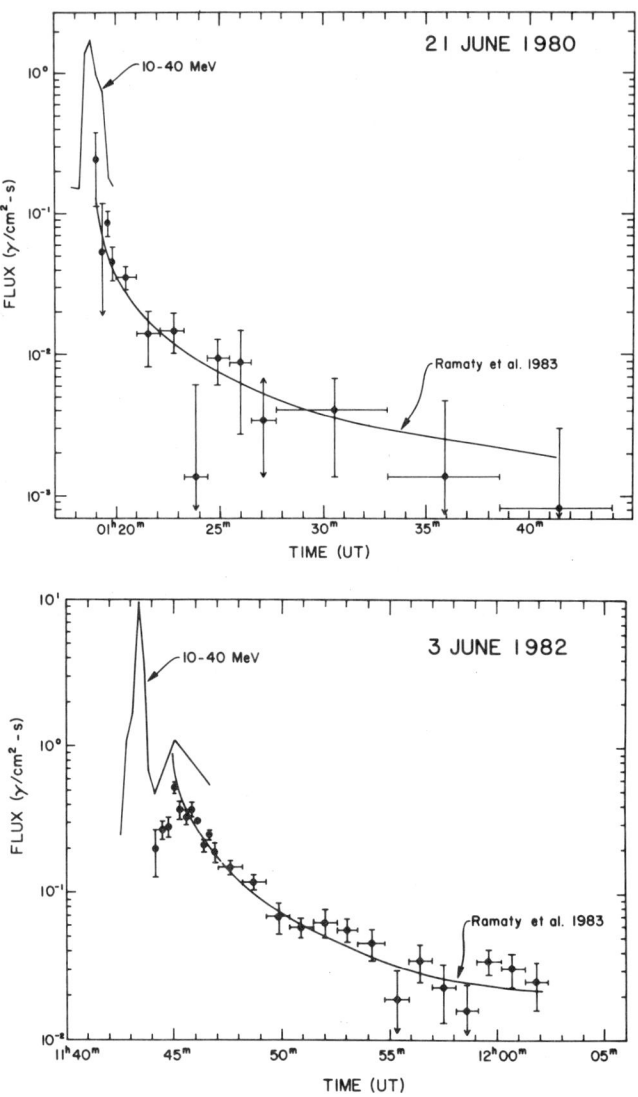

Fig. 3 Histories of the annihilation line for the two flares. (The scales of the 10-40 MeV intensities are relative.)

Plotted in Figure 3 are histories of the solar 0.511 MeV line during the two flares. These were obtained by fitting the background corrected spectra from 430 keV to 570 keV with a power-law continuum and a gaussian-shaped line. We did not attempt to fit the line during the intense portions of the impulsive phase. Shown for comparison are the rates (relative scale) in the 10-40 MeV band plotted at 16.38 s resolution. These rates can be used to estimate the interaction history of the accelerated ions. We have not plotted the rates from the spectrometer in the MeV region because of dead time and gain shift during the intense impulsive phases. Also plotted for comparison is a calculated history of the 0.511 MeV radiation assuming instantaneous acceleration and interaction of ions in a thick target (ambient densities >10^{14} n_h/cm^3).[1] The calculated history was positioned manually to fit the data.

DISCUSSION

The upper limit of 20 keV on the line width of the 0.511 MeV line indicates that the temperature in the annihilation region was less than 3×10^6 °K. The positronium continuum should be detectable for densities <10^{14} n_h/cm^3 at times when the bremsstrahlung continuum is small.[5] This condition occurred late in the 21 June flare and analysis is in progress; however, the positronium continuum may be masked by scattered 0.511 MeV radiation.[1]

We have integrated the 0.511 MeV rates in Figure 3 to obtain the observed fluences in the flares. These fluences are 14.6±3.3 γ/cm^2 for the 21 June 1980 flare and 103±8 γ/cm^2 for the 3 June 1982 flare. The fluence of the 21 June flare may have been reduced significantly by scattering because it was on the limb.

The history of the 0.511 MeV line radiation provides information on the source of positrons and characteristics of the annihilation region. The annihilation radiation of the 21 June flare peaked ≲20 sec from the start of the impulsive phase. The energies of positrons which are created in the decays of radioactive nuclei do not typically exceed a few MeV; therefore the ≲20 s slowing-down time implies that the density in the interaction region is ≳10^{12} n_h/cm^3. Ramaty et al.[1] calculated the 0.511 MeV history assuming that the interactions took place instantaneously at an ambient density >10^{14} n_h/cm^3. They also assumed that the accelerated ions had a composition similar to the solar abundances and followed a Bessel function spectrum similar to that inferred for several flares for which nuclear gamma rays have been detected. The calculated profile fits the data quite well and indicates that the history of the 0.511 MeV line is primarily determined by the half lives of the radioactive nuclei.

The 0.511 MeV line history of the 3 June 1982 flare is more difficult to interpret. Its intensity appears to increase, or is relatively constant, during the first minute after the intense impulsive peak and falls gradually over the next 15 minutes. This fall is reasonably well fit by the decay of radioactive nuclei[1], as can be seen in the figure. However, this fit assumed that the interactions occurred instaneously at 11h44m50s. Having the interactions follow the shape of the high-energy secondary peak will produce a better fit to the early data points, but leaves us with a serious problem. Where are all the 0.511 MeV photons produced during the >10X more intense primary peak? A possible explanation is that the spectrum of the ions accelerated during the primary peak was considerably harder than used in the calculations. This harder spectrum would have produced a large number of positrons from π^+ decays in addition to those from decay of radioactive nuclei. If the density in the interaction region was $\lesssim 5 \times 10^{12}$ n_h/cm^3, then these π^+ positrons would have slowing-down times close to 60 s and could account for the early 0.511 MeV history. Alternatively, the annihilation line could have been absorbed early in the flare. This would have required densities exceeding 10^{18} n_h/cm^3 in the interaction region and a mechanism by which the radiation could escape at later times.

Consistent with the theme of this conference, it is of interest to compare these solar observations of annihilation radiation with observations of gamma-ray bursts and the galactic center emission. This comparison is summarized in Table II below.

Table II

Observations of annihilation radiation from astrophysical sources

	Flares	Galactic cntr[11]	γ-ray bursts[12]
Peak intensity (γ/cm^2-s)	5×10^{-1}	2×10^{-3}	10^0
Emissivity (ergs/s)	2×10^{21}	2×10^{37}	4×10^{36} (200 pc)
Duration (s)	$10^2 - 10^3$	10^7	$10^0 - 10^2$
Line width (keV, FWHM)	<20	<2	~100

REFERENCES

1. R. Ramaty, et al., NASA Tech. Mem. 84969, in Recent Advances in the Understanding of Solar Flares, Solar Physics (1983) in press
2. R. Ramaty, et al., Space Sci. Rev. 18, 341 (1975).
3. R. Ramaty, in The Physics of the Sun, eds., T.E. Holzer, et al. (Stanford Univ. Press, Stanford 1983) in press.
4. H.T. Wang, R. Ramaty, Astrophys. J. 202, 532 (1975).
5. C.J. Crannell, et al. Astrophys. J. 210, 582 (1976).
6. E.L. Chupp, D.J. Forrest, et al., Nature 241, 333 (1973).
7. D.J. Forrest, E.L. Chupp, et al., Solar Phys. 65, 15 (1980).
8. G.H. Share, et al., Bull. of AAS 14, 875 (1982).
9. D.J. Forrest, this volume (1983).
10. E.L. Chupp, et al., Astrophys. J. Lett. 263, L95 (1982).
11. G.R. Riegler, et al., Astrophys. J. Lett. 248, L13 (1981).
12. E.P. Mazets, et al., Nature 290, 378 (1981).

GAMMA RAY BURSTS

K. Hurley
Centre d'Etude Spatiale des Rayonnements
(CNRS-UPS)
B.P. 4346, 31029 Toulouse Cedex, France

ABSTRACT

The time histories, size spectrum, spatial distribution, and repetition rates of gamma ray bursts are reviewed briefly. Evidence for a neutron star origin for gamma ray bursts may be found in many of these aspects of bursters. New results from optical searches are described. Substantial progress has been made recently in the optical identification of the 1978 November 19 burst.

INTRODUCTION

Almost a decade has passed since the discovery of gamma ray bursts was announced in the literature (1). Information on over 200 bursts has now appeared (2,3), and new events are currently being detected at the rate of one every four days. Despite the fact that only one or two tentative identifications of gamma ray bursters have been made, our understanding of the phenomenon has advanced significantly. There remain today only a handful of theories which attempt to explain gamma ray bursts, and most of them involve neutron stars. Conversely, it is generally felt that most gamma ray bursts are generated on or about neutron stars. Theories and observations of gamma ray bursters have been discussed at two recent meetings: The Symposium on Gamma-Ray Astronomy in Perspective of Future Space Experiments (4) and the Workshop on Accreting Neutron Stars (5). The purpose of this paper is to review the observational status of gamma ray bursts, and point out the relations between these observations and theories involving neutron stars. The most well known relation is perhaps that between the observation of redshifted electron-positron annihilation radiation in burst energy spectra, and the mass-to-radius ratio which it implies for the emitting object. Indeed, that is one of the best reasons for including the subject of gamma ray bursts in this conference. However, as the subject of energy spectra is to be covered in another paper, this review will concentrate on three other aspects of bursts: time histories, statistical properties (i.e., the log N-log S relation, the spatial distribution, and the repetition rates), and the searches for counterparts.

TIME HISTORIES

The shortest gamma ray burst observed lasted about 50 ms; the longest continued for over a minute. Inside this large dynamic range of durations, a wide variety of events exists, but several categories can be identified (6,7,8,9): as a minimum, they are very short events ($\simeq$100 ms), 1 s long events, and long complex events. Examples are shown in Figures 1, 2, and 3. In addition to the total duration of bursts, at least two other time scales are of interest: the rise and decay times, and the periods of oscillations.

Thermonuclear models for gamma ray bursts (10,11) appear to be able to accomodate all three categories: detonation models, in which a detonation wave propagates rapidly (9000 km/s) about a neutron star, igniting a helium layer as it goes, would best explain events lasting <1 s. In convective burning models, on the other hand, the ignition is governed by a slowly moving convective burning front; if one assumes that the helium is present in puddles on the surface of a neutron star, the length of the gamma ray burst will be governed by the geometry of the puddles, and can be longer than 10 s. The rise times tend to be quite short (< 1 ms) in the detonation models, but can be much longer in the convective burning case. It should be noted that the only gamma ray burst known to have a rise time shorter than 5 ms is that of 1979 March 5, for which the rise time is about 0.2 ms (12). Comet or asteroid collisions with neutron stars also predict very fast rise time, short duration bursts, similar to March 5 (13).

In the context of accreting neutron star models, it has been pointed out that the 0.2 ms rise time of the March 5 event is consistent with dynamical time scales near the surface of a neutron star (14), while 100-200 ms rise times (which are characteristic of the 1s long bursts) are about what would be expected for the free-fall time from the magnetospheric radius of a neutron star to its surface (15).

The very short events, which have decay time constants of about 100 ms, are interesting for another reason. The damping time constant of neutron star vibrations via gravitational radiation is of this order: this was first pointed out in the context of a model for the 1979 March 5 event (16), but could equally well apply to other short events.

All the models of gamma ray bursts which invoke neutron stars can accomodate the relatively long time constant periodicities, such as 8 s for the 1979 March 5 event (17,18,19) or 4.2 s for the 1977 October 29 event (20). These are sometimes assumed to be due to the rotation of the neutron star, although in the 1979 March 5 case, the age of the N49 supernova remnant, $<10^4$ y, appears to be far too small to be consistent with such slow rotation (18,21). Another explanation for periodicities this long is that they are due to precession of the neutron star (21). This is attractive, since the neutron star may then be assumed to be rotating much faster, with a period on the order of 10's of ms, and the underlying energy source for the gamma ray burst may be the rotational energy (21). This in turn suggests that periods of the order of 10's of ms may be present in gamma ray bursts. There is in fact some evidence for the existence of quasi-periodicity in the March 5 time history (Figure 4) (22). Besides rotation, however, another interpretation of 20 ms oscillations may be torsional vibrations of a neutron star (23), or interactions between the plasma and the magnetic field of the neutron star (10).

THE LOG N-LOG S CURVE, SPATIAL DISTRIBUTION, AND REPETITION RATES

Localizations of various accuracies now exist for 71 gamma ray bursts (2,3,24). The distribution of sources in galactic coordinates is shown in Figure 5. A curious feature of the distribution of events has been noted, namely, that a "clustering" seems to exist at latitudes around $b^{II}=-15°$ (25). This is seen more clearly in Figure 6, which shows the b^{II} distribution of events. These figures are based on 19 events localized by the international network, and 52 KONUS events; however, the same result is obtained if only the KONUS events are considered (25). While inconsistencies have been noted between the KONUS localizations and those of the international network (3), they are not sufficient to explain this effect. It should be noted that if only the distribution in $|b^{II}|$ is considered, these sources appear to be distributed isotropically. No correlation between source intensity and source latitude has been found (26). The clustering effect is not very significant, statistically speaking, even though its appearance is striking; an acceptable χ^2 is obtained for an isotropic fit to the data, and the excess is significant at only about the 2σ level. A possible explanation for the clustering has been proposed (27). It has been shown that the KONUS burst positions are strongly correlated with the earth-Venera spacecraft vector. As this vector was frequently directed towards galactic latitudes around $-12°$, the apparent clustering may be explained as a spatial selection effect.

Considerable progress has been made on the log N-log S curve of gamma ray bursts over the past few years. Both the KONUS results and recent balloon results have been added to the results of previous experiments to arrive at the compilation of Figure 7 (26). The interpretation of these results poses two problems. The first is that the KONUS results do not appear to be compatible with other experiments. This, however, may be due to the different energy thresholds of the experiments involved (28). Early attempts at renormalization of the results used the exponential spectrum approximation. If renormalization is done using the more recently observed spectral shapes, however, a reasonable agreement can be found. The second problem involves the interpretation of these results in terms of source models, which have become considerably more sophisticated over the past two years.

Here it should be noted that Figure 7 is based on only 38 gamma ray bursts (the two Mazets curves were included only for completeness, and were not used to evaluate the models). Of the 38, 32 are contained in the IMP and VELA curves, and thus pertain only to high S, while the remaining 6 at low S are all unconfirmed. This is clearly an unfortunate situation, since models tend to be accepted or rejected on the basis of their behavior at low S. On the other hand, since this situation is unlikely to improve dramatically before the advent of very large area, satellite burst experiments (e.g., the Burst and Transient Source Experiment aboard GRO), the interpretation of the the log N-log S curve is an exercise which should be carried out now, realizing that the current data may be subject to large corrections later. The interpretation of the curve at low S has been treated in detail recently (47).

Two of the first detailed studies to appear considered specific halo and disc source geometries, but used the simplifying assumption of a "standard candle", or monoluminosity burst (29,30). Both concluded that disc distributions with scale heights larger than 300 pc fit the available data best. Two recent discoveries, however, have prompted a re-evaluation of the monoluminosity assumption: first, the observation of repeating bursts from two sources. In one case, the luminosities span the range 750:1 (7). Second, the absence of any source clustering about the galactic plane (neglecting the clustering about $b^{II}=-15°$, which does not seem to have any explanation in terms of a reasonable physical model) and the lack of correlation between intensities and galactic latitudes are difficult to accomodate in disc models. With the inclusion of an intrinsic distribution of burst luminosities, both disc and halo spatial distributions are found to be acceptable (26); in fact, with this assumption, the log N-log S curve is difficult to interpret in terms of spatial distributions, since most reasonable models may be accomodated by altering the luminosity function. Two examples are shown in Figure 7: the number of bursts as a function of luminosity has been taken to be a power law in L, αL^{-1}, and the range of luminosities is given by the parameter $\zeta=L_1/L_2$. The problem of the inconsistency with the observed spatial distribution persists, however, even for the halo model (31): a significant anisotropy between the galactic center and anticenter regions should be observed, but is not.

This model makes specific predictions about source strengths and repetition rates. For example, halo models having an L^{-2} luminosity distribution should have source strengths between 5.4×10^{38} and 9.0×10^{42} ergs, with repetition rates of 5.2×10^{-10} pc^{-3} y^{-1} for isotropically radiating sources, and so on. Note that source luminosities are thus above the Eddington limit, and that in order to observe sources which repeat over time scales of say, 1-10 y, relatively rare objects are required, since their densities must be on the order of 10^{-9} to 10^{-10} pc^{-3}.

It can thus be seen that the repetition rate of burst sources plays an important role in these models. Although two cases of repeating sources have been found (7), these observations appear to be irrelevant to the basic question for several reasons. First, one repeating source is the 1979 March 5 event, which must be considered an exceptional object by any standard. Second, for both sources, the repetition rates are on the order of days to months, and only several repetitions are observed. Finally, the observed repetition concerns fewer than 2% of the total number of bursts. It seems safer to conclude that, neglecting these two exceptional sources, no convincing evidence for source repetition exists in the data of 1978-1980 (see, e.g., ref. 32), nor in the data of 1969-1981, although the earlier localisations tend to be much less accurate (3). Thus the present data are consistent with repetition periods greater than 1-10 y. The results of archival searches, as discussed below, do not alter this conclusion. The current generation of gamma ray burst instruments aboard Venera 13, Venera 14, SMM, PVO, and ISEE-3 may shed more light on this question. Thermonuclear models for bursts can accomodate repetition periods >0.3 y. Ironically, theories involving the accretion of a solid body onto a neutron star (13,32) tend to be modeled after the 1979 March 5 event, yet do not specifically predict repeating bursts on a several day to ∿2 month time scale, and too little is known about the statistics of cometary and asteroid clouds about neutron stars to draw any significant conclusions about longer period repetition rates.

COUNTERPARTS

It was originally felt that once ∿arcmin size error boxes were derived from the localization data of the international network, the discovery of optical counterparts to gamma ray burst sources would follow. However, despite the fact that some 15 observing runs have been devoted to 5 small error boxes over the past 3 years on the 1.5 and 3.6 m ESO telescopes using both a CCD camera and photographic plates, with limiting magnitudes reaching beyond about 23, the "missing link" between gamma ray and optical identification has come from an unexpected source.

Figure 8 illustrates one problem which can occur in searching for optical counterparts to bursters. This figure shows the localization of the 1979 June 13 event, a 0.7 arcmin^2 error box, superposed on a POSS plate of limiting magnitude ∿21 (48). As was the case for the 1979 April 6 event (49), the box contains no objects down to the plate limit. This phenomenon has given rise to the widespread, but erroneous idea that all gamma ray burst error boxes are "empty". Although two error boxes have indeed been found which contain no objects down to the survey plate limits, this is largely beside the point: there are many other error boxes which contain tens of objects (none particularly suspicious), and in any case, when "empty" error boxes are observed down to fainter magnitudes, objects are invariably found (50). The problem is that, based on statistical arguments, perhaps ½ or more of these objects are extragalactic and therefore probably quite unrelated to the gamma ray burster.

A breakthrough may finally have come as a result of the discovery of an optical burst on an archival plate taken in 1928 (51): the error box of the 1978 Nov 19 event was found to contain an intense (perhaps 2nd mag.) transient (perhaps 1 s long) optical source. Two faint, variable sources have been found whose positions are consistent with that of the transient (52). One of these observations might therefore constitute the first detection of a gamma ray burster at optical wavelengths. Although further study of such objects will be complicated by their extreme faintness (about 24th magnitude), the importance of the discovery cannot be overstated: it is significant first, because it indicates that simultaneous optical emission accompanying gamma ray bursts may be fairly easy to detect, despite the negative results of earlier searches (53), and second because it indicates that future deep searches for burster counterparts should probably concentrate on very faint, variable objects.

Some other consequences of this discovery should be noted. The optical to gamma ray luminosity ratio for this burst appears to be in the range $L_\gamma/L_{op} \approx 800$; this is large, and comparable to the X ray to optical luminosity ratio of X ray bursters (54). Perhaps significantly, X ray bursters are thought to be powered by thermonuclear processes. A detailed analysis indicates that the optical emission in the 1978 Nov 19 case probably arose in an accretion disc around a neutron star, and not in the same region where the gamma ray emission was generated (55). On the other hand, the X ray to optical luminosity ratio of the steady state emission is in the range 6-60, which is rather low (56). Finally, it should be noted that rather simple statistical arguments can be used to show that this discovery does not significantly constrain the repetition rate of burst sources.

Further archival searches carried out in conjunction with the Sonneberg observatory have failed to detect optical emission from three gamma ray burst positions (57). A search for simultaneous optical emission in conjunction with gamma ray bursts has been in progress at the Pic-du-Midi (France) and Ondrejov (Czechoslovakia) obseratories using wide-field cameras; again, the results have been negative to date, but the number of gamma ray bursts occurring in the field of view of the cameras has been quite small, and only a portion of the data have been searched at present.

CONCLUSION

The new international network has been in operation since November 1981, detecting gamma ray bursts at the rate of about one every four days. Many of these will be localizable to extremely precise positions. Deep optical searches should prove to be extremely rewarding for these positions. The launch of EXOSAT, expected for late 1983, should also contribute substantially to the identification of gamma ray bursters. Satellite localization using the arrival time analysis method will continue throughout the ISPM era. Beyond that date, however, the sharp curtailment of planetary exploration programs will mean that further work will have to rely on other methods.

REFERENCES

1. Klebesadel,R. et al., 1973, Ap. J. Lett. (182) L85
2. Mazets, E.P. et al., 1981, Ap. and Space Sci. (80),3
3. Klebesadel, R.,et al., 1982, Ap. J. Lett. (259), L51
4. Proceedings of the 24th COSPAR, Ottawa, Canada, 1982, to be published by Pergamon Press
5. Workshop on Accreting Neutron Stars, Max-Planck Institut für Physik und Astrophysik, Garching, Germany, MPE Report 177, 1982
6. Barat, C. et al., Ap. and Space Sci. 1981 (75), 83
7. Mazets, E.P. et al., 1982, Ap. and Space Sci. (84),173
8. Desai, U., 1981, Ap. and Space Sci. (75),15
9. Klebesadel, R., to appear in Ap. J. Supp.
10. Woosley, S. and Wallace, R., Ap. J., 1982, (258), 716
11. Fryxell, B. and Woosley, S., 1982, Ap. J. (258), 733
12. Cline, T. et al., Ap. J. Lett., 1980,(237), L1
13. Colgate, S., and Petschek, A., 1981, Ap. J. (248), 771
14. Lamb, D., AIP Conference Proceedings No. 77, p. 249, 1982
15. Lamb, F., Ap. J., 1977, (217), 197
16. Ramaty, R., et al., 1980, Nature, (287), 817
17. Mazets, E.P. et al., 1979, Sov. Astron. Lett.(5),163
18. Barat, C. et al., 1979, Astron. Ap. Lett. (79), L24
19. Terrell, J. et al., 1980, Nature, (285),383
20. Wood, K. et al., 1981, Ap. J. (247), 632
21. Brecher, K., 1982, AIP Conference Proceedings No. 77, p. 293
22. Barat, C. et al., 1982, in preparation
23. Van Horn, H.M., 1980, Ap. J. (236), 899
24. Ozel, M.E. et al., 1982, accepted, Astron. Ap.
25. Vedrenne, G., 1981, Phil.Trans.R.Soc.Lond. A, (301), 645
26. Jennings, M., 1982, Ap. J. (258), 110
27. Laros, J. et al., Ap. Space Sci., in press (1982)
28. Barat, C. et al., 1982, Astron. Ap. Lett. (109), L9
29. Fishman,G.J., 1979, Ap. J. (233), 851
30. Jennings, M. and White,S., 1980, Ap. J. (238), 110
31. Jennings, M., 1982, AIP Conference Proceedings No.77, p.107
32. Newman, M. and Cox, A., 1980, Ap. J.(242) 319

33. Cline, T. et al., 1977, Nature (266), 694
34. Herzo, D. et al., 1976, Ap. J. Lett. (203), L115
35. Nishimura, J. et al., 1978, Nature (272), 337
36. Bewick, A. et al., 1975, Nature (258), 686
37. Share, G. et al., Bull. Am. Phys. Soc., 25(4), 527, 1980
38. Beurle, K. et al., Ap. Space Sci., (77), 201, 1981
39. Agrawal, P. et al., in (COSPAR) X-Ray Astronomy, ed. W.A. Baity and L.E. Peterson, Pergamon (Oxford), 1979
40. Fishman, G. et al., 1978, Ap. J. Lett. (222), L13
41. Mazets, E.P. et al., 1978, Sov. Astron. Lett. (4), 188
42. Mazets, E.P. et al., 1980, Sov. Astron. Lett. (6), 318
43. Cline, T. and Desai, U., 1981, Ap. Space Sci., (42), 17
44. Strong, I. et al., 1974, Ap. J. Lett., (188), L1
45. Klebesadel, R. et al. (1982), Ap. J. Supp. (in press)
46. Mazets, E.P. et al. (1979), preprint 599, Ioffe Institute, Leningrad, U.S.S.R.
47. Jennings, M. 1982, preprint
48. Barat, C. et al., 1982, in preparation
49. Laros, J.G. et al., 1981, Ap. J. Lett. (245), L65
50. Chevalier, C. et al., 1981, Astron. Ap. Lett. (100), L1
51. Schaefer, B., 1981, Nature (294), 722
52. I.A.U. Circulars 3711, 3734, and 3752, 1982
53. Grindlay, J. et al. 1982, Nature, in press
54. Bradt, H. and McClintock, J., 1983, Ann. Rev. Astron. Astrophys., in press
55. Ricker, G. and Schaefer, B., 1982, Nature, in press
56. Grindlay, J. et al. 1974, Ap.J. Lett. (192), L113
57. Wenzel, P.C., 1982, private communication

ACKNOWLEDGEMENTS

This work was supported by CNES Contract 82-212.

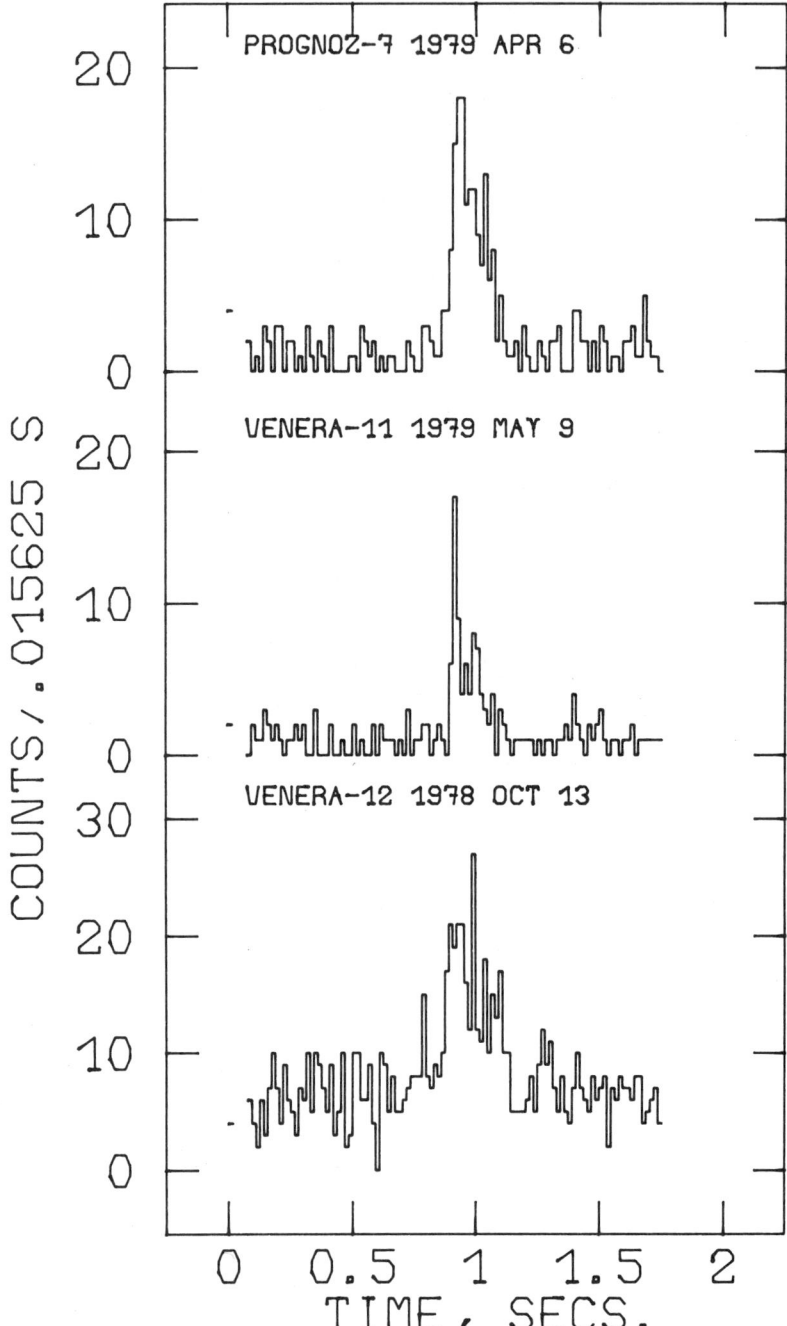

Figure 1. ~200 ms. long gamma ray bursts from the Franco Soviet SIGNE experiments. Energy ranges are >100 keV.

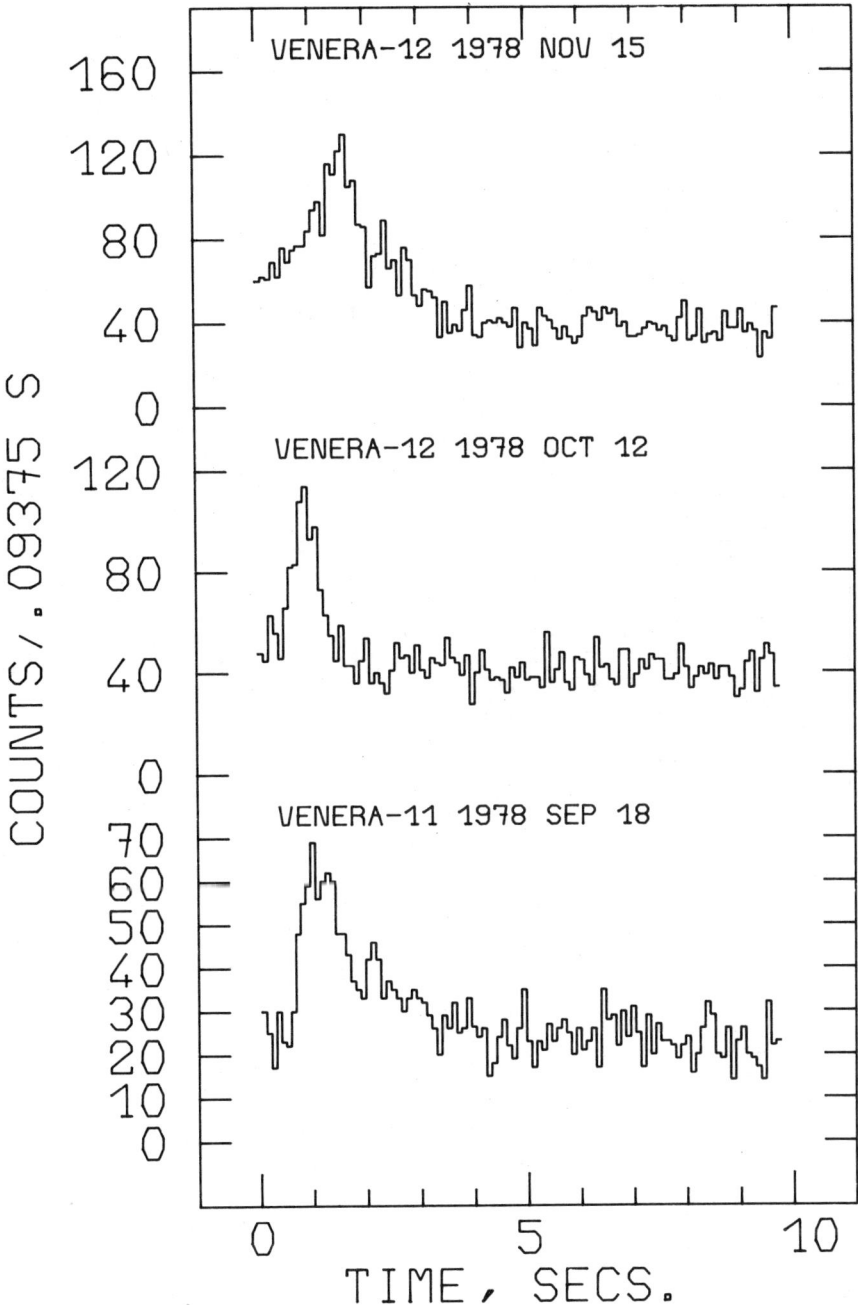

Figure 2. ~1s long gamma ray bursts from the SIGNE experiments.

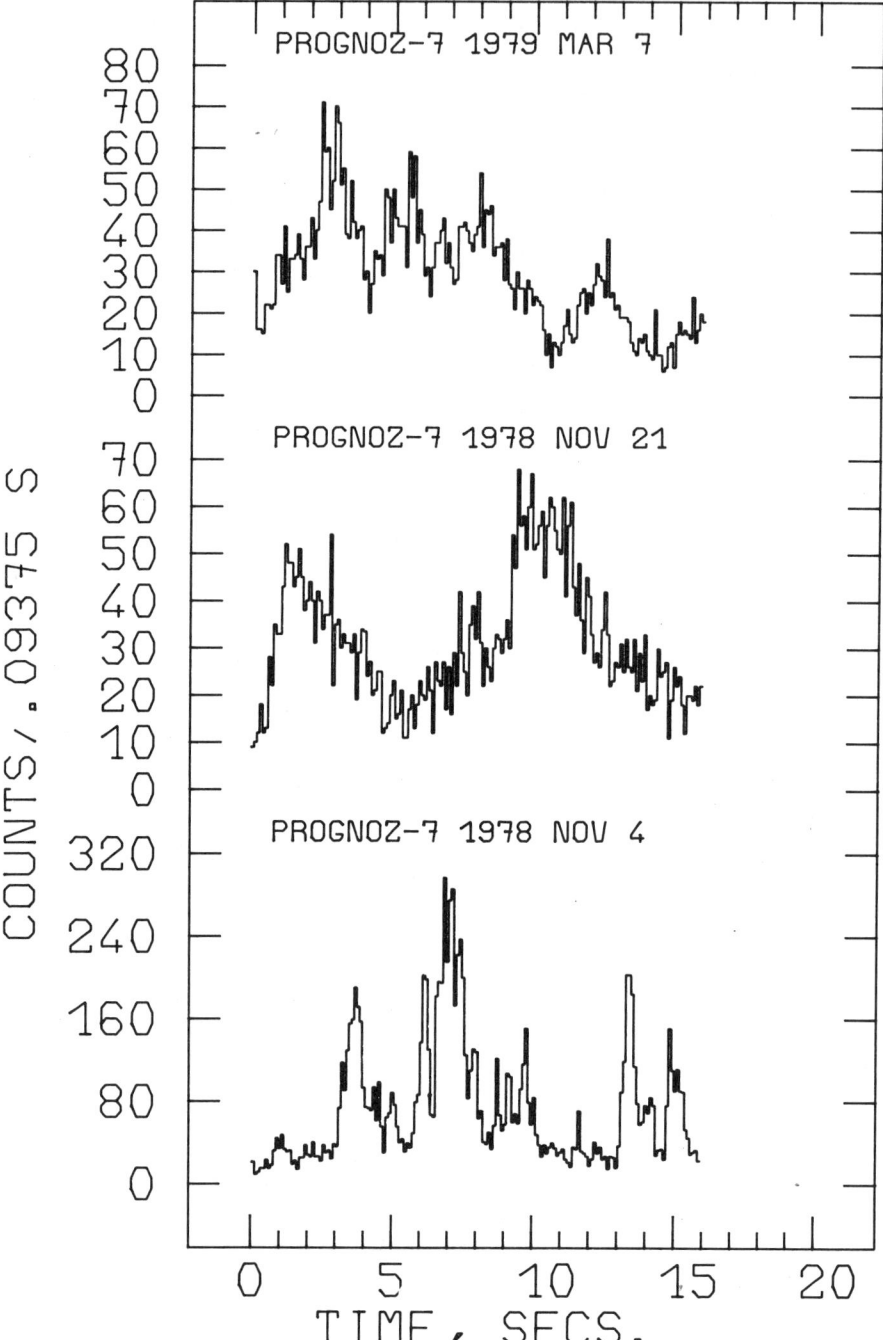

Figure 3. Examples of long, complex bursts from the SIGNE experiments.

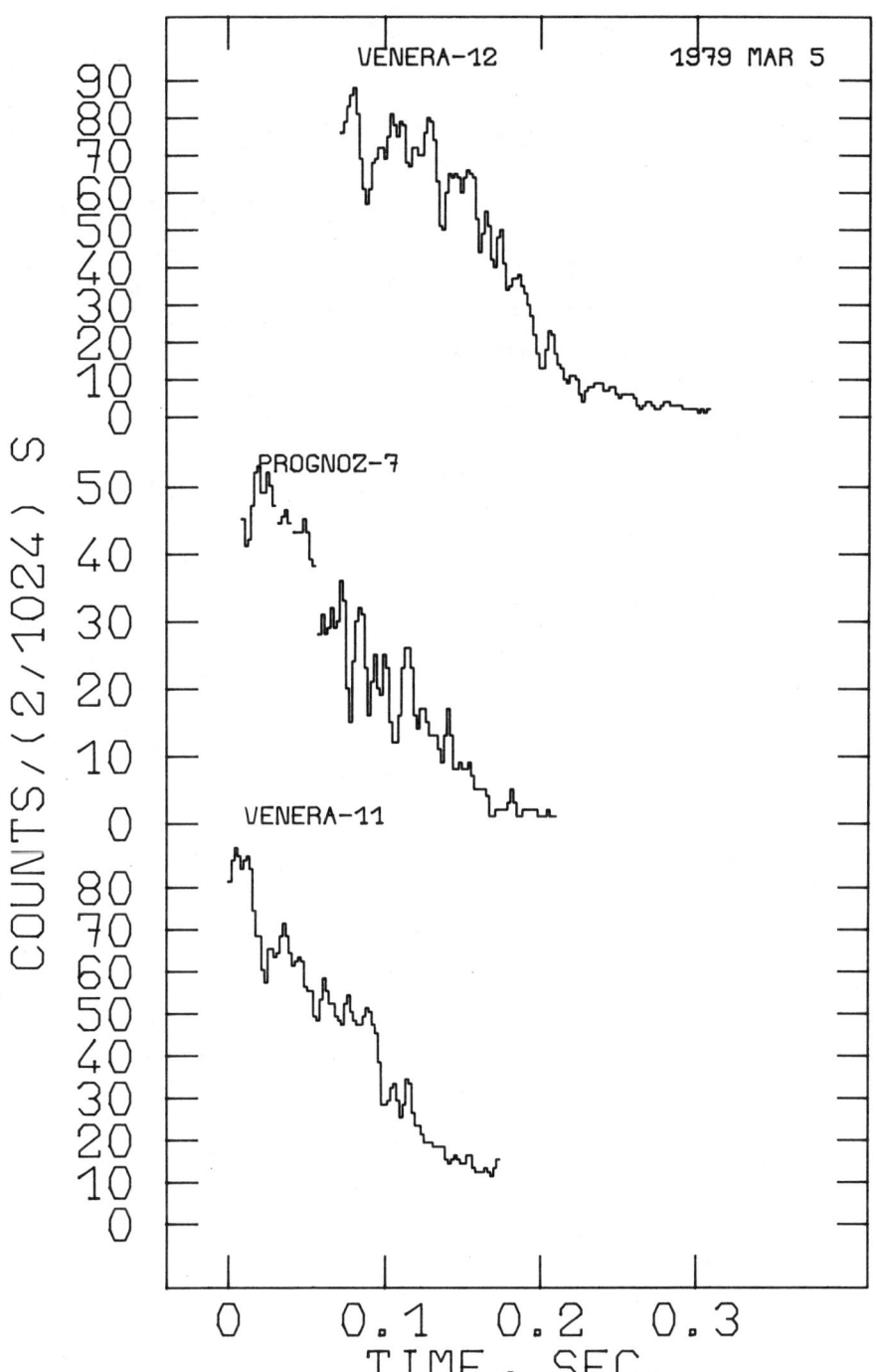

Figure 4. 2 ms plot of the 1979 Mar 5 time history.

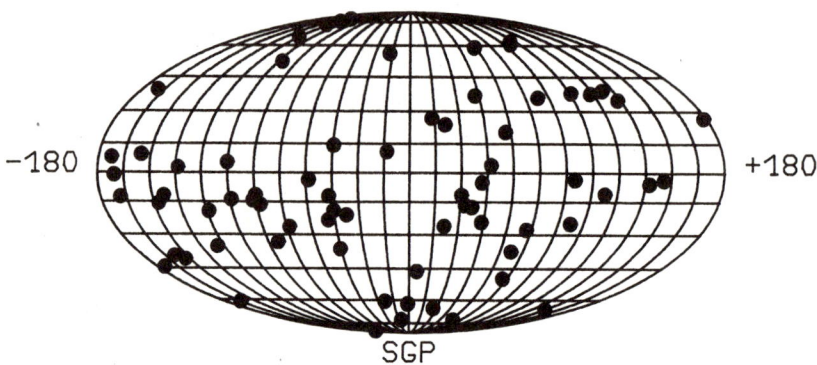

Figure 5. Distribution of 71 gamma ray bursts in galactic coordinates.

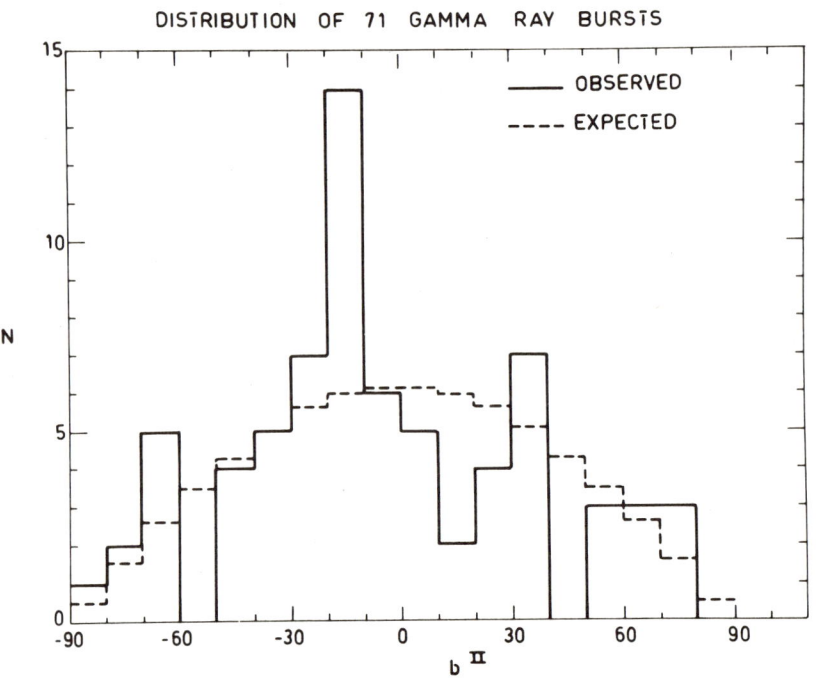

Figure 6. $|b^{II}|$ distribution of 71 gamma ray bursts.

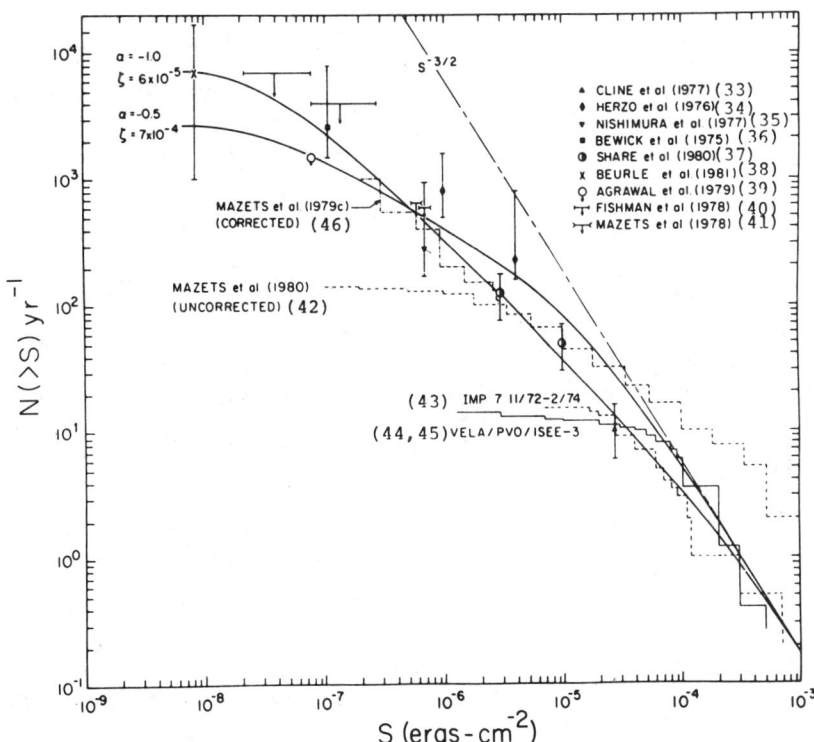

Figure 7. Log N-log S curve of gamma ray bursts, from (26)

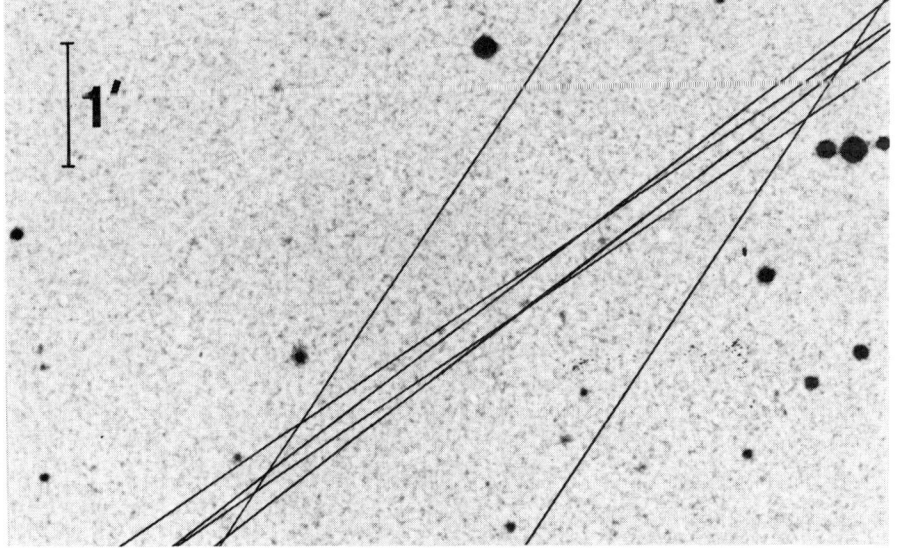

Figure 8. Error box for the 1979 June 13 gamma ray burst superposed on a POSS plate.

ENERGY SPECTRA OF THE COSMIC GAMMA-RAY BURSTS

E.P.Mazets, S.V.Golenetskii, Yu.A.Guryan, R.L.Aptekar,
V.N.Ilyinskii, V.N.Panov
A.F.Ioffe Physical-Technical Institute,
194021 Leningrad, USSR

ABSTRACT

Preliminary results of a spectroscopic study of gamma bursts in the KONUS experiment on Venera 13 and 14 are surveyed. A brief description is given of the new instrumentation and data treatment procedure used. The characteristics of continuous spectra and of the absorption and emission features are considered. Evidence is presented for a fast spectral variability and a strong positive correlation between the hardness of the spectrum and gamma-ray intensity which suggests that the instantaneous values of the source luminosity are determined by the temperature in the emitting region.

INTRODUCTION

Since the discovery of the gamma bursts [1] their investigation has been progressing in the following three directions:
(1) Precise source localization and a study of the localization regions by methods of the optical, radio and X-ray astronomy;
(2) Gamma burst spectroscopy;
(3) A study of statistical distributions.
Parallel with observations, theoretical studies of

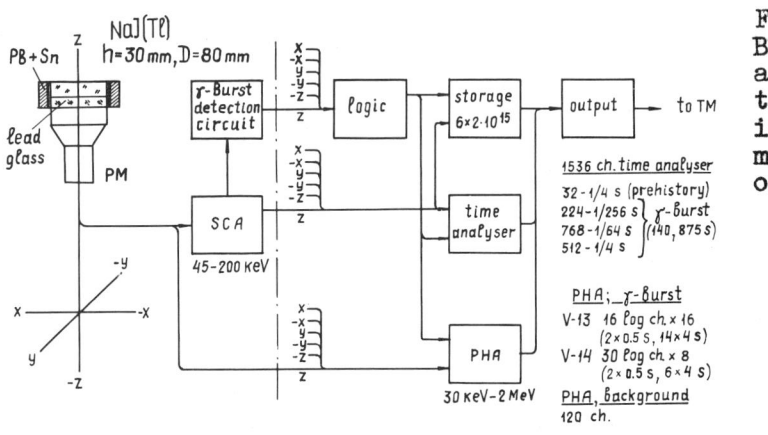

Fig.1. Block diagram of the KONUS instrumentation.

the problem of the gamma burst origin have been developing intensively. The results obtained up to date (see, e.g., reviews [2-4]) suggest strongly magnetized neutron stars as candidates for the burst sources. The actual mechanism of burst production remains, however, unclear. The most valuable information in this respect may come from a detailed study of energy spectra of the gamma bursts.

The available observational data on the gamma burst spectra have been discussed in many papers and reviews (see, e.g., [5,6]). The present work contains preliminary results of a study of the gamma burst spectra in the new KONUS experiment on the Venera 13 and 14 spacecraft in 1981-82.

INSTRUMENTATION

The schematic of the instrumentation used is shown in Fig.1. It comprises a sensor system of six scintillation detectors with a close-to-cosine angular sensitivity pattern arranged along the six axes of the spacecraft's Cartesian frame, six devices for gamma burst detection, a system to measure the gamma burst arrival time, scalers to measure the count rates in each detector channel, time and pulse height analysers, the associated logics and some other auxiliary devices.

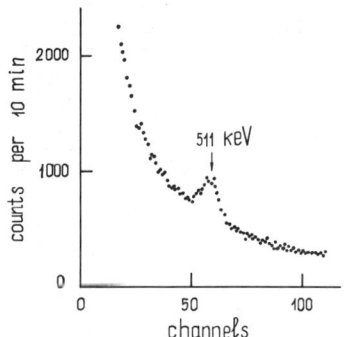

Fig.2. A background gamma spectrum with the 0.511 MeV line used for in-flight monitoring of detector gain.

In the new KONUS instrumentation the time and energy resolution was improved substantially compared to the instruments onboard Venera 11 and 12. The number of the time analyzer channels was increased from 320 to 1536 at a maximum time resolution of 1/256 s instead of 1/64 s. The duration of a time profile measurement was increased from 66 to 140.875 s. In the PHA on Venera 14 the number of the energy channels was raised from 16 to 30. while the 16-channel PHA on Venera 13 is now capable to perform 16 consecutive spectrum measurements rather than 8 as before.

A special detection device of an analog type in which readings of two count rate meters with time constants of 0.25 and 1.5 s are compared with those of a reference rate meter with a 30-s time constant provides a high threshold sensitivity to both short gamma bursts with a fast rise time and to long events with a slowly rising intensity.

The equipment operates in a triggering mode. The radiation flux level in space is monitored periodically by measuring once every 20 min with each of the six detectors the background gamma ray count rate in the 45-200 keV range, and separately the charged particle background. In addition, the background energy spectra in the 30 keV - 2 MeV range are obtained once every 4 hrs by means of a 120-channel PHA. For in-flight calibration the 0.511 MeV annihilation line always present in the background gamma-rays is used (Fig.2). A possibility is provided for an in-flight detector gain correction by a radio command from Earth.

When a gamma burst occurs, the output of the burst detection device is analyzed by a logic circuit. As a result of this, the time and pulse height analyzers are connected to the detector which is most favorably oriented with respect to the gamma burst source. At the same time the gamma-ray count rate is measured by each detector in 0.5 and 8 s intervals and stored in the experiment memory. These data are used to determine the burst arrival direction and for subsequent source localization. The time analyzer stores the 8 s burst prehistory recorded with a 1/4 s resolution and measures the burst time profile for 0.875 s with a resolution of 1/256 s, then during 12 s with a 1/64 s resolution and for 128 s with a 1/4 s resolution. The detection devices, the scalers and the time analyzer operate in the nominal energy interval $\Delta E = 45-200$ keV.

The PHA on Venera 13 converts to 16 quasilogarithmic channels and obtains 16 energy spectra in 16 consecutive time intervals (2 intervals of 0.5 s and 14 of 4 s duration). The 30-channel PHA on Venera 14 carries out 8 consecutive spectrum measurements (2 of 0.5 s and 6 of 4 s duration).

OBSERVATIONS

In the period November 1981 through December 1982, about 180 cosmic gamma bursts and more than 600 solar flares were detected altogether in the KONUS experiment. The new gamma burst observations on Venera 13 and 14 support the main conclusions drawn on the basis of the KONUS experiment on Venera 11 and 12 [2,5,7]. The average frequency of the gamma burst occurrence is 2 bursts in 5 days, just as this had been in the 1978-80 period. The pattern of the time profiles of the bursts is extremely diverse. One observes bursts of both simple and complex shape consisting of separate, sometimes overlapping peaks; there are long($>$ 2.5 min) and very short

(< 15 ms) events. Energy spectrum measurements reveal in most cases strong temporal evolution in the course of a burst, manifesting itself primarily in that the variation of the emission temperature follows that of the time profile. Some bursts exhibit spectral features which may be attributed to the cyclotron absorption and annihilation line emission in the sources.

Continuous monitoring of detector gain is of primary importance for reliable spectral measurements. Regular in-flight calibrations of the energy scale by the 0.511 MeV line in background spectra shows the gain drift to be small. Fig.3 exemplifies the calibration of one detector of the Venera 13 detector set. Even though small, such drift was taken into account in the spectral data treatment for each burst.

In one of the detector channels on Venera 14, however, a fault in electronics brought about a decrease of the PM tube voltage, which resulted in a drop of the gain. As a consequence, the energy interval in which this detector records the time profiles shifted to $\Delta E = $ 150-700 keV. This did not, however, affect significantly the quality of the data obtained since the corresponding detector on Venera 13 is functioning normally.

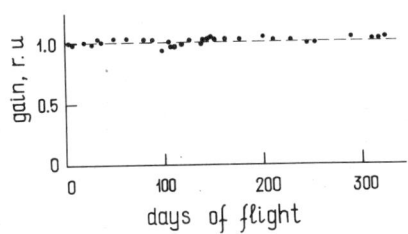

Fig.3. In-flight detector gain stability monitoring.

Furthermore, it permitted us to obtain additional important information on the spectral variability of the gamma bursts which will be discussed in the last Section.

SPECTRAL DATA TREATMENT

The instrumental energy loss spectrum in a scintillation detector, $N(E')$, is related with the incident photon spectrum $F(E)$ by the well known expression

$$N(E') = \int_{E'}^{\infty} R(E',E) F(E) dE \qquad (1)$$

The function $R(E',E)$ characterizes the efficiency and spectral properties of the detector. The relationship $R(E',E_o)$ at $E = E_o = $ const represents the detector response function to monochromatic radiation of energy E_o, the function $R_c(E',E)$ at $E' = $ const being the counting efficiency. Because of the finite width of the instrument energy channels, integration is in practice replaced by summation, the problem being considered in the matrix representation, $n(k) = R(k,j)F(j)$. Basically, the solution of this set of equations can be reduced to finding the vector $\vec{F}$ by means of the inverse matrix $F(j) = R^{-1}(k,j)n(k)$. We cannot dwell here on the well known practical and principal difficulties inherent in this approach to the solution of the inverse problem. Instead, we will limit ourselves to describing briefly the method used by us.

A library of spectra representing essentially a family of response functions $R(E',E_0)$ were compiled for our detector from the available tabulated data, our laboratory measurements and calculations. It was used to compute a family of the detector counting efficiencies $R_c(E',E)$. These data, in their turn, could now be employed to calculate energy loss spectra $n(E)$ in the detector for a broad set of standard incident photon spectra of the forms $F(E) \propto E^{-\gamma}$, $F(E) \propto \exp(-E/E_0)$ and $F(E) \propto E^{-\gamma}\exp(-E/E_0)$, as well as, in case of need, for spectra of any other desired shape.

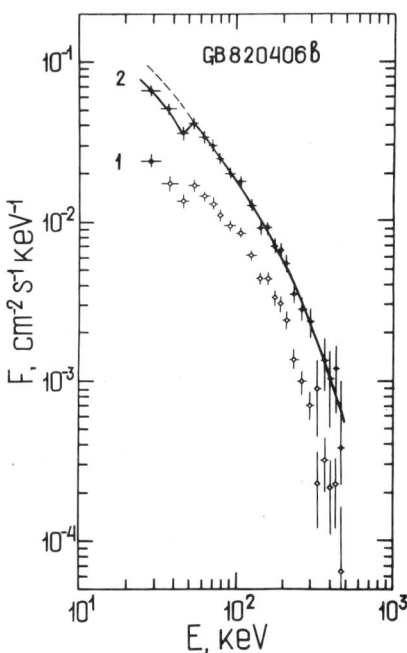

Fig.4. The instrumental energy loss spectrum (1) and reconstructed incident photon spectrum (2) for the GB820406b event. For convenience the spectrum (1) is shifted down by a factor of two.

Any instrumental spectrum $N(E')$ we want to reconstruct can be represented as a superposition of such calculated spectra, $N(E') = \sum a_i n_i(E')$. Then the incident photon spectrum will be obtained by summing the standard photon spectra

$$F(E) = \sum a_i F_i(E). \qquad (2)$$

In our reconstruction of the experimental gamma burst spectra it was usually sufficient to take one spectrum of the form $E^{-1}\exp(-E/E_0)$ with a fixed value of $E_0 = kT$.

The real procedure used was somewhat more complex.

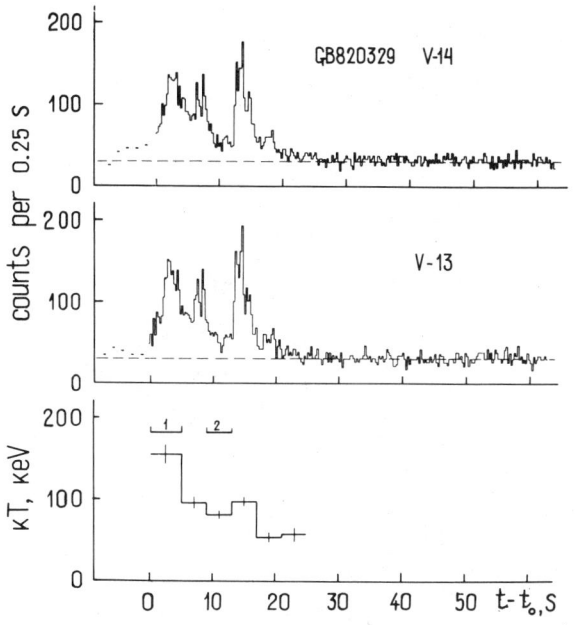

Fig.5. An event of moderate duration GB820329: time profiles from Venera 13 and 14 (top) and emission temperature (below).

As a matter of fact, measurement of the instrumental spectrum is done with the corresponding statistical error in each channel. The parameters of the relationship which fits the shape of the continuum should be determined by the least squares method. The calculated energy loss spectra cannot be represented in a simple analytical form. Therefore it is preferable to refine the parameters basing on the photon spectrum. The relation between the calculated energy loss spectrum and a standard incident photon spectrum $n(k) = \varepsilon_k F_i(k)$ is known for each energy channel. This permits one to use the iterative technique. We take as an illustration a single-component spectrum. We use as a zero approximation the standard spectrum $F_o(E) \propto E^{-\gamma_o} \times \exp(-E/kT_o)$. Multiplying $N(k)$ by ε_k^{-1}, we obtain the first approximation for the photon spectrum $F_1(k)$. Next, new values of $\gamma_1 \pm \Delta\gamma$ and $kT_1 \pm \Delta kT$ are determined by the least squares technique for the spectrum $F_1(k) = A_1 E^{-\gamma_1} \exp(-E/kT_1)$. If they differ markedly from the values of γ_o and kT_o, the procedure is repeated. The iteration is usually completed at the first or second step. If the spectrum has features, for instance, an absorption band, then the corresponding additional incident photon spectrum (with a negative intensity) is introduced also into consideration.

As an illustration, Fig.4 presents an instrumental energy loss spectrum for the GB820406b event and the corresponding reconstructed photon spectrum.

CONTINUOUS SPECTRA

Our new observations are in a full agreement with the results of the gamma burst studies on Venera 11 and 12. In most cases the shape of the spectra at energies below 1 MeV can very well be fitted within experimental accuracy with a simple expression $dN \propto E^{-\gamma}\exp(-E/E_0)$. As a rule, the index $\gamma \approx 1$. When deriving the two parameters γ and E_0 by the least squares techniques, the index γ may be found to differ somewhat from 1. However the χ^2 test for this approximation and the $\gamma = 1$ representation does not give grounds to consider the difference of the index γ from 1 as statistically significant. Thus from a formal standpoint the continuous gamma burst spectra correspond to the thermal bremsstrahlung radiation from an optically thin plasma, their shape being determined only by one parameter, $E_0 = kT$. It is possible, however, that this relationship provides only a good analytic approximation of the experimental data and does not imply that the radiation is dominated by free-free transitions [5,6]

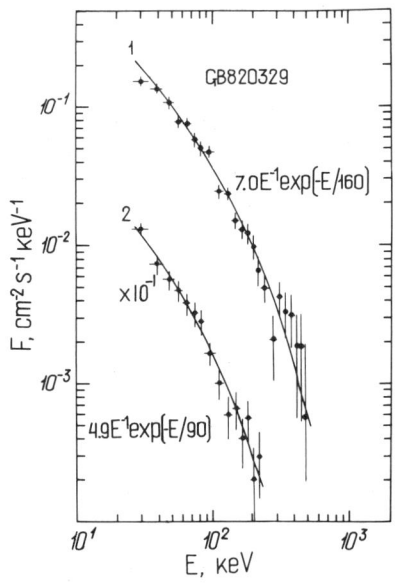

Fig.6. Energy spectra of the GB820329 event measured in intervals (1) and (2) (see Fig.5). The best fit parameters for them are: (1) - $A = 7.0 \pm 0.6$; $kT = 160 \pm 16$ keV; (2) - $A = 4.9 \pm 0.4$; $kT = 90 \pm 8$ keV.

(see also the last Section).

In the course of a burst the energy spectrum, as a rule, exhibits a fast evolution in time. The spectral variability may be considered as resulting from a variation of the emission temperature kT.

Fig.5 presents the time profile of a typical gamma burst of medium duration, GB820329. The graph shows the dependence on time of the count rate in the energy range 45-200 keV obtained with a 1/4 s resolution. The time t is reckoned from the time t_0 at which the burst detection device had been triggered. The points in the graph at $t < t_0$ represent a recording of the burst pre-

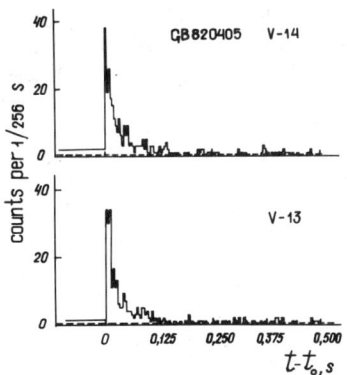

Fig.7. Time profile of very short GB820405 event shown with a 1/256 s resolution.

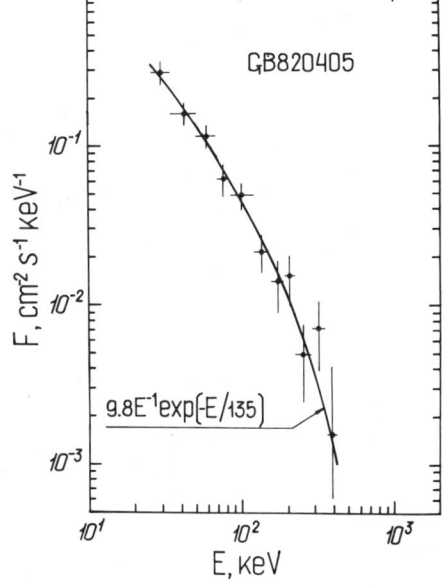

Fig.8. Energy spectrum of GB820405 event stored in a 0.5 s time interval. The best fit parameters are: A = 9.8 ± 0.9; kT = 135 ± 18 keV.

history obtained by the same detector for which the solid histogram further shown. The horizontal dashed line is the background level measured before and after the gamma burst. The graph below illustrates the variation of the emission temperature. The spectra measured at the times specified by bars 1 and 2 are presented in Fig.6.

Figs. 7 and 8 display the time profile and the spectrum of the GB820405 event. It is one of the shortest events observed by us thus far. The width of the pulse at half maximum is only 12-15 ms. Nevertheless, the gamma-ray spectrum of this short burst can be fitted by the thermal bremsstrahlung law as well as is the case with the spectra of 10^3 times longer events.

One of the most important results of the gamma burst spectral studies has been recently obtained on SMM [8,9]. It appears that gamma burst spectra can in some cases have hard tails extending to a few MeV or even higher. Such hard tails in the spectra originate apparently in the region of the emission features with an energy of 400-500 keV (see the next Section).

CYCLOTRON AND ANNIHILATION LINES

As follows from the observations on Venera 11 and 12, the spectra of many bursts contain absorption

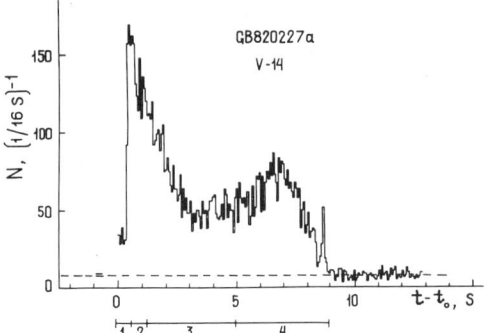

Fig.9. Time profile of the GB820227a event shown with a 1/16 s resolution. Below: spectral measurement time intervals (see Fig.10), 1,2 - 0.5s long, 3,4 - 4 s long each.

Fig.10. Evolution of the absorption feature in the GB 820227a spectrum. Values of continuum temperature kT are presented for each spectrum.

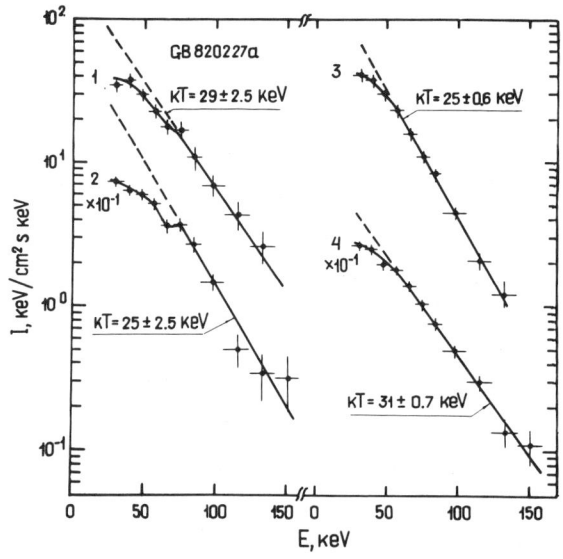

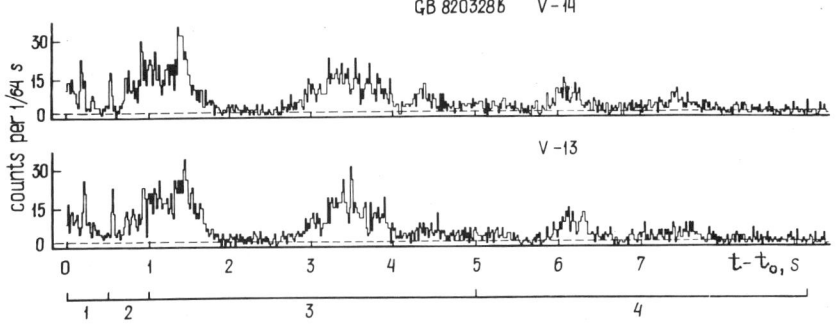

Fig.11. Time profile of the GB820328b event shown with a 1/64 s resolution. Below: spectral measurement time intervals (see Fig.12), 1,2 - 0.5 s long, 3,4 - 4 s long each.

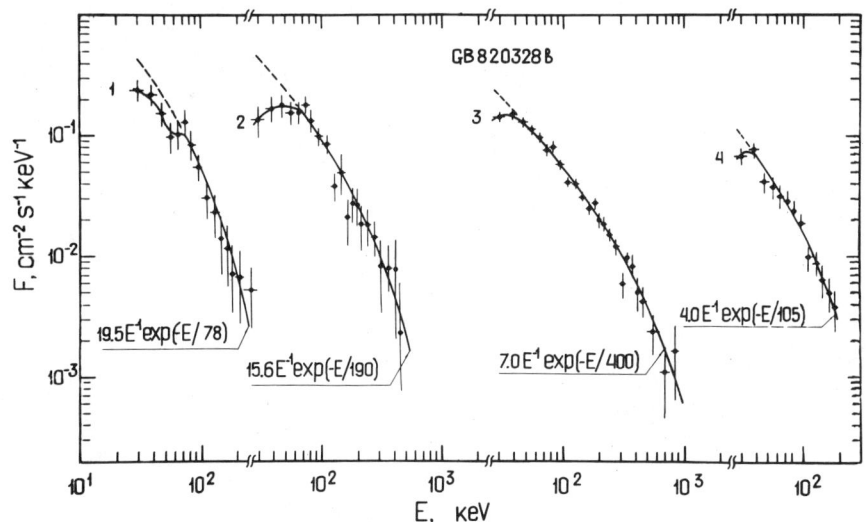

Fig.12. Energy spectra of the GB820328b event.

($E_\gamma < 100$ keV) and emission ($E_\gamma \approx 350-450$ keV) features which have been interpreted, respectively, as broad cyclotron scattering and redshifted annihilation lines [10-12]. The number of reported observations revealing such features keeps increasing [13,14]. Thus the spectra of the bursts provide one of the most convincing proofs for their being generated in the neutron stars.

Our new observations increase the number of recorded spectra with spectral features by at least a factor of two. Some examples are presented below. Figs. 9 and 10 show the time profile and spectra of GB820227a, and Figs.11 and 12, those of GB820328b. The absorption features are most clearly pronounced in the initial stages of the bursts. The last example shows clearly that the available time resolution in spectral measurements precludes a study of fast changing details in the spectral variability both in the continuum and in the absorption band. In connection with this we would like to emphasize here the urgent need in including fast spectroscopy in the programs of future gamma burst experiments.

The GB820406a event is particularly interesting (Figs.13 and 14). The features in the spectrum of the main burst peak may be considered as a manifestation of the cyclotron absorption at the fundamental and second harmonics.

As already pointed out [5], emission features are observed in the burst spectra much rarer than the absorption ones. Figs.15 and 16 show the observational data

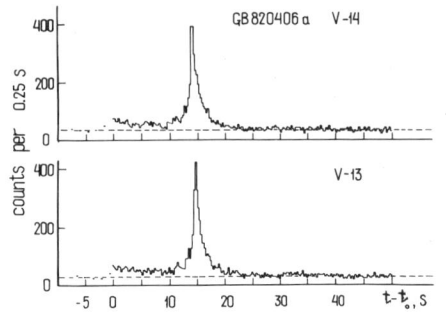
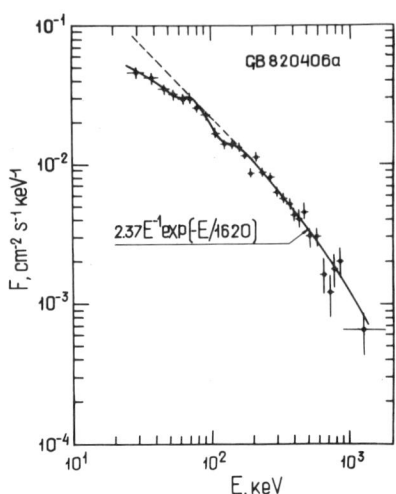

Fig 13. Time profile of the GB820406a event.

Fig.14. Two absorption bands in the main pulse spectrum of the GB820406a.

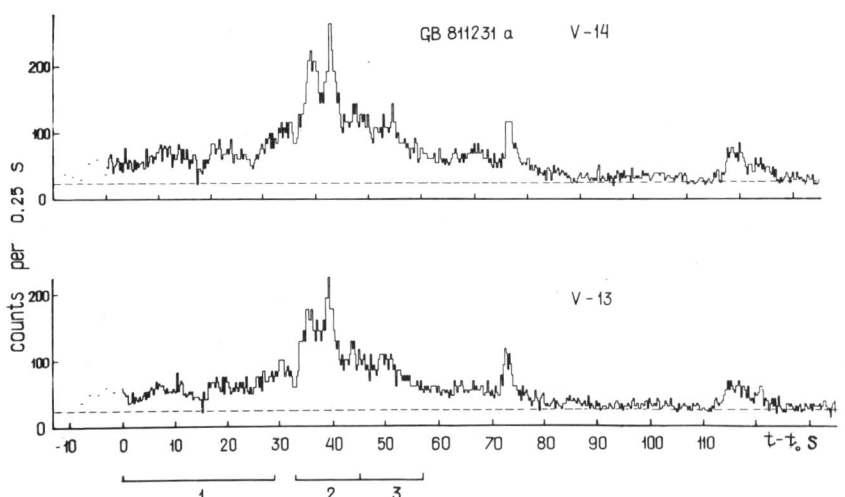

Fig.15. Time profile of the very long event, GB811231a. 1,2,3 - the time intervals, in which spectra shown in Fig.16 had been measured.

on GB811231a. The emission feature in this burst reveals an intensive hard wing. This source, GBS1620+10, is also remarkable in that its emission feature becomes noticeable in the main peak rather than in the beginning of the burst.

Fig.17 presents the time profile of GB820320, while Fig.18 shows the gamma-ray spectra for the first strong and the beginning of the second peak. The duration of the burst exceeds that of our spectral measurements. The spectral evolution in this event is very strong. Spect-

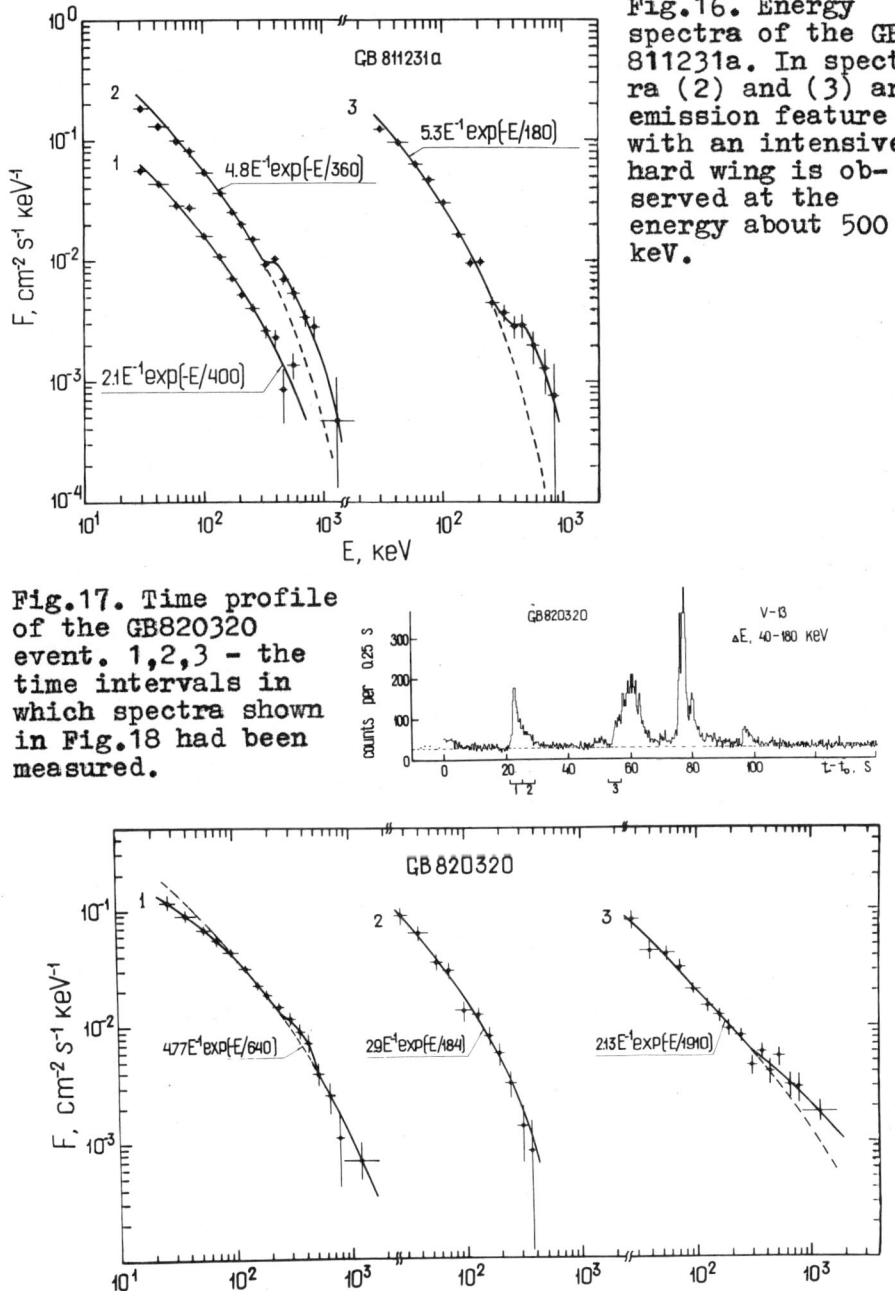

Fig.16. Energy spectra of the GB 811231a. In spectra (2) and (3) an emission feature with an intensive hard wing is observed at the energy about 500 keV.

Fig.17. Time profile of the GB820320 event. 1,2,3 - the time intervals in which spectra shown in Fig.18 had been measured.

Fig.18. Energy spectra of the GB820320. In spectrum (3) the annihilation emission feature turns into an extensive hard tail.

rum 1 contains an absorption and an emission feature which are no more present in the next spectrum 2. In spectrum 3 obtained at the beginning of the second peak the emission feature transforms into an extended hard tail.

This burst from the source GBS0610+20 was recorded also on SMM [9] where a hard tail could be followed up to about 40 MeV. Thus it appears that the hard tails in the gamma burst spectra originate in the region of the annihilation features.

FAST SPECTRAL VARIABILITY; LUMINOSITY AND TEMPERATURE

By reducing the accumulation time in the measurement of the first two spectra down to 0.5 s we have ensured a possibility not only of measuring spectra of short bursts, but also of studying in more detail the initial phase of long events. GB820827c may serve as an interesting illustration. Fig.19 shows 5 spectra measured consecutively on Venera 13. The spectral variability of this burst is very clearly pronounced. The first spectrum averaged over 1 s to improve accuracy reveals a strong absorption and an emission line. The photon continuum has a peculiar shape and fits to an exponential with a characteristic energy $E_0 = 325$ keV. No emission line is seen in the subsequent spectra obtained with a 4 s accumulation time. The absorption band takes

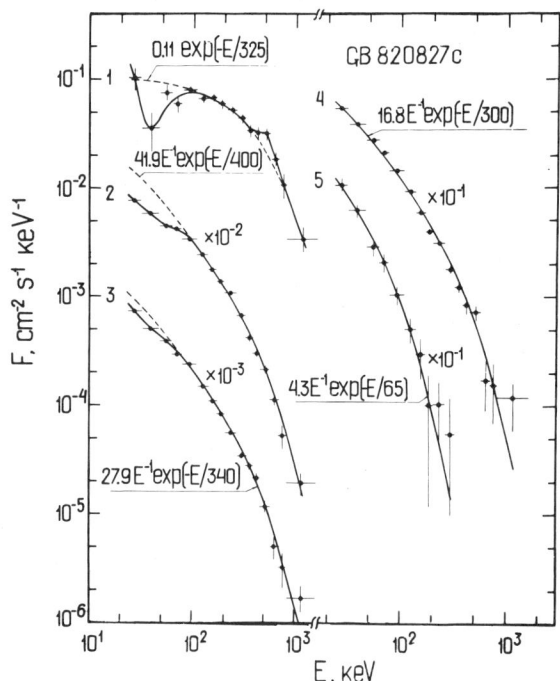

Fig.19. Energy spectra of the GB820827c exhibit a strong evolution of both the absorption and emission features.

more time to disappear. The continuous spectra acquire the usual shape typical for thermal bremsstrahlung. The temperature falls off gradually. This pattern of evolu-

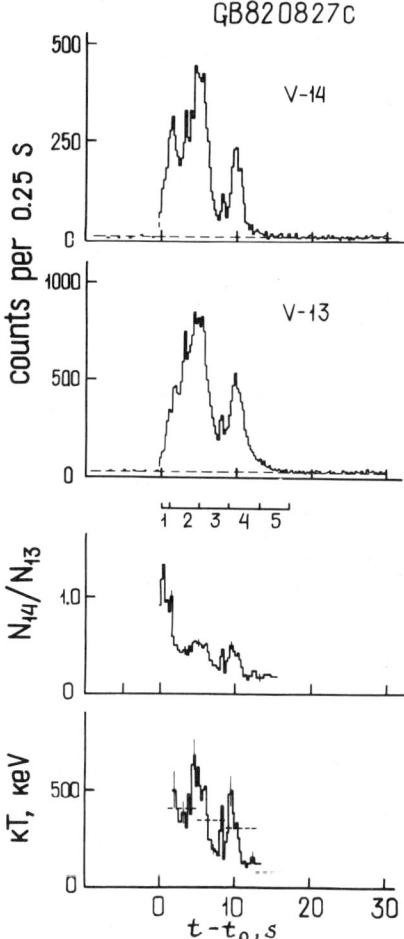

Fig.20. GB820827c event, from top to bottom: time profile (Venera 14 and 13); count rate ratio N_{14}/N_{13}; emission temperature determined from this ratio. Dashed line shows the temperature obtained from spectral measurements in time intervals 2-4 (compare with Fig.19).

tion is observed in many events. The insufficiently high resolution of 4 s does not permit us to follow the spectral variability on a finer time scale. We have succeeded, however, in obtaining very interesting data on fast spectral variations for this gamma burst and some other events. Such a "fast" spectroscopy was carried out making use of a change in the gain of one detector on Venera 14 (see above). This detector continues to detect and record gamma bursts in a shifted energy window $\Delta E = 150-700$ keV when the source lies within the region of its maximum angular sensitivity. Among such events is GB820827c. Fig.20 shows the time profiles of the burst recorded on Venera 13 and 14 in different energy windows. The differences in profile shape originate from spectral variability. These differences reflecting the hardness of the spectra can be characterized by the ratio of the count rates N_{14}/N_{13} measured in the corresponding 0.25 s time intervals. Assuming that after 1 s the spectrum in each interval has the form

$$dN \propto E^{-1} \exp(-E/kT) dE,$$ we can evaluate the temperature kT from the ratio N_{14}/N_{13}. The result thus obtained is shown in the lower graph of Fig.20. We see that the pattern of fast variability differs strongly from the temperature behavior derived from the spectra of Fig.19 by averaging over 4 s time periods. When displayed on a fast time scale, the correlation between the radiation intensity and temperature manifests itself much more clearly. This correlation can be

studied in a quantitative way. From the count rate in the $\Delta E = 40-180$ keV range (with a better statistical precision) and the corresponding value of kT one can derive the flux of gamma-rays with energy $E_\gamma > 30$ keV (or $E_\gamma > 0$ assuming the spectrum to maintain its shape at low energies) for each interval of the time profile. The result of such a treatment is presented in Fig.21 where the flux $F(>30)$ (i.e. the source luminosity $L \propto F$) is plotted as a function of kT.

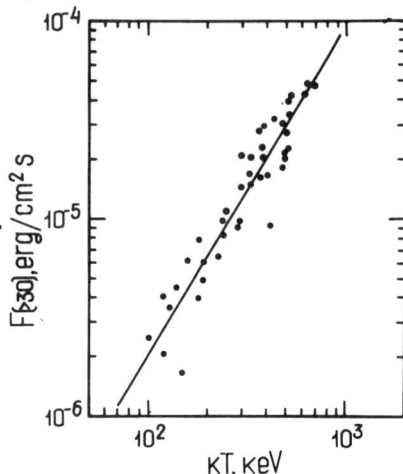

Fig.21. Gamma-ray flux $F(>30)$ vs. temperature kT for the GB820827c event.

As follows from the graph, the luminosity and temperature in the source are strongly related by a close to functional relationship. The scatter in the plot is determined primarily by the experimental errors. The regression treatment carried out for this plot yields $L \propto (kT)^{1.65 \pm 0.10}$ with a coefficient of correlation $\rho = 0.93$.

In a general case the luminosity of a source should be a function of temperature, volume and concentration of the plasma in the emitting region, $L(T,V,n_e)$. The clearly pronounced dependence of luminosity on temperature obtained by us implies that the other characteristics of the emitting region remain approximately constant during the burst. One cannot, however, exclude an alternative possibility that these parameters vary with temperature in a rigidly correlated way. Anyhow, the observed direct relationship between the luminosity and the source temperature clearly indicates that the temporal structure of a burst is a consequence of its spectral variability.

Turning now back to the form of the experimental relationship $L \propto (kT)^\gamma$ with $\gamma \approx 1.6$, we see that it differs strongly from the expression for the luminosity of an optically thin plasma, $L \propto n_e^2 V T^{1/2}$, which is valid, however, only for the nonrelativistic region, $kT \ll mc^2$. It is possible, that taking into account the relativistic corrections and the contribution of the electron-electron bremsstrahlung [15], when extended to the case of moderately relativistic plasma, $kT \sim 100-1000$ keV, will result in a reduction of this discrepancy. Nevertheless, the data obtained may imply that thermal bremsstrahlung is not a dominant process for the gamma bursts.

In a source with a strong magnetic field, thermal synchrotron radiation is an efficient mechanism of emission from a hot plasma. This possibility of interpretation of the gamma bursts has been widely discussed in the literature [16-18]. Liang et al. [19] have demonstrated that the observed gamma burst spectra can be fitted with the spectra of synchrotron radiation from a moderately relativistic ($kT \lesssim mc^2$) optically thin plasma [20]:

$$dN \propto \exp(-4.5\omega/\omega_c)^{1/3} d\omega , \qquad (3)$$

where $\omega_c = eBT^2/mc$, and $T = kT_e/mc^2$.

This interpretation, provided it is valid, permits one to derive from the observed spectra only a combination of the parameters, BT^2. To evaluate B and T separately, one has to invoke additional assumptions [19], which are apparently not the only ones possible [21]. When applied to the GB820827c event, the representation of the spectra as due to thermal synchrotron radiation implies that the spectral variability may be determined by the variation of the quantity $(B/10^{12}) \times (kT/mc^2)^2$ within the range 0.1-2, with the luminosity varying during the burst as $L \propto (BT^2)^{1.5 \pm 0.2}$, which also disagrees with the expected relationship $L_{syn} \propto nVB^2T^2$ [18,22].

The same analysis is being carried out at present for a few more gamma bursts, with the results to be discussed in a separate publication.

CONCLUSION

The new cosmic gamma burst observations in the KONUS experiment on Venera 13 and Venera 14 have fully confirmed our previous findings and are providing new valuable information on the spectral behavior of the bursts. The spectra of many bursts exhibit cyclotron absorption features at energies $E < 100$ keV, and in some cases also emission annihilation lines at $E \approx 350-450$ keV. These data provide the most convincing evidence for the gamma bursts being produced by strongly magnetized neutron stars. Both the continuum emission and the spectral features are characterized by a fast variability. Convincing indications have been obtained that the time scale of the spectral variability in the continuum is closely related to the intensity variations in the time profile of the burst. This may imply that the gamma-ray intensity during the burst (i.e. the source luminosity) is governed at each moment of time primarily by the temperature of the emitting region.

We have shown the hard tails observed on SMM in the spectra of some bursts at an energy of about 10 MeV to originate near 400-500 keV, i.e. in the region where the annihilation features are seen. This may imply that at sufficiently high temperatures the plasma becomes pair-dominated, the emission processes in such a plasma acting as an additional source of hard gamma-rays.

It is obvious from the observational point of view that further studies of the origin of the gamma bursts require progress in the fast and high precision spectroscopy over a wide energy range extending from ~ 1 keV up to $\geqslant 10$ MeV.

REFERENCES

1. R.W.Klebesadel, I.B.Strong and R.A.Olson, Ap.J. 182, L85 (1973).
2. E.P.Mazets and S.V.Golenetskii, Astrophysics and Space Physics Reviews, v.1 (Harwood Academic Publishers, N.Y., 1981), p.205.
3. G.Vedrenne, Phil. Trans. R. Soc. Lond. A301, 645, (1981).
4. R.W.Klebesadel, E.E.Fenimore, J.G.Laros and J.Terrell, Gamma Ray Transients and Related Astrophysical Phenomena, AIP Conf. Proc. No 77, 1 (1982).
5. E.P.Mazets, S.V.Golenetskii, V.N.Ilyinskii, Yu.A.Guryan, R.L.Aptekar, V.N.Panov, I.A.Sokolov, Z.Ya.Sokolova and T.V.Kharitonova, Astrophys. Space Sci. 82, 261 (1982).
6. B.J.Teegarden, Gamma Ray Transients and Related Astrophysical Phenomena, AIP Conf. Proc. No 77, 123 (1982).
7. E.P.Mazets, S.V.Golenetskii, V.N.Ilyinskii, V.N.Panov, R.L.Aptekar, Yu.A.Guryan, M.P.Proskura, I.A.Sokolov, Z.Ya.Sokolova, T.V.Kharitonova, A.V.Dyatchkov and N.G.Khavenson, Astrophys. Space Sci. 80, 3 (1981).
8. G.H.Share, J.D.Kurfess, S.Dee, E.L.Chupp, J.M.Ryan, D.J.Forrest, J.Lanigan, E.Rieger, G.Kanbach and C.Reppin, Gamma Ray Transients and Related Astrophysical Phenomena, AIP Conf. Proc. No 77, 45 (1982).
9. E.Rieger, C.Reppin, G.Kanbach, D.J.Forrest, E.L. Chupp and G.H.Share, Accreting Neutron Stars, MPE report 177 (Garching, 1982) p.229.
10. E.P.Mazets, S.V.Golenetskii, V.N.Ilyinskii, R.L.Aptekar and Yu.A.Guryan, Nature 282, 587 (1979).
11. E.P.Mazets, S.V.Golenetskii, R.L.Aptekar, Yu.A.Guryan and V.N.Ilyinskii, Nature 290, 378 (1981).

12. B.J.Teegarden and T.L.Cline, Ap. J. **236**, L67 (1980).
13. B.R.Dennis, K.J.Frost, A.L.Kiplinger, L.E.Orwig, U.Desai and T.L.Cline, Gamma Ray Transients and Related Astrophysical Phenomena, AIP Conf. Proc. No77, 153 (1982).
14. C.J.Hueter and D.E.Gruber, Accreting Neutron Stars, MPE report 177 (Garching, 1982), p.213.
15. R.J.Gould, Ap. J. **238**, 1026 (1980).
16. R.Ramaty and R.E.Lingenfelter, Astrophys. Space Sci. **75**, 193 (1981).
17. D.Q.Lamb, Gamma Ray Transients and Related Astrophysical Phenomena, AIP Conf. Proc. No 77, 249 (1982).
18. E.P.Liang, Nature **299**, 321 (1982).
19. E.P.Liang, T.Jernigan and R.Rodrigues, Stanford Univ. Inst. Plasma Res. report No 943 (1982).
20. V.Petrosian, Ap. J. **251**, 727 (1981).
21. J.P.Lasota and B.M.Belli, Non-Absorption Dips in Spectra of Gamma-Ray Bursts, Istituto Astrofisica Spaziale preprint (Frascati, 1982).
22. G.Bekefi, Radiation Processes in Plasmas (John Wiley & Sons, N.Y., 1966).

FINE TIME RESOLUTION SPECTRAL ANALYSIS OF
THE 1978 NOVEMBER 4 AND 19 GAMMA-RAY BURSTS

C. Barat
Centre d'Etude Spatiale des Rayonnements (CNRS/UPS) B.P. 4346 -
31029 Toulouse cedex - France

ABSTRACT

The French SIGNE experiments aboard the Soviet Venera probes allow a spectral analysis with a time resolution of 250 ms. From a study of the intense 1978 November 4 and 19 gamma-ray bursts, evidence is presented for a) a significant evolution in the continuum spectra, b) the appearance on a time scale of several hundred ms of spectral features such as emission lines around 400 keV, absorption lines (E< 200 keV), and a high energy cutoff (E>1 MeV).

INTRODUCTION

In revealing the existence of annihilation lines around 400 keV and cyclotron resonance features around 60 keV in the spectra of cosmic gamma-ray bursts, Mazets et al.[1] have shown that these phenomena are probably associated with neutron stars. But the emission mechanism and the nature of the energy sources remain poorly understood. Thus, the shape of the continuum may be described either by free-free (bremsstrahlung) emission [2,3] or by synchrotron emission from a hot and optically thin plasma[4]. However, these models are for a continuum measured over several seconds, and their validity may be attenuated by the fact that rapid spectral variation exists on a 250 ms scale, as shown here.

This paper presents results of our analysis of the 1978 November 4 and 19 events, observed by the SIGNE experiments aboard Venera 11 and 12. A description of these experiments is given by Chambon et al.[5] and Barat et al.[6].

OBSERVATIONS

a) The 1978 November 4 event

Figure 1 shows four spectra measured over the 250 ms intervals indicated on the time history. Note that spectrum 1 was obtained for the leading edge of the first peak, and that spectra 3 and 9 correspond to the most intense portions of the two main peaks in this event.

Spectrum 1 is well fit by a power law with an index around 1 between 90 and 1230 keV. No excess counts are observed for the 1230-1810 keV and 1810-2350 keV channels, even though detection in the 575-1230 keV channel was at 7.6 sigma over background. This cutoff disappears 250 ms later in spectrum 2; a positive detection is made at 6.9 sigma for 1230-1810 keV and 3.7 sigma for 1810-2350 keV. A broad ($\Delta E \sim 260$ keV FWHM), significant (12-16 counts/cm^2 s, assuming a power law continuum with an index 1.45) line feature

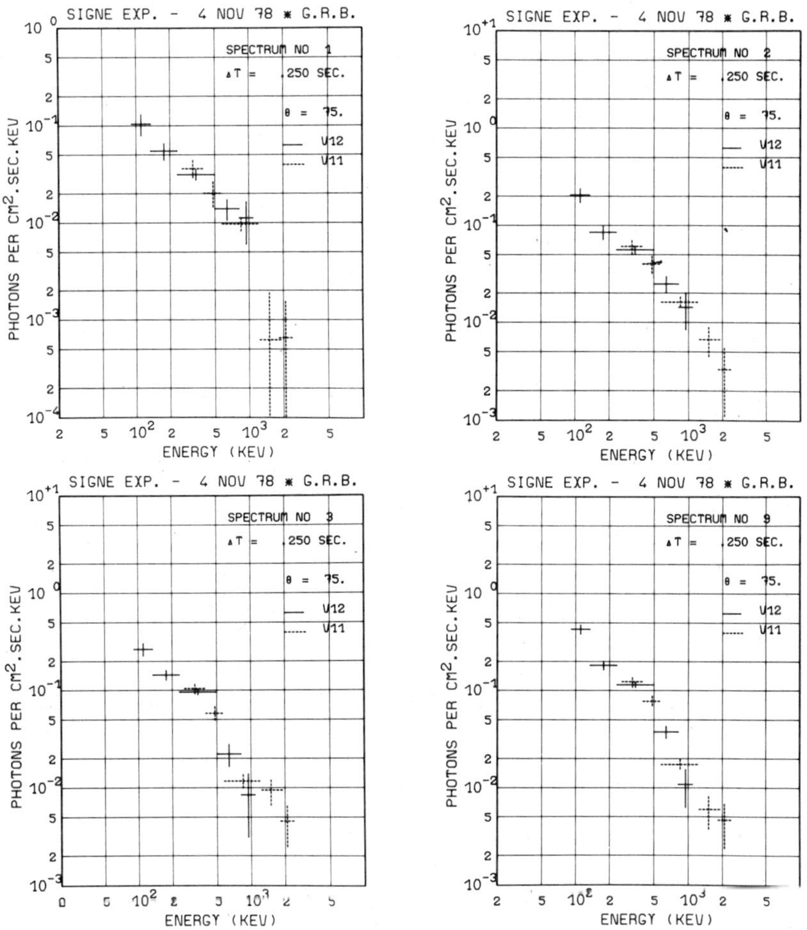

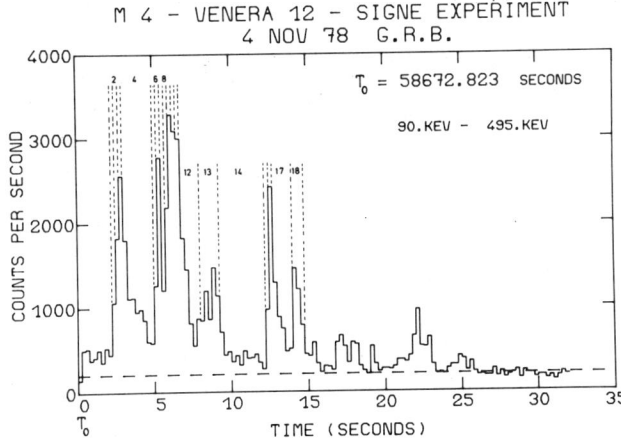

Fig. 1. Four spectra and time profile of the 1978 November 4 event.

appears around 380±20 keV in spectrum 3. This feature, probably an annihilation line, disappears in spectrum 4, which is well fit by a power law with an index around 1.7 (not shown). The same phenomenon occurs at the most intense part of the second peak, with a comparable width around E = 400 ± 20 keV and intensity for the line which lasts through spectrum 10 (Δt=250 ms), although with a smaller intensity, and then disappears.

The spectrum integrated over the first interval of four seconds[7] is in agreement with the corresponding spectrum of Mazets et al.[2], and therefore in disagreement with the result of Fenimore et al.[8]. We do not observe any significant feature in this integrated spectrum.

b) The 1978 November 19 Event

Four spectra, taken over time intervals less than or equal to 250 ms for the first three, and equal to 3.25 s for the fourth, are shown in Figure 2. They correspond to the intervals indicated on the time history.

Spectrum 1, taken at the start of the event, shows an absorption feature around 220±20 keV ($\Delta E \sim 60$ keV FWHM), significant at the 3.7 sigma level in counts, and probably an emission line around 420±20 keV ($\Delta E \sim 250$ keV FWHM), in agreement with the value found by Teegarden and Cline with a germanium detector[9]. An absorption is also present for the first channel (36-53 keV), despite the large error bar in the photon spectrum which is due to the propagation of errors from the Compton contribution and the small detector efficiency at these energies. In fact, the excess count rate in this channel (1.7 sigma above the background level) comes exclusively from the Compton contribution of photons with energies above 150 keV. In other words, there is no escess observed in this channel which is due to incident photons with energies between 36 and 53 keV.

The absorption lines at 45 and 220 keV do not appear in the following spectra, but an absorption line with an energy of 65±10 keV does appear in spectrum 2, with a width ΔE<30 keV FWHM (the detector resolution is 16 keV FWHM at 65 keV). This line lasts several seconds, and has a variable intensity, which reaches 6 sigma in counts in spectrum 3, although it does not appear in the spectrum published by Mazets et al.[2] As far as the SIGNE experiment is concerned, this is not an artifact of the instrument, since the background measured in the same channels does not present any discontinuity ; moreover, the experiment has two independent pulse height analysis systems, for measuring the background, which are in agreement before and after the event. Finally, the 65 keV absorption feature disappears at the end of the event, in spectrum 15.

The annihilation line is a significant feature in spectrum 2 (E=420±20 keV, $\Delta E \sim 280$ keV FWHM, 8 counts/cm^2.s.). A possible excess appears around 500 keV in spectrum 3, but since it involves only one channel, it is not statistically significant.

CONCLUSION

A complete analysis and an interpretation of these results is in

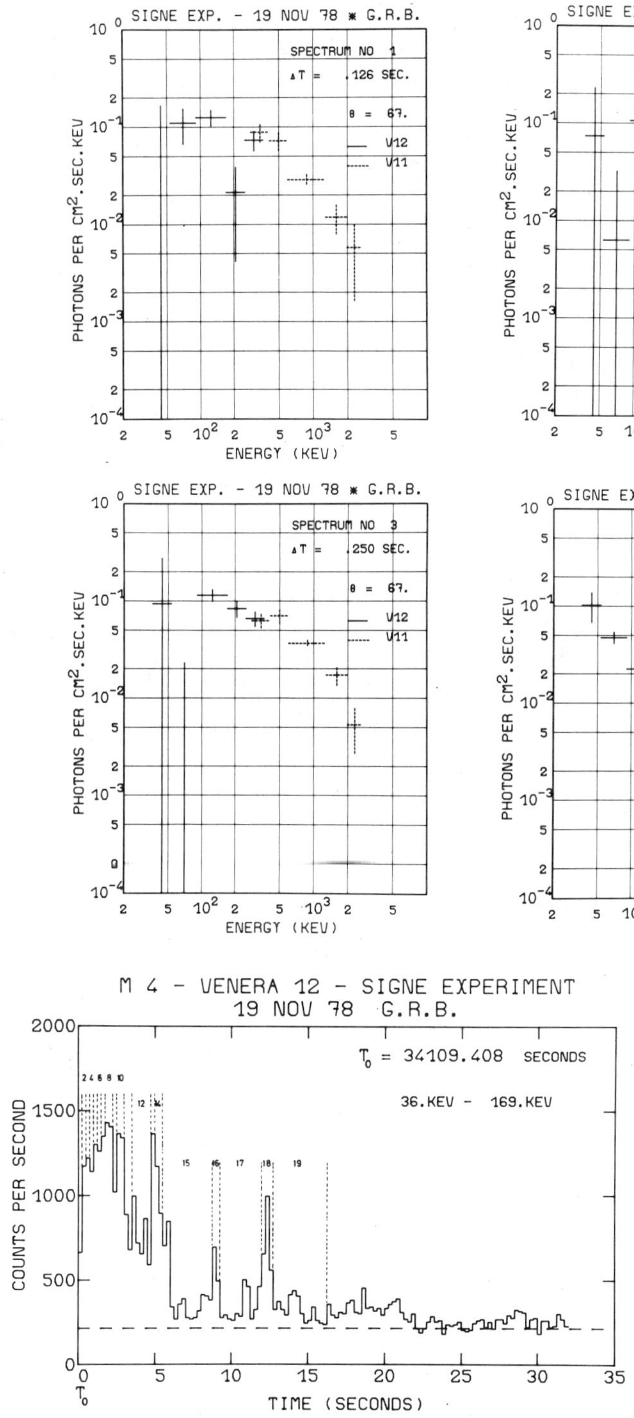

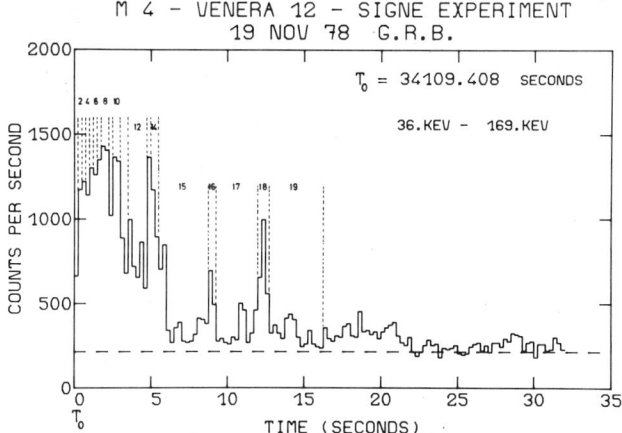

Fig. 2. Four spectra and time profile of the 1978 November 19 event.

preparation [7]. The following points may be noted here, however. First, the rapid variation of the continuum spectrum, especially for the 1978 November 4 event, means that any interpretation of this spectrum integrated over long periods of time is difficult. Second, spectral features such as the annihilation line, the high energy cutoff (1978 November 4 event), and the low energy absorption lines (e.g., spectrum 1 of the 1978 November 19 event), also vary on short time scales. Finally, the important role of the magnetic field: the low energy absorption lines may correspond to cyclotron resonance features, even though the interpretation of spectrum 1 in the November 19 event is not clear. Also, the cutoff observed in spectrum 1 of the November 4 event might be the result of electron-positron pair creation by photon splitting in a strong magnetic field [7].

REFERENCES

1. E.P. Mazets et al., Nature 290, 378 (1981).
2. E.P. Mazets et al., Ast. and Sp. Sci. 80, 3 (1981).
3. D. Gilman et al., Ap. J. 236, 951 (1980).
4. E.P. Liang, Nature 299, 321 (1982).
5. G. Chambon et al., Sp. Sci. Inst. 5, 73 (1979).
6. C. Barat et al., Ast. and Sp. Sci. 75, 83 (1981).
7. I.G. Mitrofanov et al., in preparation.
8. E.E. Fenimore et al., Gamma-Ray Transients and Related Astrophysical Phenomena (Am. Inst. Physics, New York, 1982). p. 201.
9. B.J. Teegarden and T.L. Cline, Ap. J. 236, L67 (1980).

UPPER LIMITS ON NARROW ANNIHILATION LINES IN GAMMA-RAY BURSTS

P. L. Nolan*, G. H. Share
E. O. Hulburt Center for Space Research,
Naval Research Laboratory, Washington, DC 20375

D. J. Forrest, E. L. Chupp, S. Matz
University of New Hampshire, Durham, NH 03824

E. Rieger
Max Planck Institute for Physics and Astrophysics
Garching bei Munchen, FRG

ABSTRACT

The SMM gamma ray experiment has observed 60 gamma-ray bursts from March 1980 to August 1982. Details of one burst are illustrated. There is weak evidence for a broadened spectral feature early in the burst which was also detected by the Konus experiment. Our detailed search was limited to narrow lines (< 70 keV FWHM) near 500 keV; we have found none in any of the bursts. This absence contrasts with the reported detection of several broader features. We discuss the implications for the physical conditions in the annihilation region.

INTRODUCTION

Emission lines have been observed in gamma-ray bursts by instruments on Venera 11 and 12, ISEE-3 and HEAO 1[1] [2] [3]. These lines occur in the range 400-500 keV and have a typical width of 250 keV FWHM. They have been interpreted as redshifted 511 keV lines from positron-electron annihilation. The redshift and the low-energy cyclotron features are considered to be the strongest evidence that bursts come from the surfaces of neutron stars[4]. The line width can provide important information about the temperature and magnetic field in the annihilation region[5]. In this report we have limited our search to lines narrower than 70 keV because unambiguous results could be obtained without introducing complications due to the instrument response function.

OBSERVATIONS

The observations discussed here were made using the Gamma Ray Experiment (GRE)[6] on the Solar Maximum Mission spacecraft. The SMM was launched into a low earth orbit in February 1980, and the GRE is still functioning well. There is no overlapping coverage with the observations by the Konus instruments on Venera 11 and 12[7], but there is with the new Konus instruments on Venera 13 and

*NRC/NRL Research Associate.

14, beginning in November 1981. The GRE main detector consists of seven NaI scintillator crystals, each 7.6 cm in diameter and 7.6 cm thick. The instrument is pointed within a few degrees of the sun. Active CsI shields limit the field of view to about half the sky. The effective detector area is about 150 cm² at 511 keV for events near the center of the field of view. Some intense gamma-ray bursts also penetrate the CsI shield behind the detector. The energy resolution is 7% at 662 keV. Detector gain is automatically stabilized. The time resolution for spectroscopy is 16.384 s.

We classify transient events as gamma-ray bursts, solar flares, or magnetospheric phenomena by their spectral and temporal behavior, as well as by correlation with other solar and cosmic observations when possible. We have identified 60 possible gamma-ray bursts from March 1980 to August 1982[8,9]. We are confident that our selection process is a good one because 20 of the 24 bursts detected by SMM since November 1981 have been confirmed by instruments on Venera 13 and 14.

Figures 1-3 illustrate the data we have for a strong burst on 31 December 1981, which was also observed by the Konus instruments on Venera 13 and 14[10]. The time history in several energy bands is shown in Figure 1. Its duration was more than a minute, and it was detected up to about 4 MeV. Figures 2 and 3 are the energy loss spectra, not corrected for detector response, of the first 5 seconds of the burst (interval 1 in Figure 1) and of the whole burst, respectively. A feature is visible near 475 keV early in the burst, but is not apparent in the integrated spectrum in Figure 3. It is weak, ~2.5 σ, and we can only place an upper limit of 200 keV on its width. However, it is significant that the Konus instruments detected a similar emission line in the first 0.5 second of this burst[10].

We searched for narrow annihilation lines in each of the 60 burst spectra. We fitted the data in the 300-600 keV range with a power-law continuum and a gaussian line with its centroid constrained to be in the 350-550 keV range and its total width constrained to be < 80 keV. Since the detector resolution is 38 keV, this limits the intrinsic width to <70 keV.

The 16 second time resolution of SMM does not allow us to study portions of bursts separately. About half of the 60 bursts were contained entirely within a single 16 second interval. Many of the longer bursts were too weak to allow scrutiny of individual sections. Emission lines have[1] been observed to occur mainly in the early parts of bursts, so the long 16 s integration time of SMM reduces the sensitivity of this survey.

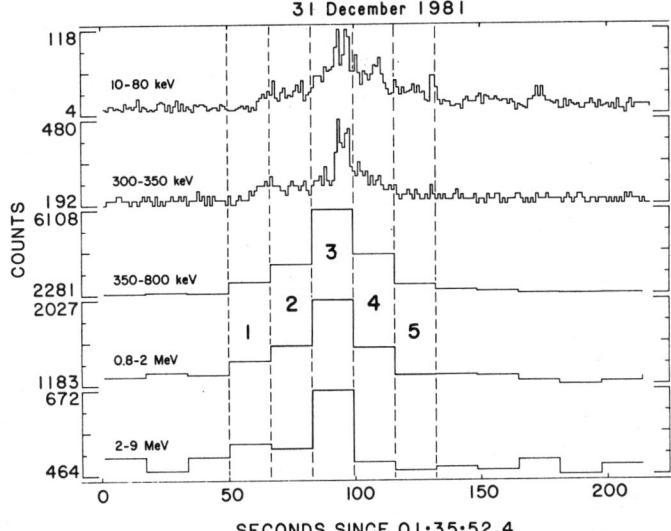

Figure 1 Time history of a select burst.

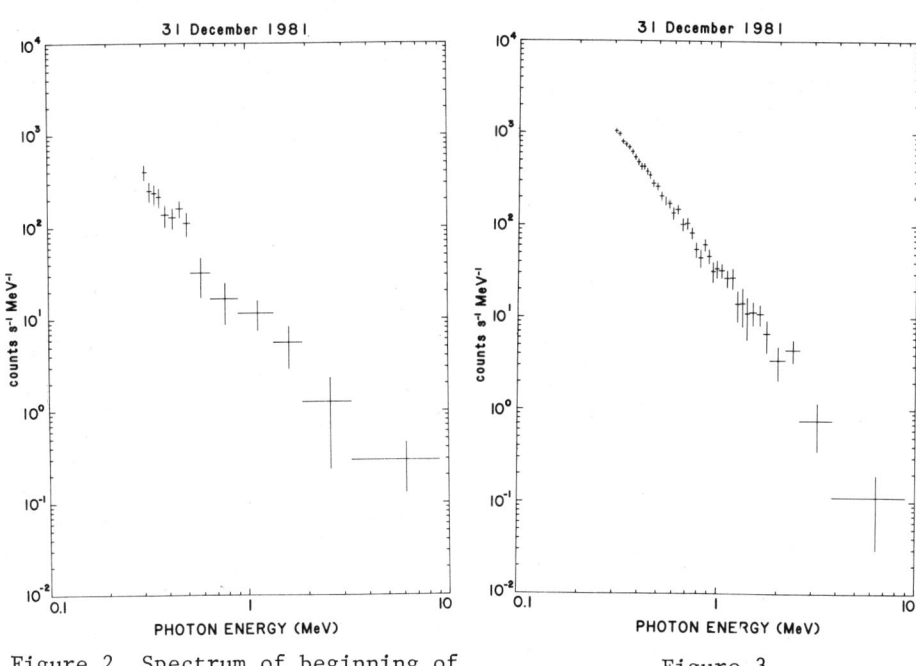

Figure 2 Spectrum of beginning of burst.

Figure 3 Spectrum of entire burst.

RESULTS

We found no narrow lines with a statistical significance more than 3.7 σ. To have 95% confidence that a line detection is real, its formal significance must be at least 4 σ. This is required because we searched about 750 independent energy resolution elements for lines in the 60 bursts. We have therefore established 2 σ upper limits to the narrow line strengths in the bursts.

Figure 4 shows histograms of these upper limits, expressed in terms of line fluence (4a) and relative line strength ([line fluence in counts/cm^2] / [interpolated continuum strength in counts/cm^2 keV]) (4b). The former is strongly affected by the amount of shielding penetrated, but the latter much less; therefore it can be compared with the results of other observations. The four small marks in the Figure indicate the fluence and estimated relative strength of the four lines detected above 4 σ by Mazets et al.[1]. We integrated the Konus continuum flux over the entire burst to obtain a relative line strength which can be compared with our limits. All four of the Konus detections were in the 24 strongest bursts. The shaded region in Figure 4b corresponds to our 12 strongest bursts, which are of comparable strength. In spite of the GRE's poor time resolution, our sensitivity to narrow lines is about as good as theirs to broad features. If lines narrower than 70 keV were as common as the broader ones observed by Konus, we should have observed ~6 in our survey. A significantly larger number of annihilation lines must be observed in bursts, resulting in an observed distribution of line widths, before a minimum line width (if any) can be firmly established and interpreted.

DISCUSSION

Data from the Gamma Ray Experiment on SMM indicate that narrow (<70 keV) annihilation lines are, at best, relatively rare features in gamma-ray bursts. However, broadened features near 500 keV have been detected. Such broadening can be produced by high temperatures, intense magnetic fields, and doppler shifts due to bulk motion and gravitational fields. It is possible that all or some combination of these effects are present in bursts which emit annihilation lines. If the broadening is purely thermal, then the <70 keV width implies a minimum temperature of ~4 x 10^{7}K in the annihilation region[11]. On the other hand, if intense magnetic fields dominate the line profile, as might be expected near neutron stars, then a minimum width can be used to infer a minimum field strength of ~2.4 x 10^{12} gauss, using the analysis of Katz[5]. This is comparable to the values obtained from cyclotron features observed in hard X-rays. If a minimum field strength exists, it would mean either that no bursts occur in weaker fields, or that those which do occur in weak fields do not produce annihilation lines.

To refine these results we plan to do the following: (1) Study only bursts for which rough positions can be established. This will allow the effective detector area to be corrected for shield penetration. (2) Extend the search to broader lines so that our results can be compared with the Konus findings directly. (3) Fit the spectra with various continuum models to obtain more information about the environment in the radiating region.

This work was supported by NASA contracts S-70926A at NRL and NAS S-23761 at UNH and by BFFT contract 010K 017-ZA/WS/WRK 0275.4 at MPI.

REFERENCES

1. E.P. Mazets et al., Nature 290, 378 (1981).
2. B.J. Teegarden and T.L. Cline, Ap. J. Lett. 236, L67 (1980).
3. G.J. Hueter and R.E. Lingenfelter, these proceedings (1983).
4. E.P. Mazets et al., Sov. Astron. Lett. 6, 372 (1980).
5. J.I. Katz, Ap. J. 260, 371 (1982).
6. D.J. Forrest et al., Solar Phys. 65, 15 (1980).
7. E.P. Mazets et al., Astrophys. Space Sci. 80, 1 (1981).
8. E. Rieger et al., "Accreting Neutron Stars," ed. Brinkmann and Trumper (MPI, Munich, 1982), p. 229.
9. G. H. Share et al., "Gamma Ray Transients and Related Astrophysical Phenomena," ed. Lingenfelter et al. (AIP, New York, 1982), p. 45.
10. E.P. Mazets, these proceedings (1983).
11. R. Ramaty and P. Mèszåros, Ap.J. 250, 384 (1981).

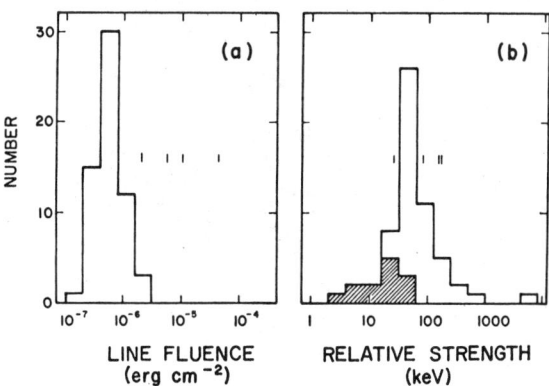

Figure 4 Upper limits on line strength.

CANDIDATES FOR A GAMMA-RAY BURSTER OPTICAL COUNTERPART*

BRADLEY E. SCHAEFER**
M.I.T., Room 37-576, Cambridge, Massachusetts 02139
PATRICK SEITZER
HALE V. BRADT***
Cerro Tololo Inter-American Observatory, La Serena, Chile

ABSTRACT

The small size of the 1928 optical error box for the 19 November 1978 gamma-ray burster (GRB) allows for a very deep search for the quiescent optical counterpart. We have used a CCD camera on the CTIO 4m telescope to search this field to a m_B fainter than 25.0. We find three objects in the 1928 error box, including one which varies by over one magnitude in under a day (see Fig. 1). This variable object corresponds in position to Pedersen et al.'s object A. Since the disappearance of their object B is confirmed, there are two faint variables in the field. It is unclear which one (if either) of the variables is the true GRB counterpart. The details of our observations and their implications are contained in a letter submitted to the Astrophysical Journal.

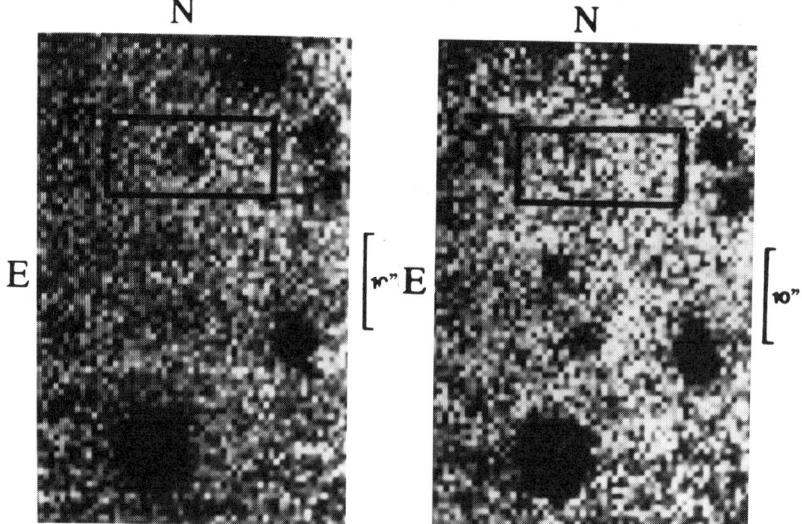

Fig. 1. The variable object AA. In the V data from 22 October 1982 (left panel), a source is seen in the 1928 optical error box (shown as a rectangle). This object is not visible in the V data obtained on the 19, 20, or 23 October 1982 (right panel).

*This work was supported in part by N.S.F. grant AST 82-14569.
**Visiting Astronomer, Cerro Tololo Inter-American Observatory supported by the N.S.F. under contract AST 74-04128.
***Visiting Resident Astronomer, on leave from M.I.T.

THEORIES OF GAMMA-RAY BURSTS

J. I. Katz
Washington University, St. Louis, Missouri 63130

ABSTRACT

Gamma ray bursts have remained an enigma for a decade. This is attributable to the difficulty of obtaining accurate positions, the low duty cycle of burst sources which prevents planned observations, and their low mean power which rules out arguments based on gross energetics. Several lines of evidence now point to an origin in neutron star magnetospheres, confirming early speculations largely based on the availability of high energy density. The evidence includes spectral features interpreted as cyclotron and gravitationally redshifted annihilation lines, and temporal periodicity interpreted as rotation. The reason for the outbursts remains as much a mystery as when they were first discovered. It is unclear whether gamma ray bursters are located in binary stars, or whether this is incidental or essential to their activity. It is not known if there is any evolutionary connection or physical resemblance between gamma ray bursters and pulsars or accretional X-ray sources. I discuss some of the problems which arise in constructing models for gamma ray bursters, with particular attention to the event of March 5, 1979, physical processes at high energy density, and the role of electron-positron pairs in producing line and continuum radiation.

INTRODUCTION

This talk is intended as a brief discussion of some theoretical problems posed by γ-ray bursts. It is not meant as a complete or systematic review of the subject, (which would necessarily quickly become outdated) but rather as a guide to some questions I think interesting and to directions I hope will be fruitful. Therefore my own interests and biases will be evident, and I apologize to those whose work and points of view are ignored or slighted. Literature citations are not intended to be complete, but only to guide the reader to the literature through a few recent papers.

The empirical description of γ-ray bursts may be summarized for theoretical purposes very briefly: They are bursts of soft γ-ray energy ($h\nu \sim 100$ KeV at the spectral peak, extending to $h\nu \sim 1$ MeV) with irregularly varying intensity, lasting ~ 30 seconds. At typical Galactic distances of ~ 300 pc the intensities of the brighter bursts correspond to luminosities of $\sim 10^{37}$ erg/sec. There is usually no apparent correspondence to any other astronomical object or evidence for repetitive or periodic behavior of any one object; this may only be because very few γ-ray bursts have accurately determined celestial coordinates. The famous March 5, 1979 event (see reference 1 for a review) is an exception to most of these statements, for it is identified with a supernova remnant in the Large Magellanic Cloud, had a peak luminosity of about 5×10^{44} erg/sec (more than the entire Galaxy, though only for 0.1 second), showed an 8 second period during its several minutes of outburst, and has repeated its outburst at a much lower intensity on several occasions since.[2] It is unclear whether conclusions drawn from this event are applicable to γ-ray bursts in general. I will usually assume that this is the case, and may be proved wrong when more data become available.

In the nearly 10 years since they were first reported the study of γ-ray bursts has been hampered by the lack of optical identifications. The reason for this was first the difficulty of obtaining accurate positions. Since the development of an extensive network of satellite-borne burst detectors some accurate positions have become available, but except for the March 5, 1979 event no optical identifications have been apparent. At this meeting Schaefer and Pedersen have discussed very deep searches conducted by their groups. In the early development of radio and X-ray astronomy optical identifications were the key to progress; in some cases (especially for radio sources) identifications with peculiar objects were apparent. Historical analogy suggests that the measurement of more accurate positions will lead to rapid progress.

The life of the γ-ray burst astronomer is complicated by the problem of dealing with transient, unpredictable events. He generally cannot go back for a second look at a given object, which breaks the cycle of observation-thought-observation. He can only do this statistically, returning to the entire population of γ-ray bursts.

The total time averaged γ-ray burst luminosity of the Galaxy is about 3×10^{34} erg/sec. This is very low, and deprives the theoretician of his favorite energetic arguments, because a great many processes are capable of supplying this power. In fact, this permits the consideration of very inefficient models, in which the γ-rays are incidental to some more energetic phenomenon. A wide field of speculation is permitted, and until recently clues were scarce.

NEUTRON STARS

A number of lines of evidence lead to the association of γ-ray bursts with neutron stars. Their strong gravitational and magnetic fields are natural sites for the release of large amounts of energy, especially in a small volume. The absence of obvious optical counterparts suggests neutron stars, especially single ones. The absence of obvious X-ray counterparts rules out binary neutron stars in which continuous mass transfer is taking place, but sets no constraints on single neutron stars. The identification of the March 5, 1979 event[1] with a supernova remnant makes the case for neutron stars persuasive, because supernova remnants are known to frequently contain young neutron stars and do not contain anything else which could conceivably produce a γ-ray burst. Observations of this event and of a burst on November 19, 1978[3] found a γ-ray emission line at an energy between 400 and 450 KeV, which is naturally attributed to two-photon positron annihilation at rest, redshifted by the gravitational potential of a neutron star surface. A similar line has been reported in the spectra of a number of bursts by one group,[2,4] but was not found by another group;[5] it is unclear whether the two measurements are directly contradictory or whether a general problem afflicts measurements of this type. These spectral observations argue for the origin of γ-ray bursts near the surfaces of neutron stars. They are supported by the observation of an 8 second periodicity in the March 5, 1979 event, which is naturally explained as a neutron star rotation period.

There is no evidence for an association of γ-ray bursts with pulsars, the known class of single (with few exceptions) neutron stars, or with pulsar glitches, the known class of neutron star impulsive events. This does not discredit the neutron star model, partly because it is unclear that these results are statistically significant arguments against association (it is likely that the observed glitches are only a small fraction of those occurring) and partly because it is entirely possible that burst sources and pulsars may represent different classes of neutron star. For example, no pulsar is as slow as the 8 second period observed March 5, 1979.

MECHANISMS

Two broad classes of mechanisms have been suggested to explain the production of γ-ray bursts on neutron stars. In one class energy is released in a near-thermal equilibrium process at high optical depth. Two subclasses have been proposed: the sudden accretion of matter,[6] (possibly originally a solid object), and the sudden release of thermonuclear energy[7] from below the neutron star surface. These are natural models, because we know that these processes occur. The problem they face is that of explaining the observed properties of γ-ray bursts.

The most acute problem is presented by the observed spectra. Ordinary X-ray sources contain accreting neutron stars, and their spectra qualitatively resemble black bodies with temperatures of order 1 KeV. γ-ray bursts have characteristic photon energies 30 to 100 times higher, and their spectra do not resemble black bodies. A higher accretion rate would tend to soften the spectrum (rather than harden it), by enveloping the source region in a thick cloud. Thermonuclear models have similar problems, because energy is released at high optical depths. In fact, thermonuclear runaways are the satisfactory explanation of X-ray bursts,[8] which have photon spectra resembling ordinary X-ray sources, rather than γ-ray bursts. One should not underestimate the ingenuity of Nature (or theorists[6,7]) in solving these problems; the spectra do not rule out these models, but are their central problem.

An additional problem is posed by containment of the radiating material. If the luminosity L exceeds the Eddington Limit L_E

$$L_E = \frac{4\pi cGM}{\kappa} \qquad (1)$$

then the outward force of radiation pressure will exceed the acceleration of gravity and the matter will be accelerated outward, turning its radiation energy into kinetic energy of expansion and gravitational potential energy. The expansion of the photophere will soften further the spectrum of whatever radiation is emitted. Most γ-ray bursts do not exceed L_E in luminosity, for ordinary values of the opacity κ (corresponding to electron scattering in ionized hydrogen), but the March 5, 1979 event exceeded it by a factor of 10^6, using the distance to the Large Magellanic Cloud to estimate L. This problem also exists for any γ-ray burst which contains electron-positron pairs, since for a pair plasma κ is 1837 times greater than for ordinary hydrogen plasma, and L_E 1837 times less. It is natural to suppose that the matter is confined by the neutron star's magnetic field, but this is a problem for accretion models in which the radiating matter has just fallen in along field lines which necessarily are open to infinity. Thermonuclear models may also suffer from this problem, because energy is released in recently accreted matter which may still be near the magnetic poles. The March 5, 1979 event poses the additional problem of a rise time of no more than 0.2 ms, which is inconsistent with diffusion of radiation from a significant depth.

The second class of mechanisms involve completely non-equilibrium processes. They are free of the problems afflicting the first class, but require the ad hoc assumption of nonthermal processes with unprecedented parameters, chosen to satisfy the observations without any theoretical basis. These are generally and loosely termed flare models, in analogy with solar flares. The analogy is qualitative, but the model is based on several resemblances. These are the irregular time-history of γ-ray burst intensity, including the submillisecond rise time of the March 5, 1979 event, the hard spectrum (including the presence of positrons, whose formation requires the presence of MeV photons or particles, and hence probably a nonthermal distribution function), and the high positron annihilation line flux inferred for the March 5, 1979 event, which implies confinement to thin sheets or other strongly clumped regions. All of these phenomena suggest the release of magnetostatic energy by plasma instabilities during field line reconnection in regions of low density and low optical depth with the energy appearing in relativistic particles. These particles radiate γ-rays, the more energetic of which produce electron-positron pairs in the neutron star's magnetic field, which are themselves accelerated ... ultimately producing a shower of pairs and γ-rays. Unfortunately no quantitative model of these complex processes exists, so they remain a speculation, attractive to some[9]. Even the question of why such processes should occur is unsolved, just as is the question of why there are solar flares. An advantage of flare models is that the observed hard spectra are natural outcomes of energy release at low optical depth. The characteristic energy of a few hundred KeV may even be explained as the result of degradation of photons and particles above the pair production threshold. In addition, there is no reason to expect energy release to occur on field lines which extend to great distances or to infinity, so the full magnetic stress can be expected to be available (and sufficient) to confine the radiating plasma.

CONTINUUM RADIATION

The continuum radiation spectrum of γ-ray bursts has been described by a bremsstrahlung spectrum of a few hundred KeV temperature. While this fits the spectral shapes, it is probably not the actual mechanism, because bremsstrahlung is not an efficient radiation process. In order to avoid the development of a strong Wien peak in the spectrum by Compton scattering, a bremsstrahlung source must be much more strongly clumped than even an annihilation line source. In the March 5, 1979 event it must have been confined to sheets 100 Å thick (a clumping factor of 10^{12})![9]

In neutron star magnetic fields cyclotron emission is much more efficient than bremsstrahlung at reasonable densities, and therefore is a more plausible source for the continuum radiation. This may be considered an additional argument in favor of neutron stars (and, in particular, magnetic neutron stars) as the origin of γ-ray bursts. Liang[10] has compared the shapes of spectra expected for thermal cyclotron sources with the observed spectra, and found it possible to obtain excellent agreement. This only establishes the possibility of a cyclotron source, because almost any function with the right qualitative form and a free parameter or two could probably be made to fit the spectral data, which have limited resolution and substantial error bars. In fact, the assumption of a thermal electron spectrum is unrealistic, because cyclotron cooling is much faster than collisional heating at reasonable densities. The problem of determining the particle distribution function requires consideration of the rates of these (and other) processes, and cannot be solved until the mechanism of plasma heating (and burst energy release) is understood. The first step will probably be calculation of plasma instabilities and pair production cascades in neutron star magnetic fields. These problems are known to pulsar physicists, who know them to be difficult.

ANNIHILATION LINE

The observations[1-4] of the positron annihilation line place constraints on the conditions in the region in which it is formed. The redshift establishes that this region is near the neutron star surface. The narrowness of the line places upper bounds on both the temperature and the magnetic field[9,11] because both thermal Doppler shifts and magnetic effects on the kinematics of annihilation will broaden it. The numerical limits depend on the angle between the field and our line of sight, but typical upper bounds are 50 KeV and 10^{13} gauss. These values are generally inconsistent with the parameters[10] of the continuum source region. This does not invalidate the model, but rather indicates that the annihilation line is produced in a location separate from the continuum, perhaps the dense, cool neutron star photosphere below the optically thin continuum source. Annihilations taking place in region of higher field or temperature will produce a broader line, possibly blending unrecognizably into the continuum.

ASTRONOMY

A variety of arguments have led to the conclusion that γ-ray bursts occur in the magnetospheres of neutron stars. This conclusion is based on the observed properties of the radiation and is independent of one's preference for the mechanism of energy release- thermonuclear, accretional, or flare. The astronomical circumstances of the responsible neutron stars remain obscure. The questions to answer are whether they are single or binary, whether they were once pulsars or accretional X-ray sources or neither, their magnetic field magnitude and orientation, their spin rate and its history, their spatial and velocity distributions, and their origin.

The March 5, 1979 event appears to be a young neutron star (because of its location in a supernova remnant). Its 8 second period implies slow rotation, too slow for it to be a pulsar. Its youth implies that it was born with essentially its present period, unless its rate of spindown was extraordinarily high. If the 8 second period can be detected during its recurrent outbursts[2] it may be possible to measure its spindown rate, which could help determine its magnetic field and rotational dynamics.

The absence of obvious optical candidates implies that the responsible neutron stars are single, or have very faint companions. This does not by itself give any useful information about burst mechanisms, for it is compatible with all of them; even accretion may occur from a very dim low mass star like the secondary in the dwarf nova WZ Sge. Further, because the distances to bursts can only be estimated using rough statistical arguments quantitative conclusions about the nature of any companions are uncertain.

Because of the rarity and unpredictability of γ-ray bursts it is much harder to obtain limits on counterparts in other wavelength levels to the bursts themselves, as opposed to the quiescent state of the burster. Information has come from photographic archives. The Prarie Network produced an upper limit to the visible brightness of a burst.[12] Schaefer[13] found in the Harvard plate collection a record of an optical outburst coincident with but 50 years before the γ-ray burst of November 19, 1978.

In order to interpret this observation it is necessary to assume some relation between the magnitudes of the two events. For lack of more information it is generally assumed that the two bursts were identical, though for an object with a distribution of outburst energies this ignores a possible statistical bias. The result is a γ-ray to visible flux ratio of 10^3. The simultaneously observed upper limit[12] is consistent with this number. A similar result is found for the luminous X-ray source Sco X-1, and is generally expected for binaries in which a luminous (L $\sim 10^{37}$ erg/sec) compact object accretes from a low mass star.[14] The flux ratio is determined by the fraction of high energy flux intercepted by the star in Roche geometry, and by the bolometric correction of the heated stellar surface. This interpretation requires that the companion, if not in contact with its limiting Roche surface, not be an order of magnitude smaller, in order to subtend the required solid angle. Compact companions are ruled out because of the short gravitational radiation lifetime of very close orbits.

Schaefer's observation[13] is naturally explained by the presence of a low mass, nondegenerate binary companion, yet very sensitive observations of the positions of the few accurately located bursts have shown that any quiescent companions must be very dim. If they are at distances ≤1 Kpc, as implied by the space distribution of γ-ray bursters,[15] the absolute magnitudes are ≥14, consistent with only the faintest red dwarfs (this argument does not apply, of course, to the March 5, 1979 event; even if it is not at the distance of the Large Magellanic Cloud the crowded field makes it harder to place upper bounds on the brightness of optical candidates). Binary models are not ruled out, even statistically, but one wonders at the absence of bursters with bright companions, such as are found among the binary X-ray sources.

An alternative possibility is that the optical flux was produced by the neutron star itself, rather than by reprocessing in the atmosphere of a binary companion. The brightness temperature implied by Schaefer's estimate of a 3rd magnitude outburst is

$$T_b = \left(\frac{D}{100pc}\right)^2 \left(\frac{A}{10^{12} cm^2}\right)^{-1} 1 \times 10^{16} \, °K \qquad (2)$$

where D is the distance and A the area of the emitting region. This clearly rules out any thermal radiation mechanism; even the γ-ray spectrum corresponds to a temperature only of order 10^9 °K. Either there are radiating TeV particles (for which there is no other evidence) or some coherent process must be responsible. For comparison, the visible pulses from the Crab pulsar are about 100 times less luminous if D = 100 pc, and about 11 apparent magnitudes dimmer. Most nonthermal or collective processes produce strongly polarized radiation, while reradiation is slightly polarized because of transfer effects in a scattering atmosphere. Polarization measurements can distinguish among these models for the origin of the visible radiation and should be a major objective of any program to search for optical counterparts of γ-ray bursts.

Schaefer and Bradt[16] and Pedersen et al.[17] have suggested identifications of the quiescent state of the November 19, 1978 burster with apparently variable stars of approximately 24th magnitude at maximum. This is consistent either with reradiation in a binary from a low level of quiescent γ-ray activity (provided the companion is sufficiently faint), or with a nonthermal origin from a single neutron star. An estimate of the simultaneous γ-ray activity would be informative, but an interesting limit (21 magnitudes below burst intensities) is far beyond present capabilities in the hard X-ray and soft γ-ray bands. Reradiation efficiencies are limited by geometry and Planck functions, while the ratio of γ-ray to nonthermal optical radiation could vary widely. Quite rough measurements of optical polarization in the quiescent state could again determine the origin of the optical radiation, and hence decide whether the neutron star is binary; these may be easier than trying to observe an optical outburst.

RECURRENCES

Schaefer[13] has inferred a repetition rate of once per year for optical counterparts to bursts bright enough to be found in photographic archives. Although based on a single event, this estimate is probably correct in order of magnitude, and I will adopt it. He has used this value to argue against the impact accretional and thermonuclear models. The former argument is based on estimates of the rate of solid body accretion events, while the latter points out that the mean accretional energy release should exceed that in γ-ray bursts by a factor of 100, and compares this number with upper bounds on steady X-ray emission from the direction of three γ-ray bursters. The recurrence rate also implies that the reported quiescent state visible flux is only about 10% of the averaged outburst flux. If the visible radiation is reradiated γ-ray energy then the quiescent and averaged outburst γ-ray powers are in a similar ratio (to order of magnitude, with the quantitative value depending on the bolometric correction); if it is produced nonthermally then no statement is possible.

ENERGETICS

It may be useful to collect a few numbers concerning the energetics of γ-ray bursts. If we assume isotropic radiation (as would be implied, at least to order of magnitude, by all models we have considered) then a typical strong burst releases about 10^{38} erg. The March 5, 1979 event, if correctly identified with the LMC, released 10^{44} erg. The Galaxy radiates[15] 1.3×10^{-4} erg/cm^{-2} yr^{-1} in γ-ray bursts, corresponding to a total luminosity approximating 3×10^{34} erg/sec.

The energy available in thermonuclear models is about 1% of the accretional (gravitational) energy released. The luminosity of compact X-ray sources (about 3×10^{39} erg/sec for the Galaxy) might be used to estimate this, except for the absence of any empirical association between them and γ-ray bursters. If γ-ray bursters are thermonuclear they almost certainly are a population distinct from the well-known accretional X-ray sources (whose thermonuclear runaways are known as X-ray bursts), and it is not possible to estimate the available energy. The required accretional energy is 3×10^{36} erg/sec, a quite modest value.

The energy available in accretional models is equally hard to estimate. The ordinary accretional X-ray sources again clearly are a distinct population, and their accretion rate is irrelevant. The required mean rate of 10^5 gm/sec per neutron star (assuming 3×10^9 in the Galaxy) is far below the Bondi-Hoyle estimate of the mean interstellar accretion rate, which is 10^9 gm/sec. The fraction of this rate which will occur as discrete events is unknown, as is the contribution from solid bodies, including those in bound orbits. It is clear that the energy conceivably available is more than sufficient, although the actual power is hard to calculate.

In flare models the released energy is magnetostatic. Using an estimate[18] of 10^{-4} kpc^{-2}yr^{-1} for the pulsar birthrate to estimate the neutron star birthrate (despite the inference from the 8 second period on March 5, 1979 that some neutron stars are born spinning too slowly to be pulsars) leads to a total release of magnetostatic energy of 1×10^{43} ergs per neutron star. In order that an exterior dipole field store this energy its equatorial value must be 6×10^{12} gauss. The energetics of the flare model are strained, though not impossible. The March 5, 1979 event, radiating 10^{44} ergs, individually poses a more serious problem.

It may be possible to regenerate magnetostatic energy from more ample sources. A number of possibilities are evident, although regeneration mechanisms are unclear. If pulsars are born with 1 second spin periods then the Galactic rate of production of their rotational energy is about 10^{38} erg/sec. Energy is lost at this rate during pulsar spindown. If they die when their periods decay to 3 seconds then their corpses are created with 10^{37} erg/sec of rotational energy; this latter energy may be more relevant because of the absence of an association between γ-ray bursters and active pulsars. Neutron stars also contain thermal energy. They are born with a great deal, but radiate most of it rapidly in neutrinos, and the remainder more slowly from their surfaces. It is unclear how to estimate the fraction conceivably available for regenerating magnetostatic energy; after 10^6 years 10^{42} ergs remain,[19] less than the required magnetostatic energy. Much younger neutron stars may contain orders of magnitude more thermal energy, particularly if their interiors do not contain pion condensates.

Thus both rotational and thermal energy may be sufficient to regenerate the required magnetostatic energy, if the original supply is inadequate. Possible mechanisms remain obscure. Greenstein[20] has suggested that energy stored in interior differential rotation may be released, perhaps breaking the crust and winding up the exterior magnetic field. Blandford[21] has suggested that thermoelectric currents may produce pulsar magnetic fields.

The energy problem of flare models is particularly acute for the March 5, 1979 event, which radiated about 5×10^{43} ergs in 0.15 second, and a similar amount of energy in the next 200 seconds.[1] A conceivable (speculative) interpretation is that the initial magnetostatic energy was dissipated in the first 0.15 second, and that the 200 second time scale is related to the

regeneration time. Regeneration from rotational energy is only
possible if differential rotation is present, because of the
constraint of the conservation of angular momentum. A significant
period change ($\sim$1-10%) is required by the magnitude of the event;
relaxation of differential rotation predicts that this change is
a spin-up, like a pulsar glitch.

CONCLUSIONS

In the early days of γ-ray burst studies dozens of models were
suggested. Most of them have been ruled out, but several possibilities remain. None of these is completely satisfactory. Just
as in radio and X-ray astronomy, obtaining optical identifications
has been necessary for progress. Much of our present understanding
depends on the identification of the March 5, 1979 event in the
LMC[1] and Schaefer's[13] discovery of a historical counterpart to
the November 19, 1978 event. We may hope that a few more accurate
positions and optical identifications will lead to equally great
progress.

By the time this paper appears I hope the question of the
optical counterpart[16,17] to the quiescent state of the November
19, 1978 event will have been resolved, and that we will know
whether the γ-ray burster has been detected, or whether some
other variable object, Galactic or extra-Galactic, is present.
I even hope that there will be some estimate of its polarization,
and hence of the radiation mechanism.

I have hardly mentioned a wide variety of theoretical problems
because only the most preliminary work has been done on most of
them. In the accretional and thermonuclear models the most important is the question of producing the hard observed spectrum.
In flare models there are questions of energy storage and release.
In all models the problems of particle acceleration, and radiation
are complex. Emissivities and opacities depend strongly on
particle energies, so that a very few energetic particles produce
nearly all the radiation, while the bulk of slower particles may
only carry currents and dissipate energy in plasma instabilities.
Relaxation processes are slow and distribution functions are non-
equilibrium. Pair production interferes with the free escape of
radiation above its threshold, though a model[9] argues that this
does not constrain burst luminosities and distances.

It would be particularly valuable to have optical observations
simultaneous with γ-ray observations. Modern photoelectric instrumentation may make this possible. A detector with 300,000
pixels, viewing the 4 steradians of sky at zenith angles less than
70° through a fisheye lens, divides the sky into cells of about
0.04 square degrees. The integrated sky brightness in such a
cell is about 7th magnitude, so that a 3rd magnitude outburst of
the type inferred by Schaefer[13] would be readily observable, even
with modest aperture. A small number of such systems, distributed
at clear dark sites around the world can provide good sky coverage.
The large background of atmospheric events (airplanes and meteors,

for example) can be eliminated using the parallax between pairs of nearby stations operating in coincidence, perhaps with orthogonal polarizers. The information processing requirements of such a system may be severe.

I thank R. Bussard and R. Walker for useful discussions, and NSF Grant AST 81-21704 for partial support.

REFERENCES

1. T. L. Cline, Comments Ap. $\underline{9}$, 13 (1980).
2. E. P. Mazets, this meeting.
3. B. J. Teegarden and T. L. Cline, Ap. J. (Letters) $\underline{236}$, L67 (1980).
4. E. P. Mazets, S. V. Golenetskii, R. L. Aptekar', Yu. A. Gur'yan, and V. N. Il'inskii, Nature $\underline{290}$, 378 (1981).
5. P. L. Nolan, G. H. Share, D. J. Forrest, E. L. Chupp, S. Matz, and E. Reiger, this meeting.
6. S. A. Colgate and A. G. Petschek, Ap. J. $\underline{248}$, 771 (1981).
7. S. E. Woosley and R. K. Wallace, Ap. J. $\underline{258}$, 716 (1982).
8. W. H. G. Lewin and P. C. Joss, Sp. Sci. Rev. $\underline{28}$, 3 (1981).
9. J. I. Katz, Ap. J. $\underline{260}$, 371 (1982).
10. E. P. Liang, Nature $\underline{299}$, 321 (1982).
11. J. K. Daugherty and R. W. Bussard, Ap. J. $\underline{238}$, 296 (1980).
12. J. E. Grindlay, E. L. Wright, and R. E. McCrosky, Ap. J. (Letters) $\underline{192}$, L113 (1974).
13. B. E. Schaefer, Nature $\underline{294}$, 722 (1981).
14. J. I. Katz, Astron. Ap. $\underline{39}$, 241 (1975).
15. M. C. Jennings and R. S. White, Ap. J. $\underline{238}$, 110 (1980).
16. B. E. Schaefer and H. Bradt, I.A.U. Circ. 2752 (1982).
17. H. Pedersen, M. Tarenghi, P. Grosbøl, and J. Danziger, I.A.U. Circ. 3711 (1982).
18. J. H. Taylor and R. N. Manchester, Ap. J. $\underline{215}$, 885 (1977).
19. G. Baym and C. Pethick, Ann. Rev. Astr. Ap. $\underline{17}$, 415 (1979).
20. G. Greenstein, Ap. J. $\underline{231}$, 880 (1979).
21. R. D. Blandford, private communication.

EMISSION MODEL OF GAMMA-RAY BURSTS

E. P. Liang
Lawrence Livermore National Laboratory,* University of California,
Livermore, Ca. 94550
and
Institute for Plasma Research,[+] Stanford University,
Stanford, Ca. 94305

ABSTRACT

This talk reviews the emission mechanisms of cosmic gamma-ray bursts. In particular, the thermal synchrotron model is discussed in detail as the most viable mechanism for the majority of the continuum emission. Within this framework various information about the source region can be extracted. The picture that emerges is that of a hot ($kT = .2 - 1.0$ mc^2), thin sheet of dense pair-dominated plasma emitting via cyclo-synchrotron radiation in a strong magnetic field ($B \sim 10^{11}$ to 10^{12} gauss). Speculations on the origin and structure of this sheet are attempted. We also briefly discuss the problem of high-energy photons above pair production threshold escaping from the source.

INTRODUCTION

Despite numerous attempts by astrophysicists over the past decade, the origin and mechanisms of cosmic gamma-ray bursts remain a total mystery. Yet a number of significant observational developments over the last few years have greatly narrowed the field of viable speculations even by the most creative theorists. (See, e.g., Ruderman[1] and Katz[2] for reviews.) The discovery of the March 5, 1979, event (Cline[3]) and the optical predecessor of the November 19, 1978, event (Schaefer[4]) plus the slightly more controversial discoveries by the Konus experiments (Mazets et al.[5]) of the presence of redshifted annihilation lines and low-energy spectral features, all help to reduce the number of viable candidates for the sites of these events. Currently, the most popular choice is the surface of strongly magnetized neutron stars. Theoretically, a strong magnetic field ($> 10^{10}$ gauss) is also needed to confine such a hot plasma, especially if it is pair dominated. In this talk I shall try to review in some depth recent attempts along these lines to understand the emission mechanism for the continuum spectrum of most gamma-ray bursts.

THE THERMAL SYNCHROTRON MODEL

Ever since the early days of their discovery, it was recognized that most of the gamma-burst spectra assume a universal exponential

---

*Operated under DOE Contract W7405-ENG48
[+]Partially supported by NASA Grant NGR 05-020-668

shape with a characteristic $kT \sim mc^2$. Thus it was suggested that they were optically thin thermal bremsstrahlung (TB) emission by mildly relativistic thermal plasmas. Unfortunately, this interpretation immediately encountered difficulty because of the inefficiency of TB emission. Combined with the lack of detectable Comptonization ($\tau_{es} < 1$), TB emission required unrealistically high aspect ratios for the emission geometry and nearby clustering of the sources (e.g., Katz[6]). If we also believe in strong magnetic fields, then the cyclo-synchrotron emission of these hot plasmas would also greatly exceed the TB emission for all reasonable situations. Moreover, if the March 5, 1979, event is indeed at the distance of N49 ($\sim$ 55 kpc), then only the synchrotron emissivity of a hot plasma in a strong field has the remote chance of accounting for the high luminosity (Ramaty et al.[7], Liang[8]). This, therefore, motivated several authors to suggest that cyclo-synchrotron emission is the natural emission mechanism of most gamma-ray bursts (Lamb[9], Katz[6], and Liang[10]). Recently, using the semi-analytic results of Petrosian[11], Trubnikov[12] and the numerical results of Lamb and Masters[13], we have succeeded in fitting most gamma-burst spectra reported to date satisfactorily with the thermal synchrotron spectrum (TS) of mildly relativistic (.2 - 1.0 mc^2) plasmas in strong fields (B $\lesssim 10^{12}$ gauss) (Liang[10], Liang et al.[14]). This is encouraging because it at least makes the strong field neutron star picture a self-consistent framework. In addition, various additional spectral features, when interpreted in this framework, provide us with valuable information about the source conditions.

Figure 1 illustrates the shape of the thermal cyclotron spectrum as the temperature is progressively increased. By the time the temperature gets up to hundreds of keV, all higher harmonics blend into a smooth continuum with only the first couple of harmonics barely visible, their peaks time-dilated to energies below the Larmor frequency (ν_L = 11.6 keV (B/10^{12} gauss)).

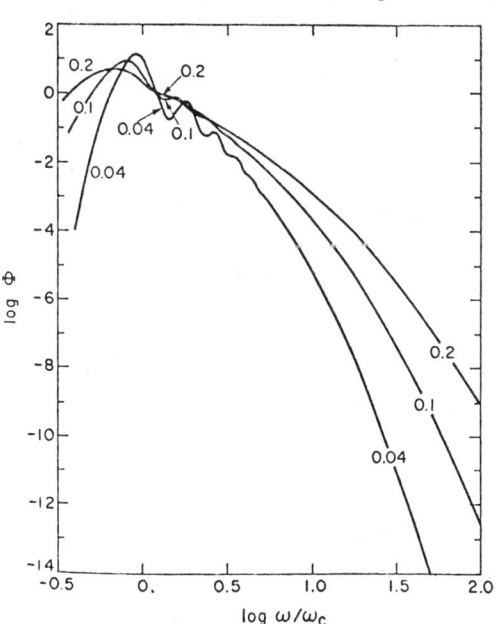

Fig. 1. Evolution of thermal cyclo-synchrotron spectra with increasing temperature (from Ref. 9). ($\Phi \propto j_\omega/\omega^2$; T is in units of $m_e c^2$.)

Above the third or fourth harmonic, the continuum emissivity is well approximated by the analytic formula (cf. Ref. 14):

$$j_\nu(\theta) = \frac{\pi e^2}{\sqrt{2}\cdot 3\, c} n_e\, \nu K_2^{-1}(1/T)\, \exp(-(4.5\nu/\nu_c \sin\theta)^{1/3}),\ \nu_c \equiv \nu_L T^2\ ,\quad (1)$$

where $T \equiv kT/m_e c^2$ and K_2 is a modified Bessel function.

Figure 2 shows how well this shape fits the typical gamma-ray burst spectra.

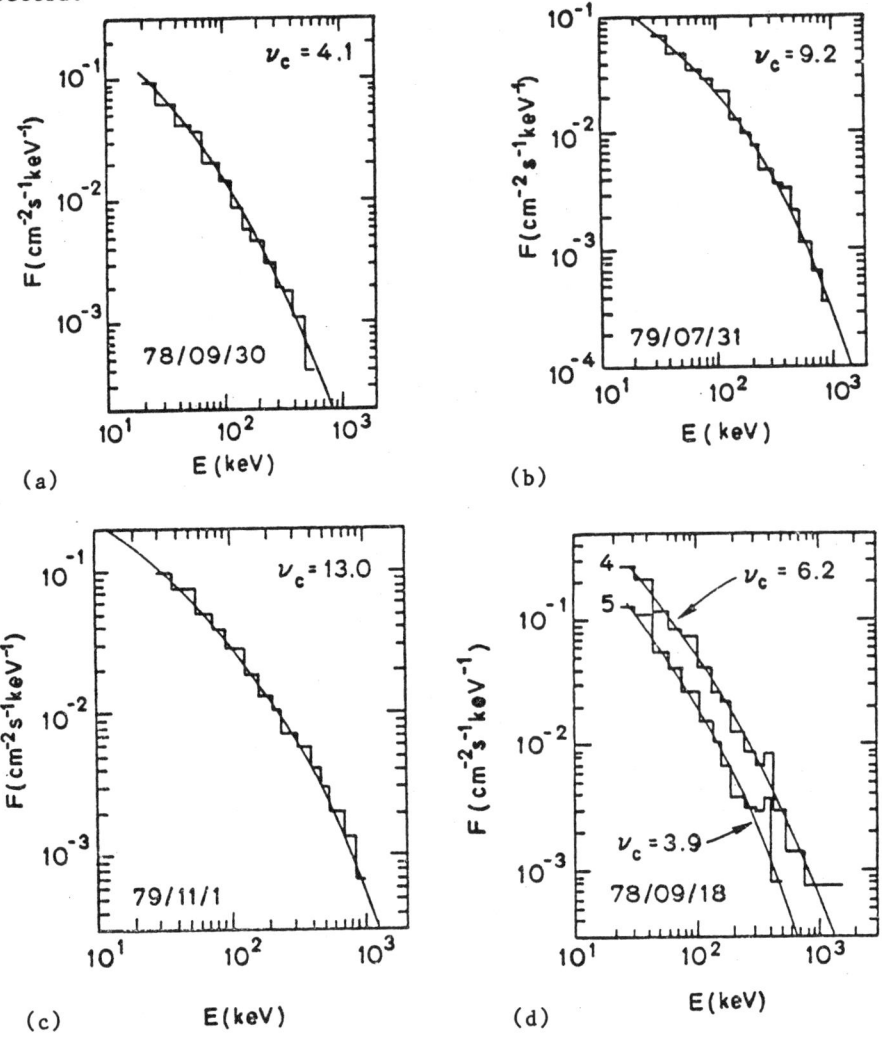

Fig. 2. TS fits to some typical gamma-bursts spectra. Value of ν_c is in keV (from Ref. 14).

When the emission column density is too large, the low-energy part of the spectrum becomes self-absorbed and turns over into a Rayleigh Jeans spectrum. The location of the turnover, ν_m, determines (cf. Bekefi[15]) the emission column density and therefore the optically thin flux. These are given by the formulas:

$$n_e h = 3.8 \times 10^{19} \nu_m T K_2(T^{-1}) e^{x_m} \text{ cm}^{-2}, \quad x_m = (4.5 \nu_m/\nu_c)^{1/3} \qquad (2)$$

$$F_{syn} \cong 3.6 \times 10^{26} T \nu_{m\,keV} \nu_{c\,keV}^2 \sum_{i=0}^{5} x_m^i/i! \text{ erg} \cdot \text{cm}^{-2} \cdot \text{sec}. \qquad (3)$$

DATA FROM THE KONUS CATALOGUE

Recently Liang et al.[14] have completed a detailed analysis of the entire Konus Catalogue (Mazets et al.[16]), which represents the largest collection of recorded gamma-burst spectra, using the TS model. Some of the key results are summarized here.

(a) The characteristic frequency ν_c has a distribution peaking near 3 keV and cutting off sharply above 12 keV. For a nominal field of 2×10^{12} gauss, this means a temperature distribution of around $.4 \text{ mc}^2$ (Fig. 3).

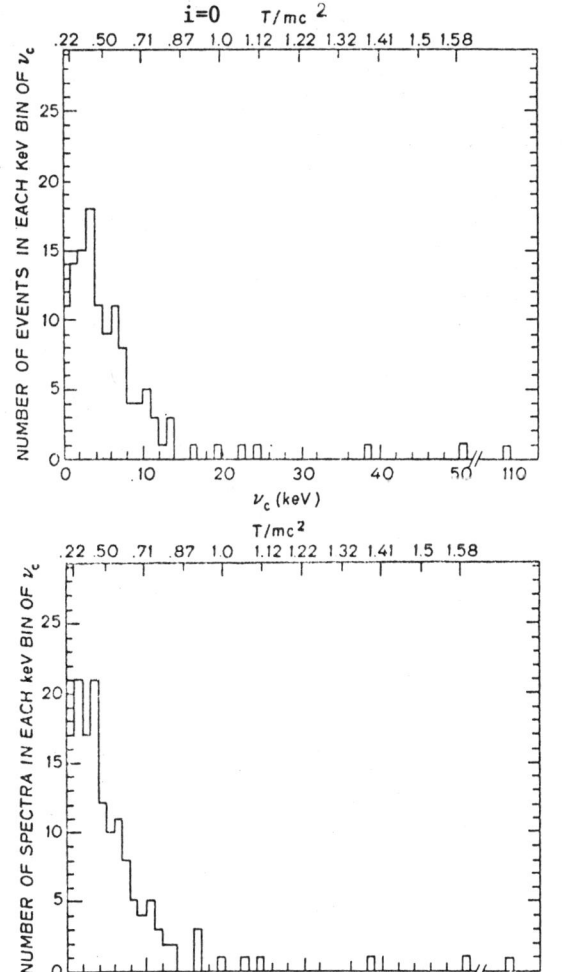

Fig. 3. Distribution of ν_c for the spectra of the Konus catalogue (from Ref. 14).

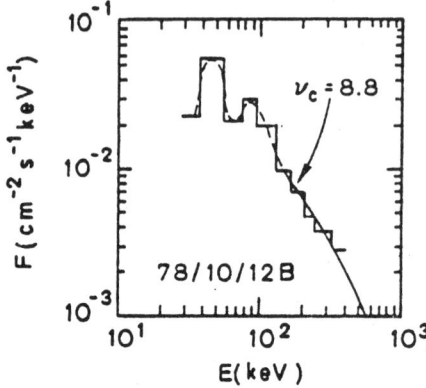

Fig. 4. Example of Konus spectrum showing possible first and second harmonic emissions (from Ref. 14).

(b) Over half a dozen events show double peak features at low energies (< 70 keV) with the second peak sitting at twice the frequency of the first, suggestive of fundamental harmonics (Fig. 4). Using the temperature deduced from the fit to the continuum, we can try to estimate the ratio of the first to the second harmonic peak flux. Figure 5 compares the theory prediction and the observed flux ratio. The result is clearly very encouraging. Future observations should concentrate on the search of the harmonics, in the X-ray energies, if possible.

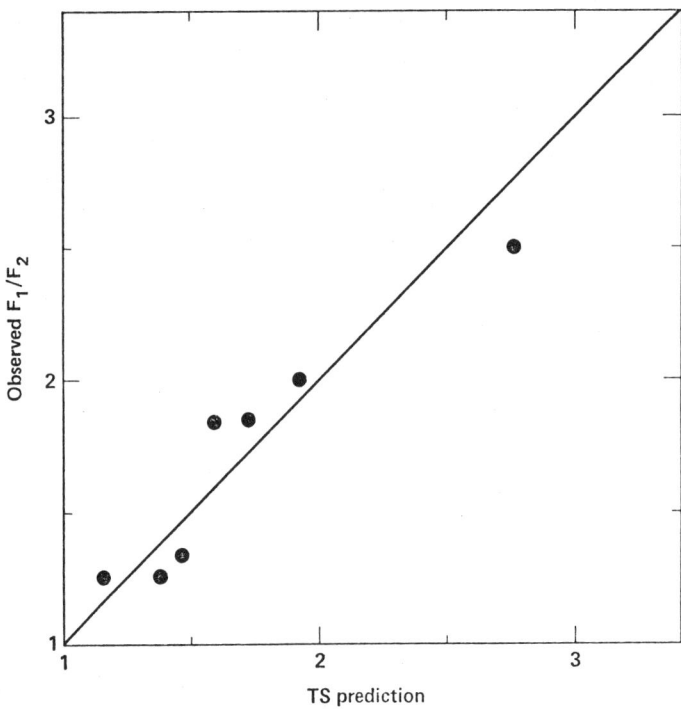

Fig. 5. Comparison between theoretical predictions and the observed ratios of first to second harmonic peak fluxes for several Konus spectra.

(c) Most of the Konus data, which cover only the range up to several hundred keV to 1 MeV, really cannot distinguish between TS, TB, or inverse Compton (IC) spectral fits (proposed by Fenimore et al.[17]). (See Fig. 6a.) However, preliminary data from the SMM (Nolan et al.[18], Fig. 6b) up to 10 MeV seem to suggest that single temperature TB or IC fits would fall short at high energies due to the exponential cutoff, whereas the TS spectrum has no problem because of its hardness (cf. Eqn. (1)).

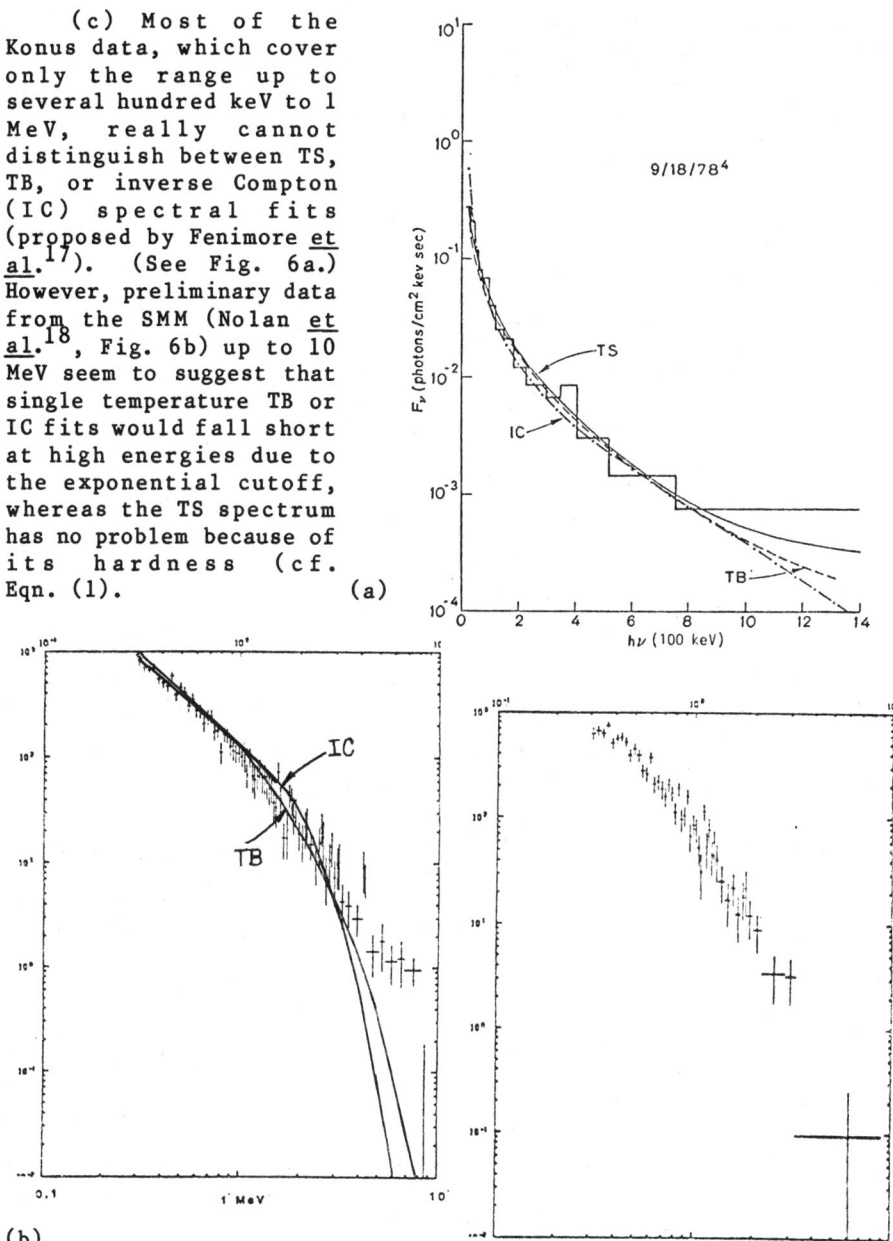

Fig. 6. (a) TS, TB and IC fits to the observed spectrum of GB780918 show little distinction (from Ref. 14). (b) SMM gamma-burst spectra (raw data) out to 10 MeV show that TB and IC fits fall short at higher energies (from Ref. 18; the high-energy data are expected to move up after detector corrections).

(d) A small fraction of Konus spectra show low energy turnover, suggestive of self-absorption (Fig. 7).

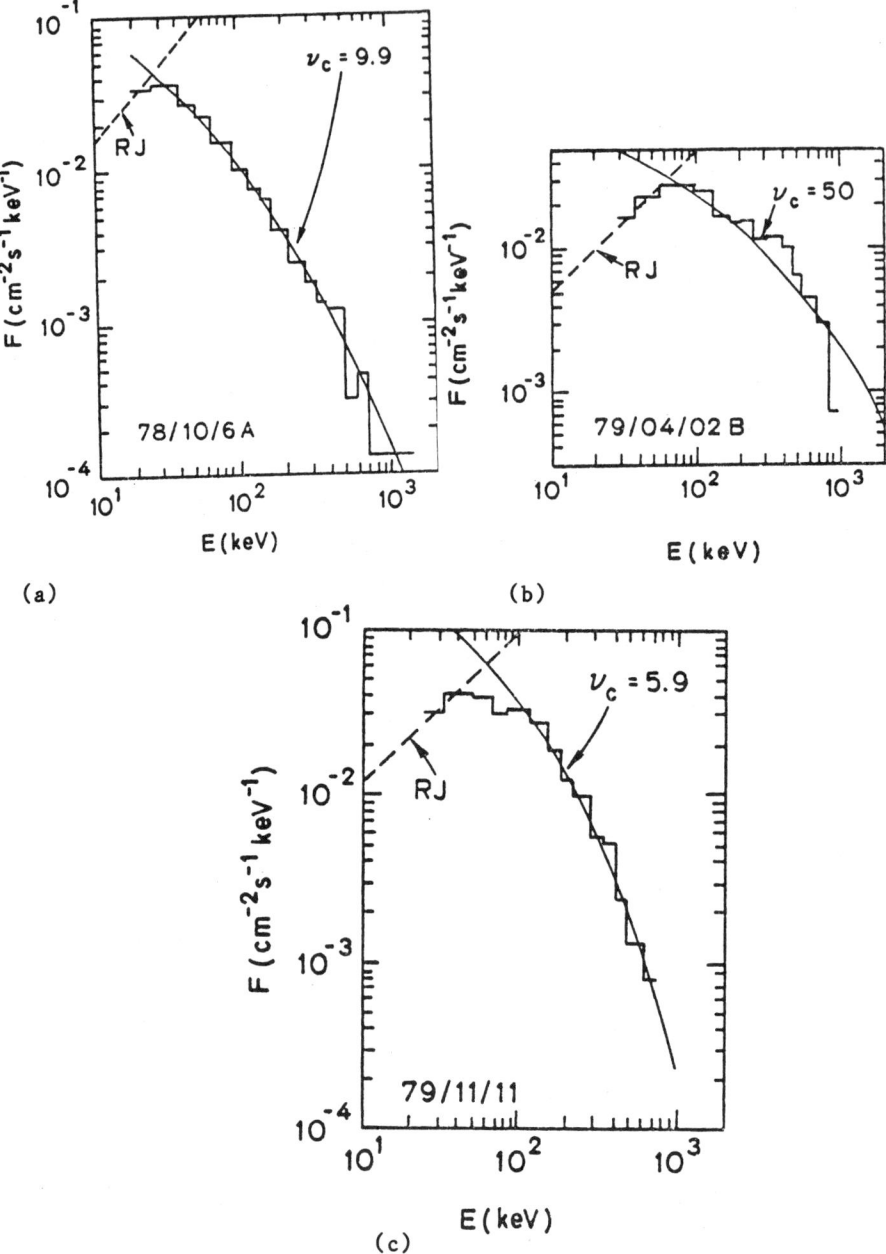

Fig. 7. Examples of Konus spectra showing self-absorption (from Ref. 14).

Table I summarizes data on these events. Note that unless the emission area is much smaller than a km^2, several events could be extragalactic, maybe even as far as the March 5 event. Note also the uniform thinness of the emission depth ($n_e h \sim 10^{20} - 10^{21}$ cm^{-2}).

Table I Spectra with self-absorption (data from Ref. 14)

Spectrum	ν_{ab} (keV)	ν_c (keV)	$n_e h/TK_2(T^{-1})$ (10^{22}/cm^2)	$T \geq$	Assuming $\nu_L^o \leq \nu_{ab}$		
					$\tau_{es}(10^{-4}) \geq$ Thompson scattering depth	$\dfrac{L_{syn}{}^a}{A(km^2)} \geq$	$\dfrac{d(kpc)^b}{A^{1/2}(km)} \geq$
10-06-78A	30	9.9	1.3	.39	1.8	1.6×10^{40}	18
11-04-78A	25	2.1	4.2	.22	.27	1.1×10^{39}	6
11-07-78	42	6.0	3.8	.29	1.4	1.2×10^{40}	22
11-11-78	28	1.9	6.1	.20	.2	1.1×10^{39}	65
11-19-78[1]	40	110	0.5	.87	.16	1.7×10^{42}	23
11-19-78[3]	40	6.0	3.4	.29	1.2	1.1×10^{40}	6
11-21-78A	26	4.8	1.8	.31	1.0	3.9×10^{39}	11
2-14-79	30	4.1	2.8	.28	0.8	3.9×10^{39}	7.7
4-02-79B[1]	60	50	1.4	.56	7.7	6×10^{41}	29
4-06-79	45	38	1.0	.56	5.5	2.7×10^{41}	30
6-13-79	50	1.9	25.7	.17	.26	3.3×10^{39}	7
10-14-79[1]	40	5.9	3.5	.28	.9	1.0×10^{40}	9
11-11-79[1]	50	5.9	5.5	.24	.5	1.3×10^{40}	18

[a] Multiply all numbers in this column by the factor 2.9

[b] Multiply all numbers in this column by the factor 1.7

(e) Many of the Konus spectra have candidate annihilation lines at 400 - 450 keV. Some of these appear in conjunction with low-energy self-absorption, in which case the pair density at the source can be estimated if we assume that the annihilation line source coincides with the TS source and that the annilation region is approximately one annihilation depth thick. Table II summarizes the data on these spectra. Note that n_+ lies in the range 10^{23} to 10^{26}. Combined with the nh limits, we are forced to consider very thin emission layers with $h \sim 10^{-3}$ to 10^{-4} cm.

Table II Data on Konus spectra exhibiting redshifted pair annihilation lines

Event (GB)	511 flux (ph/cm²·s)	ν_c (keV)	$\frac{d(kpc)}{A^{1/2}(km)}$*	n_+ (cm⁻³) (assume $n_-\sigma_T h_{annih} \sim 1$)	$\frac{F^{511}}{F_{>threshold}}$
78-09-18⁴	.18	6.2			.14
78-09-18⁵	.08	3.9			.26
78-10-06A sa	.01	10.0	≥ 30	≥ 6.0 × 10²³	.04
78-11-19¹ sa	5.0	110.	≥ 38	≥ 3.0 × 10²⁶	.12
78-11-19³ sa	.15	6.	≥ 38	≥ 9.2 × 10²⁴	.35
79-01-16	.05	5.			.15
79-03-05**	1.7	0.5	~ 1	≥ 5.0 × 10²⁶	.42
79-04-02¹ sa	.3	50.	~ 50⁺	≅ 4.8 × 10²⁵	.08
79-04-02³	.15	13.			.21
79-04-12	.12	19.			.08
79-06-22	.10	13.			.13
79-10-14 sa	.23	6.	≥ 15	≥ 3.6 × 10²⁴	.30
79-11-11¹ sa	.07	6.	~ 30⁺	≅ 4.2 × 10²⁴	.09
79-12-30	.06	5.8			.35

          ↑   ↑
      harder   lie in range
      than average $10^{23}-10^{26}$ => $h_{syn} \leq 10^{-3}-10^{-4}$ cm

*for self-abs. cases
**assumed at 55 kpc
⁺estimated from presence of harmonics at later times

To summarize, the TS interpretation of gamma-burst spectra requires the emission region to be hot (kT ≃ .2 - 1.0 mc²), optically very thin (nh ≲ $10^{20} - 10^{22}$ cm⁻²) with typical synchrotron flux $10^{29} - 10^{32}$ erg/sec. cm². Events showing pair-annihilation lines suggest that the emitting plasma may be dominated by pairs ($n_+ \sim 10^{23}-10^{26}$ cm⁻³). Hence the big questions are: What is the origin of such an unusual emission layer, and how is it maintained?

STRUCTURE OF THE THIN EMISSION SHEET

Figure 8 illustrates conceptually one possible configuration of the emission layer. The surface of the neutron star, threaded by 10^{12} gauss field lines, is heated by energy fluxes streaming down along the field lines. To avoid shielding the outgoing gamma rays, the downward energy flux cannot be in the form of particles (ions, electrons or pairs). More likely it is in the form of electromagnetic or Alfvén waves. These waves dissipate within a skin depth of the surface, generating suprathermal electrons and creating pairs. These then pitch-angle scatter and thermalize within a column density corresponding to a pitch-angle scattering depth. This is also the region where the hot thermal synchrotron radiation is emitted. The cooled pairs then annihilate over a thicker layer, corresponding to about one annihilation depth for positrons. The emission region is confined sideways by the magnetic field. Vertically it is held down by the momentum flux (or "RAM pressure") of the same waves that are heating it. It is also known that the standard coulomb scattering between protons and pairs cannot maintain a thermal distribution due to the much faster synchrotron decay rate (Langer[19], Bussard[20], Bussard and Lamb[21]). It seems likely that either collective processes, which operate at close to one tenth of the electron plasma frequency, or coulomb scattering with heavy ions (e.g., Z = 26) must dominate to maintain a thermal population of higher Landau levels. In fact, one can conceive of a self-adjusting mechanism in which the pair density is maintained at a level in which the decay rate matches the pitch-angle scattering rate. When the pair density is too low, synchrotron cooling is inefficient because it is governed by the pitch-angle scattering rate, and the heated

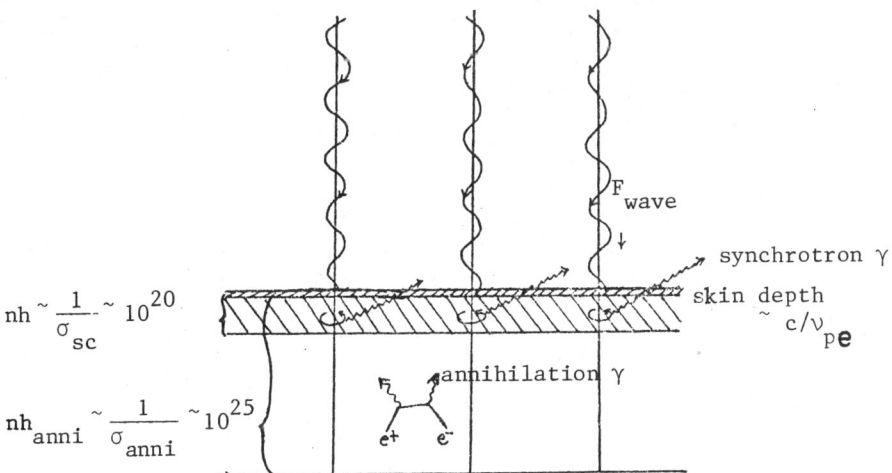

Fig. 8. Schematic diagram illustrating a possible configuration of the emitting layer of the gamma rays.

region can cool only by creating more pairs. On the other hand, when there are too many pairs, thermal and radiation pressure would exceed the wave pressure, and the layer would expand and decrease the density. While this scenario requires more detailed investigation, one can at least derive the steady-state structure from simple first-principles arguments. It turns out that the three structure variables -- n (assumed equal to $2n_+ = 2n_-$), h and T -- are uniquely determined by three steady-state requirements (for details see Liang and Antiochos[25]):

(a) energy balance:

$$F_{wave} \sim nh \dot{\varepsilon}_{syn}, \text{ where } \dot{\varepsilon}_{syn} \text{ is syn cooling rate}$$

(b) momentum balance:

$$\frac{F_{wave}}{c} = \text{pressure}_{wave} \sim n\, mc^2 T$$

(c) thermalization requirement:

$$\lambda_{pitch\ scatt} \ll \lambda_{syn\ cool} \Rightarrow h \sim (\lambda_{pitch\ scatt}\, \lambda_{syn\ cool})^{1/2}$$

For example, assuming that pitch-angle scattering is dominated by plasma collective modes, we find:

$$T \sim 0.4\ (F_{wave}/10^{30})^{.2}(B/10^{12})^{-.8};\ nh \sim 2.5 \times 10^{21} \left(\frac{F_{wave}}{10^{30}}\right)^{.6}\left(\frac{B}{10^{12}}\right)^{.4}$$

etc.,

which falls in the general ballpark of the observed parameters. These results also lead to additional predictions about correlations between the above variables and the field strength. This could be tested with future observations.

MAGNETIC FLARE MODELS

Where do the energies come from that heat the above conjectured layer? The most natural candidate seems to be reconnection of closed magnetic loops. The above sheets would then be sitting at the footpoints of the reconnecting flux tubes. Because of the high field strength, most of the reconnection energy release would likely be in the form of field perturbations rather than particles. At a flux of $\sim 10^{30}$ erg/cm$^2 \cdot$ s, a perturbation of a few percent of a 10^{12} gauss field would be adequate to account for the majority of the bursts.

What could be causing the flux tubes to develop non-potential stress fields? There are at least three conventional sources of primary energy: transient accretion (e.g., Colgate and Petschek[22]) including impact by comets or satellites; surface thermal nuclear explosion (Woosley[23]); and internal disturbances, including vibration (Ramaty et al.[7]) and differential rotations. At present we have no idea how any one of these energy sources couples to the field.

However, both the accretion and explosion pictures involve primary energy sources which are highly optically thick and therefore must be accompanied by cooler X-ray burst precursors. The hypothesis of internal energy sources does have the advantage of being capable of bypassing any optically thick phase by coupling mechanical energy of the star directly to the field in a low-density environment. For example, twisting of the footpoints due to relative motion or restructuring of the crust could produce stressed fields. In fact, it is highly likely that gamma-ray burst sources may involve a totally different class of neutron stars from radio or X-ray pulsars, which are believed to have rigid dipolar fields.

It should be mentioned that at least the majority of gamma-ray bursts, excluding the subclass which Mazets et al.[24] called the short bursts, have time structures analogous to solar flares, with a typical duration of tens of seconds broken into many spikes of subsecond duration and rise times of milliseconds or less. In the magnetic reconnection model of solar flares, most authors tend to associate the overall duration with the linear growth time of the resistive tearing mode given by the geometric mean of the Alfvén time and magnetic diffusion time, while the spikes are associated with the nonlinear Petschek type growth time equal to 10-100 Alfvén time scales. It is interesting that at least in the case of gamma-ray bursts, these two time scales also fall into the general range of the observed time scales (see Liang and Antiochos[25]).

ESCAPE OF THE HIGH-ENERGY GAMMAS

The latest SMM data (cf. Fig. 6b) shows that up to 10 MeV the gamma rays seem to emerge unattenuated by pair production self-absorption. Some authors try to argue that this puts a strong constraint on the source distance. However, as Katz[6] has pointed out, this is not necessarily the case since reannihilation may compensate for the removed photons. However, no detailed transport calculation has been attempted, and it is not clear how the original single-temperature, optically thin spectrum will be altered. In the case of TS emission, we might be saved by the fact that the highest energy gammas are all emitted close to 90 degrees from the field orientations, and therefore the relative angle with which gamma-gamma collision can occur, at least for the gamma along the line of sight, would be smaller than for isotropic sources.

A more severe difficulty is presented by the apparently unavoidable gamma-B collisions. Since gamma-B pair production cross-section depends exponentially on the parameter $h\nu/m_e c^2 (B/4.4 \times 10^{13}$ G$)$ (see Erber[26]), a 1 MeV photon would be completely wiped out by a $3 \cdot 10^{12}$ gauss field orthogonal to its path but emerge untouched from a $2 \cdot 10^{12}$ gauss field. Similarly, a $3 \cdot 10^{11}$ gauss field is opaque to a 10MeV photon, but a $2 \cdot 10^{11}$ gauss field is transparent. Unless there is something we totally miss here, observation of unattenuated spectra up to 10 MeV can only be possible if the field is quite weak or we are viewing at small angles with respect to the field lines, in which case the emission temperature must be quite high. In any case, this whole area is currently under investigation.

REFERENCES

1. M. Ruderman, Ann. N. Y. Acad. Sci. $\underline{262}$, 164 (1975).
2. J. I. Katz, these proceedings.
3. T. L. Cline, AIP Conf. Proc. No. 77, 17 (AIP, N. Y., 1982).
4. B. Schaefer, these proceedings.
5. E. P. Mazets, S. V. Golenetskii, V. N. Il'inskii, R. L. Aptekar', and Yu. A. Guryan, Nature $\underline{282}$, 587 (1979).
6. J. I. Katz, Ap. J. $\underline{260}$, 370 (1982).
7. R. Ramaty, R. E. Lingenfelter, and R. W. Bussard, Ap. Sp. Sci. $\underline{75}$, 193 (1981).
8. E. P. T. Liang, Nature $\underline{292}$, 319 (1981).
9. D. Q. Lamb, AIP Conf. Proc. No. 77, 249 (AIP, N. Y., 1982).
10. E. P. T. Liang, Nature $\underline{299}$, 321 (1982).
11. V. Petrosian, Ap. J. $\underline{251}$, 727 (1981).
12. B. A. Trubnikov, Phys. Fluids. $\underline{4}$, 195 (1961).
13. D. Q. Lamb and A. R. Masters, Ap. J. (Letters) $\underline{234}$, L117 (1979).
14. E. P. Liang, T. E. Jernigan, and R. Rodrigues, Ap. J., Aug. 15 issue, to appear (1983).
15. G. Bekefi, Radiation Processes in Plasmas (Wiley, N. Y., 1966).
16. E. P. Mazets et al., Ap. Sp. Sci. $\underline{80}$, 7 (1981).
17. E. E. Fenimore et al., COSPAR Symp. Proc. (Ottawa, Canada, 1982).
18. P. Nolan et al., these proceedings.
19. S. H. Langer, Phys. Rev. D$\underline{23}$, 328 (1981).
20. R. W. Bussard, Ap. J. $\underline{237}$, 970 (1980).
21. R. W. Bussard and F. K. Lamb, AIP Conf. Proc. No. 77, 189 (AIP, N. Y., 1982).
22. S. A. Colgate and A. G. Petschek, LANL preprint, submitted to Ap. J. (1982).
23. S. E. Woosley, AIP Conf. Proc. No. 77, 273 (1982).
24. E. P. Mazets, S. V. Golenetskii, Yu. A. Guryan, V. N. Ilyinskii, PTI (Leningrad), preprint, 738 (1981).
25. E. P. Liang and S. Antiochos, IPR (Stanford) preprint, submitted to Nature (1983).
26. T. Erber, Rev. Mod. Phys. $\underline{28}$, 626 (1966).

A FIREBALL MODEL FOR THE MARCH 25, 1978 GAMMA RAY BURST

Geoffrey J. Hueter and Richard E. Lingenfelter
Center for Astrophysics and Space Science
University of California, La Jolla, CA 92093

ABSTRACT

We suggest that the bulk of the gamma ray emission in the March 25, 1978 burst comes from a fireball of photons and $e^{\pm}$ pairs expanding from a jet off the polar cap of a neutron star and that the duration of the burst is defined by the expansion and dissipation times of the fireball. This model enables us to determine the total energy, $\sim 10^{39}$ ergs, and the distance, $\sim$ 1kpc, of the burst and should be applicable to a large class of gamma ray bursts.

INTRODUCTION

We present a fireball model for the time history and hard spectrum of the gamma ray burst which was seen on March 25, 1978 by the UCSD/MIT Experiment on HEAO-1. Over 80% of the burst energy is in a hard spectral component between $\sim$ 0.25 and 6 MeV. Requiring that the photon-photon pair production opacity in these hard photons be of the order of unity or less imposes a strong constraint on the minimum size of the hard emission region. This size greatly exceeds the diameter of a neutron star, unless the star is uncomfortably close (i.e. <0.1 pc). Thus, if the radiated energy is released in any smaller volume, the hard photons will produce $e^{\pm}$ pairs and expand as a fireball[1] until it becomes optically thin and the radiation breaks away. We suggest that the expansion and dissipation times of such a fireball in fact define the duration of the burst. If so, the observed duration provides a unique determination of the total energy released in hard radiation, which, together with the measured fluence, gives a direct determination of the distance.

Since similar hard components have been measured[2,3] in many gamma ray burst spectra, we propose that this model offers an important diagnostic for determining the intrinsic luminosities and distances of gamma ray burst sources.

The soft (<0.25 MeV) component of the burst spectrum has a strong absorption feature at 55 ± 5 keV, suggesting that this component is produced in an intense magnetic field ($\sim 5 \times 10^{12}$ gauss) close to the surface of a neutron star. In such an intense field the plasma which forms the $\sim$ MeV fireball responsible for the hard emission must first escape from the surface in a jet aligned with a polar magnetic field.

A polar cap origin of the burst also favors a thermonuclear explosion of accreted gas rather than a solid body impact as the source of the burst energy.

OBSERVATION

The March 25, 1978 gamma ray burst was discovered[4] in an ongoing search for gamma ray bursts in the data of the UCSD/MIT HEAO-1 High Energy X-ray and Low Energy Gamma Ray Experiment. The integrated flux, or fluence, between 15 keV and 6 Mev is 1.5×10^{-5} ergs/cm^2. The source direction, determined from the correlation of the responses of the two 1.7° × 20° FWHM detectors, lies in Ursa Minor in an error circle of 0.8° about RA=237.5°, Dec=76.2°, or l^{II}=112°, b^{II}=37°.

The time profile of the burst flux is shown in Figure 1, which gives the 0.03–6 MeV counting rate. The time variation of the >0.25 MeV flux is consistent with that of the entire instrument passband down to the minimum spectral accumulation interval of 5.12 seconds.

The spectrum of the emission peak at UT 06:28:52 to 06:29:02 is also shown in Figure 1. The absorption feature at 55 keV has an equivalent width of 13 ± 3 keV, consistent with the values reported[5] for the KONUS bursts. Thus the existence of these features in gamma ray bursts is confirmed. The soft component (<0.25 MeV) and its absorption feature can be understood[6] in terms of gyrosynchrotron emission and cyclotron absorption by a plasma in an intense magnetic field of as much as 5×10^{12} gauss near the surface of a neutron star. Details of the spectral evolution of the burst will appear in a forthcoming paper[7].

Here we focus our attention on the hard component between ∼ 0.25 and 6 MeV, which accounts for 80% of the observed emission from the burst.

THE FIREBALL MODEL

Schmidt[8] has used the opacity of gamma rays to photon-photon pair production to set an upper limit of a few kpc on the distance to a typical burst source which emits MeV photons. He determined the distance from the observed flux, assuming an absolute luminosity derived from a fixed source size of 10^9 cm, which corresponds to the light travel distance for the shortest fluctuations (∼ 30 ms) seen in typical bursts.

We use the opacity constraint in a different way. We assume that the hard photon emission comes from a fireball and that the total burst energy is deposited in a time short compared to the observed duration of the burst. We then determine the total energy in ∼ MeV photons and $e^{\pm}$ pairs in the fireball from the observed burst duration, assuming that the duration is defined by the time required for the fireball to expand to unit opacity and dissipate as the radiation breaks out and the pairs annihilate. From the inferred fireball energy and the measured fluence we can then directly determine the distance.

We extrapolate the observed radiation density back to a distance and hence a source size at which the radiation becomes opaque to photon-photon pair production. We assume that the energy is released in a smaller volume and expands as an opaque fireball of photons and $e^{\pm}$ pairs in which pair production and annihilation are in equilibrium and the average photon and lepton energies are equal $E_\gamma = \gamma m_e c^2$. If the hard photon energy in the burst is W, then the total number of photons and leptons in the fireball is

$$N \simeq W/\gamma m_e c^2.$$

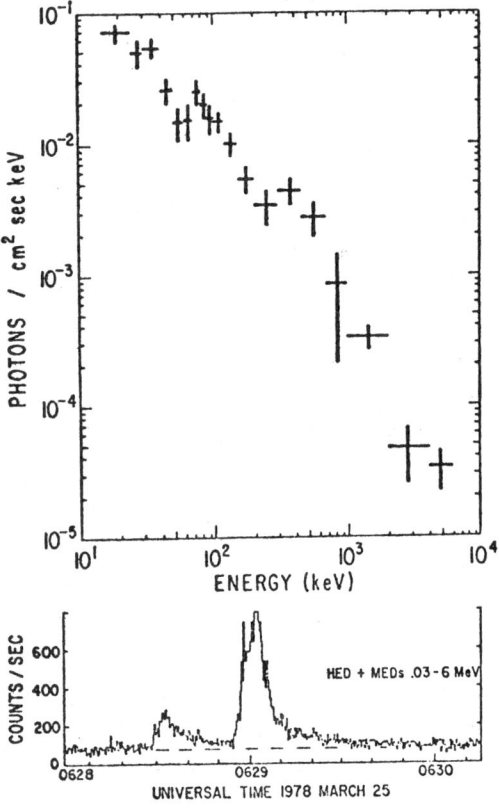

Figure 1. March 25, 1978 gamma ray burst counting rate (0.03−6 MeV) and peak energy spectrum at 06:28:52−06:29:02 UT, showing the absorption feature at 55 keV and the hard component at >0.25 MeV.

The cross sections for photon-photon pair production and for Compton scattering at these energies are comparable; approximately, $\sigma_{\gamma\gamma} \simeq \sigma_T \gamma^{-2}$, where $\sigma_T = 6.7 \times 10^{-25}$ cm^2 is the Thomson cross section. If we consider that in the fireball there is equipartition between pairs and photons, i.e., $n_\gamma \simeq n_{e\pm}$, Compton scattering and pair production contribute equally to the photon opacity. If the size of the emission region, which we assume for simplicity to be spherical, is l, then it will have unit optical depth when

$$l \simeq (n\sigma_{\gamma\gamma})^{-1} \simeq \gamma^2/n\sigma_T,$$

where

$$n = N/\text{vol} = 6W/\pi\gamma m_e c^2 l^3.$$

Substituting for n,

$$l = 4 \times 10^{10} W_{39}^{1/2} \gamma^{-3/2} \text{ cm},$$

where W_{39} is the burst energy in units of 10^{39} ergs.

If the energy is deposited in any smaller volume than this, then the fireball must necessarily expand until it reaches this size and becomes optically thin. The expected risetime for the burst is thus the time for the fireball to expand to a size l, i.e.,

$$t_r = l/\beta_e c = 1.3 W_{39}^{1/2} \gamma^{-3/2} \beta_e^{-1} \text{ sec}.$$

In a fireball of photons and pairs we would expect the relative expansion velocity β_e to be roughly on the order of 1/3, the sound speed of an $\sim$ MeV plasma. This is probably an upper bound on β_e, since the expansion will be slower if some matter (e.g. protons) is swept up in the fireball.

The burst decay timescale is determined by the annihilation time of the remaining pairs and the continued expansion of the pair plasma. The time for annihilation of a positron in a relativistic electron plasma is[9]

$$\tau_{\text{ann}} = 1.1 \times 10^{14} \gamma^2/n_{e^-},$$

where $n_{e^-} = n/4$. Substituting n_{e^-} at the time the fireball becomes optically thin,

$$\tau_{\text{ann}} = 12 W_{39}^{1/2} \gamma^{-3/2} \text{sec}.$$

Since this time is longer than the expansion time for the uncontaminated fireball, subsequent emission by pair annihilation may be suppressed by the continued reduction of the pair density due to expansion. It is possible, however, that pair annihilation could proceed much more rapidly by stimulated annihilation, which might result from the population inversion set up because the escaping photon population is depleted on a timescale shorter than the annihilation time of the lepton population. If so, such "grasar" action[10] could produce the apparent peak in the hard component at $\sim$ 0.4 MeV.

In view of the complexity of the fireball expansion, breakout and dissipation, we simply approximate the duration of the burst emission by twice the nominal risetime. Thus

$$t_{\text{FWHM}} \simeq 2t_r = 2.6 W_{39}^{1/2} \gamma^{-3/2} \beta_e^{-1} \text{ sec}.$$

Observationally, an appropriate definition of the burst duration for this model is

$$t_o \equiv S/P,$$

where S is the fluence and P is the peak flux. Since $S = W/4\pi d^2$, the source distance d is then

$$d \simeq 1.1 \times 10^{-3} \beta_e \gamma^{3/2} t_o S^{-1/2} \text{ kpc},$$

where S is measured in erg/cm^2 and t_o in seconds.

For the March 25, 1978 burst we find a duration $t \simeq 9$ sec and an average photon energy $E_\gamma/m_e c^2 = \gamma \simeq 1.4$, which gives a fireball energy $W \sim 4 \times 10^{39}$ ergs for a clean fireball ($\beta_e \simeq 1/3$) or around 4×10^{38} for a dirty one (e.g. $\beta_e \simeq 0.1$). The measured fluence of the hard component $S = 1.0 \times 10^{-5}$ erg/cm^2 then gives a source distance d of 1.7 to ~ 0.6 kpc, depending on how dirty the fireball is. At a galactic latitude b^{II} of 37°, this places the burst between a few hundred parsecs and a kiloparsec above the galactic plane, which is not inconsistent with a scale height comparable to that of pulsars[11]. Thus the estimated distance suggests a galactic population of gamma ray burst sources generally and a population of fast neutron stars in particular.

We now consider the constraints which the observed high energy component imposes on the expansion of the fireball before it becomes transparent to pair production. Free expansion does not reduce the average energy of the photons and leptons, it only cools their random motion. Unless there is a large amount of entrained material in the fireball, gravitational energy loss will also be insignificant because the escape energy of leptons from a neutron star is much less than an MeV and the gravitational redshift of photons is only on the order of 10%. A fireball can be cooled, however, by photon producing processes, such as bremsstrahlung and synchrotron emission, which reduce the average energy by increasing the number of photons that share the total energy. Thus these two processes place constraints on any fireball that must remain as hot as an MeV.

In particular, we must reconcile the absorption feature in the soft component, which implies emission in an intense magnetic field ($\sim 5 \times 10^{12}$ gauss) close to the surface of a neutron star, with the photon-photon pair production opacity limit on the hard component, which requires emission at a large distance above the surface of such a star. Since the synchrotron cooling time for MeV electrons moving across a magnetic field of 5×10^{12} gauss is $\sim 10^{-16}$ sec, allowing them to travel $<10^{-5}$ cm, we suggest that the only way that the MeV fireball responsible for the hard emission can escape from the surface of such a magnetic neutron star is in a jet aligned with the nearly radial field lines emanating from the magnetic poles. This strongly suggests that the site of energy release in this burst must in fact be at the polar caps. That in turn obviously favors as the source of the burst energy explosive thermonuclear burning of accreted gas, which would be naturally focused onto the polar caps of a magnetic star, rather than gravitational energy from a solid body impact, which should occur more randomly over the surface. We also would expect thermonuclear burning of matter on such a highly magnetized neutron star to lead to highly anisotropic expansion aligned with the magnetic field, since that is the only degree of freedom that the gas would have.

A jet, however, is not sufficient to explain the observed radiation by itself because the observed absorption feature implies that we are looking at a large angle to the neutron star magnetic field and hence the jet. Although the mechanism by which the jet is transformed into a more isotropically expanding fireball is uncertain, it is possible that the process could be initiated by the rapidly diverging field near the magnetopause. For a polar cap size of ~ 1 km^2, consistent with that expected[12] in the thermonuclear model for an energy release of $\sim 10^{39}$ ergs, the magnetopause radius is $\sim 10^9$ cm. Synchrotron losses at that distance are negligible since the cooling time for electrons ($\sim 10^2$ sec) is much longer than the local expansion time. More generally, the absence of significant bremsstrahlung cooling would constrain the minimum size at which a fireball can become isotropic to roughly a tenth of the size at which it becomes transparent to pair production and Compton scattering. Thus an MeV fireball which becomes transparent at $\sim 10^{10}$ cm cannot have been formed from a jet at less than $\sim 10^9$ cm unless its average initial energy was much greater than an MeV.

Finally, a fireball of thermonuclear origin might also explain the two separate outbursts in the event (Figure 1), if surface material is swept up into the fireball. After the dissipation of the fireball through annihilation of the pairs, entrained matter that falls back onto the surface of the star should be focused onto the poles by the magnetic field. Shock heating and compression by the sudden impact of matter reaching the opposite pole could trigger a secondary burst from the star, if sufficient nuclear fuel had accreted onto that pole. The $\sim$ 30 second interval between the two bursts peaks (Figure 1) is consistent with the freefall times of 3 to 100 seconds expected for matter falling from heights of $\sim 10^9$ to $\sim 10^{10}$ cm onto a neutron star. We expect the cap with the greater accumulation to ignite first, so that the initial outburst should be more intense than the second. Since the first peak in the March 25 burst is in fact less intense than the second by a factor of four, we attribute this difference to relativistic effects due to the net motion of the ejected fireballs relative to our line of sight; the first fireball moves away from us and the second toward us.

CONCLUSION

The major features of the fireball model are that it naturally accounts for the observed duration of the burst, predicts an intrinsic luminosity and hence a distance to the burst, and may discriminate between possible sources of burst energy. Moreover, we suggest that the fireball model is applicable to a large class of gamma ray bursts, since hard emission components are a common feature of burst spectra.

ACKNOWLEDGEMENTS

We wish to thank James C. Higdon for helpful discussions. This work was supported by the National Aeronautics and Space Administration under grants NAS 8-27974 and NSG 7541.

REFERENCES

1. G. Cavallo and M.J. Rees, M.N.R.A.S. *183*, 359(1978).
2. E.P. Mazets et. al., Astrophys. Space Sci. *80*, 3(1982).
3. G.H. Share, et. al., Gamma Ray Transients and Related Astrophysical Phenomena (AIP, New York, 1982), p.45.
4. G.J. Hueter and D.E. Gruber, Accreting Neutron Stars (Max Plank Inst., Munich, 1982), p.213.
5. E.P. Mazets et. al., Nature *290*, 378(1981).
6. E.P. Liang, Nature *299*, 321(1982).
7. G.J. Hueter, et. al., in preparation.
8. W.K.H. Schmidt, Nature *271*, 525(1978).
9. R. Ramaty and P. Meszaros, Ap. J. *250*, 384(1981).
10. R. Ramaty, J.M. McKinley and F.C. Jones, Ap. J. *256*, 238(1982).
11. R.N. Manchester and J.H. Taylor, Pulsars (Freeman, San Francisco, 1977), p.133.
12. S.E. Woosley, Accreting Neutron Stars (Max Planck Inst., Munich, 1982), p.189.

POSITRONS FROM GAMMA BURSTS

Stirling A. Colgate and Albert G. Petschek
University of California
Los Alamos National Laboratory
Los Alamos, New Mexico 87545
and
New Mexico Institute of Mining and Technology
Socorro, New Mexico 87801

ABSTRACT

We have proposed that the mechanism for the production of the hard spectrum of photons observed in the typical gamma burst is associated with the charge separation and electron-photon heating of Eddington limited accretion onto a neutron star. The charge separation occurs because of the radiation stress on the electrons and the gravitational stress on the ions and amounts to some 100 to 1000 volts cm^{-1} depending upon whether the accreting matter is dynamically decelerated by radiation or is accreted in quasi-static, steady flow. A positron produced in pair formation by the high energy part of the photon spectrum will be accelerated by both the radiation stress and the electrostatic field. Since for electrons these two stresses are equal and opposite, a positron will attain an energy significantly greater than the purely radiation driven free expansion limit where for these conditions $1/\sqrt{1-\beta^2} = \gamma_{rel} \simeq [(m_p MG/r)/m_e c^2]^{1/4} \simeq 4$. This value of γ_{rel} is near the threshold for pair production on additional in-falling matter. Only electrostatic acceleration can produce significantly greater velocities. Hence the number of positrons produced will be much less than the number of high energy photons, but may contribute to the limitation of the photon heating by shorting out the electric field.

INTRODUCTION

The spectra of γ bursts show lines,[1] indicating that the radiating region is optically thin. The overall shape of the spectra also requires an optically thin radiating region. Fenimore[2] assumes a Comptonized low energy spectrum and requires a Compton mean free path or so ($\sim 10^{24}$ cm^{-2}) of electrons. Liang[3] and Lamb[4] have proposed a synchrotron radiation process and require an even thinner region as does the ingenious proposal of Ramaty et al.[5] The characteristic temperatures of the radiating region are a few hundred keV. If γ-bursts originate uniformly over the surface of neutron stars the total energy in the radiating electrons is $10^{30} - 10^{31}$ ergs. On the other hand, assuming a source at 50-100 pc, the energy in photons observed is $10^{37} - 10^{38}$ ergs. Thus the electron energy must be renewed on a microsecond time scale. In an attempt to avoid the severe problems posed by this requirement, we have suggested a gravitational energy source. In this model a layer of appropriate thickness falls onto the neutron star, either after having been previously lofted by some sort of explosion on the surface, or while being accreted. Effectively, the available energy per electron is increased from ~300 keV to the potential energy of a proton, ~100 MeV. The characteristic time scale for the infall is a millisecond and the gravitational power available is sufficient for the observed burst provided that matter continues to fall in for the duration of the burst and provided that 10^{24} electrons cm^{-2} or more is allowed by the radiating mechanism. The time history of the bursts, which shows considerable high frequency structure, tends to support a mechanism involving repetitive events with some variation in interval. As the layer of material falls onto the neutron star, the electrons will be pushed away from the star by the emergent photon flux. This produces a charge separation and consequently an electric field that forces the electrons to fall with the nuclei. The electrons are heated by friction with the photons or by the electric field, depending on one's point of view. The photons are heated by compression between the infalling matter and the star surface, but do not heat by cooling the electrons, since the heat capacity of the latter is negligible. Most of the gravitational energy goes into heating photons. There is insufficient time to increase the number of photons. If positrons are produced, by pair production by γ-rays or other positrons, they will be accelerated outward not only by the photon flux but also by the electric field.

ENERGETICS

A layer of matter τ Compton mean free paths thick falling onto a star of radius R has a mass

$$\delta M = 4\pi R^2 \; A\tau/(ZN_A \sigma_c) \qquad (1)$$

if A is the atomic weight, Z the atomic number, N_A Avogadro's number and σ_c the Compton cross-section. The fall time, from a height h small compared to R is

$$\delta t = (2 h/g)^{1/2} \qquad (2)$$

the gravitational energy is $\delta M\,gh$ and the power hence is $\delta M\,gh/\delta t$. Putting in appropriate numbers (neutron star mass = 1 solar mass, radius = 10 km) gives a power 2×10^{34} $h_{cm}^{1/2}$ τ ergs s^{-1} while a typical γ burst produces 2×10^{-5} ergs cm^{-2} s^{-1} which, if the distance is 100 pc, corresponds to 2×10^{37} ergs s^{-1}. Hence, h = 10 km/τ^2 furnishes enough power. The effects of spherical geometry are unimportant for $\tau > 1$. We shall argue that the efficiency of the process is large so that our assumption of unit efficiency is acceptable. We shall require a photon flux below the Eddington limit, since otherwise the fall time will be stretched out.

PHOTON AND POSITRON PRODUCTION

The order of magnitude of the electron density is $10^{24} \tau h^{-1}$ or 10^{18} electrons cm^{-3}. The order of magnitude of the photon density is $L_E(4\pi R^2 h\nu c)^{-1}$ where L_E is the Eddington luminosity, $h\nu$ the average photon energy and c the speed of light. This is $\sim 10^{21}$ photons cm^{-3}, much larger. Hence, the heat capacity of the electrons is negligible. The Bremsstrahlung rate is $1.7 \times 10^{-25} Z^2 N_e^2 T_e(eV)^{1/2}$ ergs cm^{-3} s^{-1} or, at 300 keV and with $N_e = 10^{18}$, $N_e h = 10^{24}$, 3×10^{18} ergs cm^{-2} s^{-1} which is $\sim 10^{-7}$ L_E and hence negligible. Another source of photons is double Compton collisions, which is more important than Bremsstrahlung in the ratio of photon density to electron density.[6] Since this ratio is only 10^3 to 10^5, this source of photons is also insignificant. Hence the infalling matter interacts only with photons radiated by the surface of the star. We have taken this number of photons from the Eddington limit because, if the infalling matter is not stopped, it will heat the surface which will maintain itself at the Eddington limit temperature by hydrodynamic expansion.

Near threshold, the pair production cross-section is[7] $3/8 \, \alpha \, Z^2 \sigma_C (h\nu/mc^2 - 2)^3$. For an exponential spectrum, $\exp(-h\nu/h\nu_0)$ with $h\nu_0 \sim mc^2$, the integral is $3/8 \, \alpha \, Z^2 \sigma_C \times 6 \exp(-2 mc^2/h\nu_0)$. Thus for infalling hydrogen about one Compton mean free path thick the number of positrons is less by the exponential factor and the fine structure constant α than the number of photons. For infalling iron, the number of positrons and photons might be comparable if, as observed, $h\nu_0 \sim mc^2/2$.

FATE OF POSITRONS

The positrons will be accelerated outward both by the electric field and radiation pressure until they reach the outside of the accreting layer where they will tend to reverse the field and hence pick up an electron and continue outward. Their ultimate speed will be limited by photon drag.[8] Their energy, including the pair energy, comes from the gravitational energy and reduces the energy available for photons.

About 100 γ-bursts per year[9] of 2×10^{-5} ergs cm^{-2} each implies an average photon flux of 1.4×10^{-4} photons cm^{-2} s^{-1}. If the positron flux is comparable (Fe infalling), the rate of annihilation radiation should be comparable to the rate observed from the galactic center during its high state. For H infalling, it should be much less.

ACKNOWLEDGEMENT

Los Alamos National Laboratory is operated under the auspices of the Department of Energy. Work at New Mexico Tech was supported in part by the Astronomy Program of the National Science Foundation.

REFERENCES

1. E. P. Mazets, S. V. Golenetskii, Yu. A. Aptekar, R. L. Gur'yan, and V. N. Il'inskii, Nature 290, 378 (1981).

2. E. E. Fenimore, R. W. Klebesadel, J. G. Laros, R. E. Stockdale, and S. R. Kane, Nature 297, 665 (1982)

3. F. E. Liang, Nature 229, 321 (1982).

4. D. Q. Lamb, in Gamma Ray Transients and Related Astrophysical Phenomena, Am. Inst. Phys. Conference Proceedings #77, edited by R. E. Lingenfelter, H. S. Hudson, and D. M. Worrall, p. 249 (1982).

5. R. Ramaty, R. E. Lingenfelter, and R. W. Bussard, Astrophys. Space Sci. 75, 193 (1981).

6. A. P. Lightman and D. L. Band, Ap. J. 251, 713 (1981).

7. J. M. Jauch and F. Rohrlich, Theory of Photons and Electrons, (Addison Wesley, 1955).

8. P. D. Noerdlinger, Ap. J. 192, 529 (1974).

9. K. Hurley, this conference.

RADIO PULSARS: INTENSITY, POLARIZATION, AND ROTATION FLUCTUATIONS

J.M. Cordes
NAIC and Cornell University, Ithaca, NY. 14853

ABSTRACT

In this review of radio emission phenomena emphasis is given to observables that may constrain the particle content in pulsar magnetospheres. These include intensity variations within single pulses (time scales from 1 µs to tens of milliseconds); pulse-to-pulse variations (one period to years) which are both 'continuous' (drifts of subpulse features through pulse phase and other quasiperiodic and aperiodic variations) and 'discontinuous' (pulse nulling, average pulse shape changes, changes in drift rates); polarization transitions between orthogonal modes; relationship of drifting subpulses to nulling; constraints on emission altitudes; and rotation noise due to internal and magnetospheric torque fluctuations.

INTRODUCTION

Although electron-positron plasmas are generally associated with high-energy photon emission, in the case of pulsars, current thinking holds that radio emission requires pair production followed by counterstreaming of some combination of ion and pair beams to effect the large brightness temperature (10^{24} to 10^{31} °K) radiation. The evidence for pairs being involved is circumstantial but compelling: pulsar spindown rates, corroborated by cyclotron X-ray emission from Her X-1, are consistent with surface magnetic fields of order 10^{12} Gauss; combined with rotation periods between 1.55 ms and 4.3 s, the resultant electric fields imply particle energies large enough to sustain pair production avalanches near the neutron star surface.[1,2] The distribution of pulsars in P and dP/dt is consistent with the cessation or diminution of pair production at an average period of 1 s, which is notably close to the peak of the pulsar period distribution. Apart from this logic, however, no features in the radio emission can unequivocally be associated with the details of pair production. A pessimist might argue that most of the rich variety of structure in radio pulses is due to radiative transfer effects in a magnetopair medium and/or to a long list of possible plasma effects analagous to those relevant to solar bursts. But the trademark of such effects is strong frequency dependence (e.g. quantities $\propto \nu^x$, $x > 1$) which is not seen in the shapes, amplitudes, spacing, and polarization of individual pulse features. Consequently, an optimistic view is taken here, namely that some radio features reflect the details of pair avalanche processes very close to the neutron star surface. If true, that view implies that a tremendous amount of information exists on the time behaviour of pair avalanches and that we need to be more clever in piecing available facts into constraints on pulsar theory.

This paper is predominantly empirical but the discussion is organized around the notion that emission originates near the magnetic polar region and well within the light cylinder of a rotating neutron star. A posteriori self consistency tests are discussed later.

INTENSITY FLUCTUATIONS

Radio intensities fluctuate on known time scales of 1 µs to years. Figure 1 is a pulse sequence showing the broad range of quasiperiodic and intermittent structure revealed by high time resolution; the average of 500 pulses is nearly converged to a shape typical for this pulsar and is weakly frequency dependent. Figure 2 shows (with poorer time resolution) drifting subpulses and pulse nulling (no emission) from another pulsar. In the following we categorize the kinds of fluctuations seen.

<u>Intrapulse fluctuations</u>, revealed best by autocorrelation analyses of the intensity and/or Stokes parameters,[3,4] include narrow spikes, <u>micropulses</u>, with durations from 1 µs - few milliseconds[5,6] which also show quasiperiodicities (as in Fig. 1) of 0.1-few ms[6-8]. Broader features, <u>subpulses</u>, ($\Delta t \approx 1$-100 ms), sometimes appear as envelopes of micropulses, sometimes as modulations of amorphous,

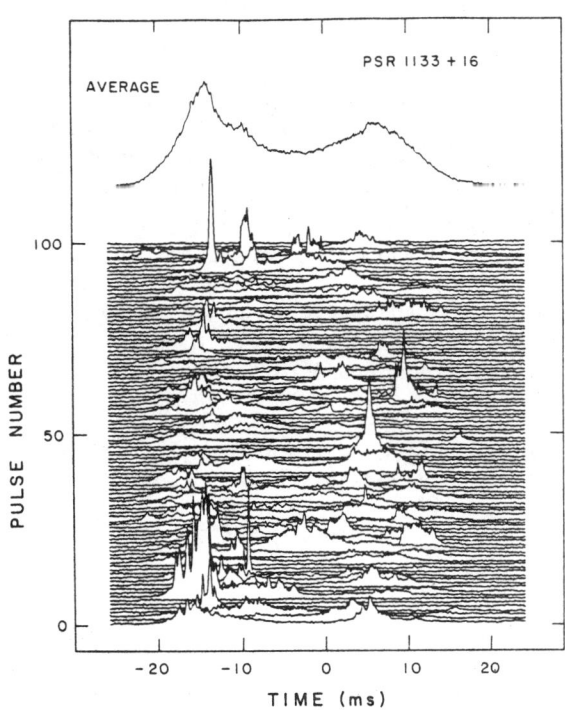

Fig. 1 A sequence of pulses from PSR1133+16 obtained at Arecibo Observatory with a linearly polarized feed at 600 MHz. Peak flux densities are 10^3 Jy corresponding to 10^7 photons per resolution interval of 100 µs. The average waveform is of 500 pulses. Arecibo Observatory (NAIC) is operated by Cornell University under contract with the National Science Foundation.

Fig. 2. Pulses from PSR 1944+17 at 1.4 GHz in total intensity. Drifting subpulses and pulse nulling (no emission) are evident. Null pulses, when averaged, are consistent with zero pulsar flux. Data are from reference 34. Horizontal extent is 54 ms.

noiselike structure ('sometimes' refers to different times for the same pulsar and differences between pulsars). Subpulses generally have weakly frequency dependent widths and separations ($\Delta t \propto \nu^x$, $x \approx 0.0$ to 0.5)[9] whereas micropulses have essentially no frequency dependence of their widths and separations ($x<0.03$)[10,11]; micropulses tend to be more prominent at frequencies below 1 GHz.

A fundamental probe into the nature of intrapulse fluctuations results from an autocorrelation analysis of the intensity obtained with a narrowband receiver. A model[12] for the intensity is $I(t)=A(t)N(t)$ where $A(t)$ describes micro/subpulse modulations that vary much more slowly than the noiselike $N(t)$, whose time scale is the inverse receiver bandwidth, typically $1/\Delta\nu \approx 1$-10 μs. The intensity autocorrelation function (ACF) is (after correcting for additive receiver noise)

$$R_I(\tau) = R_A(\tau)[1 + \beta\Delta(\tau)] / (1 + \beta) \qquad (1)$$

where $R_A(\tau)$ = ACF of $A(t)$, $\Delta(\tau)$ is a narrow spike with $\Delta(0)=1$ and width $1/\Delta\nu$, and β depends on the fluctuations of $N(t)$. We can write $N(t) \equiv |\varepsilon(\tau)|^2$ where $\varepsilon(t)$ is the complex narrowband field selected by the receiver and antenna polarization. Empirical tests on high time resolution data yield a determination of β which is the ratio $\beta \equiv \langle|\epsilon|^4\rangle/\langle|\epsilon|^2\rangle^2$. Results[13,14] are consistent with $\beta = 1$, the value appropriate for ϵ being complex Gaussian noise. By the central limit theorem, one concludes that a large number of independent emitters contribute to the signal in a resolution time of 1-10 μs, so even the narrow micropulses are incoherent composites of many coherent emissions.

Pulse-to-pulse fluctuations can be divided into 'continuous' and 'discontinuous' kinds. Continuous variations include subpulse drift (e.g. Fig. 2), quasi-periodicities in some regions of pulse phase (sometimes but not always related to obvious subpulse drift), and sundry aperiodic fluctuations on time scales from one period to years. Discontinuities occur with rise time less than one period whereby pulse nulling occurs (cessation of all radio emission), the subpulse drift rate can discontinuously switch between two or more preferred values, or a waveform 'mode change' occurs (the average shape switches between two or more preferred shapes).[15] Discontinuous changes occur over a broad frequency range (from simultaneous multifrequency observations) and may include all of the detectable radio emission.[16]

Table I summarizes the above discussion and gives examples of pulsars that display particular phenomena. It should be emphasized that some pulsars show none of the discontinuous phenomena, others show all of them. PSR 0809+74 displays an interesting interaction of nulling and drifting: during a null, the phase of the drift pattern is remembered such that the first subpulse after a null has nearly the same pulse phase as the last subpulse before the null.[17] The memory is not perfect.[18] This result implies a relationship between the physics of coherent radiation (two stream instability?)[2] and the drift physics ($\vec{E} \times \vec{B}$ drifts?)[2]. Like many pulsar phenomena, however, this effect has not been established for other objects, such as PSR 1944+17 (Fig. 2). But objects with frequent nulling, well-defined drift paths, and large signal-to-noise ratio are uncommon, so detailed studies on many objects have not been possible.

The time signature of a null (rise and fall times of the intensity) is not well known. If nulls occur via step functions in the intensity, then the probability of seeing a discontinuity when the pulsar beam points toward the earth is just the pulse duty cycle $\approx 5\%$. Investigation of 100 nulls from PSR 1944+17 should have yielded ≈ 10 transitions; none were seen.[19] A slow fadeout of the intensity (e.g. over many rotation periods) can also be ruled out by investigating amplitudes of pulses. The conclusion (for this pulsar) seems to be that the null transition time is comparable to or less than a rotation period. One possibility is that a null is a cessation of subpulse formation, not a modulation that is independent of the subpulse emission.

Subpulse drift usually occurs towards earlier pulse phase in successive pulses, although the opposite sense of drift occurs in some objects. Drift rates for three objects appear to jump between two or more preferred values on time scales less than P.[20]

Table I Pulsar Intensity Variations

Kind	Description	Δt	Properties	Examples
INTRAPULSE	i. micropulse	1μs-few ms	quasiperiodic: $0.1\text{ms}<P_\mu<4$ ms; prominent for $\nu<1$ GHz and small dP/dt.	0809+74, 0950+08, 1133+16, 1944+17, 2016+28.
	ii. subpulse	1ms-100 ms	width, spacing $\propto \nu^x$, $-0.2<x<0.5$; drifting subpulses for small dP/dt.	≈30%
PULSE to PULSE				
a. continuous				
	i. subpulse drift	$P_2 \approx$ 5-20 ms $P_3 \approx$ 3-100P	drift rate P_2/P_3; nonlinear drift paths; quantized drift rates.	0031-07, 0809+74, 1822-09, 2319+60.
	ii. quasiperiods	$\approx P_3$		many
	iii. aperiodic	P to years	bursts, trends, white noise.	most
b. discontinuous				
	i. nulling	durations: ≈1-5000P; rise/fall times < P.	no radio emission; correlated with $\dot{P}P^{-x}$, $2<x<3$.	0809+74, 1944+17.
	ii. waveform mode change	"	broadband; changes in pulse component heights and spacings.	0329+54, 1237+25, 1822-09, 2319+60.
	iii. drift rate changes.	"	quantized drift rates.	0031-07, 1919+21, 2319+60.

Note: P is the pulsar rotation period. See text for references.

Waveform mode changes are known in 7 objects and are evident as dramatic changes of the relative amplitudes accompanied by smaller fractional changes in the relative spacing of pulse components. For two objects the waveform changes are accompanied by transitions of the subpulse drift rate between preferred values.[20] As discussed below, alterations of component spacings may be effected by changes in emission altitude of particles that are collimated by a roughly dipolar magnetic field. Drift rate changes are interpreted in some models[2,21] as due to changes in $\vec{E}\cdot\vec{B}$ near the stellar surface which could also cause particle density fluctuations and altitude fluctuations.

PULSAR BIRTH AND DEATH

The exact prerequisites of observable radio emission are not known. Counterstreaming of particle beams is oft invoked, but several possibilities exist for the kinds of particles involved and those kinds may vary during a pulsar's lifetime.[22,23] Pulsars may evolve through a variation of composition of neutron star surface (He, Fe, p) which determines kinds of ions available, and through variation of accelerating potential and temperature which determine the extent to which the magnetosphere is replenished by those particles and/or by pairs. It is by no means certain that coherent radiation is produced immediately upon birth but it is clear that radiation is strongly diminished, on average, when the period becomes greater than 1 sec. Of course, the period is not the sole determinant; magnetic field strength is almost certainly involved and other 'hidden' variables may enter in.

The period histogram of 331 objects[24] (Fig 3) shows evidence of such cutoff by the paucity of objects with P>1 sec. Also shown are

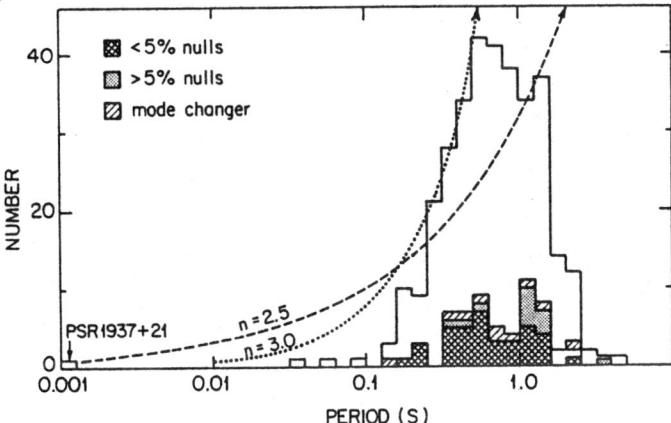

Fig. 3. Histograms of pulsar periods for 331 objects and for objects with and without frequent pulse nulling and with waveform mode changes. Theoretical histograms are shown for braking indexes n=2.5 and n=3 assuming objects are born with the same initial P and dP/dt.

theoretical distributions [calculated by assuming 1) birth of all objects with the same initial P and dP/dt (hence, same magnetic field, mass, and moment of inertia); 2) no variation of radio luminosity, L_{radio}, with P; and 3) a time independent birth rate] which are $dN/dP \propto P^{n-2}$ for $P_0 < P < P_{max}$ where n is the 'braking index' in the spindown relation $d\nu/dt \propto \nu^n$, P_0 is the initial period and P_{max} is the period attained by the oldest pulsar (P_{max} = 92 sec for a pulsar born with Crab pulsar parameters 10^{10} years ago). Clearly, such assumptions do not fit the observed pulsar distribution. In addition to the paucity of long period objects, the leading edge of the distribution is not well matched by a model assuming birth with millisecond periods and identical magnetic fields. It is possible that the structure of the distribution at short periods is more a reflection of the distribution of initial P and dP/dt than of the spindown law.[25]

There have long been associations of observed phenomena with various combinations of $P^\alpha \dot{P}^\beta$, suggesting that the period histogram for P>0.7 s is highly affected by a cessation of pair production.[1,2] The incidence of subpulse drifting correlates with small values of $\dot{P}$,[26] pulse nulling correlates with $\dot{P} P^{-x}$ with 2<x<3,[27] and L_{radio} correlates with $P^{-1.8} \dot{P}^{0.9}$.[25,28] Models invoking sustained pair discharges due to $\gamma + \gamma' \rightarrow e^+ + e^-$ where γ' is a virtual photon of the magnetic field require[1,2,23,29] large values of Ω and B. Using $B \propto (P\dot{P})^{1/2}$, these models predict a threshold in $\dot{P}P^{-y}$, 2<y<3, a quantity that bears great similarity to the empirical one for frequent pulse nulling. Although such correlations are not perfect (i.e. one cannot predict solely from P and dP/dt that a given pulsar will display nulling) they support the idea that nulling is a predictor of imminent pulsar death.[27] The point has been made,[30] however, that the well studied pulsars (i.e. the intense nearby ones) are perhaps atypical of the bulk of objects, which are on average 100 times weaker in the radio.

BEAMING

Average pulse waveforms for most objects converge to stable shapes when 10^3-10^4 pulses are synchronously averaged. Shapes of well studied objects have been stable over the 15 years since pulsars were discovered. Waveforms correspond to duty cycles of 3-80% and show single, double, triple, and five-component shapes.[15] Recent work[31] indicates that the underlying generic beam shape is a hollow cone (or cones) with a central core, the relative amplitudes of which are frequency dependent and vary during waveform mode changes. Core emission is prevalent at low frequencies and is associated with circular polarization that tends to switch sense halfway through the core component for many objects.

As can be seen in Figures 1 and 2, a substantial amount of longitude fluctuation (random jitter or systematic drift) occurs from pulse to pulse (longitude=pulse phase). In this regard, the average waveform is not a radiation beam in the same sense as a beam from a dipole antenna. Waveforms measure both the frequency of occurrence of the pulse longitude of individual subpulses and the likelihood of seeing an intense subpulse at each longitude. That is, the instantaneous intensity I at longitude ϕ is described by a non-separable two-dimensional probability density function (pdf) $g(I,\phi)$ and the average waveform is the expectation of I over g. Waveform mode changes correspond to the pdf alternating between two or more functions.

It is not known a priori whether the longitude fluctuations corresponding to the pulse jitter are physically the same as the cause of subpulse durations. Both jitter and durations could be <u>temporal</u> variations observable in the co-rotating frame. However, it appears that both are <u>angular</u> variations of radiating plasma across the magnetic polar region.[11,32] This conclusion holds for subpulses, but not for micropulses, and follows from the frequency dependence of widths and spacings of subpulses and waveforms, and the variation of polarization angle across pulse longitude. Micropulses appear to have frequency independent spacings[10,11] and appear to be true temporal modulations.

Lorentz factors of particles flowing along open field lines are thought to be large (e.g. $\gamma = 10^2$-10^3) so that emission is highly beamed in directions tangent to the local magnetic field lines. Although the radio emission mechanism is not known, the anisotropy imposed on particle flow by the ambient magnetic field implies that polarization angles Ψ are at least initially determined by the magnetic field. Measurements of polarization across pulse longitude show that $\Psi(\phi)$ is consistent with such a viewpoint except that the instantaneous angle can take on one of two more or less orthogonal angles (see below). Apart from this $\pi/2$ ambiguity, the variation of $\Psi(\phi)$ is consistent with $\approx$ dipolar fields and radiation at a given frequency arising from a small range of radii, although depth of field effects and gradients of the emission altitude across the open field line region have not been fully explored. With dipolar geometry and constant radius, the signature of $\Psi(\phi)$ is determined purely by the angles $\alpha = \cos^{-1}(\hat{\Omega}\cdot\hat{\mu})$ and $\beta = \cos^{-1}(\hat{\Omega}\cdot\hat{n})$ where $\hat{\Omega},\hat{\mu},\hat{n}$ are unit vectors in the directions of rotation axis, magnetic moment, and line of sight at zero longitude (defined as the rotational phase for which the unit vectors are coplanar). Given the abovementioned core/cone model for the waveform beam, a correlation between waveform shape and polarization signature is expected and indeed holds for many pulsars. The value of $\Delta\Psi \equiv \Psi(\phi_{max})-\Psi(\phi_{min})$ where ϕ_{max}, ϕ_{min} define the longitude range of observable radiation is determined by α,β and the radiation beam width. Recent work[33] indicates that there are more pulsars with small values of $\Delta\Psi$ than expected if the radiation beam is circularly symmetric and if α and β are uniformly distributed over all angles. The conclusion is that beams are elongated by $\approx$ 3:1 with major axis in the plane containing $\hat{\Omega}$ and $\hat{\mu}$ (i.e. in rotational latitude). Such elongations are therefore orthogonal to those sometimes discussed in magnetospheric models[22,29] and are not understood.

POLARIZATION ORTHOGONALITY

As mentioned before, the ambient magnetic field appears to determine the orientation of the polarization ellipse, but with an ambiguity of $\pi/2$: the instantaneous angle often attains a bimodal distribution with modes separated by $\approx \pi/2$. The relative frequency of occurrence of the modes is a strong function of both pulse longitude and pulsar but is surprisingly only weakly dependent on observation frequency.[34] The modes are actually elliptically polarized such that one angle correlates with one sense of circular polarization, the orthogonal angle correlates with the other sense.

As with pulsar intensities, no theory exists by which observables can be inverted into estimates of particle densities and Lorentz factors. This arises in part because the radiation transfer is likely to be nonlinear because deduced radiation energy densities are comparable to theoretically estimated particle energy densities (ref 32). Secondly, one does not know the emissivity into the normal modes of the plasma in the emission region. Attempts have been made to understand the observables in terms of linear propagation processes such as birefringence[35] and conversion[36] of linearly polarized into elliptically polarized radiation. The puzzle is that the 'cleanliness' of histograms of the position angle (see ref 34) suggests on the one hand that they reflect the 'naked' emissivities into the two normal modes (which are linearly polarized for $r \ll r_{LC} \equiv c/\Omega$ and large Lorentz factors[35]); on the other hand, the appearance of strong circular polarization (sometimes > 50%) implies conversion somewhere along the propagation path. Apart from the appearance of circular polarization, however, pulsar magnetospheres do not appear to be magnetoactive (no generalized Faraday rotation is evident, for example).

EMISSION ALTITUDES

Radio emission altitudes can be constrained by considering pulse widths and times-of-arrival at different frequencies, by using interstellar scintillations to show that pulse components originate from locations with transverse separations smaller than 10^3 km, and, of course, by theoretical argument. Arguments based on energy densities and pulse waveform stability favor emission altitudes somewhere between the neutron star and the light-cylinder radius r_{LC}. Light cylinder models[37,38] invoke azimuthal co-rotation velocities near c as the kinematical determinant of radiation whereas polar cap models invoke relativistic streaming along field lines connecting the neutron star and extra-light-cylinder space. There are no proofs that radio emission at the light cylinder does not occur. However, light cylinder models appear increasingly less plausible when the bulk of observational results are compared with such geometry, including waveform symmetries, absence of large differential aberration effects, radiation energy densities, and the P-Ṗ distribution. The biggest embarrassment is that the magnetic field energy density varies by 20 orders of magnitude at r_{LC} (assuming a dipolar r^{-3} variation of B). But emission from the neutron star surface (or a fixed multiple of R_{NS} for all objects) does not fit the empirical pulse width-period distribution.[39] It appears that emission occurs at altitudes $r \ll r_{LC}$ that vary both from object to object and possibly as a function of observation frequency, as we shall now discuss.

Magnetospheric theory invokes the Goldreich-Julian[40] scale for the magnetic polar cap,

$$\theta_{PC} \approx 2(R_{NS}/r_{LC})^{1/2} \approx 1.6 P^{-1/2} \text{ degrees} \qquad (2)$$

(with $R_{NS} \equiv$ neutron star radius $\approx$ 10 km), as the scale for 'action' at the stellar surface. Particles emanate from or are produced near the polar cap because field lines that thread it are not equipotentials. Observable radio emission arises when two conditions are satisfied: 1) amplification at frequency ν occurs due to conversion of free streaming energy into plasma waves which directly or indirectly couple to transverse EM waves; 2) radiation can propagate out of the magnetosphere. The physics of these processes have not been worked out. Radiation energy densities for micropulses are comparable to predicted particle energy densities, so it is surprising that the observations appear as if radiation occurs at some local plasma frequency, ω_p. If we assume so, i.e. $\nu_{obs} \propto \omega_p \propto n_\pm^{1/2}$ where $n_\pm$ is the number density of a pair plasma and $n \propto r^{-3}$ in a dipolar field, then $r \propto \nu^{-2/3}$ and the radiation beam width is approximately

$$\theta_{rad} = (3\theta_{PC}/2)(r/R_{NS})^{1/2} \qquad (3)$$

and $\theta_{rad} \propto \nu^{-1/3}$. This dependence can be compared with that observed, $\theta_{waveform} \propto \nu^x$, $0.1 < x < 0.5$ for $\nu < \nu_c$ and $-0.1 < x < 0$ for $\nu > \nu_c$ where ν_c ranges between 0.3 and 3 GHz.[41] One interpretation is that low frequency radiation ($\nu < \nu_c$) is narrowband at or near some plasma frequency, but high frequency radiation, for some reason, is broadband. Other interpretations are possible, however, including broadband emission occurring solely at one radius but with a <u>spectrum</u> that varies, owing to a variation of $n_\pm$ across the magnetic polar region. The truth is probably somewhere between these extremes of emission frequency varying solely in radius and solely in angle.

Interesting limits on emission altitudes can be made by comparing observed pulse widths, θ_{pulse}, with θ_{rad}. In doing so, one assumes that θ_{pulse} reflects the beam width, unaltered by geometrical effects. Such will be true so long as $\hat{n}\cdot\hat{\mu} \approx 1$ and $\hat{\Omega}\cdot\hat{\mu} \ll \cos\theta_{rad}$. For 'cone' waveform components, ususally $\theta_{cone} \gg 3\theta_{PC}/2$, implying emission altitudes of 10-$100 R_{NS}$. For core components, however, $\theta_{core} \approx 3\theta_{PC}/2$, implying that not only is the emission altitude not much greater than the stellar radius for core emission, but also that neutron star radii cannot be much greater than 10-20 km. Most effects that alter the polar cap size, such as distortion of the assumed dipolar field, increase it, so there is not much freedom in altering this conclusion except by proposing ad hoc that core emission does not involve the entire flux tube threading the polar cap.

The above estimates of cone emission altitudes are weak because of the dipolar field assumption. The field may be distorted very near (multipoles) and very far (induced currents) from the star. Another approach, still model dependent, is to measure or place limits on nondispersive time delays in pulse times-of-arrival. Cold plasma dispersion in the interstellar medium produces time delays $\Delta t \propto DM\Delta(\nu^{-2})$ where DM is the column density of electrons. Additional delays between two frequencies occur if the aforementioned radius-to-frequency mapping holds because of differential aberration and propagation times; for $r \ll r_{LC}$, this time delay is $\Delta t_{ab} \propto \Delta(\nu^{-2x}) \approx \Delta(\nu^{-0.2}$-$1.0)$. Limits on Δt_{ab} from multifrequency observations imply limits on emission altitudes of $r_{max} \approx 0.01$-$0.08 r_{LC}$ for $\nu \approx 0.1$ GHz and $r_{min} \approx 5$-$10 R_{NS}$ for $\nu \approx 0.4$-1.4 GHz. Recent possible detections of the effect[11,43] suggest $r \approx 10 R_{NS}$ for one object at 0.3 GHz and $r \approx 13$-$50 R_{NS}$ for another at 1 GHz.

Interstellar scintillations (ISS) also place useful limits on transverse separations of emission regions, which, in turn, imply altitude limits in the context of an assumed geometry. ISS causes a point source to become smeared out and produces a diffraction pattern whose transverse scale is $s_I \approx 1000$ km at meter wavelengths. By reciprocity, two point sources will scintillate differently if their transverse separation is greater than s_I. At kiloparsec distances, this implies an angular resolution < 0.1 micro-arc sec (the size of an amoeba in Los Angeles viewed from New York). ISS measurements indicate that the emission regions corresponding to different waveform components (as in Fig. 1) are identical, within errors, implying a transverse separation < 1000 km when the emitters are pointed at the Earth. As a consequence of curved magnetic field lines, this implies an emission altitude < $0.06 r_{LC}$ for PSR0525+21 at 0.4 GHz.[44]

PARTICLE DENSITIES

The derived altitude limits and the apparent altitude variations associated with waveform mode changes can be used to estimate the product $\gamma_{\pm} n_{\pm}$ and its variations for a pair plasma. One assumes the radiation is tied to the local plasma frequency ν_p of the pair plasma and that its energy spread, $\Delta\gamma_\pm mc^2$, is not large. One finds

$$\gamma_\pm n_\pm \approx \pi \nu^2 m/e^2 = 10^{10.1} \nu_{GHz}^2 \text{ cm}^{-3} \qquad (4)$$

which applies to deduced radii (or limits) of $\approx 10 R_{NS}$ at $\nu = 1$ GHz. Whether one can additionally constrain $\gamma_\pm$ depends on assuming that field line curvature plays the same role in the coherent radiation process as it would for single particle emission. The hollow cone component of the radiation beam (which the aberration limits apply to) may derive its structure from the dependence of radiation on field line curvature but it may also reflect the structure of potential drops near the stellar surface. Assuming that field line curvature is important, an additional requirement for significant amplification is that $\nu_p < \nu_c$ where $\nu_c = 3\gamma_\pm^3 c/4\pi a$ is the cutoff frequency of the single particle curvature spectrum and a is the field line radius of curvature. Results[43] imply $\gamma_\pm > 500$ and $n_\pm < 10^{7.4}$ cm^{-3}, which are not inconsistent with estimates of Lorentz factors calculated theoretically in pair cascades. The density of pair plasma (extrapolated back to the polar cap) is comparable to estimates of ref (36) which assume current is carried mostly by ions, not positrons.

Waveform components change their spacing by up to 20% during mode changes[16]. Assuming $\Delta\theta \propto r^{1/2}$ for a dipolar field and $n_\pm \propto r^{-3}$ we find that component spacing changes of $\Delta\theta_2 = 0.8\Delta\theta_1$ imply changes $r_2 \approx 0.7 r_1$ and, at fixed altitude, $n_{\pm 2} = (\Delta\theta_2/\Delta\theta_1)^6 n_{\pm 1} \approx 0.2 n_{\pm 1}$. Note that if this interpretation is correct, the minimum component spacing allowable for $r \approx 10 R_{NS}$ is $\Delta\theta_2 = \Delta\theta_1 (R_{NS}/r_1)^{1/2} \approx 0.3 \Delta\theta_1$. None of the observed mode changes show such factor of 3 decreases, so the interpretation is viable. Similar results obtain if one assumes that changes in subpulse drift rate correspond to changes in accelerating potential which would be accompanied by changes in $n_\pm$.

ROTATION FLUCTUATIONS

An empirically deduced model for rotational phase is

$$\phi(t) = \phi_S(t) + \phi_G(t) + \phi_{TN}(t) + \phi_M(t) \qquad (5)$$

where the first three terms are intrinsic to the pulsar and the fourth is measurement error; ϕ_S is the phase due to a deterministic spindown function, ϕ_G and ϕ_{TN} are stochastic with ϕ_G corresponding to 'glitches', discontinuities in P and/or dP/dt, and ϕ_{TN} is 'timing noise' intrinsic to the pulsar that behaves as nonstationary clock noise not dissimilar from that seen in terrestrial clocks. Spindown laws of the form $d\nu/dt \propto \nu^n$ predict

$$\nu(t) = \nu_{now}(1 + t/\tau_S)^{-1/(n-1)} \qquad (6)$$

where the spindown time is $\tau_S \equiv \nu_{now}/(n-1)\dot{\nu}_{now}$ and n is the braking index. For most pulsars $\tau_S \gg 1000$ years so pulse timing measurements of the rotational phase over an astronomer's lifetime should be adequately described by a spindown law

$$\phi_S(t) = \phi_0 + \nu_{now}t + \dot{\nu}_{now}t^2/2 + [\ddot{\nu}_{now}t^3/6]. \qquad (7)$$

The bracketed term has been identified only in the timing data on the Crab pulsar for which $\tau_S \approx 1200$ yr and the braking index has been estimated as $n \approx 2.5$ (ref 45). Magnetic dipole radiation from a perfectly conducting billiard ball surrounded by vacuum gives n=3. Departures of the Crab's braking index from this value has been variously attributed to loading of the magnetosphere by plasma,[39] ohmic decay of the field[27], and internal torques on the neutron star crust from a more rapidly rotating superfluid.[46]

Glitches were first observed in the Crab and Vela pulsars when they had the shortest known periods. These appeared as spinups with unknown (but < 1 day) risetime; amplitudes $\Delta\nu/\nu \approx 10^{-6}$ (Vela) and 10^{-8} (Crab); long decay times: ≈ 100 d (Vela) and 5-15 d (Crab); temporary increases in dP/dt; and no accompanying changes in observed pulse shapes[47,48]. Table II lists the objects which have shown glitches, including very long period objects.

Timing noise appears in the majority of pulsars as evidently nonstationary noise which contributes an rms residual $\sigma_\phi \propto T^x$ with $0.5 < x < 2.5$. For T = 5 yr the phase stabilities of pulsars are typically $\sigma_\phi/\phi_{total} \approx 10^{-13}$ to 10^{-9}.[49]

Timing activity, whether glitches or timing noise, correlates with dP/dt at the 50% level, is weakly anti-correlated with P, and is uncorrelated with radio luminosity, altitude above the Galactic plane, and the occurrence of pulse nulling and waveform mode changes. These results support models for rotation fluctuations involving interactions of neutron star crust with internal superfluid rather than torque fluctuations associated with the magnetosphere. In crust/superfluid models, dP/dt enters as a brake on the crust which, in turn, slows down the superfluid in a noisy, rather than smooth, way.[50] Torque fluctuations associated with magnetospheric variability may also be expected to manifest themselves in pulse waveform changes because the magnetic polar cap size is related to the global state of the magnetosphere. These are ruled out on very long time scales (e.g. > 1 hr) but cannot be ruled out on shorter times which might be of interest for timing noise.[51]

Table II Pulsar Glitches

Pulsar	P (s)	dP/dt (10^{-15} s/s)	Glitch	N	Reference
0525+21	3.745	40.1	$\Delta\nu/\nu \approx 10^{-9}$	2	52
0531+21	0.033	422.4	$\Delta\nu/\nu \approx 10^{-8}$	2	54
0823+26	0.531	1.7	$\Delta\nu/\nu \approx 0.03$	1	52
0833-45	0.089	124.7	$\Delta\nu/\nu \approx 10^{-5.7}$	6	47
1325-43	0.533	3.0	$\Delta\nu/\nu \approx 10^{-7}$	1	54
1641-45	0.455	20.1	$\Delta\nu/\nu \approx 10^{-6.7}$	1	54
2224+65	0.683	9.7	$\Delta\nu/\nu \approx 10^{-6.8}$	1	53

SUMMARY

A survey of pulsar radio phenomena has been given along with interpretations in the context of polar cap models and an inversion of some observables into estimates or constraints on physical quantities in the magnetosphere. Suggestions are made as to how some intensity fluctuations may be related to particle injection mechanisms near the magnetic polar cap, including pair production.

REFERENCES

1. P.A. Sturrock, Ap.J., 164, 529 (1971).
2. M.A. Ruderman and P.A. Sutherland, Ap.J., 196, 51 (1975).
3. T.H. Hankins, Ap.J.Letts., 177, L11 (1972).
4. J.M. Cordes and T.H. Hankins, Ap.J., 218, 484 (1977).
5. N. Bartel and T.H. Hankins, Ap.J.Letts., 254, L35 (1982).
6. J.M. Cordes, Sp.Sci.Rev., 24, 567 (1979).
7. V. Boriakoff, Ap.J.Letts., 208, L43 (1976).
8. J.M. Cordes, J.M. Weisberg, and T.H. Hankins, preprint.
9. J.H. Taylor, R.N. Manchester, and G.R. Huguenin, Ap.J., 195, 513, (1975).
10. V. Boriakoff and D. Ferguson, in Pulsars (Sieber and Wielebinski, eds.), (Dordrecht, Reidel, 1981), pp. 191-196.
11. B.J. Rickett and J.M. Cordes, in Pulsars, (Sieber and Wielebinski, eds.), (Dordrecht, Reidel, 1981), pp. 107-108.
12. B.J. Rickett, Ap.J., 197, 185 (1975).
13. J.M. Cordes, Ap.J., 208, 944 (1976).
14. T.H. Hankins and V. Boriakoff, Nature, 276, 45 (1978).
15. R.N. Manchester and J.H. Taylor, Pulsars (Freeman, San Francisco, 1977); F.G. Smith, Pulsars (Cambridge Univ. Press).
16. N. Bartel et al., Ap.J., 258, 776 (1982).
17. S. Unwin et al., M.N.R.A.S., 182, 711 (1978).
18. A. Filippenko et al., these proceedings.

19. unpublished analyses.
20. G.R. Huguenin, J.H. Taylor, and T.H. Troland, Ap.J., 162, 727 (1970); L.A. Fowler and G.A. Wright and D. Morris, Ast.Ap., 93, 54, (1981); D.C. Backer, Ap.J., 182, 245 (1973).
21. M.A. Ruderman, Ap.J., 203, 206.
22. M.A. Ruderman, in Pulsars, (Sieber and Wielebinski, eds.) (Dordrecht, Reidel, 1981), pp. 87-98.
23. F.C. Michel, Rev.Mod.Phy., 54, 1 (1982).
24. R.N. Manchester and J.H. Taylor, A.J., 86, 1953 (1981). D.C. Backer et al., Nature, 300, 615 (1982).
25. M. Vivekenand and R. Narayan, J.Ast.Ap., 2, 315 (1981).
26. G.R. Huguenin, R.N. Manchester, and J.H. Taylor, Ap.J., 169, 97 (1970).
27. F.S. Fujimura and C.F. Kennel, Ap.J., 236, 245 (1980).
28. A.G. Lyne, R.T. Ritchings, and F.G. Smith, M.N.R.A.S., 171, 579 (1975).
29. E.T. Scharlemann, J. Arons, and W.M. Fawley, Ap.J., 222, 297 (1978); P.B. Jones, M.N.R.A.S., 200, 1081 (1982).
30. A.G. Lyne, in Pulsars, (Sieber and Wielebinski, eds.) (Dordrecht, Reidel, 1981), pp.423-426.
31. J.M. Rankin, preprint (1982).
32. J.M. Cordes, in Pulsars, (Sieber and Wielebinski, eds.) (Dordrecht, Reidel, 1981), pp. 115-131.
33. R. Narayan and M. Vivekenand, Ast.Ap., 113, L3 (1982).
34. D. Stinebring, et al., Ap.J., submitted (1983); D. Stinebring and J.M. Cordes, these proceedings.
35. D.B. Melrose and R.J. Stoneham, Proc.Ast.Soc.Aust., 3, 120 (1977); D.B. Melrose, Aust.J.Phys., 32, 61 (1979).
36. A.F. Cheng and M.A. Ruderman, Ap.J., 229, 348 (1979); A.F. Cheng and M.A. Ruderman, Ap.J., 235, 576 (1980).
37. T. Gold, Nature, 221, 25 (1969).
38. D.C. Ferguson, in Pulsars, (Sieber and Wielebinski, eds.) (Dordrecht, Reidel, 1981), pp. 141-152.
39. D.H. Roberts and P.A. Sturrock, Ap.J., 172, 435 (1972).
40. P. Goldreich and W.H. Julian, Ap.J., 157, 869 (1969).
41. W. Sieber, R. Reinecke, and R. Wielebinski, Ast.Ap., 38, 169 (1975).
42. J.M. Cordes, Ap.J., 222, 1006 (1978).
43. N.S. Kardashev et al., Ast.Ap., 109, 340 (1982).
44. J.M. Cordes, J.M. Weisberg, and V. Boriakoff, Ap.J., in press (1983).
45. E. Groth, Ap.J.Suppl., 29, 453 (1975).
46. J.M. Cordes, Ap.J., 237, 216 (1980).
47. G.S. Downs, Ap.J., 249, 687 (1981).
48. P. Boynton et al., Ap.J., 175, 217 (1972).
49. J.M. Cordes and D.J. Helfand, Ap.J., 239, 640 (1980).
50. P.W. Anderson et al., in Pulsars, (Sieber and Wielebinski, eds.) (Dordrecht, Reidel, 1981), pp. 299-300.
51. J.M. Cordes and G. Greenstein, Ap.J., 245, 1060 (1981).
52. G.S. Downs, Ap.J.Letts, 257, L67 (1982); private communication.
53. P.R. Backus, Ph.D. Thesis, University of Massachusetts, 1981.
54. R.N. Manchester, in Pulsars, (Sieber and Wielebinski, eds.) (Dordrecht, Reidel, 1981), pp. 267-276.

THE EFFECT OF NULLS ON THE DRIFTING SUBPULSES IN PSR 0809+74

Alexei V. Filippenko, A. C. S. Readhead, and M. S. Ewing
Department of Astronomy, California Institute of Technology,
Pasadena, CA   91125

ABSTRACT

The manner in which nulls affect the drifting subpulses of PSR 0809+74 has been investigated. Immediately following a null, subpulses reappear at approximately the longitudes expected for subpulses in the first nulled pulse, as previously reported. The higher quality of the new data, however, has made it possible to demonstrate that the "phase memory" is not perfect; there is a small systematic phase shift in the direction of normal drift.

This result is probably inconsistent with the hypothesis that during a null, electron-positron pair production in the polar gap is absent and local "hot spots" identify the regions over which discharges last occurred. If nulling is the manifestation of temporally steady discharge within discrete regions in the polar gap, on the other hand, a slight phase shift is expected, but the observations do not confirm the prediction that its magnitude should be proportional to null length. In addition, this scenario has difficulty explaining the abnormally slow subpulse drift following long nulls and the absence of radio radiation during nulls associated with large phase shifts.

INTRODUCTION

Shortly after the discovery of pulsars it was realized that a null can affect the rate at which individual subpulses drift through a window defined by the average pulse envelope[1,2]. This observation, together with the mere existence of drifting subpulses and nulling, provided important restrictions in theoretical studies of pulsars. Perhaps the most severe constraints, however, were imposed by the discovery made by Unwin et. al. (hereinafter URWE)[3] that in PSR 0809+74, subpulses which appear immediately after a null have nearly the same longitudes (positions within the pulse envelope) as those expected for subpulses in the first nulled pulse. This demonstrates that nulling cannot be caused by an obscuration or deflection of the radiation beam, but rather must be closely linked to the processes which cause the emission of radiation and the subpulse drift.

A much more detailed analysis of PSR 0809+74 has been made in an effort to increase our understanding of these phenomena. The effect of nulls on the drifting subpulses is briefly discussed in this article, and a complete treatment will be presented elsewhere[4].

OBSERVATIONS AND ANALYSIS

PSR 0809+74 was observed with the 140-foot telescope at the

National Radio Astronomy Observatory for a total of over 20 hours during 6 consecutive days in October and November, 1978. The data were sampled with four channels at all times. Signals in two of these were recorded on magnetic tape; 64 bins, each representing either 2 or 3 ms, comprised a window of 0.128 or 0.192 s duration (~ 0.1 P, 0.15 P) centered on the pulse arrival time. The other two channels were used to search for the strongest signal by varying the central frequency (~ 160 MHz) and bandwidth (~ 1 MHz). In addition, the signal in each of the four channels was continuously plotted with low temporal resolution on a strip chart recorder.

Initially, the strip chart and the plots having high temporal resolution were examined, and all nulls identified. The data from 48 consecutive pulse windows are displayed in Fig. 1, where each bin represents 3 ms. A null consisting of 10 missing pulses disrupts the drift pattern, and to first order the subpulse behavior near the null agrees with that described by URWE.

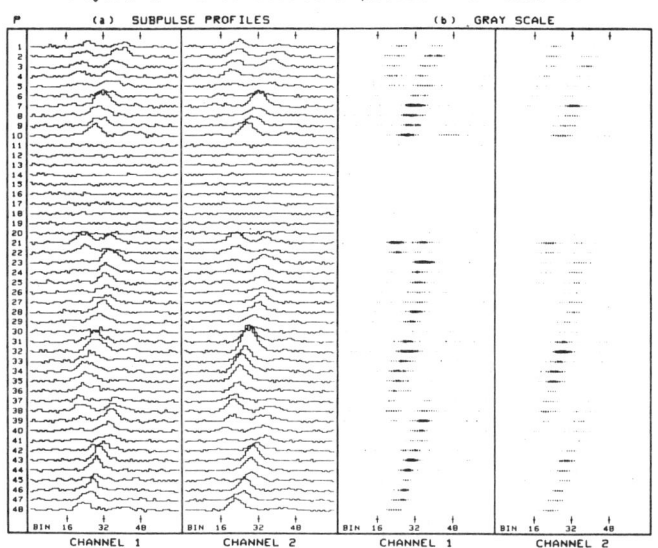

Figure 1: 48 consecutive pulses, PSR 0809+74

Fig. 2 shows the subpulse centroids for the same set of data. The systematically decreasing longitude (or increasing "phase") of subpulses in each band is evident, and immediately following the null the drift rate is slower than usual. After long nulls such as this the drift is initially very lethargic, but it becomes significantly faster within a single band as the subpulses relax to their normal pattern. Moreover, the longitude of each subpulse is somewhat smaller than that expected (in the absence of nulling) for the subpulses of the first nulled pulse, implying that a slight drift is present during the null. This increase in phase (marked by a horizontal segment and an arrow) is quite large for one of the two subpulses, leading to a temporary change in the band spacing.

The small phase shifts can be used to test several hypotheses

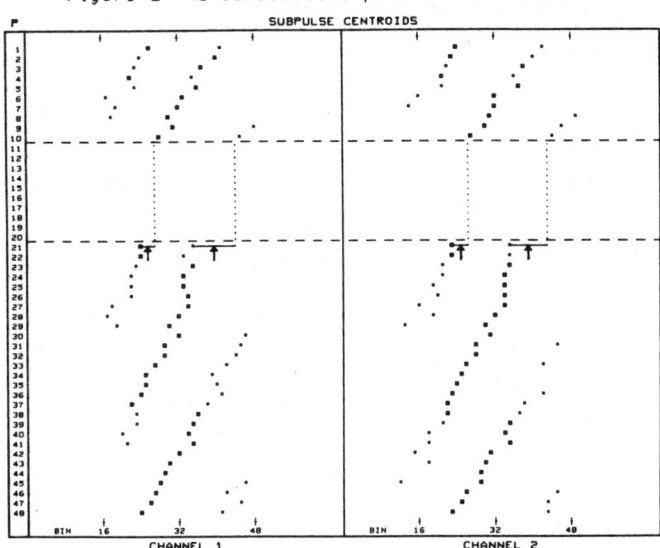

Figure 2: 48 consecutive pulses, PSR 0809+74.

concerning subpulse drift. In order to quantify the shifts, every null in data having relatively high signal-to-noise ratios has been examined. Only nulls consisting of two or more consecutive missing pulses were considered, since great variations in pulse intensity can produce apparent single nulls. For each subpulse band, the phase at which the null began and ended was estimated by inspection or by a least-squares fit through the positions of subpulses in the band, and the phase shift was defined as the difference (ms) between the observed longitude of reappearance and the longitude which would have been expected in the <u>first null</u>ed pulse (rather than in the last visible pulse).

Fig. 3 illustrates the subpulse phase shift as a function of null length. A dashed line represents the expected shift if no memory mechanism were present, whereas zero shift indicates perfect memory. The average shift is nonzero for all null lengths, so the "memory" is not perfect. (Note that the same conclusion may be drawn from Figs. 6 and 7c in URWE.) This behavior is currently being investigated in more detail, but it is clear that the average shift (~ 4 - 5 ms, marked with an arrow) is almost independent of null length and perhaps slightly exceeds the drift which normally occurs during one pulsar rotation period (~ 4.2 ms).

Nulls of a given observed length display shifts having many different sizes, and the direction of several is opposite that of the usual subpulse drift. This can be partially explained by the fact that a fixed observer can measure only integral values of null length, whereas a continuous distribution is physically more plausible. Thus, a null which consists of two consecutive missing pulses may actually be between one and three rotation periods long. In addition, the shifts were often difficult to measure, and the relative error increases with decreasing null length. Nevertheless, a portion of the scatter must be intrinsic to the pulsar.

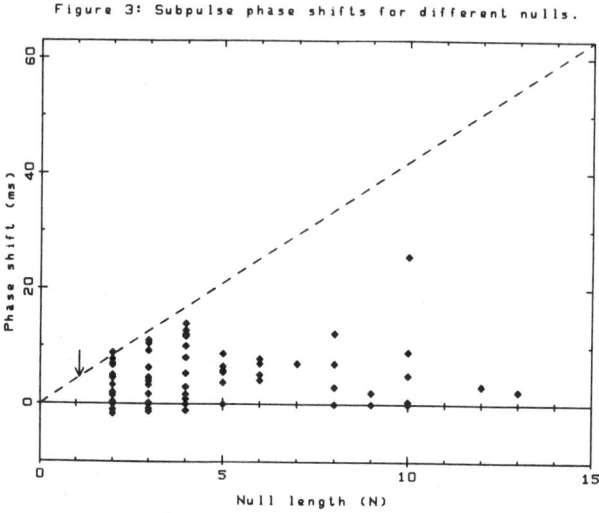

Figure 3: Subpulse phase shifts for different nulls.

INTERPRETATION

The phase "memory" during nulls which was first reported by URWE has received considerable attention during the last few years. Several hypotheses have been proposed to explain it, but the most plausible of these[5] may be summarized as follows:

In the standard pulsar model[6-9], local heating of the neutron star due to a flow of ultrarelativistic electrons creates an area whose temperature is higher than that of the surrounding surface. This excess should persist to some extent for relatively long periods of time, so during a null ions and thermal X-rays may be emitted by a surface "hot spot." Ions are accelerated along magnetic field lines to relativistic speeds by the intense electric field, and X-rays subsequently form electron-positron pairs in the Coulomb field of the ions. "Lorentz-boosted pair production" may therefore reinitiate the avalanche, and sparking is most likely to start at the positions which the localized discharges occupied immediately prior to the null because the probability of ion emission is a very sensitive function of surface temperature.

Although it has been argued[10] that this scenario cannot produce the memory in old pulsars, the new observations provide perhaps the best evidence against it. If this mechanism were indeed operating, it would provide _too_ good a memory, and following a null the localized discharges should recommence exactly where they last occurred.

The existence of a slight subpulse phase shift associated with nulls has recently been predicted[10] in some modifications to the standard theory. A configuration may exist such that the potential drop across the polar gap discharges in a temporally steady manner, unlike the case in the usual interrupted avalanche. This may happen in old pulsars because the gap thickness must be large before sparking can be initiated — and since the electric field increases slowly with gap height at large distances from the stellar surface, occasionally conditions may favor a discharge which exactly balances the growth rate of the potential drop. Such a condition is not possible in young pulsars: the gap size is small and the drop continues to grow quickly for a short time after the birth of a spark, thereby leading to an exponential growth in the production of electron-positron pairs.

The lack of modulation of the rate at which plasma is injected into the magnetosphere, together with the relatively small particle energies and small dispersion in energy, lead to the absence of the mechanism which normally produces coherent radio radiation, so a null is seen. The steady pair production occurs at the level of a dying avalanche discharge, and since the average potential drop is roughly a factor of 5 -- 10 smaller than normal, the drift rate (which is proportional to the drop) is greatly reduced during nulls. Therefore the phase of a subpulse should be slightly greater after a long null than it was immediately prior to the null, as observed.

This hypothesis has difficulty accounting for several features associated with nulls and drifting subpulses:

(i) Fig. 3 implies that the average phase shift is almost independent of null length, whereas it is predicted to be proportional to the duration of a null.

(ii) The drift rate after long nulls is initially very sluggish, demonstrating that slow drift of localized discharges is not always accompanied by a lack of observable radio radiation. Moreover, several long nulls display uncomfortably large phase shifts; hence, rather rapid drifts can occur under certain circumstances without producing intense coherent radiation.

These observations suggest that nulls and partial phase memory may be caused by mechanisms which differ from those proposed in the steady discharge scenario. "Hot spots" are also unsatisfactory, however, so a new approach to the problem is probably necessary.

Informative discussions with S. Unwin and V. Radhakrishnan are appreciated, and A.V.F. is grateful to the Fannie and John Hertz Foundation for a graduate fellowship. NRAO is operated by Associated Universities, Inc. under contract with the NSF.

REFERENCES

1. T. W. Cole, Nature, 227, 788 (1970).
2. J. H. Taylor and G. R. Huguenin, Ap. J., 167, 273 (1971).
3. S. C. Unwin, A. C. S. Readhead, P. N. Wilkinson, and M. S. Ewing, M.N.R.A.S., 182, 711 (1978).
4. A. V. Filippenko, A. C. S. Readhead, and M. S. Ewing, in preparation (1983).
5. A. F. Cheng and M. A. Ruderman, Ap. J., 235, 576 (1980).
6. P. Goldreich and W. H. Julian, Ap. J., 157, 869 (1969).
7. P. A. Sturrock, Ap. J., 164, 529 (1971).
8. M. A. Ruderman and P. G. Sutherland, Ap. J., 196, 51 (1975).
9. A. F. Cheng and M. A. Ruderman, Ap. J., 214, 598 (1977).
10. A. V. Filippenko and V. Radhakrishnan, Ap. J., 263, 828 (1982).

DISCOVERY OF A MILLISECOND PULSAR

Shrinivas R. Kulkarni
Department of Astronomy,
University of California, Berkelely, Calif. 94720

ABSTRACT

We have discovered a pulsar with a period of about 1.556 milliseconds. The discovery of the pulsar has cleared up the mystery surrounding the enigmatic source, 4C21.53. An exposition of the history of the search process, results from our timing program, the pulse waveform, the distance and the origin of the pulsar are discussed.

INTRODUCTION

I am reporting the discovery of a millisecond pulsar and some follow-up work. I would like to clarify that this is a report on behalf of a team of people of which I am one member. Some of the results have already been published and some are in the press. The discovery was made by Backer and his colleagues and first reported in mid-November, 1982 in an IAU telegram [1]. A detailed description of the history of the search process with preliminary timing observations has been published in Nature [2]. I will also present results of two other projects - intensive timing of the pulsar [3] and a determination of distance to the pulsar by HI absorption [4].

HISTORY OF THE SEARCH PROCESS

The radio source 4C21.53 (G57.5-0.3) has been an enigma for many years. It was first noticed to be peculiar by Readhead and Hewish [5] in their 81.5 MHz interplanetary scintillation (IPS) survey data. At such low frequencies compact extragalactic sources are broadened by interstellar scintillation (ISS) to an angular size $\theta = \theta_0(\sin b)^{-1/2}$ where θ_0 is the broadening toward b=90°, estimated to be 0.15" at 81.5 MHz [5]. This results in the reduction of the observed IPS modulation. In particular, sources close to the plane ($|b| < 10°$) are not expected to show much IPS unless they are nearby, i.e. compact galactic sources in which case $\theta = \theta_0(d/H)^{1/2}$ where H is the scale height of the thermal electron layer and d is the distance to the object. Based on the IPS data, Readhead and Hewish [5] (the actual figures are from Purvis [6] who has remeasured and improved all the previous Cambridge measurements) estimated the angular size of 4C21.53 to be less than 0.7".

Rickard and Cronyn [7] on re-examining the results of the Cambridge IPS group [5,8] concluded that θ_0 was quite overestimated. Despite this and primarily based on pulsar pulse-decay time data, Rickard and Cronyn [7] concluded that either the expected ISS broadening was grossly overestimated or that about 42 low-latitude sources had sizes inconsistent with predictions of ISS theories and gave them the generic name "scintars" to denote a new class of compact, non-thermal galactic radio sources. 4C21.53 was especially noted as being peculiar in that it seemed to have an extremely steep spectrum at meter-wavelengths (ν^{-2}) and to lie right in the the galactic plane but with $\theta < 0.7"$ instead of the expected 3"-10".

The next step was taken by Backer [9] who realized from a literature search that the steep spectrum IPS source was coincident with a flat spectrum ($\nu^{-0.1}$), extended (~ 60") object located to the west of the nominal 4C right ascension by approximately one 4C interferometer lobe. This superposition of an extended, flat spectrum source with a compact, steep spectrum source was so reminiscent of the radio properties of the Crab nebula and its pulsar that it led Backer to conclude that 4C21.53 was another Crab type pulsar-embedded-in-synchrotron nebula object. However no pulsar was detected in this direction in the sensitive Arecibo survey [10]. This meant that the purported pulsar had to have either a DM greater than the Arecibo survey limit of 1280 cm^{-3}pc or a period shorter than 30 milliseconds, or at the frequency of the Arecibo survey (430 MHz) $t_s > P$ where t_s is the delay of the scattered wavefront from ISS and P is the pulse period. The upper limit on θ, the observed angular size, by the IPS observations [5] of course constrained t_s and DM. However the distance to the scattering screen is not constrained by the IPS size; in particular, by moving the scattering screen toward the pulsar one can increase t_s keeping the observed angular size fixed provided the scattering angle in the screen is increased. Thus one explanation for the non-detection of the pulsar in the Arecibo Survey was that the extended object caused most of the scattering and was quite near the pulsar. Consequently Backer and colleagues mounted a pulsar search at centimeter wavelengths (to reduce pulse-decay caused by ISS) and looked specifically for short-period pulsars (P > 10ms) from the Owens Valley Radio Observatory and the Arecibo Observatory, but without success.

Matters rested at this point until Erickson [11] showed that there was yet another steep spectrum, compact object, 4C21.53E, displaced by one 4C sidelobe to the east of the nominal 4C right ascension. This observation confused the issue for a while and provided data against the pulsar-nebula superposition hypothesis. Interest in 4C21.53 returned when Cronyn [12] showed that at least one component in the 4C21.53 "complex" exhibited IPS at 34 MHz implying a brightness temperature in excess of the incoherent electron synchrotron temperature of 10^{12} K. VLA observations by Backer [12] of 4C21.53E showed it to be an 0.8" separation double source and thus most probably an ordinary extragalactic double source. Thus at this point it became quite clear that the scintillating object was actually in 4C21.53W [14], giving the superposition hypothesis a fresh lease on life.

Parallel to these largely American developments, Purvis [6] using the Cambridge IPS array, the 1-mile and the 5-km telescopes came to the same conclusion. In December 1981, he conducted a pulsar search at 81.5 MHz but failed to find any pulsed signal because his search was only sensitive to the period range 10-200 ms and also the effects of ISS are rather severe at that low frequency.

The crucial observation showing that the scintillating source was not coincident with the extended source came from the Westerbork Synthesis Radio Telescope (WSRT). In January 1982, observations of 4C21.53W at 609 MHz showed that 4C21.53W could be divided into two components - 1937+215, the extended, flat spectrum object and 1937+214 the compact, steep spectrum object. In March 1982, at the suggestion of Backer, Boriakoff carried out a pulse search with optimal sensitivity down to about 4 ms without success. By August, a higher resolution 20cm WSRT map of 4C21.53W was obtained which clearly revealed 1937+214 to be a separate component. In addition 1937+214 was found to be highly polarized : about 14% at 20 cm and 28% at 609 MHz, thus confirming that 1927+214 was certainly a pulsar.

We conducted a new pulsar search at 20 cm on Sept 25, 1982 from the Arecibo Observatory and detected two harmonics of 642 Hz. The pulsed signal was present for only 4 minutes in one frequency channel and for about 10 minutes in another channel 10 MHz away in a 20-minute scan. Scans over the next 90 minutes and five days later at 1412 and 2380 MHz did not reveal any pulsed signal. We interpreted the fading of the pulsed signal to result from ISS. It still is not clear why we did not detect any pulsed signal on the second observing run. We intensified our search in November, 1982. On the assumption that 1937+214 was a compact, unpulsed object we carried out an ISS search at 20 cm; the deep ISS modulation with a temporal coherence scale of 5 min and frequency coherence scale of 2 MHz confirmed the pulsar-like compact nature of the source. The next day we confirmed the millisecond periodicity of the pulsar.

PARAMETERS DEDUCED FROM TIMING DATA

Based on our September 25 and early November data we reported a rather large $\dot{P}$ of 3×10^{-14} ss^{-1} in our discovery announcement [1]. In mid-November we discovered that our September 25 data had some sampling problems and we immediately retracted our initial estimate of $\dot{P}$ [14].

We timed the pulsar intensively from November to the first week in January, 1983. Currently Mike Davis is carrying out timing observations about once a week at Arecibo. Our timing observations to date (~ April, 1983) have yielded the following:

RA (timing)	$19^h 37^m 28.7474^s \pm 0.0002^s$
Dec (timing)	$21° 28' 01.369" \pm 0.006"$
RA (VLA)	$19^h 37^m 28.720^s \pm 0.013^s$
Dec (VLA)	$21° 28' 01.3" \pm 0.2"$
Period	1 557 806 448 820 fs $\pm$ 3 fs
Period derivative	$(11.047 \pm 0.063)\times 10^{-20}$ ss^{-1}
Epoch (JED)	2 445 303.263 443 9
Dispersion Measure	71.065 ± 0.004 cm^{-3}pc

These results are consistent with the measurements of Ashworth, Lyne and Smith [16], after taking into account that P and $\dot{P}$ are coupled. The quoted uncertainty in the position, period and the

period derivative are the statistical uncertainties. The systematic
uncertainties could be larger by as much as a factor of 10.
Ashworth, Lyne and Smith [16] claimed that the difference between the
interferometric position and timing position could be as large as
0.7" in both the coordinates. If this claim were true then the
uncertainties that I have quoted above are meaningless and $\dot{P}$ could
well be zero. This confusion seems to have settled down now with the
retraction of this claim [17]. The uncertainties in P and $\dot{P}$ arising
out of the uncertainties in the VLA position are about 110 fs and
2.5×10^{-20} ss^{-1} respectively.

The pulse arrival time is remarkably regular; no glitches have
been observed so far. In Figure 1, I have shown the residuals of the
pulse arrival time for all our data of November 29, 1982 after
fitting a P and $\dot{P}$ for all our data through January 1, 1983. The mean
and the rms for this stretch of data are -0.4 µs and 1.8 µs,
respectively. With more data our estimates for P and $\dot{P}$ have improved
such that our typical rms for any day (in April) is about 1.3 µs. In
Figure 2, I have plotted the timing residuals averaged over one day's
observation, which typically is about 2-3 hours long, against the day
number. The data spans from 30 November, 1982 to 6 January, 1983.
Other than an arbitrary offset in the mean, we see that there are no
systematic variations in the residuals, ruling out a binary system
and precession. The very small variations in the residuals speak
extremely highly of the pulsar as a stable clock.

PULSE WAVEFORM

In Figure 3, I have plotted a very high signal to noise ratio of
the pulse waveform. The main pulse and the interpulse are separated
by about 173°; the widths of the pulses are dominated by instrumental
broadening. Ashworth, Lyne and Smith [16] and Stinebring [18] have done
some high resolution polarimetry on this pulsar. Stinebring [18]
reports FWHM for the main and the interpulse at 1415 MHz to be 65 µs
and 56 µs respectively. The pulse/interpulse morphology is quite
similar to that of the Crab. Also note that this waveform has a
little emission before the main pulse and the interpulse, very much
like the "precursor" pulse of the Crab pulsar waveform. However
improper it is to generalize on statistics of two, it is quite
tempting to go ahead and generalize that all short period pulsars
would have similar waveforms.

DISTANCE TO THE PULSAR

We have just finished a project to determine the distance to
the millisecond pulsar and 1937+215 [4]. The absence of polarized flux
in the WSRT maps and the flat spectrum of 1937+215 led us to believe
that 1937+215 (I will refer to this object as either 1937+215 or
4C21.53W) must be an HII region. Consequently we searched and
detected the H166α recombination line centered at +2 kms-1 with a
FWHM of 25 kms^{-1}. The line temperature is about 1% of the continuum
flux and thus in all respects appears to be a normal HII region. The
proximity of the HII region and the pulsar suggests a possible

physical connection. Clearly a determination of the distances to the two objects would resolve the issue. The absorption spectra toward the various components of the 4C21.53 "complex" are shown in Figure 4.

The recombination line velocity of +2 kms^{-1} and the absorption spectrum in Figure 2 places the HII region on the solar circle at the far distance of 10.7 kpc. The free-free luminosity of this HII region is comparable to that of the Orion nebula and thus it may be a site of active star formation. From the HI maps of Kulkarni, Blitz and Heiles [19] we see that, in fact, the HII region is located in a rather prominent HI concentration in the Perseus Arm. Blitz, Fich and Kulkarni [20] find a good correlation between active star formation sites, CO emission and bright HI emission. Based on this we predict that one should be able to detect CO from this HII region.

Obtaining the pulsar absorption spectrum was quite an instrumental challenge. I will however skip all the interesting instrumental details and present only the final result (Figure 4b). The pulsar absorption spectrum (Figure 4b) does not show as much absorption as the spectrum toward the HII region. In particular the pulsar spectrum does not show much absorption between 0 and 10 kms^{-1}; the prominent +16 kms^{-1} visible in the spectrum of the HII region is quite weak in the pulsar spectrum and finally the pulsar does not show as much absorption at 40 kms^{-1} as do those of 4C21.53E and 4C21.53W. Since this (last) feature is located at the tangent point, we conclude that the pulsar is in fact a little closer to us than the tangent point which is located 5 kpc away. The narrow absorption feature at -10 kms^{-1} seen in the spectra of all three components is attributed to a nearby dust cloud.

Our 5 kpc distance to the pulsar is in good agreement with the distance implied by our dispersion measure if we assume that the average electron density in that direction is about half the normal value of 0.03 cm^{-3}. Indeed, Ables and Manchester [21] find that in this general area the electron density is lower than the nominal value by a factor of two. Weisberg [22] while rejecting some of the distances determined by Ables and Manchester [21] identifies the longitude range 60° to 70° as being deficient in electrons. This has been attributed to the fact that this line of sight passes through an interarm region.

The HI absorption spectrum of 4C21.53E comes as no surprise; it shows more absorption at all velocities than either the HII region or the pulsar and hence is farther away. From the morphology we conclude that 4C21.53E is an extragalactic double source.

Knowing the distance to the pulsar one can make some estimate of the expected flux at higher frequencies. The total loss of energy from the pulsar, deduced from the measured $\dot{P}$ and nominal neutron star parameters is about 2×10^{36} ergs s^{-1} which can be compared to 7×10^{36} ergs s^{-1} for the Vela pulsar and 5×10^{38} ergs s^{-1} for the Crab pulsar. Both the Crab and Vela emit pulsed γ-ray and optical radiation. Arons [23] predicts a γ-ray luminosity of 3×10^{34} ergs s^{-1}, comparable to that of the Vela pulsar. Thus the expected γ-ray flux here on earth is about 100 times smaller than that of Vela (assuming a similar beaming factor) and probably undetectable in the COS-B data.

OPTICAL EMISSION

Optical pulses have been recently detected from the pulsar [24]. Crab and Vela are the only other pulsars from which optical pulsations have been detected. A straightforward application of Pacini's formula for the optical flux [25], $L_{opt} \propto B_0^4 P^{-10}$, with our distance estimate leads to an apparent magnitude of about 26 without accounting for extinction. Here B_0 is the surface magnetic field strength. It appears from [24] that the pulsar is about 100 times weaker than the 20±1 magnitude star identified by Djorgovski [26] as a possible pulsar candidate which has now been identified as a K-giant at roughly the same distance as the pulsar with a visual reddening of 8 magnitudes [27]. Assuming standard reddening parameters [28] of 0.61 E(B-V)/kpc and R_v=3, the visual extinction toward the pulsar is about 9 magnitudes, in fair agreement with extinction for the nearby K-giant. Thus the pulsar has an absolute magnitude of roughly 2.5 in the optical part of the spectrum.

With the availabilty of three optical pulsars, we have just enough data to derive an empirical equivalent of Pacini's formula for L_{opt}. Using the published optical absolute magnitudes for the Crab and the Vela pulsars with our estimate (above) of the absolute magnitude for PSR1937+214 (part of work done with D. C. Backer and J. Barnard, soon to be published) we derive :

$$M/2.5 = -0.97 + 10.27 \log(P/ms) - 2.07 \log(\dot{P}/10^{-19}) \quad (1)$$

i.e. $L_{opt} \propto B_0^4 P^{-12}$, where M is the optical absolute magnitude, P is the period in milliseconds and $\dot{P}$ is the period derivative in units of 10^{-19} ss^{-1}. This relation is in better agreement with the observed secular decrease of the optical flux from the Crab pulsar [28] than is the theoretical formula of Pacini [25]. This exceedingly steep dependence on P means that it may be profitable to search for millisecond pulsars in one of the optical bands rather than in the usual radio bands [29].

ORIGIN OF THE PULSAR

Since the discovery of PSR1937+214, there has been a spate of theoretical papers explaining possible origin-mechanisms of short period pulsars [30,31,32,33,23]. In the "binary origin" hypothesis [30,31,32] it is postulated that such pulsars are born in long-lived mass transfer binary systems in which one of the stars is a neutron star accreting mass from the companion . In such systems, the surface magnetic field of the neutron star is much smaller than the surface field strengths of typical radio pulsars ($\approx 10^{12}$ G) owing to the decay of the magnetic field with time. Thus the Alfven radius of the neutron star is much diminished . The accretion stops when the neutron star is spun up to the angular velocity of the inner edge of the Keplerean disk which is at the Alfven radius . Alpar et al [32] and Fabian et al [33] prefer low mass close binaries such as the X-ray bulge sources. In the binary scenario for the origin of short period pulsars [30,31], an isolated short period pulsar such as PSR1937+214 is formed when the companion eventually evolves and undergoes a supernova explosion which disrupts the binary system. I must admit that it seems a bit too much to expect all of these steps to proceed in the expected fashion. On more general grounds, all binary origin hypotheses have been critcized for invoking very high accretion rates [23]. In the competing theories [33,23], an isolated fast pulsar is simply explained as the end product of a weakly magnetized star with a rapid initial rotation. The weak initial magnetization helps preserve the angular momentum all the way to the final collapse. Such theories [33,23] satisfactorily explain the occurence of PSR1937+21 right in the galactic plane which is difficult to understand in a simple way in the scenario of Alpar et al [31].

SEARCH FOR OTHER FAST PULSARS

Including the classical binary pulsar, PSR1913+16, in the class of spun-up and isolated short period pulsars one can conclude that the galactic plane is a particularly rich area for fast pulsars which have escaped detection in all the previous sky-surveys. Alan Purvis (Cambridge) and I will be searching for such pulsars sometime this year. Our basic strategy is to look toward "scintar candidates" that Purvis has catalogued from his recently completed sensitive IPS survey. As remarked in an earlier section, selected nearby regions can be profitably searched in the optical bands. John Barnard (U. C. Berkeley) pointed out to me that since very different physical mechanisms generate the radio and the optical emission it is concievable that one could find pulsars which pulse in the optical but not in the radio. I think this is an exciting prospect and hope to exploit it in the near future.

ACKNOWLEDGEMENTS

I would like to express my gratitude to John Barnard for suggesting several ideas regarding optical emission and generally being patient with all my questions ; to Mary Stevens for painstakingly proof-reading the manuscript and Don Backer for numerous discussions; to Carl Heiles for financial support and encouragement.

REFERENCES

1. D. C. Backer, S. R. Kulkarni, C. Heiles, M. M. Davis, and W. M. Goss, IAU Circular no. 3743, November 12 (1982a)
2. D. C. Backer, S. R. Kulkarni, C. Heiles, M. M. Davis, and Goss, W. M., Nature 300, 615 (1982c)
3. D. C. Backer, S. R. Kulkarni, and J. H. Taylor, Nature 301, 314 (1983)
4. C. Heiles, S. R. Kulkarni, M. Stevens, D. C. Backer, M. M. Davis, and W. M. Goss, submitted to Astrophys. J. Letters (1983)
5. A. C. S. Readhead and A. Hewish, Mem. R. A. S. 78, 1 (1974)
6. A. Purvis, M. N. R. A. S. 202, 605 (1983)

7. J. J. Rickard and W. M. Cronyn, Astrophys. J. 228, 755 (1979)
8. P. J. Duffet-Smith and A. C. S. Readhead, M. N. R. A. S. 174, 7 (1976)
9. D. C. Backer, priv. comm. (1979)
10. R. A. Hulse and J. H. Taylor, Astrophys. J. 201, L55 (1975)
11. W. C. Erickson, Bull. A. A. S. 12, 799 (1981)
12. W. M. Cronyn, referred to in Ref. 14
13. D. C. Backer, priv. comm. (1981)
14. W. C. Erickson, Astrophys. J. 264, L13 (1983)
15. D. C. Backer, S. R. Kulkarni, and C. Heiles, IAU Circular # 3746, (1982)
16. M. Ashworth, A. G. Lyne, and F. G. Smith, Nature 301, 313 (1983)
17. A. G. Lyne, priv. comm. to D. C. Backer (1983)
18. D. Stinebring, submitted to Astrophys. J. Letters (1983)
19. S. R. Kulkarni, L. Blitz, and C. Heiles, Astrophys. J. 259, L63 (1982)
20. L. Blitz, M. Fich and S. R. Kulkarni, Science (in press, 1983)
21. J. G. Ables and R. N. Manchester, Astron. Astrophys. 50, 177 (1976)
22. J. M. Weisberg, Thesis, Univ. of Iowa, Iowa City (1978)
23. J. Arons, Nature 302, 301 (1983)
24. R. N. Manchester, B. A. Peterson, P. T. Wallace, IAU Circular # 3795, April 29, 1983
25. F. Pacini, Astrophys. J. 163, L17 (1971)
26. S. Djorgovski, Nature 300, 618 (1982)
27. J. Middleditch, D. Cudaback, C. Pennypacker, B. Oliver, G. H. Rieke, M. J. Lebofsky, J. T. McGraw, D. Dearborn, preprint, 1983
28. L. Spitzer, Physical Processes in the Interstellar Medium (Willey-Interscience, 1978), p155 and p160
29. F. Pacini and M. Salvati, sub. to Astrophys. J. (1983)
30. V. Radhakrishnan and G. Srinivasan, Curr. Sci., Dec. 5 (1982)
31. M. A. Alpar, A. F. Cheng, M. A. Ruderman, and J. Shaham, Nature 300, 728 (1982)
32. A. C. Fabian, J. E. Pringle, F. Verbunt and R. A. Wade, Nature 301, 222 (1983)
33. K. Brecher and G. Chanmugam, Nature 302, 124 (1983)

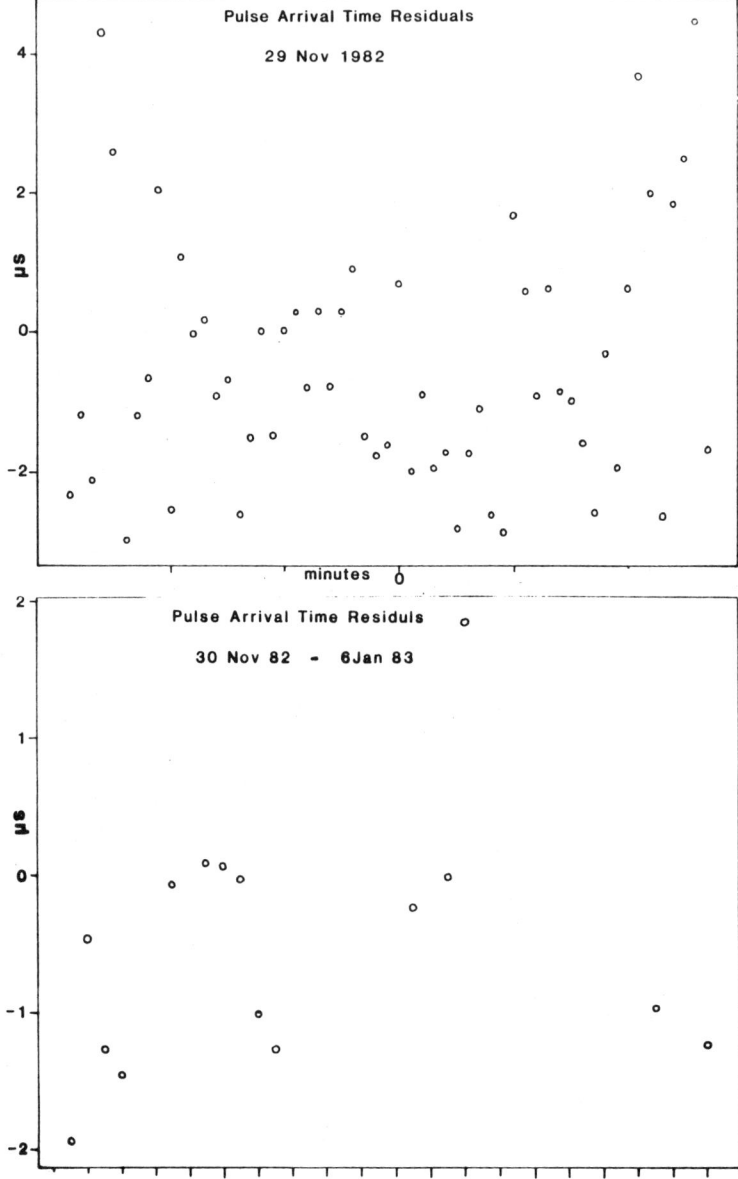

Figure 1 : (top) Timing residuals for one day's observations. Each dot in the plot is two minutes of coherent integration. Source transit is at x = 0 minutes. Reduced signal to noise at high zenith angles may explain the increased scatter in the residuals at the extremes of the plot.

Figure 2 : (bottom) Timinig residuals over 38 days. Each tick on the x-axis is two days of time. Source was observed for about 2-3 hours on any given day.

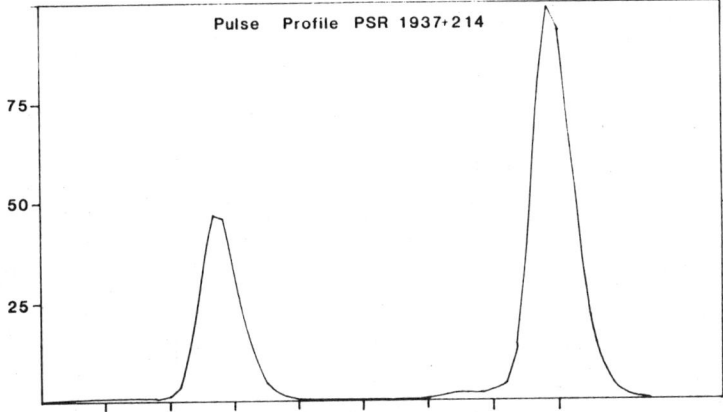

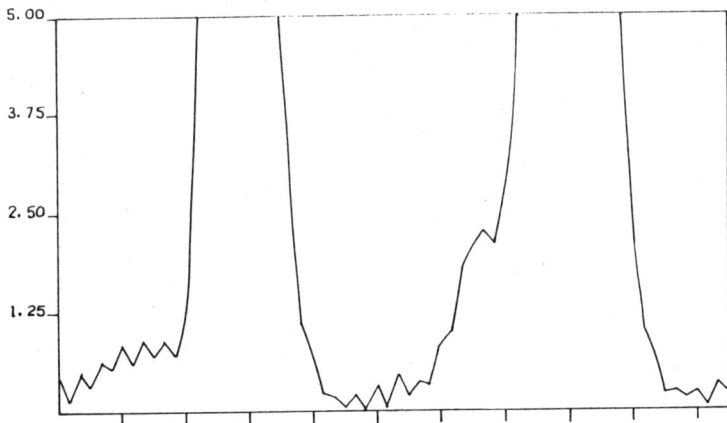

Figure 3 : (top) Plot of the pulse waveform at 1408 MHz. Most of the width of the pulse and the interpulse is of instrumental origin. Note the faint emission preceeding the main and the interpulse (bottom) Same plot as above except the vertical is now restricted to 5% of the peak value.

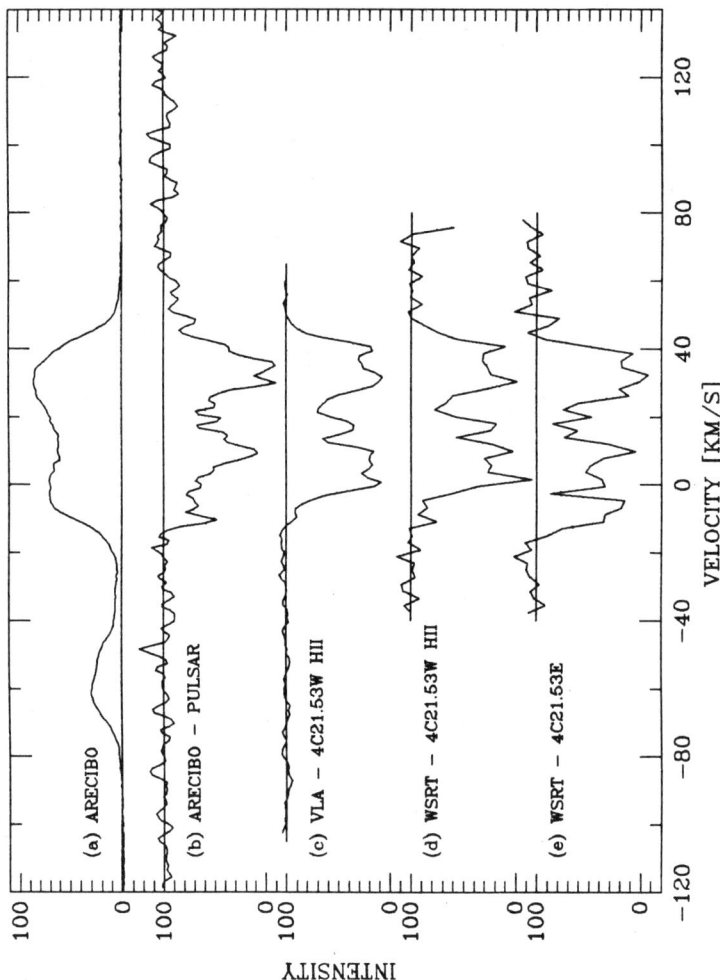

Figure 4 : Emission and absorption spectra toward the various components of 4C21.53. The emission and the absorption spectra toward the pulsar from Arecibo are presented in (a) and (b). The emission spectrum has units of K and the scale of all the absorption spectra is in units of percentage of continuum flux. The absorption spectra for the HII region, 4C21.53W are given in (c) and (d) based on synthesis observations with the VLA and the WSRT, respectively; (e) shows the absorption spectrum of the extragalactic source 4C21.53E from the WSRT.

HIGH ENERGY OBSERVATIONS OF PULSARS

Gottfried Kanbach

Max-Planck-Institut für Physik und Astrophysik,
Institut für extraterrestrische Physik,
8046 Garching, Fed. Rep. of Germany

ABSTRACT

High energy emissions have so far only been definitely observed from two pulsars: Crab (PSR 0531 + 21) and Vela (PSR 0833-45). In these sources, however, the luminosity at gamma-ray energies surpasses the radio luminosity by approximately 10^6 which demonstrates the basic importance of this emission for the pulsar radiation budget. Deep observations performed with the COS-B instrument in the 0.1 to 1 GeV gamma-ray energy range have revealed several previously unknown details: The Vela pulsar gamma-ray spectrum changes with light-curve phase such that the second peak of emission exhibits a harder spectrum than the first peak. In the Crab pulsar a secular change of the relative strength of the two peaks of emission has been detected with high probability in the course of observations spread over 5 years. It is remarkable that concurrent observations of the X-ray (2-12 keV) pulse profile of Crab do not show this variation.

The search for gamma-ray emission from further radio pulsars has up to now been inconclusive. From the estimated flux limits it is however possible to derive limits to the contribution of pulsars to the high energy luminosity of the Galaxy. It is expected that the next generation of telescopes on board the Gamma Ray Observatory will discern several more gamma-ray pulsars.

INTRODUCTION

Pulsar research has traditionally been and continues to be the domain of radio astronomy. With about 330 pulsars detected according to the latest surveys[1,2] the power of a radio telescope to find these strange objects in the frequency bands around several hundred megahertz is unsurpassable. The stronger radio pulsars have mean flux densities exceeding about 100 mJy at 400 MHz and are easily detected even at the level of individual pulses. The radiospectra observed show typical nonthermal behaviour with steep power laws (average index -1.5) and the brightness temperatures indicate that coherent emission processes are required. It is clear that the coherence of the emission has to break down when a certain minimal size corresponding to a maximal frequency of emission has been reached. The straightforward extension of pulsar radio spectra into the infrared, visible, X-ray and gamma-ray parts of the spectrum under the assumption that the same radiation process is at work would therefore impose quite unrealistic demands on pulsar models and at the same time one would, as an example, reach flux densities in the 100 MeV range of the order of 10^{-22} Jy which would put pulsars out of reach of any conceivable gamma-ray instrument. Fortunately nature provided

0094-243X/83/1010129-12 $3.00 Copyright 1983 American Institute of Physics

in several cases new phenomena in this so called high energy regime of pulsars, which can be taken to include the total spectrum beyond the radio range. The first pulsar found to be such a high energy emitter was of course the Crab pulsar in the optical range[3]. This observation was followed by numerous detections of PSR 0531 + 21 at X-ray and low and high gamma-ray energies. In 1975 the second high energy pulsar, PSR 0833 - 45 in the Vela remnant, was reliably detected at 100 MeV in data from the SAS-2 instrument[4]. Both pulsars were confirmed by COS-B measurements[5]. The overall signatures of the Crab and Vela lightcurves are shown in figure 1. The well known

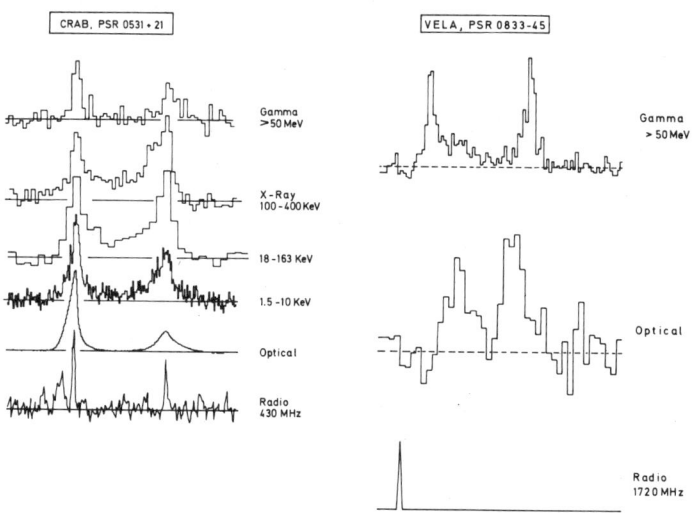

Fig. 1: Lightcurves for Crab and Vela at wavelengths where emission was observed.

morphological similarities of these two pulsars in the gamma-ray region contrast with the differences evident in the emission at other wavelengths. The power contained in the gamma-ray channel reaches nearly one percent of the available braking power in the case of PSR 0833 - 45. Both findings indicate the basic importance of gamma-ray emission for young pulsars.

The situation in the search for gamma-ray emission from other, generally older, radio pulsars does not seem to be very encouraging at present. None of the claims or indications derived from the SAS-2 and COS-B observations[6-9] were confirmed in both data sets. A contribution to this symposium by the COS-B team[30] describes in detail the search procedures used in a concerted effort of radio observations and gamma-ray measurements. The result indicates that none of the available older pulsars emits above 50 MeV at a level exceeding 10^{-6} photons/cm^{2}s.

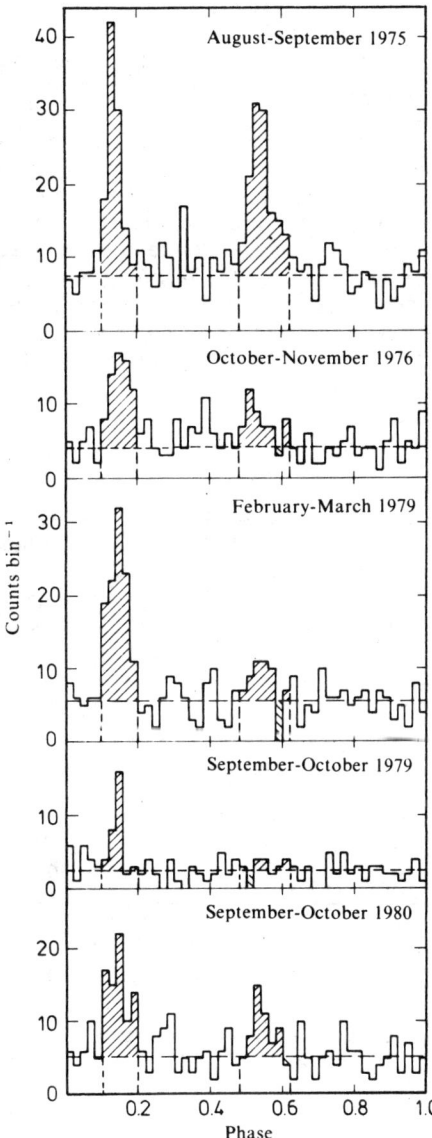

Fig. 2: Gamma-ray lightcurves (>50 MeV) of PSR 0531 + 21 at the epochs indicated[15].

CURRENT OBSERVATIONAL TECHNIQUES AND LIMITATIONS

Cosmic high energy radiation at X-ray and gamma-ray frequencies is observed using techniques originating from nuclear physics laboratories however with instruments that are adapted to the remote operations on balloon and satellite platforms. The principal interactions that lead to the detection of these photons are with increasing energy the photoeffect, the Compton scattering on electrons and the pair creation interaction in the field of a target nucleus. The released secondary electrons are measured according to the particular need to determine their energy and direction in solid state detectors, ionisation and scintillation counters, spark chambers or Cerenkov detectors. It goes without saying that an astronomical instrument must have a capacity to resolve the direction of incidence of the detected radiation and if pulsar studies are done a good timing system is necessary. A schematic review of currently operating or planned instruments is given in table I together with characteristic angular and energy resolutions and with sensitivities achievable with feasible technical sizes[10-12].

TABLE I

Energy range	10keV-10MeV	0.5MeV-30MeV	30MeV-10GeV
Typical Instruments	GRO-OSSE[12] (Balloon[10])	GRO-COMPTEL[12] (Balloon[11])	GRO-EGRET[12] (SAS-2[13], COS-B[14])
$\frac{\Delta E(FWHM)}{E}$	8% (0.6 MeV) (20%, 60 keV)	5 - 8% (10%)	15% (50%)
Angular Res.	10 arc min (3°)	8 arc min (4°)	6 arc min (1°)
Maximum Effective Area	2300 cm^2 (2000 cm^2)	50 cm^2 (40 cm^2)	2000 cm^2 (50 cm^2)
Source Sensitivity	3x10^{-5}cm^{-2}s^{-1}	5x10^{-5}cm^{-2}s^{-1}	5x10^{-8}cm^{-2}s^{-1} (1x10^{-6}cm^{-2}s^{-1})

A rough upper limit for the flux to be expected from a rotating neutron star can be written as

$$\varphi = \frac{4\pi^2 I \dot{P} \eta \beta}{\Omega d^2 \langle E \rangle P^3} \quad cm^{-2} s^{-1} \tag{1}$$

where I is the moment of inertia, d the distance and it is assumed that the pulsar expends a fraction η of its rotational energy loss in the form of photons of average energy $\langle E \rangle$ into the solid angle Ω. The duty cycle β accounts for φ being the time averaged flux.

In useful dimensions equation (1) can be given for the integral flux above 100 MeV, assuming an E^{-2} spectral shape:

$$\varphi(>100 \text{ MeV}) \approx 7\times 10^{-9} \, I_{45} \, \eta \, \beta \, \Omega^{-1} d_{kpc}^{-2} \, \dot{P}_{-15} \, P^{-3} \text{ cm}^{-2} \text{ s}^{-1} \quad (2)$$

with $I = 10^{45} \cdot I_{45}$ g cm^2, d in kpc and $\dot{P} = 10^{-15} \, \dot{P}_{-15}$ s/s.

The low photon fluxes expected in the gamma-ray range from pulsars and the limited size of available detectors make long observation periods necessary in order to collect a statistically meaningful number of events. The long observations, for COS-B typically one month, demand extremely accurate knowledge of the pulsar parameters which are used to convert the arrival times of the photons at the detector into a phase value of the pulsar's periodic cycle. For example the period of a moderately fast pulsar observed in the course of a month must be known to an accuracy of about $\Delta P/P \approx 10^{-9}$. This nearly rules out the search for unknown periodicities in the presently available gamma ray data independent of guidance from radio, optical or X-ray results. The excessive number of independent trials necessary to cover a frequency interval in a gamma-ray search invalidates even the seemingly most improbable effects. The coordination of observations at different wavelengths is therefore very important to obtain the maximum results from future missions like the Gamma Ray Observatory, which is due to be launched in 1988.

PSR 0531 + 21, THE CRAB PULSAR

The anticenter of the Galaxy was observed with COS-B six times during the mission which started in August 1975 and ended in April 1982. Five observations have so far been analysed for the pulsed emission from PSR 0531 + 21 and gamma-ray light curves in the energy range 50-3000 MeV were derived[15] for each epoch. As shown in figure 2 the general features of the gamma-ray lightcurve are present in all observations: two maxima of emission separated by 13 ms, which is equivalent to 0.43 (155°) of one period. It is however apparent that during the first observation (1975) the relative strength of the two pulses looks different with respect to the later observations. The lightcurves given in figure 2 have been constructed with only a fraction of the detected photons from the pulsar namely those with a measured arrival direction within 7° of the source for energies below 150 MeV and within 4° for higher energies. For a quantitative analysis of the above hinted secular variation of PSR 0531 + 21 the full instrument point spread function was fitted to the data and the total numbers of pulsed counts above background in the lightcurves were derived[15].

The analysis shows that during the first observation the second pulse is indeed much stronger relative to the first pulse than it is at later times. The effect is consistently present in a low energy (50 - 150 MeV) and a high energy band (150 - 3000 MeV). Previous measurements on SAS-2[16] (E > 35 MeV, 1972-1973) and with a balloon experiment[17, 18] (E > 240 MeV and E > 800 MeV, 1971, 1973)

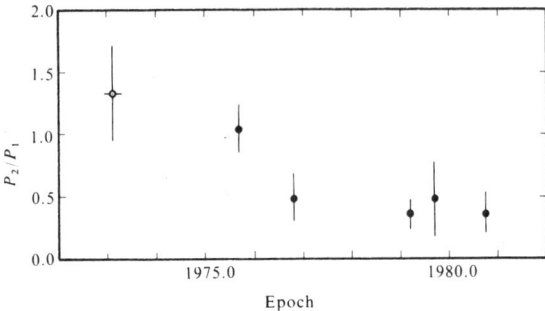

Fig. 3: Variation with epoch of the ratio of pulse strengths for the second pulse to the first pulse[15]. The open circle indicates the SAS-2 measurement[13] (>35 MeV). The full circles are from COS-B (>50 MeV).

derived light curves for PSR 0531 + 21, albeit with less counting statistics, that can be analysed similarly. In figure 3 the ratio of the second pulse to the first is shown for the COS-B analysis and the SAS-2 result. The balloon results[17,18] for P_2/P_1 range from 1.5 to 2.0 with an uncertainty of $\pm$ 1.0. The overall picture suggests with high probability (>99%) that the Crab pulsar changed its high energy emission pattern sometime around 1976.

The X-ray monitor on COS-B has allowed simultaneous observations of 2-12 keV emissions of the Crab pulsar. The authors[15] find the shapes of the X-ray lightcurves in 1975, 79 and 80 to be unchanged. A tentative analysis of the X-ray phase histogram using pulse definitions similar to the gamma-ray curves results in a ratio $P_2/P_1 = 1.10$. Optical lightcurves of Crab have been examined for variability[19] between observations in 1970 and 1977 thus spanning the gamma-ray change in 1976. Only variations on a level less than 1% were found. Again a rough estimate of P_2/P_1 was made and a value of ~ 0.4 was found for the above phase definitions. Further pulse strength ratios are quoted in reference 20 and the general picture is that radiation below and including the optical range has $P_2/P_1 \lesssim 0.6$ while X-rays and previously measured gamma-ray values indicate $P_2/P_1 \gtrsim 1.1$. The present result for E > 50 MeV radiation therefore suggests a transition from a classical "high-energy profile" to a "low-energy profile".

The sum of all COS-B observations of PSR 0531 + 21 is depicted in figure 4. This represents the statistically best lightcurve above 50 MeV yet obtained. The sharp peaks which point to rather compact emission regions are evident. The peak widths (FWHM) were measured to be 1.6 $\pm$ 0.4 ms and 2.0 $\pm$ 0.5 ms for the first and second peak respectively. Furthermore positive evidence for pulsed emission in the region between the two peaks has been found at a level of intensity of (15 $\pm$ 4)% with respect the total pulsed emission (contained in the phase interval 0.10 to 0.62). For a more detailed discussion on the significance of this interpulse emission the reader is referred to reference 21. The pulsed spectrum of PSR 0531 + 21 above 50 MeV was found[22] from the COS-B measurements to be proportional to $E^{-2.1}$ and the pulsar's signature was detected at least up to energies of 1 GeV. With the increase in sensitivity planned for the GRO-EGRET instrument it will be possible to follow the Crab spectrum into the region of 5 GeV and it will be interesting to observe the spectral continuation towards the ultra high energy region covered by air Cerenkov techniques[23].

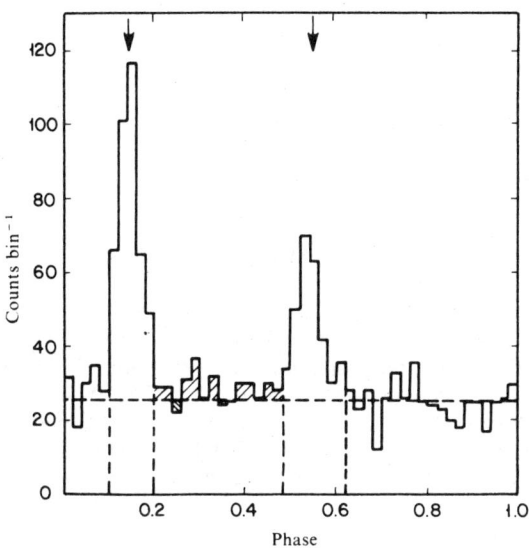

Fig. 4: Summed >50 MeV lightcurve[15] of PSR 0531+21. The arrows indicate the radio peaks. The shaded area shows the interpulse emission.

PSR 0833-45, THE VELA PULSAR

The Vela pulsar is the most conspicuous gamma-ray source in a >50 MeV galactic survey because it is the brightest object known in the gamma-ray sky. This made it possible to study PSR 0833-45 in greater detail[5,22,24,25] than any other high-energy gamma-ray source. COS-B observed the Vela/Puppis region eight times between 1975 and 1979 and PSR 0833-45 has consistently been detected at different aspect angles to the telescope axis. An analysis similar to the one described for Crab has been applied to the PSR 0833-45 lightcurves[24]. The relative strength of the second peak to the first peak shows no variation with time. The value of $P_2/P_1 = 0.96$ for Vela compares well with the pre-1976 lightcurves of Crab. As a side remark one should note that the constancy of P_2/P_1 found for Vela, where we will see that the two pulses have a different spectrum, proves that the COS-B instrument does not suffer from an energy dependent sensitivity change over the course of nearly 7 years. The

Crab result can therefore not be attributed to an instrumental effect. A lightcurve above 50 MeV with 0.5 ms binning was derived[25] for PSR 0833-45 and is shown in figure 5. The peak separation is

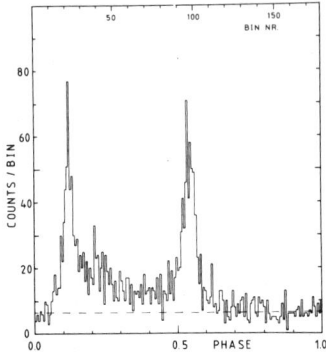

Fig. 5: Fine time resolution (0.5 ms) lightcurve[25] for PSR 0833-45 at E > 50 MeV.

0.420 ± 0.007 in phase and the widths (FWHM) are 3 ± 0.5 ms and 6 ± 1 ms for the first and second peak respectively. Again the sharpness especially of the first peak is noteworthy. The pronounced emission between the peaks ("interregion") and emission following the second pulse ("trailer", 0.58 < phase < 0.77) have been resolved. If phase resolved skymaps from the Vela region are analysed it is de-

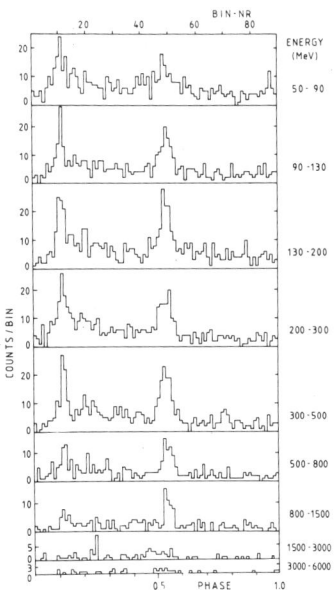

Fig. 6: PSR 0833-45 lightcurves in energy intervals[25]

rived that the source above 100 MeV must be to more than 90%
pulsed. This is in contrast to the Crab case[22] where the pulsed fraction below 400 MeV is only about 55%. In figure 6 the PSR 0833-45
lightcurves are shown for successive energy bands from 50 to
6000 MeV. The pulsed signal is significantly present even in the
highest energy interval from 3 GeV to 6 GeV. Phase resolved spectra
were derived for the lightcurve components relative to the first pulse
which suppresses instrumental effects completely. The phase boundaries of the components as given in table II are of course somewhat
arbitrary.

Table II

Phase definitions for PSR 0833-45 lightcurve components

Background	0.77-0.05
First pulse	0.05-0.17
Interregion	0.17-0.49
Second pulse	0.49-0.58
Trailer	0.58-0.77

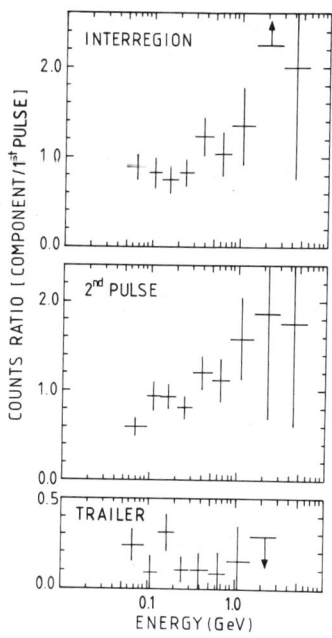

Fig. 7: Ratio of counts in lightcurve components[25] relative to counts in the first pulse of PSR 0833-45.

The result for the pulsed count ratios is given in figure 7. For the interregion and in particular the second pulse a clear trend is visible: these components exhibit a harder spectrum than the first pulse. A recent theoretical work[26] explains this difference in terms of an inverse Compton scattering model where the first pulse is produced near the light cylinder while the second peak comes from the polar region near the star.

The total pulsed flux from Vela above 50 MeV is 2.4×10^{-5} photons cm^{-2} s^{-1} and the spectral shape can be well fitted with a power law of index -1.89. If we assume a pencil beam type of emission pattern[8] the luminosity of Vela above 50 MeV is about 4×10^{34} erg/s which corresponds to 0.6 percent of the available braking power.

SEARCH FOR HIGH ENERGY EMISSION FROM OTHER RADIO PULSARS

Several theoretical works[27, 28] have suggested that the efficiency to convert rotational energy loss into gamma radiation increases with age for pulsars. This is mainly supported by the fact that Crab (age $\sim 10^3$ years) converts about 3×10^{-4} of the available energy and Vela (age $\sim 10^4$ years) about 5×10^{-3} into gamma radiation above 100 MeV. If we follow the derivation[28] for the efficiency and normalize to the Crab and Vela values we obtain

$$\eta = 10^{-3} \, P^{1.3} \, \tau_{years}^{0.5} \qquad (3)$$

Inserted into (2):

$$\phi(>100 \text{ MeV}) = 10^{-4} I_{45} \, \beta/\Omega \, d_{kpc}^{-2} P^{-0.7} \tau_{year}^{-0.5} cm^{-2} s^{-1} \qquad (4)$$

Let us assume a beaming pattern in rough analogy to the Crab and Vela lightcurves with two pencil beams each of half opening angle α. Then $\beta/\Omega = \alpha/2 \pi^2 (1-\cos\alpha)$ and with a typical value of $\alpha = 18°$ one obtains from Eq. (4):

$$\phi(>100 \text{ MeV}) = 3 \times 10^{-5} I_{45} d_{kpc}^{-2} P_s^{-0.7} \tau_{year}^{-0.5} cm^{-2} s^{-1} \qquad (5)$$

For average radio pulsars[20] with $P = 0.65s$ and $\tau \sim 10^6$ years and $I_{45} = 1$ one estimates

$$\phi(>100 \text{ MeV}) = 4 \times 10^{-8} d_{kpc}^{-2} cm^{-2} s^{-1} \qquad (6)$$

This shows that at COS-B and SAS-2 sensitivities only old pulsars closer than about 200 pc would have been detectable. This was however not the case[9, 30, 29]. The consequences for the theoretical models[30] can be stated in terms of the efficiency: while it was believed that η increases up to several times 10^6 years it is now likely that gamma ray production is not anymore efficient beyond an age of about 10^5 years.

An upper limit to the efficiency was derived[30] at $\sim 6\%$ for a collection of older radio pulsars. This result indicates that old pulsars contribute only a negligible amount to the high-energy emission of the Galaxy.

Young pulsars, yet undetected in radio surveys, may however present a different story. Even when one accounts for a limited number of young pulsars (i.e. younger than Vela, $\gtrsim 10^3$) in the Galaxy which would result from a limited although not well understood birthrate of these objects, several sources of the COS-B catalogue[31] could be such young and extremely fast pulsars. The case of the 1.5 ms pulsar [32] in 4C21.53 is a good example that radiosearches for fast pulsars are still a rewarding exercise. The position of this PSR 1937 + 214 in the COS-B skymaps incidentally does not show any enhancement (limit $\sim 10^{-6}$ cm^{-2}s^{-1}) in gamma-ray intensity. This is however no surprise if one regards the extremely low $\dot{P} = 10^{-19}$ s/s and an estimated distance of more than 2.5 kpc.

REFERENCES

1. R.N. Manchester, J.H. Taylor, Astron. J. $\underline{86}$, 1953 (1981).
2. R.N. Manchester et al., Mon. Not. R. Astr. Soc. $\underline{202}$, 269 (1983).
3. W.J. Cocke, M.J. Disney, D.J. Taylor, Nature $\underline{221}$, 525 (1969).
4. D.J. Thompson, C.E. Fichtel, D.A. Kniffen, H.B. Ögelman, Astrophys. J. Lett. $\underline{200}$, L79 (1975).
5. K. Bennett et al., Astron. Astrophys. $\underline{61}$, 279 (1977).
6. H.B. Ögelman et al., Astrophys. J. $\underline{209}$, 584 (1976).
7. G. Kanbach et al., Proc. 12th ESLAB Symposium, Frascati, ESA SP-124, 21 (1977).
8. R. Buccheri, IAU-Symposium No. 95, Pulsars, p. 241, Dordrecht, Reidel Publ. Co. (1981).
9. D.J. Thompson et al., GSFC preprint (1983).
10. C. Reppin et al., Proc. ESRANGE Symp. Ajaccio, ESA SP-135, 293 (1978).
11. V. Schönfelder et al., Astron. Astrophys. $\underline{110}$, 138 (1982).
12. GRO Science Working Team (D.A. Kniffen, GSFC, Chairman) The Gamma-Ray Observatory Science Plan, Sept. 1981.
13. C.E. Fichtel et al., Astrophys. J. $\underline{198}$, 163 (1975).
14. L. Scarsi et al., Proc. 12th ESLAB Symposium, Frascati, ESA SP-124, 3 (1977).
15. R.D. Wills et al, Nature $\underline{296}$, 723 (1982).
16. D.J. Thompson et al., Astrophys. J. $\underline{213}$, 252 (1977).
17. B. McBreen et al., Astrophys J. $\underline{184}$, 571 (1973).
18. K. Greisen et al., Astrophys. J. $\underline{197}$, 471 (1975).
19. D.H.P. Jones, F.G. Smith, J.E. Nelson, Nature $\underline{283}$ 50 (1980).
20. R.N. Manchester, J.H. Taylor, Pulsars, W.H. Freeman Co., 1977.
21. F.K. Knight, Astrophys. J. $\underline{260}$, 538 (1982) and contribution to this symposium (1983).
22. G.G. Lichti et al., Non-Solar Gamma Rays, Adv. in Space Exploration Vol. 7 (eds. Cowsik and Wills, Pergamon), 49 (1980).
23. G.K. Gupta et al., This conference (1983).
24. R.D. Wills et al., Proc. of the 17th Int. Cosmic Ray Conf., Paris, Vol. 1, 22 (1981).
25. G. Kanbach et al., Astron. Astrophys. $\underline{90}$, 163 (1980).
26. M. Morini, Mon. Not. R. Astr.Soc. $\underline{202}$, 495 (1983).
27. R. Buccheri et al., Nature $\underline{274}$, 572 (1978).
28. A.K. Harding, Astrophys. J. $\underline{245}$, 267 (1981).
29. U. Graser and V. Schönfelder, submitted to Astrophys. J. (1983).
30. R. Buccheri et al., submitted to Astron. Astrophys. and contribution to this conference (1983).
31. B.N. Swanenburg et al., Astrophys. J. $\underline{243}$, L69 (1981).
32. D.C. Backer et al., Nature $\underline{300}$, 615 (1982).
33. H.A. Mayer-Hasselwander et al., Astron. Astrophys. $\underline{105}$, 164 (1982).

Properties of the Crab Pulsar Inferred
from the Phase-Averaged Spectrum

F. K. Knight
NRC/NRL Research Associate
E. O. Hulburt Center for Space Research
Naval Research Laboratory, Washington, DC 20375

ABSTRACT

The Crab pulsar emission can be described adequately by using the phase-averaged spectrum, which is continuous from the near infrared to the γ-ray range. Using most of the published observations over this energy range, a power law fit to the emission requires at least two power laws with a break of 0.74±0.02 in spectral index at 39±3 keV. If this emission is synchrotron radiation, then the emitting electron spectrum is continuous over about five decades and has a break of ∼1.5 in spectral index at γ = E/mc^2 = 1900±200 $(10^6 \text{ gauss}/B_\perp)^{1/2}$. With $B_\perp = 10^6$ gauss, the electrons emitting in the infrared have $\gamma = 10$. Because a fit with three power laws has approximately the same χ^2, a gradual turnover is consistent with the data, as one would expect from synchrotron emission.

INTRODUCTION

The study of the Crab pulsar spectrum is important because it is the best diagnostic for the particles and magnetic field near the pulsar. Presumably, the emission comes from particles emitted from the neutron star surface and their secondaries produced within the light cylinder, interacting with the ordered magnetic field of the pulsar. Because the Crab pulsar emission is well above instrumental sensitivities, we can characterize it well and learn about the emitting particles and the magnetic field in the emitting region.

Here I discuss the evidence for examining the pulsar spectrum using its phase-averaged properties, and fit this spectrum using simple power law models. More realistic models, including self-absorption at low energies and a gentle curve in place of sharp breaks assumed in the multiple power-law spectra, will be examined using the existing data in future work.

THE PHASE-AVERAGED SPECTRUM

Although the Crab pulsar light curve is basically similar from optical to γ-ray energies, there is a variation of the spectrum with pulse phase, particularly evident in high-energy X-rays. Figure 1 shows the variation of spectral index as a function of pulse phase in the energy range 13-250 keV. The data from the

UCSD/MIT instrument on HEAO-1[1] and the data of Hasinger et al.[2] (MPI) show significant variations of spectral index with pulse phase, while the data of Strickman, Johnson and Kurfess[3] (NRL) show a trend of the same nature, although it is not significant. Although these observations disagree in detail, it appears that the high-energy X-ray emission coming from the the two main pulse peaks is softer than the emission from the region between the peaks, what I call the interpulse region.

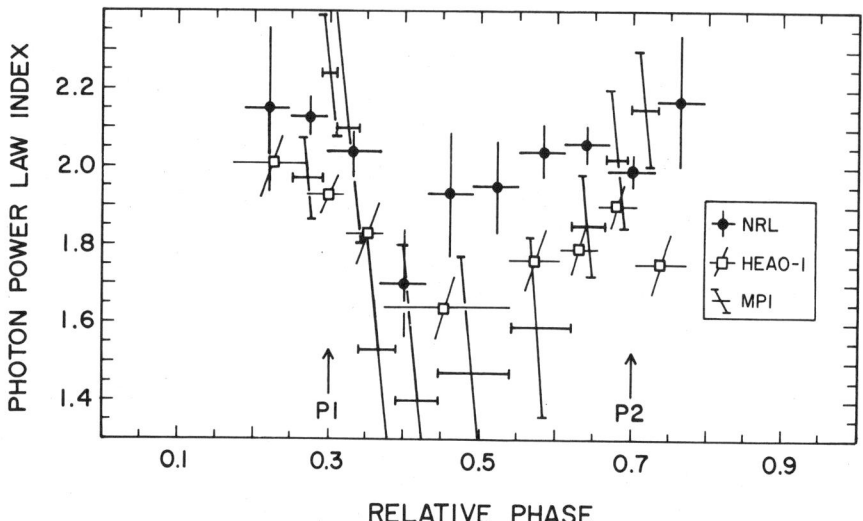

Fig. 1. Photon power law index versus pulse phase. Results from three observations show similar trends: a hardening of the high-energy X-ray spectrum in the phase region between the primary peak (P1) and the secondary peak (P2).

Figure 2 shows the interpulse emission spectrum as measured by OSO-8[4] and HEAO-1[1] in the energy range 4-700 keV. Because this spectrum shows a curvature while the spectrum common to the peaks is a power law in this energy range, I proposed[1] two "thermal" models to account for the emission. These models use a thermal electron population separate from the one producing the emission in the two main peaks to produce the emission by bremsstrahlung or Comptonization. However, these models are not compatible with new observations. There is no way to obtain the observed 30% optical polarization in the interpulse[5], or the significant infrared flux from the interpulse[6] with these models. Also, the summed observations of COS-B[7] show a 3.8σ detection of the interpulse at energies above 100 MeV with an intensity that is orders of magnitude above the exponential tail of the assumed "thermal" emission.

The alternative to these "thermal" models is to assume that the variation of the spectral shape with pulse phase in high-energy

X-rays is not due to a second electron population. Instead, it is due to some unexplained effect, possibly a variation of the pitch angle distribution versus energy for the single electron population in the high-energy X-ray range. These remarks indicate that there is no compelling reason to assume more than one electron population. Hence, the phase-averaged spectrum - as a whole - can characterize the emission.

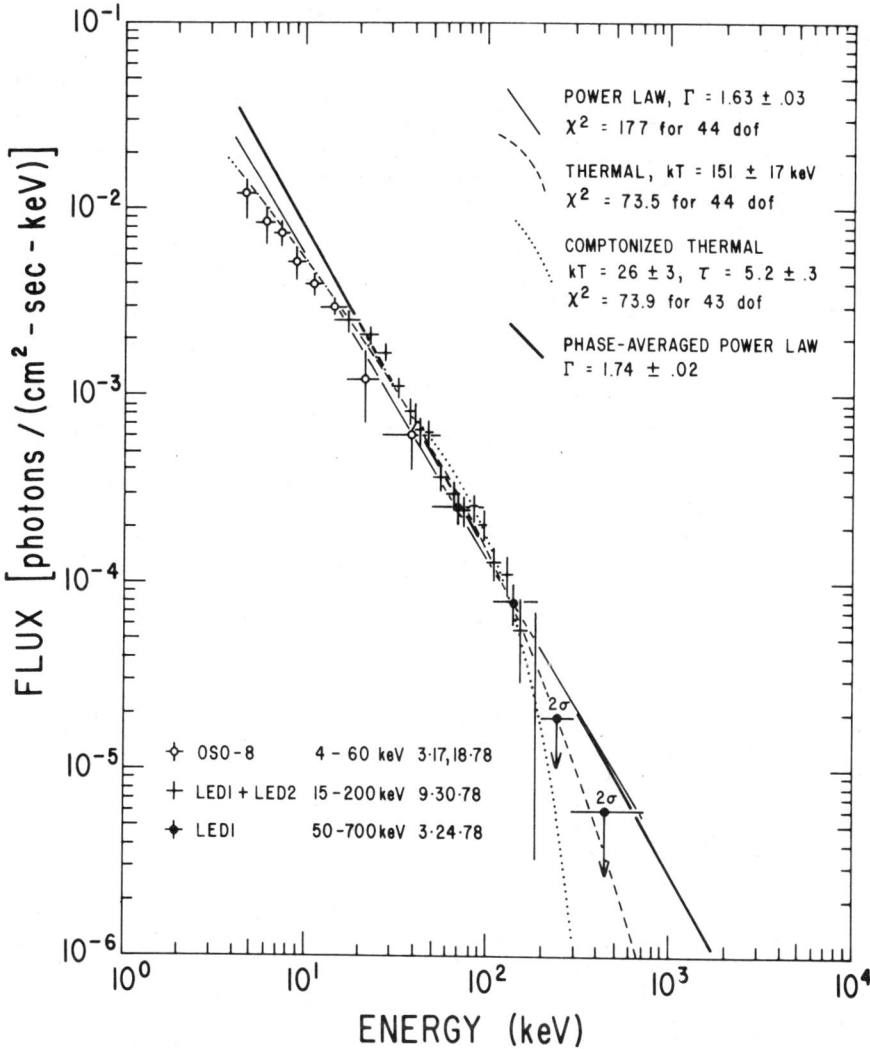

Fig. 2. Interpulse spectrum. Using the data from the region of phase 0.37 to 0.59 in Figure 1, two "thermal" models (marked thermal for thermal bremsstrahlung and Comptonized for Comptonization) give significantly better fits than a power law model, which is adequate for the spectra of the primary and secondary peaks.

FITS TO THE PHASE-AVERAGED SPECTRUM

Figure 3 shows most of the data from infrared to γ-ray energies expressed as energy flux. Motivated by the fact that the emission may be synchrotron radiation, I fit these data with the power law models, summarized in the Table. A single power law is inadequate to describe the data. A double power law appears adequate, even though χ^2 is about 3 per dof. There is no significant reduction in χ^2 with the addition of a third power law. For the double power law, the break occurs at 39 ± 3 keV. An electron population emitting this spectrum as synchrotron radiation needs a break at $\gamma = 1900\pm200$ $(10^6$ gauss$/B_\perp)^{1/2}$. A perpendicular magnetic field of 10^6 gauss allows the spectrum to be produced entirely by synchrotron radiation. If the power law extends down to $\gamma \leq 10$, the infrared flux can be explained. If the power law extends up to $\gamma \geq 10^6$, the γ-ray flux can be explained.

TABLE: POWER LAW FITS TO PHASE-AVERAGED SPECTRUM[a]

	single[b]	double[b]	triple[b]
χ^2	6068	347	314
dof	120	119	117
χ^2/dof	50.6	2.92	2.68
p1[c]	10.3±2.3	2.38±0.09	4.1±19.1
p2[keV]	100	100	100
p3	1.77±0.02	1.44±0.01	1.38±0.43
p4[keV]	∞	38.6±2.5	2.99±37.3
p5		2.18±0.01	1.63±0.04
p6[keV]		∞	57.4±6.1
p7			2.20±0.02

[a] Parameters without errors are fixed. Quoted errors increase χ^2 by 1 with other parameters held fixed.
[b] p1 $(E/p2)^{-p3}$ E < p4
  p1 $(p4/p2)^{-p3}$ $(E/p4)^{-p5}$ p4 < E < p6
  p1 $(p4/p2)^{-p3}$ $(p6/p4)^{-p5}$ $(E/p6)^{-p7}$ E > p6
[c] Flux units for p1 are 10^{-4} keV $(cm^2$-s-keV$)^{-1}$

The perpendicular magnetic field of 10^6 gauss requires either that the emission region be close to the light cylinder where the field strength is approximately 10^6 gauss or that the electrons be directed along the field lines close to the neutron star where the field strength approaches 10^{12} gauss. Primary electrons moving along the field lines and secondary electrons produced with initial trajectories along the field lines lose their energy via synchrotron radiation before they move far enough along the field lines to introduce a significant perpendicular field component due simply to the field line curvature.

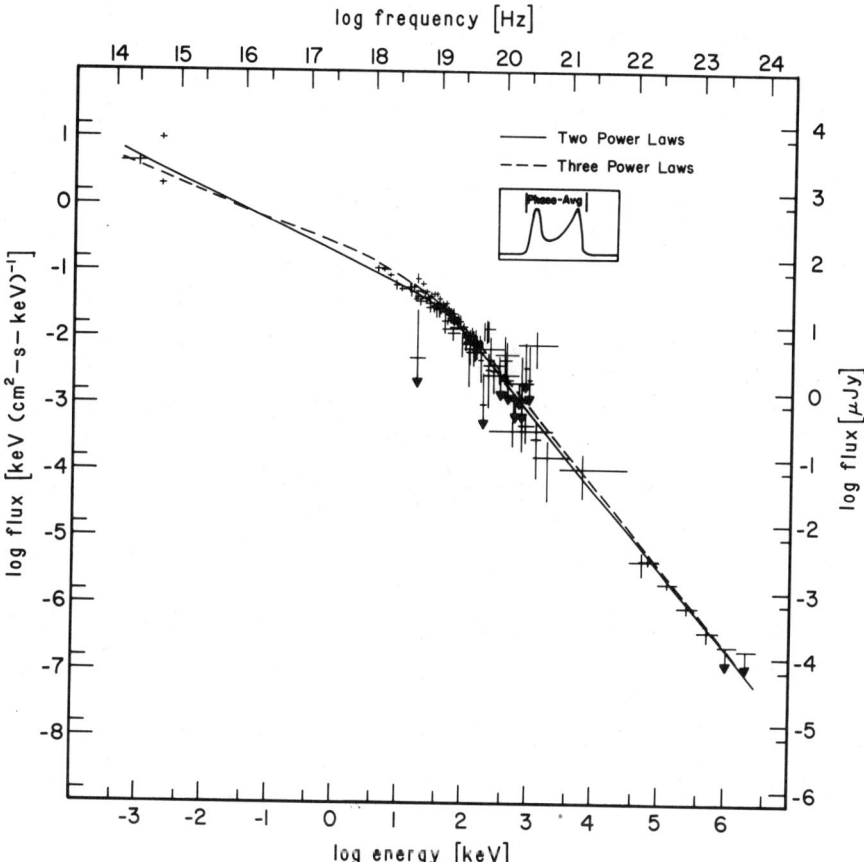

Fig. 3. Power law models for the phase-averaged spectrum. Two or three power laws give similar χ^2 when compared to the data in the range plotted. Data are from references 1,4,6,9-20.

Testing the emission with such a simple model is no doubt unrealistic. The energy-conversion processes for electrons within the pulsar magnetosphere involve many other mechanisms[8]. Because the data extend over more than eleven decades in energy, it is possible to test more realistic models, a study for future work.

SUMMARY

In summary, the Crab pulsar emission can be characterized using the phase-averaged spectrum. This spectrum extends from infrared to γ-ray energies as a double power law with a break at 39±3 keV. An electron population emitting this radiation as synchrotron radiation in a magnetic field with perpendicular component of 10^6 gauss would extend from γ of 10 to γ of 10^6 with a break at γ of 1900±200.

ACKNOWLEDGEMENTS

I thank Mark Strickman for valuable discussions and the National Academy of Sciences for providing my support.

REFERENCES

1. Knight, F.K., Ap. J., 260, 538 (1982).
2. Hasinger, G., Pietsch, W., Reppin, C. Trümper, J., Voges, W., Kendziorra, E., and Staubert, R., in Accreting Neutron Stars, W. Brinkmann and J. Trümper, eds. (1982), p.130.
3. Strickman, M.S., Johnson, W.N. III, Kurfess, J.D., Ap. J. (Letters), 253, L23 (1982).
4. Pravdo, S.H. and Serlemitsos, P.J., Ap.J., 246, 484 (1981).
5. Smith, F.G., in IAU Symposium 95, Pulsars, W. Sieber and R. Wielebinski, eds., (Dordrecht, Reidel, 1981), p.221.
6. Pennypacker, C.R., Ap. J., 244, 286 (1981).
7. Wills, R.D., et al., Nature 296, 723 (1982).
8. Michel, F.C., Rev. of Mod. Phys., 54, 1 (1982).
9. Graser, U. and Schönfelder, V., Ap. J., 263, 677 (1982).
10. Kurfess, J.D., Ap. J. (Letters), 168, L39 (1971).
11. Laros, J.G., Matteson, J.L. and Pelling, R.M., Nature (Phys. Sci.), 231, 109 (1973).
12. Lichti, G.G., et al., in Non-Solar Gamma-Rays, R.Cowsik and R.D. Wills, eds., (Oxford, Pergamon Press, 1980), p.49.
13. Nather, R.E., Warner, B. and MacFarlane, M., Nature, 221, 527 (1969).
14. Oke, J.B., Ap. J. (Letters), 156, L49 (1969).
15. Orwig, L.E., Chupp, E.L. and Forrest, D.J., Nature (Phys. Sci.), 231, 171 (1971).
16. Strickman, M.S., Johnson, W.N. III, Kurfess, J.D., Ap. J. (Letters), 230, L15 (1979).
17. Thompson, D.J., Fichtel, C.E., Hartman, R.C., Kniffen, D.A. and Lamb, R.C., Ap. J., 213, 252 (1977).
18. Walraven, G.D., Hall, R.D., Meegan, C.A., Coleman, P.L., Shelton, D.H. and Haymes, R.C., Ap. J., 202, 502 (1975).
19. Wilson, R.B., et al., in Proc. 15th Int. Conf. Cosmic Rays, Plovdiv, Bulgaria, vol. 1 (1977), p.24.
20. Wilson, R.B. and Fishman, G.J., Ap. J., 269, in press (1983).

COS-B UPPER LIMITS ON GAMMA-RAY EMISSION FROM RADIO PULSARS
==

R. Buccheri
IFCAI-CNR, Palermo, Italy
on behalf of the Caravane Collaboration for the COS-B satellite

INTRODUCTION

The number of known radio pulsars has now increased up to 333 (Manchester and Taylor, 1981; Manchester, Tuohy, D'Amico, 1982; Backer et al., 1982; Boriakoff et al., 1983). Among them only three have a low timing age $\tau = P/2\dot{P}$ such to be considered "young" pulsars. These are PSR0531+21 (the Crab pulsar) aged $\sim$ 1000 years, PSR1509-58 aged $\sim$ 1500 years and PSR0833-45 (the Vela pulsar) aged $\sim$ 11000 years. The Crab and Vela pulsars are strong gamma-ray emitters being the most and the third most intense sources of gamma-ray emission of the 2CG catalogue (Swanenburg et al., 1981). PSR1509-58 is located inside the error box of a gamma-ray source contained in the list published by Wills et al., 1980.

Gamma-ray emission from older radio pulsars has been indicated several times (Thompson et al., 1976; Ögelman et al., 1976; Kanbach et al., 1977; Pinkau, 1979) but never confirmed (see for example Buccheri, 1981). In this paper I will report on the results of a search for pulsed gamma-ray emission from old radio pulsars in the COS-B data in the energy range 50 MeV to 2 GeV. The analysis has been done by using either pulsar parameters obtained in simultaneous radio and gamma-ray observations or parameters which could be reliably derived by interpolation from radio observations spanning the COS-B observations.

ANALYSIS AND RESULTS

The values of P and $\dot{P}$ used in this analysis are derived from simultaneous radio and gamma-ray observations in the cases of PSR0540+23, PSR0611+22, PSR0740-28 (twice) and PSR1822-09, whereas in 140 other cases these values are taken from literature (Gullahorn and Rankin, 1978; Newton, Manchester and Cooke, 1981; Manchester et al., 1982). The total number of 145 independent pulsar observations refer to 117 radio pulsars some of which were repeatedly observed by COS-B during five years from August 1975 up to April 1980. In all cases the rms phase residual derived from radio measurements was less than 0.02 of the pulsar period and therefore sufficient for a reliable analysis of the gamma-ray photon arrival times collected by COS-B. The arrival times, after reduction to a phase value ϕ in the interval (0,1) using the values of P and $\dot{P}$ as determined by radio measurements, were analyzed by means of the statistical variable

0094-243X/83/1010147-05 $3.00 Copyright 1983 American Institute of Physics

$$z_2^2 = \frac{2}{N_i} \sum_{k,1}^{2} \left(\sum_{j,1}^{N_i} \cos k \phi_j \right)^2 + \left(\sum_{j,1}^{N_i} \sin k \phi_j \right)^2$$

where N_i is the total number of photons used for the i-th pulsar observation (i=1,145) and ϕ_j (j=1,N_i) are the phase values obtained by folding the N_i arrival times of the gamma-ray photons.

The 145 values of z_2^2 so obtained indicate absence of signal from any individual pulsar. In addition to that the sum

$$N_\sigma = \frac{\sum_{j,1}^{16} n_{\sigma j}}{\sqrt{16}} = -0.11$$

(where $n_{\sigma j} = \frac{z_2^2 - 4}{\sqrt{8}}$ is the no. of standard deviations of z_2^2 above the expected average 4, as measured in each single pulsar observation) shows absence of cumulative signal from the 16 pulsars expected to have the highest signal-to-noise ratio according to their distance and rotational energy loss and in dependence of their location in the galactic disc. Details on the method of analysis can be found in Buccheri et al., (1983).

The list of the 16 higher priority pulsars is given in Table 1 which shows also the value of the pulsar parameters $P(s)$ and $\dot{P}(10^{-15} s/s)$, the distance d (Kpc) and the 2σ upper limit on their gamma-ray flux (in terms of 10^{-6} ph/cm^{2}s).

Fig. 1 shows the upper limits on the intrinsic luminosity of the 16 pulsars of the priority list compared with the luminosity of Crab and Vela pulsars. In computing these values it has been assumed a beaming factor $4\pi f = 2.3$ equal to the Crab and Vela case.

The upper limit on the average value of the product ηI between the conversion efficiency η from $\dot{E}$ into gamma-ray luminosity and the moment of inertia I results in

$$\overline{\eta I} < 6 \times 10^{43} \text{ g.cm}^2$$

This value can be used to estimate the contribution of old pulsars to the total galactic gamma-ray emission above 50 MeV. This is

$$L_p < \langle \eta \dot{E} \rangle \times N = 2.6 \times 10^{40} \text{ ph/s}$$

where $\langle \dot{E} \rangle$ is referred to the N = 100,000 pulsars of the galaxy.

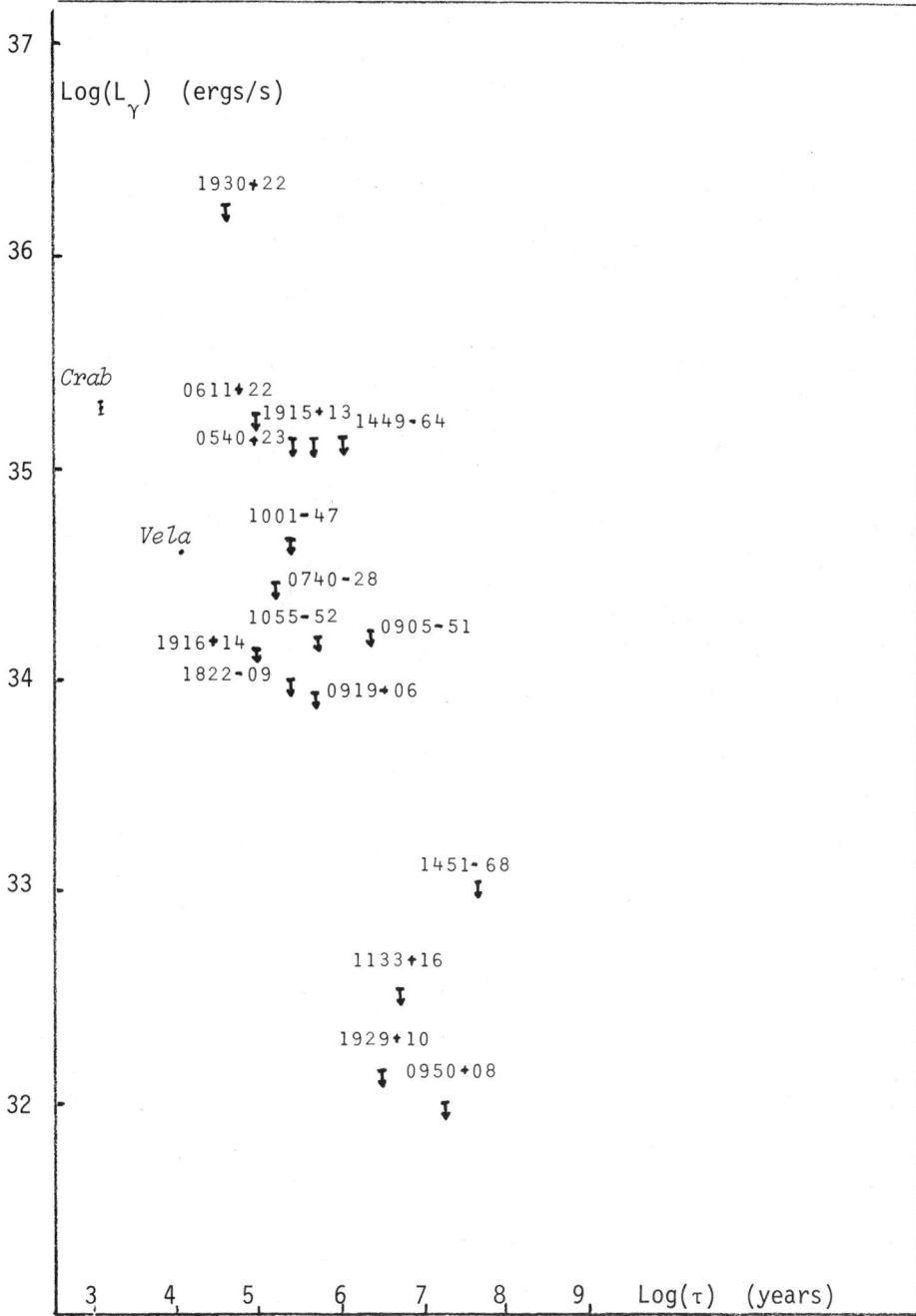

Fig.1. Upper limits on gamma-ray luminosity for the pulsars of the priority list versus the timing age (logarithmic scale).

TABLE I

PSR	P(s)	$\dot{P}(10^{-15}s/s)$	d(Kpc)	FLUX $(10^{-6} ph/cm^2 s)$
0540+23	0.246	15.4	2.60	2.4
0611+22	0.335	59.6	3.30	1.9
0740-28	0.167	16.8	1.50	1.5
0905-51	0.253	1.8	0.86	2.7
0919+06	0.431	13.7	1.00	1.0
0950+08	0.253	0.2	0.09	1.4
1001-47	0.307	22.1	1.60	2.1
1055-52	0.197	5.8	0.92	2.2
1133+16	1.188	3.7	0.15	1.8
1449-64	0.179	2.7	2.20	3.4
1451-68	0.263	0.1	0.23	2.4
1822-09	0.769	52.3	0.56	3.7
1915+13	0.195	7.2	2.40	2.9
1916+14	1.181	211.4	0.86	2.8
1929+10	0.226	1.2	0.08	2.6
1930+22	0.144	57.8	7.00	4.1

Considering that the galactic luminosity above 50 MeV is 2.4×10^{42} ph/s (Caraveo and Paul, 1979 and assuming an E^{-2} spectrum), the contribution by old pulsars in this energy range is less than 1%

REFERENCES

1) D.C.Backer, S.R.Kulkarni, C.Heiles, M.M.Davis, W.M.Goss
   Nature, 300, 615, 1982

2) V.Boriakoff, R.Buccheri, F.Fauci, D.C.Ferguson
   Astrophys.J., in preparation

3) R.Buccheri
   Proc. of IAU Symp. n.95 "PULSARS" Bonn, p.241, 1981

4) R.Buccheri, K.Bennett, G.F.Bignami, V.Boriakoff, J.V.G.M.Bloemen, P.A.Caraveo, W.Hermsen, G.Kanbach, J.L.Masnou, R.N.Manchester, H.A. Mayer-Hasselwander, M.E.Özel, J.A.Paul, B.Sacco, L.Scarsi, A.W.Strong
   Astron.Astrophys., in preparation

5) P.A.Caraveo, J.A.Paul
   Astron.Astrophys., 75, 340, 1979

6) G.E.Gullahorn, M.J.Rankin
   Astron.J., 83, 1219, 1978

7) G.Kanbach, K.Bennett, G.F.Bignami, G.Boella, M.Bonnardeau, R.Buccheri, N.D'Amico, W.Hermsen, J.C.Hidgon, G.G.Lichti, J.L.Masnou, H.A.Mayer-Hasselwander, J.A.Paul, L.Scarsi, B.N.Swanenburg, B.G.Taylor, R.D.Wills
   Proc. 12th ESLAB Symp., ESA SP124, 21, 1977

8) R.N.Manchester, J.H.Taylor
   Astron. J., 86, 1953, 1981

9) R.N.Manchester, I.R. Tuohy, N.D'Amico
   Astrophys. J., 262, L31, 1982

10) L.M.Newton, R.N.Manchester, D.J.Cooke
    Mon.Not.R.Astr.Soc., 194, 841, 1981

11) R.N.Manchester, L.M.Newton, W.M.Goss, P.A.Hamilton
    Mon.Not. R.Astr.Soc., 202, 269, 1983

12) H.B.Ögelman, C.E.Fichtel, D.A.Kniffen, D.J.Thompson
    Astrophys.J., 209, 284, 1976

13) K.Pinkau
    Nature, 277, 17, 1979

14) B.N.Swanenburg, K.Bennett, G.F.Bignami, R.Buccheri, P.A.Caraveo, W.Hermsen, G.Kanbach, G.G.Lichti, J.L.Masnou, H.A.Mayer-Hasselwander, J.A.Paul, B.Sacco, L.Scarsi, R.D.Wills
    Astrophys.J., 243, L69, 1981

15) D.J.Thompson, C.E.Fichtel, D.A.Kniffen, R.C.Lamb, H.B.Ögelman
    Astrophys. Lett. 17, 175, 1976

16) R.D.Wills, K.Bennett, G.F.Bignami, R.Buccheri, P.A.Caraveo, N.D'Amico, W.Hermsen, G.Kanbach, G.G.Lichti, J.L.Masnou, H.A.Mayer-Hasselwander, J.A.Paul, B.Sacco, B.N.Swanenburg
    Proc.COSPAR Symp. "Non Solar Gamma-Rays" Bangalore, Advances in Space Exploration 7:43, Oxford, Pergamon

THE QUEST FOR ELUSIVE GEMINGA :

A UNIQUE OBJECT PROPOSED AS THE COUNTERPART OF 2CG 195+04

P. A. Caraveo, G. F. Bignami
Istituto di Fisica Cosmica del C N R Milano - Italy

R. C. Lamb
IOWA State University, Ames, IOWA - U.S.A.

INTRODUCTION

The second COS-B catalogue of high-energy (> 100 MeV) γ-ray sources [17] lists 25 objects which appear as both significant and unresolved excesses in the γ-ray sky. The majority of them are galactic in nature, as shown by their celestial distribution, and are yet unidentified (for a review on γ-ray source properties see [1]). One obvious way to search for counterparts is to explore the COS-B error circles at other wavelengths using the opportunity of the Einstein Observatory [7] Guest Observer program. As discussed in detail by [5], a good coverage was obtained for many COS-B sources down to a flux limit of $\sim 10^{-13}$ ergs/cm^2-s in the IPC energy range of $\sim$ .1 to 5 keV.

Geminga is the brightest unidentified source in the COS-B catalogue, second only to the Vela pulsar PSR0833-45, with a flux limit of 5×10^{-6} photons (> 100 MeV)/cm^2 -s, $\sim 2.4 \times 10^{-9}$ ergs/cm^2-s in the interval 50 MeV-3GeV [13]. The source, discovered during the SAS-2 mission [6], was repeatedly observed by COS-B. No long term variability was found. Short term variability was also actively sought, with no convincing success. The long observation times devoted by COS-B to Geminga, coupled with the relatively hard spectrum of the source allow for a positional accuracy unusually high for current γ-ray astronomy: the error box, centered at $l = 195°.1$ and $b = 4°.2$, has a radius of ~ 24 arcmins. The complete γ-ray error box could thus be contained in a single Einstein IPC field, providing a unique opportunity for in-depth search in the keV region. The several sources found were investigated at radio and optical wavelenghts, leading to some identifications.

THE DATA

Three observations of the Geminga region of the sky were performed with the Einstein Observatory, two with the IPC instrument in the focal plane and one with the HRI. Since the pointing directions of the two IPC fields coincide, we have summed the two. The combined IPC image is shown in Fig. 1a, where the "unmasked" raw data can be seen.

Firstly, one notes that neither in the raw data, nor in more smoothed versions of the images, is it possible to see evidence of diffuse X-ray features. On the other hand, four sources pass the IPC "5σ" confidende test and are well visible in Fig. 1a where they are numbered from 1 to 4. As discussed in detail by[2] the sources 1 to 3 appear compatible with a serendipitous source population expected in a $\sim$10,000 sec IPC field.

Source 4 is by far the brightest in the field, so that it is possible to present for it results on its spectral properties, search for temporal variability and for spatial extent. The source spectrum is very soft, with the majority of the counts below $\sim$1 keV, and with little evidence for interstellar absorption. This impression is confirmed more quantitatively by a variety of fits which were tried for the spectral data. Acceptable fit (reduced $\chi^2 \lesssim 1$) was obtained equally with power law ($3 < \alpha < 3.5$), exponential ($.1 < kT < .4$) and black body ($kT \lesssim .1$) spectra. However, in no case the maximum allowed column density N_H to the source could exceed $2 \times 10^{20} cm^{-2}$, implying a very low interstellar absorption. In fact, for all the spectral types considered, the data were best fitted by N_H in the 10^{19}-$10^{20} cm^{-2}$ range. A special case of spectral fit to the data can be tried with a monochromatic line; in our case a good χ^2 was obtained for a line centered at $\sim$260 $\pm$20 eV.

An analysis was performed on temporal behaviour of source (1E 0630+178). The constancy of the "crude" flux in IPC1 and IPC2 is well within the instrument stability, so that a rough limit of 10% can be placed on the long term variability. Short term random variability was investigated using as input the count rates for both the source and background binned in intervals from minutes to hours. No evidence of source variation was seen. A search for periodicities was also performed using the FFT algorithm in the second to milliseconds range. The frequency interval covered was quite extensive, and in particular period values down to 2.6 msecs were included. No significant effect was found. 1E 0630+178 was also investigated with an HRI exposure as shown in Fig. 1b. The final (i.e.,"reprocessed") best position for the source is

α (1950) = $6^h 30^m 59^s.15$ δ (1950) = $17°48'33".0$

to which a 90% confidence radius of 3" should be attached.

The HRI error circle of 1E 0630+178 falls into a black field on both the red and blue POSS prints. The interstellar absorption in this direction appears to be of the order of $A_V \simeq .5$ mag.,[14] A sensitive measurement from the VLA (by RCL) also did not detect a source at the $\sim$ 1mJy level at 6 cm. A pulsar search was carried out at its position from Arecibo[3]; (V.Boriakoff, private communication). No evidence for radio pulsar was found.

At our request, two CCD images of the field containing the HRI error circle were obtained, one with the SAO CCD camera at Mt. Hopkins, Arizona, and the other with the ESO CCD instrument at the 1.5 m Danish telescope at La Silla, Chile. In the first image, only sensitive to $m_V \simeq 22$, the HRI error circle is still blank (P.Hertz and J.Grindlay, private communication). On the ESO images in the V,R and I, a faint blue object is seen (H.Sol and M.Tarenghi, private communication). In the following we shall tentatively assume a V magnitude of $\sim$ 23.5 for this object, pending more accurate estimates.

The probability of a chance occurrence of a blue object of

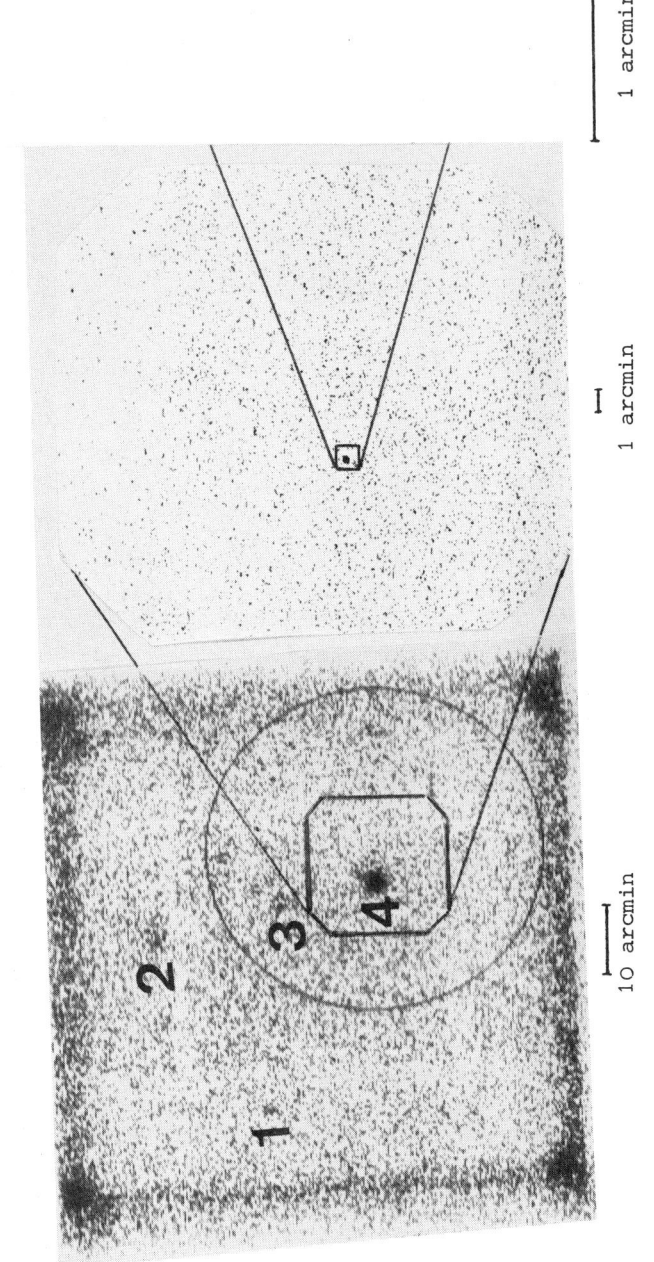

Fig. 1 : The raw X-ray data, shown in increasing resolution.

(a) The extended IPC field (N at top, E at left) sum of IPC1 and IPC2. The serendipitous sources, numbered 1 to 4, are well visible. The COS-B error box for 2CG 195+04 "Geminga" is also shown.
(b) The HRI field, showing 1E 0630+178 and no other source or extended feature in its vicinity.
(c) Detail of the HRI field showing, at full resolution, the point like nature of 1E 0630+178 ("Geminga").

$m_V = 23.5$ in a 3" radius circle can be evaluated from the data of e.g.[12,18,8], to be conservatively, $\leq 1 \times 10^{-3}$. We are thus led to associate 1E 0630+178 with the faint blue object in its error box.

DISCUSSION

A number of observed characteristics combine to render 1E 0630+178 an unique object. They are: the high L_X/L_V value, which is about 1300 in the case of the optical identification (or more otherwise), the point-like appearance of the HRI source, the absence of any known radio pulsar at its position, despite the dedicated Arecibo search, and, in fact, the absence of any radio source in the sensitive VLA survey. To our knowledge, so far no other Einstein source at this flux level and with HRI positioning has a similar set of characteristics. [16] and [8] have discussed complete optical identifications for high-latitude surveys, and found no comparable source. At low latitudes, work is still in progress[11], so that it cannot be excluded that other such objects will eventually be found. Considering the size of the COS-B error box for 2CG 195+4, the probability of a chance appearance of a $\sim 2 \times 10^{-12}$ ergs/cm²-s IPC source in it around 5%. Such probability, althought indicative, is certainly not sufficient to claim an association.

However, we believe the uniquenesses of the X-ray source properties to be a much stronger argument, and on this account we propose the association of 1E 0630+178 and of its optical counterpart with the COS-B high-energy γ-ray source. The quest for elusive Geminga, started back in 1974 after the SAS-2 discovery of the γ-ray source, comes here to an end.

As to the astrophysical nature of the object, we note that only two classes of known X-ray sources have such a high L_X/L_V value: low-mass binaries[4] and radio pulsars[9,10]. The first can be easily excluded, since their optical luminosity function is quite narrow and peaked around $M_V \sim 1$[15]. A distance modulus of $(m_V - M_V) = 22$ would place the source at ~ 250 kpc, well outside our Galaxy. Since all the best spectral fits are obtained for $10^{19} < N_H < 10^{20}$ cm⁻², one can realistically place the source at ~ 100 pc and speculate on the consequences. The soft X-ray luminosity would be a few x 10^{30} ergs/sec.

The only type of known object satisfying the observational constraints on L_X/L_V, absolute X-ray and optical magnitudes is a pulsar, or better, a neutron star. The COS-B spectrum of Geminga in the 50 MeV-3GeV interval[13] shows similarities with that of the Vela pulsar. The L_X/L_V values of the objects are also similar, and so would be the L_γ/L_X values which are ~ 500 for the Vela pulsar unresolved emission and ~ 1200 for Geminga.

In summary, althought the idea of identifying Geminga with a Vela type pulsar appears, somewhat by default, the most immediate, a peculiar, if not unique, object cannot be excluded considering that the distance estimate given above could also be too generous and that nothing in the data prevents the source being even much nearer. A more general difficulty, of course, is that of attaching the label "pulsar" to an object for which no time variability of any kind, at any wavelength, is seen.

REFERENCES

1. G. F. Bignami and W. Hermsen, Annual Rev. Astron. Astrophys. $\underline{21}$, (1983) - in press.
2. G. F. Bignami, P. A. Caraveo, R. C. Lamb, Ap.J. Submitted, (1983).
3. R. Buccheri, N. D'Amico, F. Fauci, B. Sacco, Atti del 1º Convegno Nazionale di Fisica Cosmica, Lecce,1982, p.71.
4. H. V. D. Bradt and J. E. McClintock, Annual Rev. Astron. Astroph., $\underline{21}$,(1983) - in press.
5. P. A.Caraveo, Space Science Rev.,(1983) - in press.
6. C. E. Fichtel, R. C. Hartman, D.A. Kniffen, D. J. Thompson, G. F. Bignami, H. Ogelman, M. E. Ozel, T. Tumer, Ap.J. $\underline{198}$, 163 (1975).
7. R. Giacconi et al., Ap.J. $\underline{230}$, 540 (1979).
8. R. E. Griffiths et al., Ap.J. (1983) - in press (CFA preprint No. 1733).
9. D. J. Helfand, IAU Symp.No. 95 "Pulsars" (W.Sieber and R.Wielebinski editors, 1981), p. 343.
10. D. J. Helfand, IAU Symp.No. 101 "Supernova Remnants and Their X-ray Emission", (P.Gorenstein and J.Danzinger editors, 1983), in press.
11. P. Hertz and J. E. Grindlay, B.A.A.S. $\underline{13}$, 787 (1981).
12. B. M. Lasker, Ap.J. $\underline{203}$, 193 (1976).
13. J. L. Masnou et al., 17th I.C.R.C. $\underline{1}$, 177 (Caravane Collaboration 1981).
14. Th. Neckel and G. Klare, Astr.Ap.Suppl. $\underline{42}$, 251 (1980).
15. J. van Paradijs and F. Verbunt, Space Science Rev. $\underline{30}$, 361(1981).
16. J.T. Stocke, J. Liebert, I. M. Gioia, R. E. Griffiths, T. Maccacaro, J. Danzinger, D. Kunth, J. Lub, Ap.J. (1983) - in press.
17. B. N. Swanenburg et al., Ap.J.Lett. $\underline{243}$, L69 (Caravane Collaboration 1981).
18. J. A. Tyson, Ap. J. Lett. $\underline{248}$, L89 (1981).

ENERGY SPECTRA OF VERY HIGH ENERGY γ-RAYS FROM THE CRAB AND VELA PULSARS

S.K. Gupta, P.V. Ramana Murthy, B.V. Sreekantan
S.C. Tonwar*and P.R. Viswanath
Tata Institute of Fundamental Research
Homi Bhabha Marg, Colaba, Bombay-400 005

ABSTRACT

Using the atmospheric Cerenkov radiation technique we have observed pulsed emission of very high energy, $E_\gamma \gtrsim 10^{12}$ eV, γ-rays from the Crab and Vela pulsars. Ooty Cerenkov telescope composed of 18 reflectors has been used in the 'Distributed Mode' for these observations during early 1981. In this mode the arrival direction of individual showers has been measured to an accuracy $\sim 0.3°$ which has made possible a more reliable estimation of the γ-ray energy thresholds in the direction of these two pulsars. Observations on the Crab pulsar show evidence of sporadic emission with significant variability of the pulsed flux over periods as short as tens of minutes. The measured pulsed flux of $2.2 \pm 0.5 \times 10^{-11}$ ph. cm^{-2} s^{-1} at energies $> 9 \times 10^{11}$ eV during the active phase agrees well with the extrapolation of the Cos-B spectra $2.9 \times 10^{-11} E_{GeV}^{-1.17}$ ph.cm^{-2} s^{-1}, both in amplitude and spectral slope. The time averaged flux is however considerably smaller. On the other hand the time averaged pulsed flux of $1.0 \pm 0.4 \times 10^{-11}$ ph.cm^{-2} s^{-1} at energies $> 2.3 \times 10^{12}$ eV for the Vela pulsar is about two orders of magnitude smaller than the extrapolated value from the Cos-B spectrum, $1.8 \times 10^{-7} E_{GeV}^{-1.11}$ ph.cm^{-2} s^{-1}. Also the energy spectrum for the Vela pulsar at energies $\sim 2 - 5 \times 10^{12}$ eV with a slope of -2.2 ± 0.2 is much steeper than at lower energies. These observations suggest significant differences between γ-ray emission from these two pulsars.

INTRODUCTION

Pulsed high energy γ-ray emission has been observed[1,2] from Crab and Vela pulsars by satellite borne detectors in the energy range of a few times 10^6 to 10^9 eV. Many attempts have been made since late 1960's to observe[3,4] pulsed γ-ray emission at very high energies $\sim 10^{11}-10^{13}$ eV using the atmospheric Cerenkov radiation (CR) technique. It is generally realized that observations on the structure of the pulsed flux and the energy spectrum at these energies are very significant for understanding the radiation processes in pulsars and these may provide crucial information about pulsar magnetospheres. Ground based atmospheric technique is ideally suited for the study of very high energy γ-rays since the effective γ-ray collection area is many orders of magnitude larger than the physical area of the CR detector itself. Positive detection of pulsed flux from Crab and Vela pulsars has been reported only from few of the many observations[3,4] made so far by various groups, suggesting significant variability of the flux over the periods of days.

Some of these observations, for example those reported by the Tata group[5,6] during 1977-80 (Figs. 1 and 2), have shown double peak light curves very similar to those reported by COS-B group.[2] Based on these observations it has been suggested[7] that the pulsed γ-ray spectra for these two pulsars as measured at energies ∼35 MeV - 10 GeV by the COS-B group must be steepening at higher energies. This

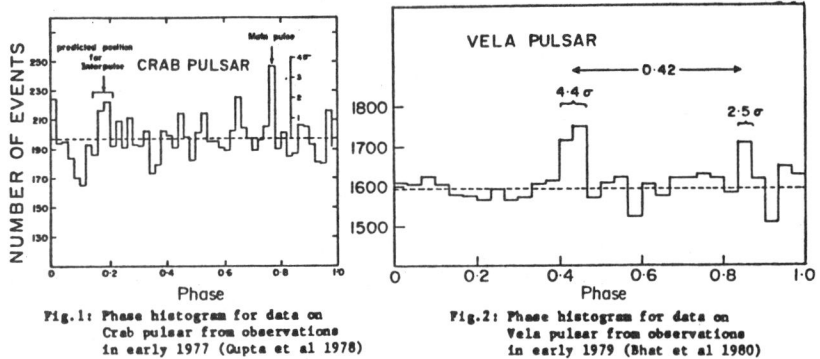

Fig.1: Phase histogram for data on Crab pulsar from observations in early 1977 (Gupta et al 1978)

Fig.2: Phase histogram for data on Vela pulsar from observations in early 1979 (Bhat et al 1980)

conclusion was based on indirect estimates of γ-ray energy thresholds of CR telescopes. The CR telescope operating at Ooty (11.4°N latitude, 2.2 km altitude) in southern India since early 1977 has been recently modified to enable a relatively direct measurement[8] of the angular aperture of the telescope which has allowed a more reliable estimate of the γ-ray detection energy thresholds and also a measurement of the energy spectrum over a relatively narrow energy range. We present here a brief discussion of the observations with the Ooty CR telescope made in early 1981 and discuss the results obtained for the flux, energy spectrum and time variability for the Crab and Vela pulsars.

OBSERVATIONS

Ooty CR telescope consists of an array of 18 reflectors, eight of area 1.8 m^2 each and 10 of area 0.64 m^2 each, mounted on independent fully steerable equatorial mounts. Each reflector has a RCA 8575 photomultiplier at its focus behind a metal mask with a circular hole in the center to define the physical aperture. Observations during the period 1977-80 were made in the 'Compact Mode' when all the reflectors were placed close to each other. This configuration allowed participation of the signals from all the reflectors in the selection of showers using a majority coincidence system. In early 1981 this array was changed to the 'Distributed Mode' as shown in figure 3. In this configuration selection of showers is made using only the signals from eight reflectors placed at the center. The 10 reflectors located at the periphery of the circle of radius 55 m are used to measure the relative arrival time of the Cerenkov light front which enables a determination of the arrival direction of individual showers to an accuracy ∼5 mr. This feature helps in rejection of showers arriving at larger angles relative to the direction of the pulsar thus improving the signal (γ-ray showers) to background (cosmic ray showers) ratio. Observations[8,9] with this new array have shown that the effective aperture of the telescope is significantly larger than the physical aperture defined by the mask in front of the photomultiplier (figure 4). Since the estimates of the threshold energies for detection of γ-rays for various source regions of the sky depend sensitively on the value of the effective aperture this measurement of the angular

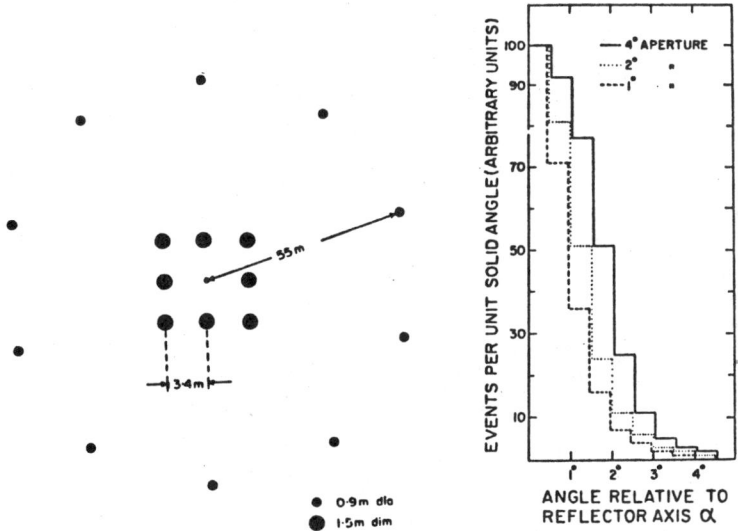

Fig.3: Ooty CR telescope array in the Distributed Mode

Fig.4: Angular distribution of showers for different physical size apertures

distribution of selected showers, obtained experimentally for the first time, has allowed a more reliable determination of the energy thresholds. It should be emphasized here that these new values of the energy thresholds are significantly higher than the values estimated earlier leading to different results for the γ-ray spectra as discussed later.

During the observational season of 1981, limited to the three month period from January to March by the requirement of cloud-less moon-less nights, data was taken for the Crab pulsar for 32 hours spread over 15 nights with a mean trigger rate of 7.5 s^{-1}. The exposure for the Vela pulsar was 16 hours spread over seven nights and the mean trigger rate was 2.33 s^{-1}. Note that the Crab pulsar is only about 10° N relative to the zenith at meridian transit compared to the angle of about 56° S for the Vela pulsar. These data have been analyzed[9] for the γ-ray emission using pulsation parameters obtained from radio or optical observations. The arrival direction was computed for each recorded shower which triggered four or more outer reflectors. It may be noted that showers with no outer reflectors triggered are mostly due to lowest energy primaries and the detection energy threshold increases with increasing number of outer reflectors triggered.

RESULTS

The phase histogram for all events observed for the Crab pulsar does not show any significant excess (> 3σ) in any bin. However, since earlier observations have suggested variability of the pulsed flux over a period of days, data from 7 out of 15 nights which individually showed excess >2σ in the same phase bins were selected for further study. The phase histogram for events from this restricted sample, with either events of lower energy 900 < E < 1100 GeV or those with arrival angle of less than 2° relative to the pulsar direction, is shown in figure 5a. The histogram for only events whose arrival angle was less than 2° relative to the pulsar direction is shown in figure 5b. Note that the separation between the two peaks in these two figures is 0.45 ± 0.02 which is rather close to the value of 0.42 seen in COS-B observations on the Crab pulsar. Therefore it is very tempting to call these two pulses as the pulse and the interpulse. The combined probability of both peaks arising due to chance is only 5.1×10^{-7} provided the initial selection effects are ignored. The number of events contained in these two peaks have been used to compute the pulsed flux at various energies and the integral energy spectrum for the pulsed flux is shown in figure 6. It is to be noted that the observed flux of

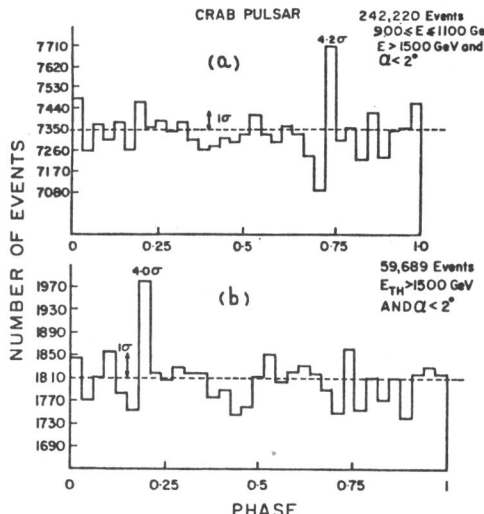
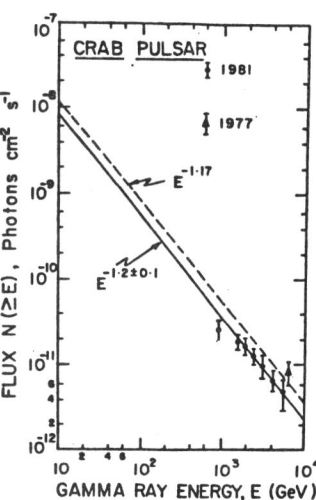

Fig.5: Phase histogram for data on Crab pulsar;
(a) All events with either lower energy or with angle < 2° relative to pulsar,
(b) Only events with angle < 2° relative to pulsar.

Fig.6: Integral energy spectra for pulsed γ-ray flux from the Crab pulsar

$(2.2 \pm 0.5) \times 10^{-11}$ ph.cm^{-2} s^{-1} at energies $> 9 \times 10^{11}$ eV agrees well with the extrapolation of the COS-B spectrum, 2.9×10^{-8} $E_{GeV}^{-1.17}$ ph.cm^{-2} s^{-1}, both in amplitude and spectral slope. The flux values shown in figure 6 are however applicable only for the active phase of emission and the time averaged flux would be about one third of these values.

The phase histogram for all events observed for the Vela pulsar is shown in figure 7a. The histogram for events with lower energy only is shown in figure 7b. It is interesting to note that the peak seen in figure 7a becomes more significant in figure 7b.

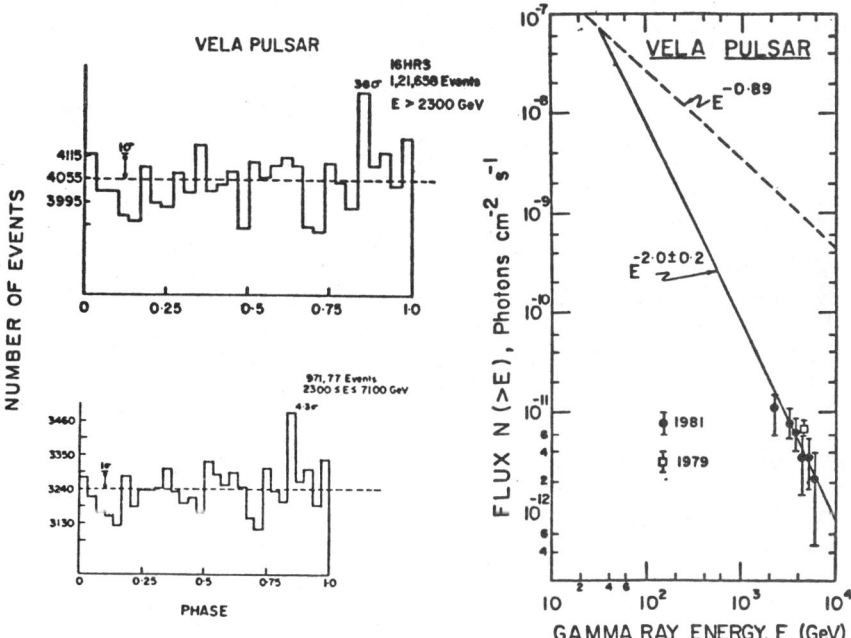

Fig.7: Phase histogram for data on the Vela pulsar; (a) All events, (b) Events with lower energy only.

Fig.8: Integral energy spectra for pulsed γ-ray flux from the Vela pulsar

It is also seen from the energy spectrum of the pulsed flux (figure 8) that the time averaged pulsed flux of $(1.0 \pm 0.4) \times 10^{-11}$ ph.cm^{-2}.s^{-1} at energies $> 2.3 \times 10^{12}$ eV for the Vela pulsar is about two orders of magnitude smaller than the extrapolated value from the COS-B spectrum, 1.8×10^{-7} $E_{GeV}^{-1.11}$ ph.cm^{-2}.s^{-1}. Also the energy spectrum for the Vela pulsar at energies ~ 2-5×10^{12} eV with a spectral index of -2.2 ± 0.2 is much steeper than at energies of few GeV.

Data for both the Crab and Vela pulsar have been studied for variability over very short periods of minutes. It is interesting to remark that both these pulsars show statistically significant variation of the flux over such short periods suggesting that the pulsed emission at these very high energies may be mostly sporadic. Details of this analysis are discussed elsewhere.[9]

CONCLUSIONS

Observations with the Ooty CR telescope in early 1981 have reconfirmed pulsed emission of very high energy γ-rays from the Crab and Vela pulsars seen earlier in few experiments. With more reliable values of the γ-ray detection energy thresholds available from these observations with the telescope in the 'Distributed Mode', it can be concluded that there are significant differences between γ-ray emission from these two pulsars at very high energies compared to the emission at energies less than a few GeV.

REFERENCES

1. D.J. Thompson et al., Astrophys. J. <u>200</u>, L79 (1975)
2. G.G. Lichti et al., Non-Solar Gamma Rays (COSPAR, Bangalore, 1979); Advances in Space Exploration <u>7</u>, 49 (1980)
3. N.A. Porter and T.C. Weeks, SAO Special Report 381 (1979) unpublished
4. P.V. Ramana Murthy, Non-Solar Gamma Rays (COSPAR, Bangalore, 1979); Advances in Space Exploration <u>7</u>, 71 (1980)
5. S.K. Gupta et al., Astrophys. J. <u>221</u>, 268 (1978)
6. P.N. Bhat et al., Astron. Astrophys. <u>8.</u>, L3 (1980)
7. P.N. Bhat et al., Pulsars - IAU Symposium 95, eds. W. Sieber and R. Wielebinski (D. Reidel, Holland), 1981, p. 251
8. S.K. Gupta et al., preprint (1983) unpublished
9. P.R. Viswanath, Proc. International Workshop on Very High Energy Gamma Ray Astronomy, Ootacamund, India, eds. P.V. Ramana Murthy and T.C. Weeks (Tata Institute, Bombay), 1982, p. 21

*Present address: Department of Physics and Astronomy, University of Maryland, College Park, MD 20742

ELECTRON POSITRON PAIRS IN RADIO PULSARS

Jonathan Arons

Astronomy Department, Physics Department and Space Sciences Laboratory
University of California at Berkeley

ABSTRACT

I outline the role of electron-positron pairs in some theoretical models of the magnetospheres of radio pulsars. While there is little direct evidence for the presence of pairs in the form of an observed recombination line, their formation leads to many promising features in theories of the electrodynamics and plasma physics of these objects' magnetospheric structure and for the interpretation of the observed photon emissions, both from pulsars themselves and from the surrounding interstellar medium.

INTRODUCTION

Cordes and Kanbach have reviewed the observational knowledge of photon emission by radio pulsars at this meeting. The interpretation of these observations in terms of the physical conditions at each pulsar is a goal still far in the future. Almost all of us agree that the origin of the emission derives from the energization of charged particles by the dynamo EMF of the rotating, very strong magnetic field rooted in each neutron star. The rotating, magnetized neutron star hypothesis is founded in the observed regularity of the short period clock defined by the stable pulse window, the "stellar" magnitude of the photon luminosity, the high degree of polarization of the radio emission, the energization of the Crab Nebula and the kinematics of the pulsar population in the galaxy. All these aspects of pulsar astronomy and the "zeroth order" pulsar astrophysics are summarized by Manchester and Taylor[1], for example. One would think that the "simple" problem of accelerating charged particles in a dynamo would be readily solved, and the emission properties of the accelerated charges would be as straightforward as the emission from a TV or radio antenna, familiar systems which have brightness temperatures at several hundred Mhz vastly in excess of the thermal energies of the particles, just as do pulsars. Unfortunately for such simple analogies, the electric field in a pulsar's magnetosphere is profoundly modified by the charged particles in ways which are much harder to understand than in an antenna with an externally applied voltage.

The circumstances in pulsar magnetospheres do have some analogies to the goings on in planetary magnetospheres. In these systems, we find an environment in which the energy source is the single mechanical degree of freedom contained in the relative streaming between the magnetized solar wind and the planetary magnetic field. Because the motional EMF in the wind differs from that in the planet, the charged particles in the magnetosphere respond to the voltage difference, accelerate, and "thermalize" the energy, primarily into large scale flow of the magnetospheric plasma with a small fraction put into high energy particle acceleration. Some of this energy in accelerated particles finally appears as incoherent auroral line emission, and some shows up in coherent radio emission. If the magnetospheric plasma density were as high as in a stellar atmosphere, one might have expected the energy derived from the motional EMF to have gone into driving currents against a collisional ohmic resistance, with eventual radiation of all the energy in truly thermal photons with a locally Planckian spectrum. That one sees nothing of the sort is a consequence of the relatively low density and high "temperature" of the system, which results in negligible collisionality (except at the boundary with the ionosphere). The magnetic stress associated with the currents driven by the motional EMF is converted into large scale, hydromagnetic flow rather than into heat, with collective processes tapping a small fraction of this free energy to give radiation to infinity. Local thermodynamic equilibrium (LTE), so useful in studies of stellar interiors, atmospheres and the interstellar medium (in this last case, only applicable to the continuum- atomic bound states have populations far from LTE) is a very poor starting place in magnetospheric studies. The result has been that in spite of this wealth of in situ observations, theoretical progress on the appearance of planetary magnetospheres to remote observations has been slow because of the zoo of plasma physical issues which must be confronted.

By contrast, the remote, accreting X-ray pulsars in binary star systems pose a somewhat simpler problem. Here also, the emission appears to have its energetic roots in the thermalization of a simpler mechanical degree of freedom, the free fall of plasma from large altitude onto a neutron star's surface. Because the density is much higher (probably in excess of 10^{20} cm^{-3} at the base of the accretion flow), the thermalization of the gravitational energy in each cubic centimeter's kinetic energy of infall really is (roughly) approximated by LTE. The zeroth order picture is then one in which the magnetosphere is passive, with the magnetic field entering only as a guide which forces the flow onto a small polar region[2-8]. Energization by current flow, which must exist because of the gravitational stress applied to the magnetic field by the infalling plasma, is literally a first order issue (the small parameter is the ratio of the stellar radius to the magnetopause = Chapman-Ferraro radius, often misnamed the Alfven radius in this field). So far, the effects of current flow beyond the ideal magnetohydrodynamic level have been neglected, in favor of more detailed radiation "hydrodynamics" studies of the emission zone implied by the zeroth order model, in an attempt to model the observed X-ray spectra and light curves with more depth than is given by the poor (but not ridiculous) level of simple LTE.

The radio pulsars reveal themselves in a manner which clearly points to a distant kinship with the planetary magnetospheres; their relation to the accreting X-ray pulsars is only through the astronomical "accident" of their being magnetized neutron stars, rather than through the specific plasma dynamics and radiation physics that leads to the observed photon emission. Of course, in the more "astronomical" questions involving the structure and evolution of these stars, including the interior currents which support the external magnetic field in the first place, the relation is much closer to the X-ray pulsars. The main energy loss is in the form of some combination of bulk flow kinetic energy and Poynting flux, as is shown by the total rate of change of rotational kinetic energy being much larger than the sum of all the photon luminosities. The exact form of this energy loss is not specified by the timing observations, nor is it uniquely determined by the "thermalization" observed in the box calorimeter formed by Crab Nebula around its pulsar. There is widespread agreement that the rotation of the star is the fundamental reservoir of the energy, and that a relativistic outflow of something is implied by the timing data and by the excitation of the Crab Nebula[9-16]. Recent work on the excitation of the Crab Nebula hints at a curious sort of relativistic wind with a mixed magnetohydrodynamic, vacuum wave structure as the carrier of the energy and angular momentum of the star[17-19]. The photons carry only a small trickle of the energy, and these are clearly due to highly nonthermal and indeed collective processes in the plasma somewhere in or near the magnetosphere. Once again, we have a system with one degree of freedom, now in the rotation of the stellar magnetic field with respect to the universe, with its associated induced (vacuum) EMF. (I omit discussion of the alternate idea expressed by Michel and Dessler[20], in which the rotation of the magnetic field with respect to an external conductor, in their case a fossil disk, creates a direct analogy to the planetary systems, especially Jupiter. As far as I am concerned, this hypothesis is not forced on us either by direct observations or by any basic theoretical necessity, as will be seen below.) As in the planetary case, the "thermalization" is weak, and the dissipation of the high quality energy of rotation is very far from LTE, presumably because the density is too low and the "temperature" is too high to support almost equilibrium energy transport. The gamma rays from the Crab and Vela pulsars point to the same conclusion, since the largest photon outputs in these objects occur in the $10^2 - 10^4$ MeV range, far in excess of the effective temperature (~ 1 keV) which would be expected if the gamma ray luminosity ($\sim 10^{34} - 10^{35}$ erg/s) were created by thermalized matter, as does occur (to zeroth order) in the accreting neutron stars.

Because we have neither the in situ observations available to the students of planetary magnetospheres nor the advantage of a zeroth order picture which is successful at the zeroth order level available to the X-ray pulsar modelers, we are forced to proceed by more sophisticated theoretical modeling even to get a feeling for the plasma configuration and the associated self-consistent electric field. As one might expect, this has led to a variety of theoretical ideas, none of which is far enough advanced to produce the observed photons (except in the form of the sophisticated propaganda published in the scientific journals) and many of which appear to occupy mutually orthogonal axes in idea space, in spite of the fact that everyone starts from a rotating magnetic field rooted in the star (usually assumed to be a centered dipole, for simplicity). Here, I will outline what I think is the most likely story, that is, the results and hypotheses which have influenced my own research. Michel[21] has given a summary of the theoretical work done through mid 1981, but with a drastically different emphasis from the line of thought I follow here. In particular, I think the creation of electron-positron pairs is an essential part of the magnetospheric dynamics as well as of the photon emission, even though no really convincing direct evidence has appeared in the form of an observable 511 keV annihilation line (pace Leventhal et al.[22]).

A THEORIST'S VIEW OF THE OBSERVATIONS

One route to understanding the physics of the magnetosphere is to concentrate on the region where the energy is "thermalized." The best studied such region is the Crab Nebula. The energization of this system was a long standing mystery, "solved" by the discovery of the pulsar with a total rate of loss of rotational energy comparable to the nonthermal synchrotron luminosity of the nebula[11]. However, the existence of relativistic electrons and magnetic fields in the nebula is not a specific clue to the means employed by the pulsar to excite its surroundings, partly due to the inability to spatially resolve the fine details of the region where the excitation of the nebula is the strongest, the inner edge of the wisp zone[23] where the particles which radiate the highest energy X-rays first appear as luminous entities[24-25]. Possibly the Space Telescope will make a direct contribution here by examining the "fine fibered structure" long known in the highest resolution optical photographs[23] of the synchrotron continuum, as will the improved X-ray imaging resolution of AXAF[26]. The theoretical models which are doing the best in modelling the existing data are the shock excitation schemes[17-19], in which some sort of relativistic wind flows out from the pulsar. These ideas owe their origin to models of the pulsar's magnetosphere, which are themselves oriented toward explaining the timing data and the trickle of photon energy in the observed pulses.

The basic facts of the timing data are that pulsars are temporally modulated radio sources with modulation periods now known in the range 1.56 msec to ~ 5 sec. The shortest period, that of PSR 1937+214[27], is close to the minimum allowed by the rotational stability of a neutron star, while the longest seem to reflect some sort of turn-off of the pulsar emission[28-31]. These periods slowly increase, with $\dot{P} \sim 10^{-15}$ s/s typical, but with a much wider range than is true of the known periods. On top of this systematic spindown, many pulsars show irregular period residuals, most likely representing some sort of noise in the moment of inertia or in the magnetospheric torque[32-34]. Almost all the known objects are galactic objects, with distances usually guessed from the dispersion measures to be $10^2 < D < 10^4$ pc.

The observed regularity and short periods of the clocks immediately led to the rotating neutron star hypothesis[11], and the observed spin-down rates led to the magnetic neutron star hypothesis, based on the textbook physics of rotating a bar magnet in a vacuum with magnetic axis oblique to the rotation axis[10-12]. This model leads to a successful spin down theory if the neutron stars have magnetic moments $\sim 10^{30}$ Gauss-cm^3, and initial theorizing "explained" this by invoking flux conservation of presupernova cores, without giving any real reason for the apparently preferred values of the precollapse flux. Indeed, there are drastic exceptions to the canonical magnetic moments--in the extreme, one finds values as low as $\sim 10^{26}$ Gauss-cm^3 in PSR 1937+214[35-36] if the conventional vacuum formula is applied to the observed[37] spin down rate of this object.

At the same time as Ostriker and Gunn[12] gave the most sophisticated formulation of the vacuum theory, Michel[13] showed that a relativistic magnetohydrodynamic wind gives rise to essentially the same torque as does the vacuum theory, while Goldreich and Julian[14] proposed a scenario in which a completely charge separated outflow would allow the magnetosphere and the wind zone to act somewhat like a force-free transmission line for the rotational energy and angular momentum, again yielding a torque formula like the vacuum prediction. All of these ideas lead to the same result because they all employ the same basic assumption about the global properties of the magnetosphere--in all cases, the electromagnetic energy density is assumed to be wholly dominant over the plasma energy, all the way from the stellar surface through the light cylinder (the cylindrical radius $R_L = c/\Omega_* = cP/2\pi$ where a corotating particle would be moving at the speed of light) and on until the outflow is thermalized at the radius where the energy flux balances the pressure of the surrounding medium. See Arons[38] for a discussion of the various radii typifying these and other magnetospheres. Magnetic moments of this magnitude imply surface field strengths $\sim 10^{12}$ Gauss, and indeed, cyclotron line features in the spectra of several X-ray pulsars[39-41] suggest surface fields, possibly dominated by non-dipolar components, to be in this range.

If one were to ask no further questions, one could conclude that a satisfactory theory of the basic energetic fact of pulsar spindown is in hand, and proceed on to other astrophysical problems without worrying about the complications of collective radio emission needed to model the pulses. However, the vacuum model is very sensitive to the presence of even a small number of charges[42-46] and does not lead to the generation of the spectrum of the X-ray emitting particles at the inner edge of the Crab Nebula. The mhd wind idea requires a high density plasma from the star in order to be valid, whose source was not addressed in the original work, while the completely charge separated, aligned magnetosphere of Goldreich and Julian faces several insuperable difficulties, all traceable to the difficulty of populating some regions of the magnetosphere with a non-neutral plasma from the surface when all motion in the corotating frame is flowing along field lines, as is assumed in their force-free scenario[21,47-48]. The resolution of these problems lies, in my opinion, in the flows of electron-positron pairs, initially hypothesized[49-50] in an attempt to explain the direct photon emission from pulsars. If pairs are implicated in the observed photon emission some broad and brush attention to the directly observed photons is necessary.

When observed with high time resolution (~ 1 msec), the individual pulses forming the basic objects of pulsar emission turn

out to be the individual subpulses, temporally narrow bursts of
radiation (width ~ 10^{-2} P) which occur within a fixed window of
rotation phase, the pulse waveform. This waveform is observed by
averaging together several thousand pulses. This forms the
characteristic fingerprint of each object, as well as giving a
template that allows the accurate timing of the pulsar's spin.
Individual pulses as strongly polarized as 100% linear polarization
have been observed in some cases[51-52]. It was once thought that the
polarization was mostly linear, but more recent work[53-55] has shown
the radiation to be in two orthogonal elliptical polarization states,
usually overlapping to some degree so that the total polarization is
less than 100%. The spectra of the subpulses are broad, piecewise
power laws. The pulses fluctuate in intensity simultaneously at all
frequencies, as if the emitter is itself broadband in one location, or
at least, maintains the phase space organization of the subpulse
plasma as it moves between different locations with different narrow
band frequency of emission in times too short to be resolved[56-57].
The superposition of the subpulse power laws yields the observed power
laws of the average spectra.

Some (but not all) pulsars exhibit microstructure, temporal
modulations of the signal on time scales between 100 μsec and ~ 1
μsec, but there are pulsars whose subpulses appear smooth[58], at least
with present sensitivity. In addition, the polarization direction of
the subpulses seems to mimic the location expected for the E vectors
from the swing of the average polarization, once the effect of the
orthogonal mode structure is removed from the data, while the
micropulse polarization appears to switch to follow the local
intensity state[59]. These aspects suggest the emission/transfer of the
basis beam of radiation is embodied in the subpulses, rather than the
subpulses being themselves the average of a group of micropulses. A
final, curious numerical aspect of the subpulse structure is that
normally only one subpulse appears within a waveform component, in
spite of the fact that several (sometimes many) subpulse beams could
be fit into the pulse window. This suggests the fluctuations of the
subpulse beam directions which go into forming the broader average
pulse window are related to the global structure of relativistic flow
within the magnetosphere, since the fluctuation time is on the order
of the relativistic transit time ~ P in the magnetosphere[34].

The radio emission is clearly "coherent" in the sense of being
the collective emission of the electric field of groups of particles
which move together in phase space on time scales which include
significant power on the 100 Mhz time scale, even though the basic
magnetospheric structure may be time steady or have at most variations
on the rotation time scale, when observed in the corotating frame.
The evidence for this is simply the lower bound of the brightness
temperature of the subpulses, > 10^{27} K, imposed by the requirement
that the emission region yield a beam of angular width less than
c x subpulse width/R_L. This intensity corresponds to a wave electric
field in excess of 10^5 V/m at 300 Mhz; the exact value depends on
one's assumptions about the emission site/transfer properties in the
magnetosphere.

We don't know whether or not the emission and transfer along a ray is intrinsically phase coherent, as well as having high intensity. It is known that that the distribution of subpulse and micropulse (when observed) intensities can be modelled as the result of large numbers of overlapping shots of emission[60-63], suggesting that the radiation received is the sum of many emitters, each of which has high brightness temperature but has no fixed spatial phase relation with respect to the other emitters. Because we do not resolve the emission/transfer regions (interstellar scintillation implies angular scales for the region where the radiation last couples to the magnetospheric plasma to be less than 10^{-7} arc sec), the observed lack of phase coherence could be due to completely phase coherent emission from many small "clouds" within the subpulse plasma, each of which is incoherently arrayed with respect to its companions, or it could be due to the random phase saturation of a plasma instability throughout the whole of the subpulse plasma (of course, the formation of the "clouds" must also be the result of some "instability" process).

One of the simplest ways of organizing a large fraction of these data in an easily remembered manner is given in the kinematic beaming model of Radhakrishnan and Cooke[64]. Here, the beaming of the emission is attributed to relativistic outflow along polar magnetic field lines of an oblique dipole field, just the geometry assumed in the spin down theory. The beaming is controlled by the magnetic geometry if the particle Lorentz factors exceed ~ 10-100, assuming the intrinsic beaming of the coherent emission is controlled by the same $1/\gamma$ width of relativistic single particle curvature emission. Of course, this assumption is irrelevant to collective emission, in principle, since the beaming can equally well manifest the direction in which the gain of the spatial amplifier is the greatest. In their idea, the polarization of the average profile (and, in the modern descendent of this idea, of an individual subpulse beam) is determined by the local direction of the magnetic field. The basic success of this idea in explaining the rotation of the polarization states observed in most pulsars (once the orthogonal modes are accounted for) has led to the widespread acceptance of the polar lighthouse view of pulsar emission, although it is by no means the only way of accounting for the kinematics of polarized emission (e.g., Ferguson[65]). It has even been applied in its simplest, undistorted dipole form to rather fine details of the polarization structure by Narayan and Vivekanand[66], although as I shall argue below, the physical realization of polar plasma streams include strong field aligned conduction currents which prevent the application of the perfectly dipolar Radhakrishnan and Cooke beaming theory to the fine details of the polarization data as was done by these authors. We shall also see that other emission sites in the magnetosphere may be thinkable, although at the moment their properties are much less well explored. In this picture, therefore, we view the pulse waveform as measuring the opening angle of the polar magnetic field at the emission altitude, at least in zeroth order models in which the effects of refraction[67-68] and the effects of current induced nodding and wobbling of the polar flux tube[34] are neglected. The subpulse beams are then the manifestation of angularly localized emitters within the polar flux tube.

Some aspects of subpulse organization (marching subpulses and mode changing, especially) point to variations in the magnetosphere which require several to many pulsar days (times > P) for their occurence. These are briefly discussed below.

In a few cases, high frequency emission gives us an energetically more significant view into pulsars' magnetospheres, although without the splendid time resolved polarimetric information possible in the radio part of the spectrum. In the Crab pulsar, near IR, optical, X-ray and gamma ray pulses coincide with the main radio pulse and interpulse, with photons in the spectrum all the way up to 10 GeV[69]. The radio precursor and the possibly variable TeV emission[70] form the only real exceptions to this temporal line up, although weak optical emission may be present through 360° of rotation phase[71]. In this object, therefore, relativistic particles certainly are present and the temporal coincidence of the pulses suggests that the same relativistic plasma is responsible for the radio emission. In the Vela pulsar, there is optical and gamma ray (100 MeV - 10 GeV) emission but no X-ray emission at the level expected by extrapolating the gamma ray emission to lower energies, and the pulses are not coincident with each other or with the radio pulse (here single), although the gamma ray pulses are similar to those of the Crab object in the same energy range. PSR 1509-58 was recently found to have a single broad X-ray pulse[72] in the 1-4 keV band and a narrow radio pulse at 1.4 Ghz[73] with the relative phase unknown.

These data are too sparse and discrepant to draw any simple direct conclusions about the model, although their importance is clear, since the non-collective, high frequency emissions involve luminosities many orders of magnitude greater than the radio luminosities and the presence of the gamma rays points directly to the existence of relativistic plasma in pulsars' magnetospheres. About the only single common feature of these three pulsars is that they are among the fastest rotators known (0.033, 0.089 and 0.15 sec rotation periods) and have the largest known values of dP/dt, indicating very strong magnetic fields. As we shall see, these aspects lead to very strong particle acceleration phenomena in the magnetospheres of these stars. They are also all objects which were not found in the normal pulsar surveys, which are insensitive to the detection of short period objects[1]. This suggests that more such objects may be lurking out there, if we could improve the sensitivity of high frequency (including mm and IR) observations of the known pulsars, which have longer periods and smaller $\dot{P}$, and/or increase our sensitivity to the discovery of short period objects, even though those with large $\dot{P}$ are not expected to linger long in the small P regime where really strong high frequency emission seems easier to achieve.

THEORY

Since the radiation characteristics of pulsars are not immediately suggestive of a simple physical process in a simple zeroth order model medium, one is forced to couple together investigations of magnetospheric plasma dynamics with studies of the radiation physics of the dynamical models, in order to reach even the lowest order stage of identifying the most likely physical processes which might give

rise to the observed emission. The most important requirement is that
the dynamical models must have an energy input into the plasma which
is larger than the observed luminosity in photons. Furthermore, this
energy input must have a morphology relevant to the observed beaming,
and include sufficient "free" energy in the plasma physicist's sense
to give rise to plasma collective processes ("instabilities") which
have some chance of explaining the radio emission. Streaming of
plasma along the polar field lines is not in itself a model of the
emission, if there exists a single Lorentz transformation to a frame
where the plasma is at rest and has a single humped distribution
function with plasma pressure $\ll B^2/8\pi$, since such a plasma has no
free energy for collective emission processes. One example of polar
flow which does have free energy, in principle, occurs if the plasma
has several components which stream relative to each other. Then the
energy of _relative_ motion is available to drive the collective
emission. The dynamical models must be able to assess these
relatively delicate aspects of the plasma structure, and they must do
so in a self-consistent manner if the "garbage in, garbage out"
phenomenon is to be avoided.

Because the observational data may point to relativistic polar
flow as the emission site, _locally_ self-consistent theories of the
polar flux tube may be the most appropriate starting place. After
discussing these, I will turn to some possible connections to the
global aspects of the magnetosphere, and then to the relevant aspects
of the radiation physics and its limited relation to the observed
pulses.

Radio pulsars are electrically isolated neutron stars. They are
not in mass transfer binary systems[74], they do not have planetary
companions so far as is known[75], and there is no direct or indirect
observational evidence for the disks or rings suggested by Michel and
Dessler[20]. The enormous Poynting flux implied by the zeroth order
theories of spindown makes accretion from the interstellar medium
unlikely[38] (impossible is a more probable description of this plasma
source) and the same Poynting flux and radiation losses excludes the
entry of ultra high energy cosmic rays into the magnetosphere in any
interesting number[76]. External gamma rays can enter and cause the
creation of a small number of electron-positron pairs[77], but most
stars are not in the opacity window which allows this process to be
interesting even as a catalyst for shower development in the
magnetosphere[78], and the total number of particles created without
shower development is electrically negligible[79]. Then a stellar
surface is the only possible source of a plasma sufficiently dense to
affect the electric field induced by the rotation of the magnetic
field.

This isolation leads to a very peculiar system. The
gravitational escape speed from the surface is $\sim c/3$, yet the observed
surface temperatures of radio pulsars are less than $\sim 10^6$ K[80].
Therefore, pressure is negligible in populating even the inner
magnetosphere with quasi-neutral plasma. For all the observed
objects, $\Omega_* R_* = 2\pi R_*/P \ll c/3$; therefore, centrifugal force is
negligible in lifting particles off the surface (even for PSR
1937+214, this effect is likely to be only a small, but not

fantastically small, correction to the basic electrical effects).
However, the component of the vacuum electric field along B vastly
exceeds the component of gravitational force along B, which causes the
magnetosphere to become populated with a non-neutral, completely
charge separated plasma whose density is such that $E_\parallel = \vec{E}\cdot\vec{B}/B$ is
greatly reduced below the vacuum value, a point first made by
Deutsch[81], who was thinking about magnetic A stars, and by Goldreich
and Julian[14] in the explicit context of pulsars, at least at the
stellar surface where the magnetosphere is in contact with the
essentially infinite density of the crust. Along field lines which
close near the star, the supposition is that these charges form an
essentially static, electrically supported atmosphere, while along
polar field lines, there is at least the possibility that the
atmosphere is a continuous flow of charge, if only because along the
field lines which would have closed beyond the light cylinder, the
centrifugal forces would continue to urge the particles outwards,
assuming the much larger electrical forces cooperate with the outflow.
As will be seen, centrifugal force is unlikely to play any major role,
but for now, assume the polar field lines do compose the flux tube of
a dipole field which would have closed beyond the light cylinder, and
assume that a non-neutral beam outflow from the stellar surface is
indeed acceptable to the outer magnetostructure.

1. ELECTRICALLY DRIVEN FLOW ALONG THE POLAR FIELD

Because all nonelectrical forces are unable to supply particles
to the magnetosphere, the only remaining possibility is that the
electric field extracts charged particles from the surface until the
vacuum $E_\parallel$ induced by the rotation of the magnetic field is
substantially reduced, at least at the stellar surface. The
orthogonal rotator $i \cong 90°$ is a special case which will not be
discussed further. The sense of $E_\parallel$ is such that the vaccuum fields
extract electrons by motion along B when $i < 90°$, while they try (and
most likely succeed) to extract ions in the case $i > 90°$. The stellar
interior has sufficiently good conductivity to ensure corotation below
the surface. Since the outer layers of neutron stars are solid[82], the
electric field inside the star is given by

$$\underset{\sim}{E} = -(1/c)\left(\underset{\sim}{\Omega}_* \times \underset{\sim}{r}\right) \times \underset{\sim}{B}, \qquad (1)$$

while ouside it is adequate[38] to write

$$\underset{\sim}{E} = -(1/c)\left(\underset{\sim}{\Omega}_* \times \underset{\sim}{r}\right) \times \underset{\sim}{B} - \underset{\sim}{\nabla}\Phi. \qquad (2)$$

Nonzero acceleration along B is entirely dependent on Φ. In the
vacuum, $\Phi \sim 10^{13}$ to 10^{16} V, while the escape energy from the star is
$\sim 10^5$ eV (electrons) and ~ 1 GeV (iron ions). Thus, once electrical
forces of even a small fraction of the vacuum magnitude operate,
gravity, centrifugal force, pressure and all the usual astrophysical
modelling assumptions that go along with them are forgotten. The
price, obviously, is that the surface only wants to supply a
completely non-neutral plasma to the magnetosphere.

The density scale of this nonneutral medium is easily found. Upon substituting (2) into Poisson's equation, we find

$$-\nabla^2 \phi = 4\pi (\eta - \eta_R) \quad (3)$$

where η is the charge density and

$$\eta_R = -\frac{\Omega_* \cdot B}{2\pi c} + \frac{1}{4\pi c}(\Omega_* \times \underline{r}) \cdot (\nabla \times \underline{B}) \quad (4)$$

is the charge density such that E is the corotation value everywhere (pace boundary conditions)[14]. Notice that this charge density is proportional to $B_z = \Omega_* \cdot B/\Omega_*$, not to B itself. Numerically, a non-neutral plasma with this charge density has density/elementary charge $\cong 10^{11}$ (B/1 TG)(1s/P) cm^{-3}. In the early days, it was suspected that the magnetosphere would fill by flow along B until $E_\parallel = 0$ everywhere[14,83], with the departures of Φ from 0 only of order the small gravitational energy/charge. Notice how gravity is huge on the scale of all the non-electrical forces available, but is tiny compared to the electric force on a nonneutral plasma.

We hypothesize the plasma on the polar field lines to be composed of a relativistic, nonneutral beam flowing out along field lines, with a density such that $E_\parallel$ is reduced below its vacuum value at the surface. Because of continuity of flow along B, the current density in the beam is $J_\parallel = J_*(B/B_*)$, where B_* is the surface magnetic field. In a relativistic, non-neutral flow, the charge density is $\eta = J_\parallel/c$. Since the surface is dense (number densities of electrons and ions of the basic lattice are expected[84] to be $\sim 10^{26}$ cm^{-3}), the surface emits sufficient particles to yield $E_\parallel = 0$ at the surface, assuming the emission (either cold-field or thermionic) is free. Ruderman[84] suggested that the binding of ions might be sufficiently strong to prevent such free emission in the $i > 90°$ case. More detailed calculations of the binding energy of an infinite, perfect lattice of strongly magnetized iron[85] showed that such binding is possible, but is sensitive to the unknown details of the electronic structure of the lattice. Variational calculations were done with the result that even the sign of the binding energy depends on the form of the trial wave functions assumed. However, the emissivity of real, compositionally pure surfaces is controlled by the roughness, since $E_\parallel$ concentrates on local peaks when the surface is a cold-field emitter, reducing the effective work function well below the binding energy (see the ancient discussion by Fowler[86] for example). In addition, all locally self consistent models of particle acceleration at the poles lead to bombardment of the surface by TeV electrons in the $i > 90°$ case[77,87-88]. These induce showers in the crust (see below), which in turn cause the emission of protons, neutrons and alpha particles (as well as gamma rays) which are probably more weakly bound than the iron ions of the basic lattice and which may have sufficient flux to sustain a space charge limited beam flow, even if the binding of iron nuclei is perfect. If this is the case, the stimulating model of Ruderman and Sutherland[77] is simply an initial transient, lasting only as long as it takes to build up an interesting layer of pollutants in the first 100 to 1000 gm/cm^2 (i.e., within the outer layer of the crust ~ 1 mm to 1 cm thick). I discuss this crustal shower problem below; for now I assume that during almost all of its life, a pulsar's polar cap is a free emitter of whatever sign of charge is attracted by the residual vacuum E field of the immediately overlying magnetosphere. Then at the surface, the current density adjusts to a value such that the charge density $\eta = J_\parallel/c$ is very close to η_R(surface)(B/B$_*$), as is required if the emission of the nonneutral beam is to be space charge limited[89]. Notice that this surface boundary condition fixes only the current <u>density</u>; the total current, which must be fixed by the global structure, depends also on the area of the polar cap, usually <u>assumed</u> to be ~ $\pi\theta_c^2 R_*^2$ with $\theta_c = (\Omega_* R_*/c)^{1/2}$ as in the force free hypothesis of Goldreich and Julian[14]. In addition, the free emissivity at the surface really fixes only the charge density in a charge separated surface; the current density is fixed only in regions where the non-neutral plasma is composed of a unidirectional flow. Both of these assumptions are changed in the global theory outlined below, but their adoption leads to a potential Φ which is <u>not</u> changed in the globally self-consistent theory, and which provides the basic element in the supply of pair plasma to the magnetosphere.

Since η varies in proportion to B while η_R varies in proportion to B_z, $E_\parallel = 0$ is possible only at one surface, here the stellar surface, in this relativistic flow. Any real (non-monopole) magnetic field has B_z/B non-constant along each field line. In particular, those field lines which bend toward the rotation axis have B_z/B increasing with increasing altitude. Then the beam in this part of the flux tube is unable to completely short out the vacuum $E_\parallel$ at all points above the surface--this "favorably curved" part of the polar flux tube is "starved" of charge, relative to the corotation value, leading to a residuum of the vacuum $E_\parallel$ (the "starvation" electric field), which is not shorted out. When the neighboring closed zone and the part of the polar flux tube which curves away from the rotation axis (the "unfavorably curved" zone) are <u>assumed</u> to be regions of $\Phi \cong 0$, the accelerating potential is, <u>in a pure dipole</u> model with dipole moment μ,

$$\Phi = (-\text{sign}\eta_R)(10^{13}\text{ Volts})\left(\frac{\mu}{10^{30}\text{ cgs}}\right)\left(\frac{0.1}{P}\right)^{5/2}\left[\left(\frac{r}{R_*}\right)^{1/2} - 1\right]f(\text{angles}) \quad (5)$$

with the numerical value and the scaling with period dependent on the assumption that the polar cap's opening angle is that of the globally force-free hypothesis. The detailed derivation of (5) is given by Scharlemann et al.[48], who did their calculations explicitly for the i < 90° case but with results which apply to the flow of freely emitted ions by simply changing the sign.

2. PAIR CREATION IN THE POLAR FLUX TUBE

The potential (5) is derived without consideration of possible pair creation. It is expected[48,91] that this flow might fit into a global model of a neutron star's magnetosphere when i >> θ_c and

$\pi - i \gg \theta_c$; in the aligned case, such relativistic, non-neutral flow is unlikely[48] since all polar field lines of simple magnetic fields are unfavorably curved. In the rest of my discussion, I assume such a global, charge separated model exists and might be the relevant model of the magnetosphere when the voltage is too small to lead to pair creation. However, it has long been known that if charged particles accelerate along the polar field lines to the energy implied by (5), the gamma rays emitted might be converted into an electrically interesting density of electron-positron pairs[49-50]. The $i < 90°$ geometry leads to a simpler model of the effects of pair creation, since every beam electron can radiate convertible gamma rays, so I use it to outline the basic ideas and processes, then return to the ion emission case.

(a) $\underset{\sim}{\Omega}_* \cdot \mu > 0$

In this case, TeV electrons of the beam extracted from the stellar surface move out along the curved magnetic field. Since this is an accelerated motion, they emit GeV gamma rays. If neutron stars were somewhat warmer, inverse Compton scattering of the thermal photons from the surface would also contribute, but for the observed radio pulsars, the curvature process dominates. Sufficiently high energy gamma rays are magnetically converted into $e^\pm$ pairs (γ + virtual photon of the magnetic field $\to e^\pm$). The creation of pairs through the process γ + thermal photon $\to e^\pm$, so prominent in quasar theories[91,92], is not an important contributor to the opacity, given the observational upper limits to the surface temperatures of most radio pulsars, either over the whole stellar surface or at the polar cap alone[80]. The magnetic conversion opacity declines exponentially with $1/B$[93,94], so if conversion occurs at all, it must happen in the region where the magnetic field strength is still close to its maximum, at radii less than $2R_*$. This assumes the rate of decrease of B with radius exceeds the rate of increase of the gamma rays' typical energy, since the opacity really depends on $1/\varepsilon B$, where ε is the photon energy. This is, in fact, the case in the potential given by (5) and any other potential so far proposed for the polar flux tube. Then if the number of pairs created by each outgoing electron is too small to modify the potential within a height $< R_*$, the number of pairs created at all radii is negligible, a circumstance described in brief as opacity bounded pair creation. However, at heights $< R_*$, the residual charge density $\eta - \eta_R$ is small compared to η itself. Therefore, only a small amount of pair creation is needed to alter the potential to that of locally force-free flow, $E_\parallel = 0$. In addition, because the opacity depends exponentially on $1/\varepsilon$ and therefore depends exponentially on $1/\Phi^3$ (for curvature emission), the number of pairs created/beam particle increases very rapidly with height, typically[95] like $\exp[-\text{const}/(\text{height})^4]$. Therefore, when pair creation is sufficiently strong to short out the starvation electric field at some height $H < R_*$, the alteration of the flow from that of the pure charge separated beam to a locally force-free region at higher altitude occurs in a thin layer, the pair formation front (PFF) whose thickness is much less than H itself. The alteration occurs through the capture of positrons in the electric field of the pair formation front just below the height H. The positrons are stopped as they form, reversed and accelerated downwards until they strike the polar cap with ~ TeV energy/particle, causing the cap to emit X-rays with effective temperature ~ $10^{5.5} - 10^6$ K, too weak to be seen with Einstein sensitivity but forming a lower bound to the expected soft X-ray emissivity of neutron star surfaces due to magnetospheric bombardment in this sort of model[95]. The density of this locally generated beam of precipitating particles is much less than that of the primary beam extracted from the surface, simply because the residual charge density is so small at H.

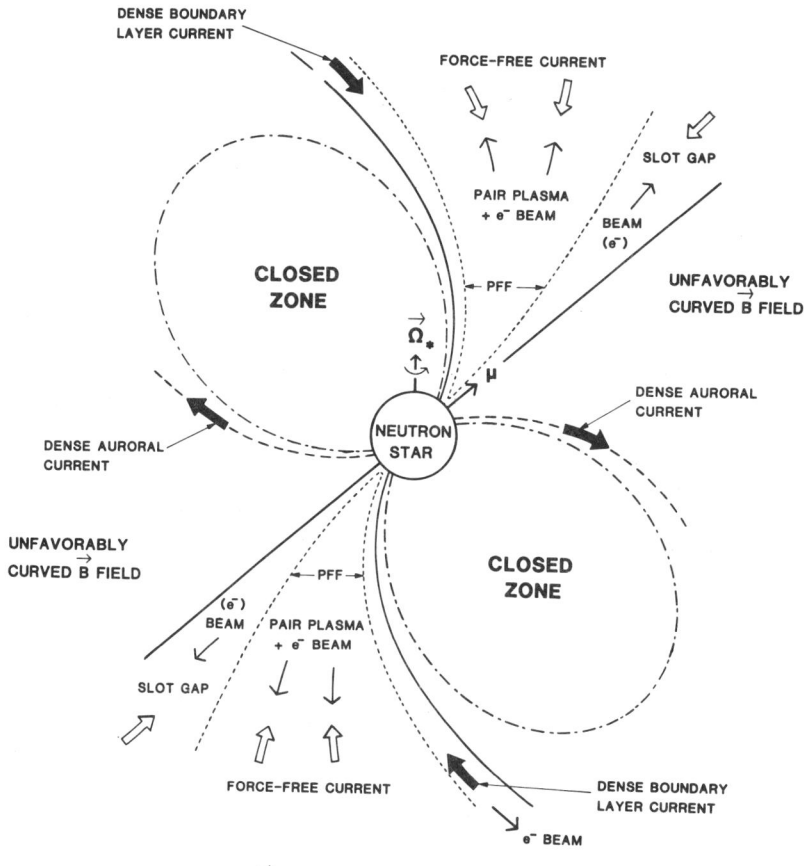

Figure 1. Low altitude beam flow structure with pair creation in the favorably curved flow zone, from ref. 88. The boundary layer and auroral currents are from the possible global structure discussed below.

The details and numerical values of all this can be found in papers by Arons and Scharlemann[96] and by Arons[95], where the height H was assumed to be a single constant. However, these calculations assume the regions bounding the favorably curved flow zone to be good conductors, in which case the flow near the boundary has a small starvation $E_\parallel$ and never has sufficient pair production to short out the parallel component of the starvation electric field in these boundary regions. Then H must curve up, becoming coincident with the field lines at a radius $\sim 2R_*$. The resulting flow pattern of this locally self-consistent model is shown in Figure 1, drawn approximately to scale for a pulsar with a millisecond rotation period if the opening angle of the polar cap is $\theta_c = (\Omega_* R_*/c)^{1/2}$. At this level of modelling, the closed field lines and the unfavorably curved open field lines are simply regarded as in perfect electrical contact with the star and therefore as corotating. In this case, the "slot gap," the annular region of nonzero starvation electric field surrounding the outflowing column of pair plasma, is filled only with the particles of the outflowing primary beam. The most important aspect of this model is that the rotational energy dissipated in this slot is more than sufficient to explain the hard photon emission from all of the radio pulsars known to emit radiation above the radio band, a virtue not shared by the low altitude zone below the pair formation front in this or the related models of Ruderman and Sutherland[77] and Cheng and Ruderman[97,98]. In addition, the beaming morphology is qualitatively promising as a site for photon emission in the boundary layers between the slot gap and the pair plasma[34]. The quantitative details of the structure of the pair formation front, and of the energetics of the non-neutral beam flow in the slot gap when dissipation in the boundary layers is neglected, are given in a recently published paper[88]. A quantitative assessment of this emission site requires modelling the structure of these boundary layers, which demands a knowledge of the distribution of pairs formed above the pair formation front as the accelerated beam particles radiate away their energy in the locally force-free zone. It also turns out to demand a knowledge of the global current flow pattern which in turn depends on the availability of the dense pair plasma. Thus, the next step in the purely theoretical aspect of the modelling is the evaluation of the vacuum cascade showers.

Above the pair formation front, each beam electron continues to emit curvature photons, although it is no longer accelerated by the starvation electric field. The number of photons emitted per unit photon energy and per unit length is proportional to $\varepsilon^{-2/3}$, up to the upper cutoff energy

$$\varepsilon_c = \frac{3}{2} \frac{\lambda_c}{\rho} \gamma_b^3 m_e c^2 \qquad (6)$$

where λ_c is the Compton wavelength, ρ is the radius of curvature of the magnetic field and γ_b = primary electron energy/mc^2. Typically, $\varepsilon_c/mc^2 \sim 10^3 - 10^4$ at the pair formation front, and declines with distance above the front as the particles lose energy to curvature emission. On the other hand, a photon emitted at some height s ≅

$r - R_*$ is absorbed (with unit probability) if the photon energy exceeds[96]

$$\varepsilon_{min} \cong \frac{8}{3} \frac{B_q}{B} \frac{R_*}{\rho \ln \Lambda} mc^2. \tag{7}$$

ln Λ comes from setting the optical depth for magnetic conversion

$$\tau = \Lambda \exp\left[-\frac{8}{3} \frac{B_q}{B} \frac{mc^2}{\varepsilon \sin \psi}\right] \tag{8}$$

equal to unity, with

$$\Lambda = 0.09 \left(\frac{e^2}{hc}\right) \frac{\rho}{\lambda} \left(\frac{B}{B_q}\right)^2 \left(\frac{\varepsilon}{mc^2}\right) \sin^3 \psi = 9 \times 10^{10} \left(\frac{\rho}{100 \text{ km}}\right) \left(\frac{B}{1 \text{TG}}\right)^2 \frac{\varepsilon}{mc^2} \sin^3 \psi. \tag{9}$$

In (8) and (9), $B_q = 4.41 \times 10^{13}$ Gauss and ψ is the angle between B and the photon momentum at the $\tau = 1$ point on a ray. If the magnetic field and the radius of curvature are assumed to scale with radius like a dipole field, ε_{min} increases in proportion to $r^{5/2}$. Thus pair creation by an individual electron continues along a field line until $\varepsilon_c(r)$ drops below $\varepsilon_{min}(r)$; above this point, the magnetosphere is optically thin to the whole curvature spectrum. This leads to the total number of pairs per primary emitting electron being

$$\kappa_c \cong \frac{\gamma_b(H) mc^2}{\varepsilon_c(H)} \left[\frac{\varepsilon_c(H)}{\varepsilon_{min}(H)}\right]^{2/3} \tag{10}$$

on a field line, where H is the height of the pair formation front on that field line. The first factor in (10) is the "usual" estimate of the amount of pair plasma created per primary[77,96], based on the thought that pair creation is negligible once the beam particles lose $\sim 1/2$ of their energy by curvature emission. The second factor comes from the fact that pair creation continues to higher altitude until the photons of energy ε_c are emitted under optically thin conditions; pair creation is opacity bounded, not energy bounded, leading to the extra factor of 10-100 over previous estimates. The power in (10) is 2/3 rather than 1 because the rate of production of photons drops in proportion to γ_b. In (10), I assumed that all the energy loss of the beam is in the curvature emission, rather than through the two stream interaction with the secondary plasma, because the longitudinal mass of the beam particles leads to a growth length for the instability much greater than the thickness of the pair creation zone. I also assumed the asymptotic form of the opacity in calculating (7), since I am mainly concerned with the boundary layer regions where H is large and the magnetic field is a factor ~ 5 weaker than the surface field even at the pair formation front; in the central regions of the flow, this can lead to a mild overestimate of the pair density created if the surface field is very strong ($> 7 \times 10^{12}$ Gauss) and H (central field line of favorably curved flux tube) $\gg R_*$, since under these conditions ε_{min} should be replaced by $2mc^2$/pitch angle of the photons[94].

A more important point is that the density of the pair plasma is controlled by the field strength, the rotation period and the radius of curvature of the field. In long period pulsars, these models only work if the surface field has a strength comparable to that of the dipole, which can be estimated from observations of P, but has a radius of curvature much smaller than that of the pure dipole field at the cap, as can be achieved if the surface field has a few low order multipoles in addition to the dipole[99]. In a pure dipole model, one would expect all observable properties to be fixed by B (dipole) and P, and therefore to correlate well with P and $\dot{P}$. In fact, the pulsar luminosity function is almost independent of both these observables[1], indicating at least one other parameter is important. The pair creation calculations suggest that the density (and also the morphology of the field lines along which plasma is injected) is a function of the radius of curvature as well as of the direct observables P and $\dot{P}$, which may be a part of the explanation for the diversity of pulsar luminosities and morphologies even when the objects have the same torque.

In addition to the curvature photons emitted by the primary beam, newly formed pairs also emit gamma rays since they are born with non-zero pitch angle with respect to the magnetic field. Suppose a photon whose momentum makes an angle $\psi \ll 1$ with respect to B makes a pair with total initial energy equal to the photon's energy ε_i. If $(mc^2/\varepsilon_i)(B_q/B\psi) > 2 - 3$, as is true in almost all of the pair creation zone above the pair formation front, equipartition between the particles' initial energies is a good approximation[94,96] and the particles each have initial energy $\gamma_i mc^2 = \varepsilon_i/2$. The particles lose their gyrational momentum instantaneously--the synchrotron loss time is $\sim 10^{-12} - 10^{-15}$ sec, depending on the pitch angle and the altitude, while the transit time is $\sim H/c$ $10^{-4} - 10^{-6}$ sec. Then the spectrum of photons created per pair[100] is proportional to $\varepsilon^{-3/2}$, up to the upper cutoff energy $\varepsilon_s = 1.5\hbar\omega_c \gamma_i^2 \psi$ with $\omega_c = eB/mc$ - cyclotron frequency. Since each pair is created where $\tau = 1$, $\gamma_i \psi = 4mc^2/3\hbar\omega_c \ln\Lambda$. Therefore the upper cutoff of the synchrotron emission created by the absorption of a photon of energy ε_i is simply

$$\varepsilon_s(\varepsilon_i) = \varepsilon_i/\ln\Lambda. \qquad (11)$$

Then one can write a simple integral equation for the synchrotron emissivity per primary emitting beam particle which takes into account synchrotron emission caused by the absorption of curvature photons and of the synchrotron photons themselves. If $Q_s(\varepsilon)$ is the number of synchrotron photons emitted per second per unit energy per primary beam particle, and Q_c is the same thing for curvature emission,

$$Q_s(\varepsilon)\varepsilon^{3/2} = \alpha \int \frac{d\varepsilon_i}{\varepsilon_i} Q_c(\varepsilon_i) F(\varepsilon/\varepsilon_i) + \alpha \int \frac{d\varepsilon_i}{\varepsilon_i} Q_s(\varepsilon_i) F(\varepsilon/\varepsilon_i) \qquad (12)$$

where α is a number ~ 1 proportional to $(\ln\Lambda)^{1/2}$ and $F(x)$ is a messy function related to the usual Bessel functions of synchrotron theory but which in the end has behavior not too far from $\exp(-x)$. (12) can be solved in general by Mellin transforms, and roughly leads to a synchrotron emission spectrum proportional to $\varepsilon^{-(\alpha+3/2)}$, quite a bit softer than curvature emission. The absorption of these synchrotron photons leads to another order of magnitude increase in κ = number of pairs/primary beam particle in the central parts of the favorably curved flow tube. Because the upper cutoff of the synchrotron emission is $\varepsilon_c(H)/\ln\Lambda$, where ε_c is the upper cutoff of the curvature emission, the synchrotron emission does not contribute to the structure of the boundary layer, where $\varepsilon_c(H) \sim \varepsilon_{min}(H)$. However, the total pair output of the pulsar can be large; in preliminary integrations over the pair creation rates, I find the output/cap to be

$$\dot{N}_\pm/\text{cap} \sim 10^{36} \, \mu_{30}^{5/3} \, P^{-7/6} \, (10 \text{ km}/R_*)^{7/3} \, s^{-1} \qquad (13)$$

assuming the opening angle of the polar field lines to be that of the force free scenario. For the Crab pulsar, this leads to a total pair emission $\sim 10^{39}$ sec^{-1} and a total number of pairs stored in the nebula $\sim 3 \times 10^{49}$, close to the number of synchrotron emitting particles which give rise to the nebular radio emission. Because these particles are relativistic, an upper limit for the flux at Earth of annihilation line photons formed by interaction of the positrons with the thermal plasma in the filaments of the nebula is $\sim 10^{-7}$ 511 keV photons/cm^2 sec. If the filaments are magnetically isolated from the surrounding plasma, the flux would be still lower.

In the extreme case possible, a pulsar with a 1 msec period and a very strong dipole moment $\mu_{30} \sim 10$, the number of pairs created might be as high as 10^{42}/s and 5×10^{48} pairs/event, with the event being the initial spin down of the pulsar lasting ~ 1 month. Total pair creation rates and variability time scales of this sort are reminiscent of recent annihilation line observations of the galactic center[101,102]. However, one needs to be able to hide the nebular emission of the expected analog of the Crab nebula, or explain why such an analog does not occur, in addition to explaining why ultrafast pulsars with strong magnetic fields are forming at a large rate in the galactic center and not elsewhere. Therefore, I think pulsars are probably a less likely candidate for the source of the galactic center pairs than a disk around a massive black hole, but their improbability is not as great as previous estimates[103] might suggest. The details of these calculations will be submitted for publication shortly.

(b) $\underline{\Omega}_* \cdot \underline{\mu} < 0$

If ions are freely available, the electrodynamics of the starvation zone below the pair formation front and of the slot gap is essentially the same as in the electron zone case, but the formation of pairs must go through more complicated channels[88]. The ions themselves do not emit convertible gamma rays, but they can trigger the creation of a low density positron beam which then plays the same role as the primary electron beam in the $i < 90°$ case. One way this can happen is through pair creation in the Coulomb field of a relativistic ion, as Doppler shifted thermal photons interact with the virtual photons of the Coulomb field[97]. The resulting low density positron beam accelerates in the same starvation zone potential as in the electron zone case (except the sign is reversed), and emits a sufficient number of curvature photons to form a well defined pair formation front. The number of pairs per positron is then the same as in the electron beam case, but since the positron beam has much lower density, the total amount of pair plasma created is ~ 3 to 4 orders of magnitude smaller than in the i < 90° case.

In addition, the bombardment of the polar cap by the electrons formed at the pair formation front may play an essential role in the model's self-consistency. These particles have a density ~ 10^{-3} n_R/Ze ~ 10^9 cm^{-3} just above the surface, with TeV energy/particle. Upon entering the dense crust (mass density^{104} ~ 10^4 - 10^5 g/cm^3), these cause old-fashioned, bremsstrahlung/Coulomb pair creation showers in the crust[87,105]. Because of the very high density, however, the Landau-Pomeranchuk-Migdal effect[106] suppresses the emission of low energy photons, causing the shower to degrade at much larger column densities than is the case for atmospheric showers, with substantial numbers of particles with energies below 1 GeV appearing only at depths in excess of 1000 gm/cm^3. Zachary (to be published) has constructed detailed solutions for the one dimensional shower equations incorporating this effect. In contrast to the models constructed by Jones[105,107], who assumed the showers propagate as if in air, these showers mainly cause the release of protons, alpha particles and neutrons (which yield more protons) in numbers which may be sufficient to supply the assumed space charge limited ion beam, even if the iron nuclei are strongly bound. Then the main difference between otherwise identical pulsars with i < 90° and i > 90° would be a reduction of the asymptotic pair density by 3 or 4 orders of magnitude in going from the acute case to the obtuse case, a difference which may appear in the morphological grouping of polarized pulsar waveforms into two distinct classes[55]. Work on this issue is in progress.

3. BOUNDARY LAYER STRUCTURE AND GLOBAL THEORY

Near the edge of the pair plasma, the height of the pair formation front approaches infinity. Since the number of pairs created per emitting beam particle comes from integrating the pair creation rate over all photon energies and from H to the point where photons emitted at the upper cutoff see the sky as optically thin, κ must decline to zero as the inner edge of the slot gap is approached. For curvature emission alone, this yields κ(field line coordinate) as shown in Figure 2--the gradient is very steep, with the plasma density dropping by many decades in a layer whose thickness is less than a tenth of the overall width of the polar flux tube. If the emissivity is confined to this boundary layer, and in addition (i) all transfer effects are absent, (ii) the emission all comes from one radius at each frequency, and (iii) the magnetic structure were completely static with respect to the star from one neutron star day to the next, every pulse component in a waveform would have this small angular width, and all subpulses would fill their components and always arrive at the same rotation phase. In fact, all of the assumptions (i)-(iii) are probably wrong, as I will now argue, and relaxing each of the

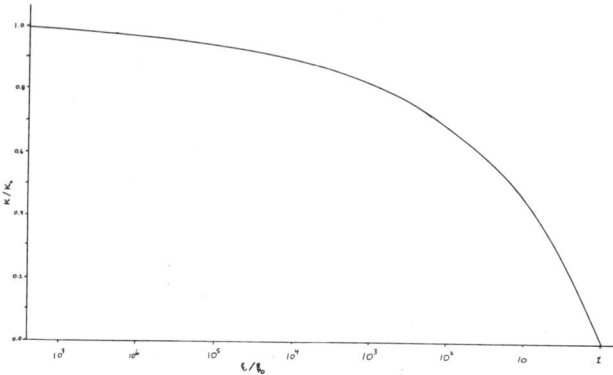

Figure 2. Transverse density profile of $e^{\pm}$ plasma injected into the asymptotic flow tube, curvature radiation only. κ is the number of pairs per primary beam particle which can radiate gamma rays, and κ_c is the maximum number of pairs per primary given by (10). $\xi = [r_\perp^{(g)} - r_\perp]/[r_\perp^{(c)} - r_\perp]$, where $r_\perp$ is the distance across B from the axis of the flow, $r_\perp^{(g,c)}$ are the transverse distance of the outer edge of the pair plasma and the edge of the polar flux tube from the flow axis, respectively; $r_\perp^{(g)}$ is the inner edge of the slot gap. $\xi_o = 0.2[\varepsilon_{min}(R_*)/\varepsilon_c(H_{min})]^{13/5} \ll 1$ sets the scale for the density gradient.

assumptions leads (qualitatively) to subpulse and waveform component structure more like the observed morphologies.

The pair creation calculations tell us something about what is the <u>injected</u> flux of pairs, but the actual flow and current structure depends on the relation of this zone to the rest of the magnetosphere. In particular, the boundary layers of the pair plasma are those in direct contact with regions of the outer magnetosphere which may have electrical properties far from the force free scenario of many previous global studies. In addition, the local model of the polar flux tube assumes a total field-aligned current $I \geqslant 2 \times 10^{11}$ $(B_*/1 \text{ TG})(1 \text{ sec}/P)$ amperes; equality assumes the force-free polar cap size is correct. If no return current flows somewhere else, the star charges up to a value very different from the corotation charge $q \sim \Omega_* \mu/c$ present in the interior[38,108,109], and the model breaks down within a single pulsar day. My remarks on this topic are speculative, based on the work in progress on the model illustrated in Figure 3 by F. Coroniti and myself.

The basic idea behind this scheme comes from the fact that conduction currents will not flow in the magnetosphere unless mechanical stress is exerted upon the magnetic field somewhere, and the observation[34] that the polar cap pair creation occurs only along

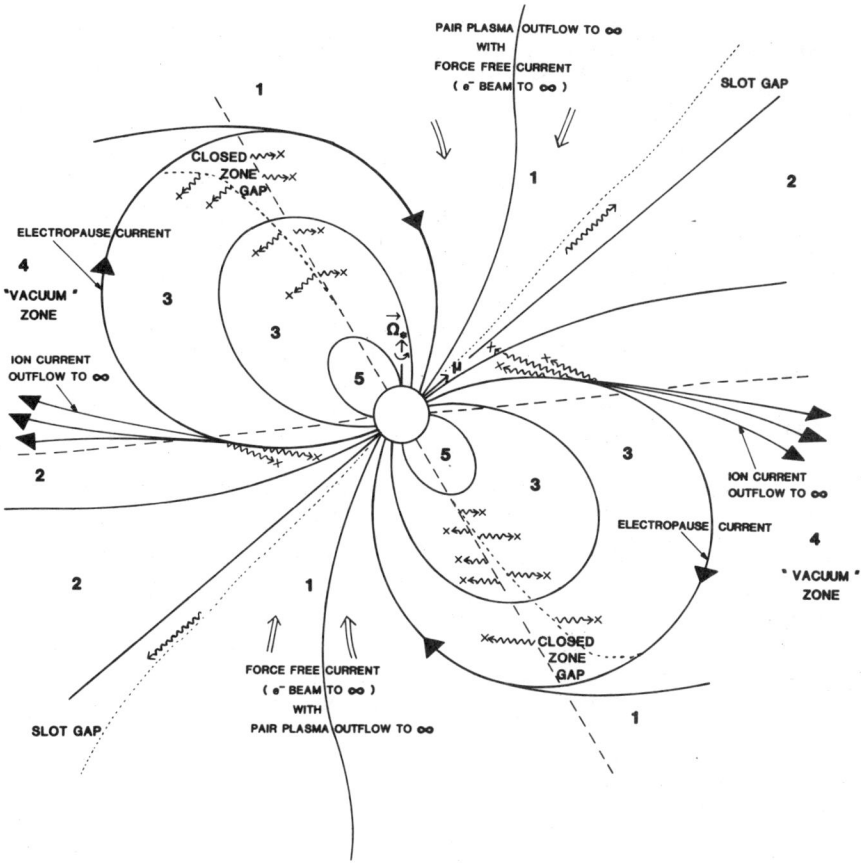

Figure 3. A cartoon of a proposed model of a radio pulsar's magnetosphere. Electric field lines (not shown) extend from the star and from the ion flow to the electropause layer and to the effectively negative charge density of the vacuum zone. The ions form a "neutral sheet" current which reverses the direction of the magnetic field in the asymptotic stellar wind.

the favorably curved field lines. The unfavorably curved polar flux tube cannot develop a unidirectional, ultrarelativistic beam flow but instead tries to fill with an electrically trapped atmosphere, at least out to the null surface where $\Omega_* \cdot B$ goes through zero[48,79]. Beyond this surface, a magnetosphere which can only move charges along B is unstable to forming a vacuum zone, if the only supply of plasma is from the surface and pair creation is absent[47]. In the geometry illustrated, the polar zone responds to the local vacuum demand by supplying electrons, while the equatorial regions at large radius demand ions. Without pair creation, this inconsistency is the essential problem in globally force free theories.

When the voltages are sufficiently large, pair creation can change this problem into a virtue, and (we hope) lead to a

self-consistent theory. In the closed region 3, a local quasi-vacuum
discharge region (which can operate in steady state) can be
constructed near one of the null surfaces, which fills the closed
region with pair plasma and allows this zone to approximately
corotate. If the rotational equator zones are the vacuum regions
shown as region 4 with total density not greater than $\Omega_*B/2\pi c e$, the
dense pair plasma in the corotating region shields itself by forming a
non-neutral surface charge (and magnetopause current) layer, in
response to the field aligned EMF which forms because the magnitude of
the jump in the perpendicular electric field varies along the
boundary. This yields a dense layer of electrons precipitating onto
the outer edge of the unfavorably curved part of each polar cap,
forming a true "auroral zone." The analogy to the terrestrial
magnetosphere is not only geometric. The density of the preciptating
electrons is large compared to $|n_R/e|$ just above the cap, which
reverses the magnetospheric electric field along B from what it would
have been in the absence of pair creation. Since the freely emitting
surface always adjusts to give a local charge density very close to
n_R, this results in the extraction of a <u>dense</u> ion beam which
counterstreams up through the precipitating electron flow, just as
occurs in the terrestrial auroral zones[110]. Similarly, the positrons
accelerated in the equatorial boundary layer precipitate onto the
outer edge of the favorably curved part of the polar cap, causing the
electron current density in this outer region to be much larger than
B/P (although since the layer is thin, the total current is not much
different than before including these global effects), with the polar
cap potential still the same as in the unidirectional flow theory,
since the charge density is unchanged. The current flow at the polar
cap is shown in Figure 4.

Upon reaching the altitude where the ion flow crosses the null
surface, the "vacuum" electric field (actually another example of a
starvation electric field) accelerates the ions to energies such that
their energy density exceeds the magnetic energy density. This "tears
open" the magnetic field lines (more precisely, the inertial drift
currents switch the closed field lines into open ones). It is
possible to show that if the jump in the electric field across the
equatorial electropause shown in Figure 3 is itself on the order of
the corotation electric field, and if the sign of the variation of the
electric field in the starvation zone outside the electropause indeed
does lead to the formation of the precipitation described, then the
total current from the star is zero, at least at the level of
<u>estimates</u> of the area occupied by these flows at the polar caps. Our
present efforts are concentrated upon approximate models of the global
electric field including these boundary layers, as well as the
quantitative determinations of the boundary layer structures which can
be done with the use of the injection rates given by the application
of the cascade theory to the polar cap flows and to the pair creation
zones in the closed region shown in Figure 3.

POSSIBLE RELATIONS TO OBSERVATIONS

While these complex plasma electrodynamical stories make for

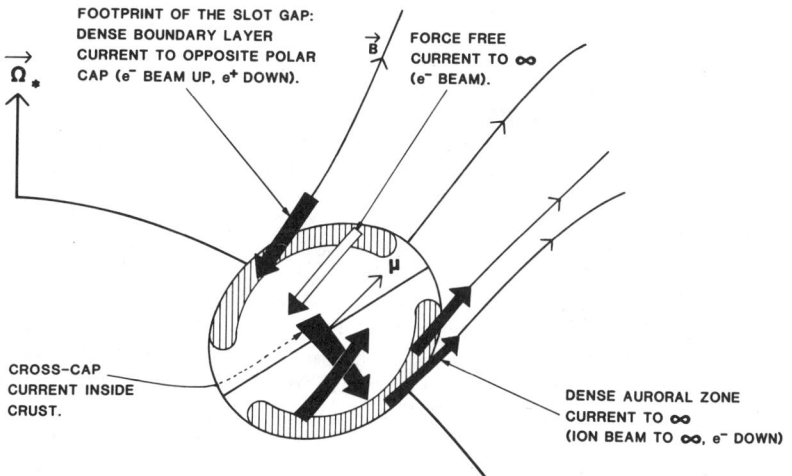

Figure 4. Current flow system at one polar cap in the proposed global theory. The footprint of the slot gap which connects to infinity lies along the boundary separating the two halves of the polar cap, on the side closest to the rotation axis (not shown).

entertaining activities by small numbers of theorists, one would obviously like to know what, if anything, they have to do with the observed world. My opinion is, quite a bit but most of the quantitative work on experimental tests of these schemes is still waiting for the completion of even the zeroth order models of the magnetospheric configuration.

The basic energetics of the magnetosphere are crudely sampled by timing measurements and by observations of the Crab nebula and possibly of other plerionic supernova remnants. The most sophisticated measurement of a pulsar's torque was done by Groth[111], who showed that the spin down of the Crab pulsar can be represented as the response to a magnetospheric torque proportional to Ω_*^n, with n = 2.512. Later work at Arecibo[112] gave n = 2.516, and both sets of observations suggest that in addition there is a fluctuating component of the torque whose individual "random" shots are not resolved by the timing data (actually, all that is measured is torque divided by moment of inertia, so that fluctuations in the star are also a possible source of noise in the spindown). In the polar cap current flow shown in Figure 4, the mechanical stress on the stellar crust is easily shown to lead to a spindown torque

$$T_\parallel \cong \frac{\mu^2 \Omega_* \sin i \cos i}{cR_*^2} \theta_c^3 \qquad (14)$$

If one adopts the force free opening angle $\theta_c = (\Omega_* R_*/c)^{1/2}$, one immediately finds[114] a torque proportional to $\Omega_*^{5/2}$, which at least suggests we are doing something right. However, the model which leads to this cannot be force free, as can be immediately observed by comparing the Poynting flux at the light cylinder to which the last closed field line extends in a force free model to the energy loss implied by (14)--they disagree by a factor of θ_c^{-1}. This is simply a direct indication that the balance of forces which determines where our electropause forms in Figure 3 must be well inside the light cylinder. Indeed, demanding that the Poynting flux leaving the star must be on the order of the energy loss implied by (14) leads to a polar cap opening angle $\theta_c \cong (\Omega_* R_*/c)^{1/3}$ and a torque proportional to Ω_*^2. If indeed this is self consistent, the larger braking index of the Crab pulsar indicates the presence of evolutionary effects such as decay of the dipole component or alignment (or indeed counter alignment) between the rotation axis and the dipole axis. It also implies that the dipole fields of pulsars are quite a bit weaker than is obtained with the usual formula advocated in all previous models, vacuum or not[15-16], and pushes us even more toward models with surface magnetic complexity.

These dense current flows also suggest the possibility that the large scale structure of the magnetosphere may be variable on time scales on the order of, or somewhat greater than the rotation period. The ion currents in Figure 3 correspond to a neutral sheet, across which B reverses, a configuration known to lead to "stormy" behavior in the terrestrial magnetosphere. In addition, a plasma with density $\sim n_R/e$ propagating in "vacuum wave" fields is known to lead to strong instabilities[114-118], with growth rates $\sim \omega_{wave} \sim \Omega_*$. If the currents fluctuate by amounts on the order of J itself in times on the order of P, one can easily show that the fluctuations in the spindown of many pulsars[32] can be explained as fluctuations in the conduction current torque[34]. It is also easy to estimate that the direction of the polar flux tube with respect to a fixed vector stuck on the star varies by amounts which can easily explain the variable phase observed for the location of the subpulses with respect to the average waveform (which is fixed with respect to the star, by construction), if the altitude of emission exceeds 500-1000 km[34]. The variable intensity of the pulses also then may be explicable by the nodding of an inhomogeneous beam of radiation which is time steady with respect to the magnetic field, but the magnetic field makes 5-10° excursions around an average direction from pulse to pulse. If, on top of this pulsar weather, there is a slow change in the climate such as alignment which slowly moves us out of the pulsar beam as the object ages, the existence of pulsar nulling without any substantial change in the star's average luminosity when the pulses can be seen is now comprehensible as the greater likelihood that we can see pulses only during the largest excursions of the beaming flux tube from its average location, as the star gets old. This suggests alignment might be a very interesting

explanation indeed of the decay of pulsar torques[29] with age and also
suggests that in a few cases in which we can see almost pole on with
rather broad pulses nulling would not be present, assuming the polar
flux tube is the emission site. Of course, the boundary layers
littered all over Figure 3 may also be interesting candidates for
emission; it is too early in the investigation of this model to be
sure that boundary layer emission from the polar flux tube is the only
candidate. It is thinkable that the most extreme exotica in pulsar
beaming morphology, the marching subpulses of PSR 0809+74 and the mode
switching between different average waveforms, may be attributable to
this sort of global weather in the magnetosphere, the first to there
being an oscillatory behavior to the current flow pattern and the
second to there being at least two different, fairly stable current
flow patterns, with rapid switching possible between them. Until we
can get a better feeling for the time-steady structure (as seen by a
corotating observer), these intriguing possibilities will remain
qualitative speculations, but I am optimistic that these long period
phenomena will turn out to be attributable to the global structure of
the current flow, rather than to the rather artificial (to me)
attempts to pin them onto time dependent, local polar cap dynamics
which has intrinsic time scales of microseconds, as in the most recent
example by Filippenko and Radhakrishnan[119].

The global scheme outlined above has some very interesting
implications for the formation of relativistic winds from these
objects. In the sectors over the rotation poles, the wind is a
relativistic, approximately force free zone of circularly polarized
electromagnetic field and relatively low energy particles. In the
rotation equator, one has high energy ions and possibly some electrons
and positrons drawn in by acceleration from the boundary layers, with
a total density not greater than the corotation charge density and
rough equipartition between fields and particles, at least near the
electropause. As mentioned above, present evidence suggests this
structure is unstable to transferring essentially all of the field
structure into particle energy, indicating a particle-dominated wind
in the equatorial sector. Kennel and Coroniti[19] have shown that just
such a particle dominated wind can lead to an adequate explanation of
the X-rays in the Crab nebula, if one assumes the relativistic wind is
thermalized in a relativistic shock wave at the radius where the wind
ram pressure balances the nebular pressure[17]. The sector structure of
the outflow described here is consistent with the inner morphology of
the nebula, where the hardest X-rays form a torus around the pulsar in
the plane of the nebula's minor axis[25,120], while the radio emission
peaks at roughly the same angular distance along the long axis of the
nebula[121]. Since the linear polarization of the optical emission lies
along the nebula's long axis, one can place the rotation axis of the
pulsar parallel to the long axis with the spun off magnetic field
toroidally wound around this axis, the particle dominated wind then
lying in the sector occupied by the hard X-ray emission torus, and the
field dominated wind flowing out along the long axis of the nebula
where the radio emission, which comes from lower energy pairs, is
strongest in the inner regions. It is of course likely that the
particle dominated wind has some residual magnetic sector structure,

since the magnetic field is here wound back in the form of a linearly polarized wave propagating with speed just under $c^{38,122}$ with a wavelength on the order of the electropause radius, perhaps as small as 1000 km. The acceleration of high energy particles with Larmor radii well in excess of this wavelength is then <u>not</u> a direct extension of the nonrelativistic theory of shock acceleration[123], since the particles are exposed to a time variable wave electric field even though the phase velocity is subluminous. Particle acceleration problems of this sort have not even been formulated, much less solved.

A final feature of the global scheme shown in Figure 3 which we hope can lead to observational tests is that it is littered with precipitating particle streams, all of which should lead to soft X-ray emission and possibly to excitation of the cyclotron line and even of the redshifted 511 keV annihilation line from the surface. These are hard to excite with the weak TeV streams incident on the surface in models so far published[95,96,98,107] where the positrons, if any, bury themselves in the crust and lose energy in the electromagnetic showers rather than give observable line emission. However, in the precipitation flows outlined above, the densities are much higher and we suspect the energies per particle are much lower than in the simple beams of the local models of polar caps now in the literature. This puts a strong emphasis on hard X-ray, soft gamma ray observations of pulsars, a spectral region where our observational knowledge is particularly weak.

I have expressed my opinion previously on what I suspect to be the dominant physics controlling the radio emission[67]; detailed research papers are in preparation. In these high density models, refraction and non-linear mode coupling play a big role in both the emergent polarization and the structure of an individual subpulse (Barnard and Arons, Arons and Barnard, in preparation); this should be contrasted to the negligible importance of such processes in the low density models considered by Stinebring[54], for example. In addition, circular polarization and the formation of notches in the beam profile[56,57] may be created by cyclotron scattering, not absorption (Arons and Barnard, in preparation), which has interesting implications for the far infrared emission from pulsars. In the high density models, microstructure is very likely to be the consequence of non-linear self modulation of the high intensity radiation beam (Arons and Konigl, in preparation), which leads to the temporal modulation of the envelope of an otherwise steady radiation beam and suggests the polarization state of a micropulse should correlate with the intensity[59]. The strongest emission process in these models seems to lie in the electromagnetic form of the non-neutral beam, slipping stream instability (ref. 124 and Arons, in preparation), since this is driven by the powerful voltage difference between dense pair plasma and a neighboring vacuum zone (the boundary between the slot gap and the pair plasma in the polar column in Figure 3, or possibly the boundary between the closed zone and the equatorial vacuum zone, although here there may be more difficulty in forming a narrow pulse window, if the magnetosphere fluctuates as outlined above). In contrast, models based on a two stream instability downstream from the pair formation front, perhaps best typified by the intriguing idea of Cheng and Ruderman[125], suffer from strong suppression due to the large momentum dispersion in the pair plasma[38,67,126] and a lack of sufficient efficiency in converting the energy in the pairs into emergent radio photons[127,128] even when the cold plasma approximation is used. The relative shear in the boundary layer can tap the essentially infinite energy reservoir of the neighboring starvation field, and can produce both orthogonal modes of polarization with effectively broad band emission in the steep density gradient of the boundary layer. Whether it is efficient enough is not yet known. Finally, there is a real possibility that incoherent inverse Compton scattering of backscattered radio photons can give rise to interesting amounts of optical, X-ray and gamma ray emission in objects with the densest pair plasmas (ref. 67 and in preparation), leading to an interesting new scheme for the high frequency emission from the Crab and Vela pulsars.

In conclusion, one can see that even though no direct evidence for electron positron pairs in pulsars' magnetospheres has yet been found at more than the 3σ level, their presence is strongly implicated by the class of dynamical models extending back to the pioneering work of Sturrock[49,50] and of Ruderman and Sutherland[77] and there may be some chance of finding them directly by improved hard X-ray, soft γ-ray searches for the annihilation line. They certainly do provide a lot of possibilities for the solution of many of the outstanding problems of pulsar physics.

I am indebted to J. Barnard, F. Coroniti, W. Fawley, A. Konigl, E.T. Scharlemann, D.F. Smith and A. Zachary for our enjoyable collaborations on pulsars, and to C.F. Kennel, C.E. Max, L. Mestel and R. Pellat for informative discussion. My reseach on pulsars is supported by NSF Grants AST 79-23243 and AST 82-15456 and by the taxpayers of California.

REFERENCES

1. R.N. Manchester and J.H. Taylor, Pulsars (W. H. Freeman, San Francisco, 1977)
2. J.E. Pringle and M.J. Rees, Astron. Astrophys., 21 1 (1972)
3. K. Davidson and J.P. Ostriker, Astrophys. J., 179, 585 (1973)
4. F.K. Lamb, C.J. Pethick and D. Pines, Astrophys. J., 184, 271 (1973)
5. J. Arons and S.M. Lea, Astrophys. J., 207, 914 (1976)
6. J. Arons and S.M. Lea, Astrophys. J., 210, 792 (1976)
7. P. Ghosh and F.K. Lamb, Astrophys. J., 232, 259 (1979)
8. D.J. Burnard, S.M. Lea and J. Arons, Astrophys. J., 266, 175 (1983)
9. J.H. Piddington, Aust. J. Phys., 10, 530 (1957)
10. F. Pacini, Nature, 216, 567 (1967)
11. T. Gold, Nature, 218, 731 (1968); 221, 25 (1969)
12. J.P. Ostriker and J.E. Gunn, Astrophys. J., 157, 1395 (1969)
13. F.C. Michel, Astrophys. J., 158, 727 (1969)
14. P. Goldreich and W.H. Julian, Astrophys. J., 157, 869 (1969)

15. P. Goldreich, F. Pacchi and M.J. Rees, Comm. Astrophys. and Space Phys., **3**, 185 (1972)
16. J. Arons, in IAU Symp. No. 94, Origin of Cosmic Rays, G. Setti, G. Spada and A.W. Wolfendale, eds. (D. Reidel, Dordrecht, 1981), 175
17. M.J. Rees and J.E. Gunn, Mon. Not. Roy. Astron. Soc., **167**, 1 (1974)
18. J. Arons, Space Sci. Rev., **24**, 437 (1979)
19. C.F. Kennel and F.V. Coroniti, preprint (1983)
20. F.C. Michel and A. Dessler, Astrophys. J., **251**, 654 (1981)
21. F.C. Michel, Rev. Mod. Phys., **54**, 1 (1982)
22. M. Leventhal, C.J. MacCallum and A. Watts, Astrophys. J., **216**, 491 (1977)
23. J.D. Scargle, Astrophys. J., **156**, 401 (1969)
24. G.R. Ricker, S.G. Scheepmaker, J.E. Ryckman, J.E. Ballantine, J.P. Doty, P.M. Downey and W.H.G. Lewin, Astrophys. J. (Letters), **197**, L83 (1975)
25. M. Oda, in Proc. International Course and Workshop on Plasma Astrophysics, T.D. Guyenne, ed., ESA SP-161 (European Space Agency, Paris, 1981), 251.
26. G. Clark, Physics Today, **35**, 26 (1982)
27. D.C. Backer, S. Kulkarni, C. Heiles, M.M. Davis, and W.M. Goss, Nature, **300**, 615 (1982)
28. J.E. Gunn and J.P. Ostriker, Astrophys. J., **160**, 979 (1970)
29. A.G. Lyne, R.T. Ritchings and F.G. Smith, Mon. Not. Roy. Astron. Soc., **171**, 579 (1975)
30. F.S. Fujimura and C.F. Kennel, Astrophys. J., **236**, 245 (1980)
31. E.S. Phinney and R.D. Blandford, Mon. Not. Roy. Astron. Soc., **194**, 137 (1981)
32. J. Cordes and D.J. Helfand, Astrophys. J., **239**, 640 (1980)
33. J. Cordes and G. Greenstein, Astrophys. J., **245**, 1060 (1981)
34. J. Arons, in IAU Symp. No. 95, Pulsars, W. Sieber and R. Wielebinski, eds. (D. Reidel, Dordrecht, 1981), 69
35. A. Alpar, A.F. Cheng, M.A. Ruderman, J. Shaham, Nature, **300**, 728 (1982)
36. J. Arons, Nature, in press (1983)
37. D.C. Backer, S. Kulkarni and J.H. Taylor, Nature, **301**, 314 (1983)
38. J. Arons, Space Sci. Rev., **24**, 437 (1979)
39. J. Trumper, W. Pietsch, C. Reppin, W. Voges, R. Staubert, and E. Kendziorra, Astrophys. J. (Letters), **219**, L105. (1978)
40. W. Voges, W. Pietsch, C. Reppin, J. Trumper, E. Kendziorra and R. Staubert, Astrophys. J., **263**, 803 (1982)
41. N.E. White, J.H. Swank and S. Holt, Astrophys. J., in press (1983)
42. C.E. Max and F. Perkins, Phys. Rev. Lett., **27**, 1342 (1971)
43. C.E. Max, Phys. Fluids, **16**, 1277 (1973)
44. C.F. Kennel, G. Schmidt, and T. Wilcox, Phys. Rev. Lett., **31**, 1364
45. C.F. Kennel and R. Pellat, J. Plasma Phys., **15**, 335.
46. E. Asseo, C.F. Kennel and R. Pellat, Astron. and Astrophys., **65**, 401 (1978)
47. N. Holloway, Nature Phys. Sci., **246**, 6 (1973)

48. E.T. Scharlemann, J. Arons and W.M. Fawley, Astrophys. J., 222, 297 (1978)
49. P.A. Sturrock, Nature, 227, 465 (1970)
50. P.A. Sturrock, Astrophys. J., 164, 529 (1971)
51. J.H. Taylor, R.N. Manchester and G.R. Huguenin, Astrophys, J., 195, 513 (1975)
52. R.N. Manchester, J.H. Taylor and G.R. Huguenin, Astrophys. J., 196, 83 (1975)
53. D.C. Backer and J. Rankin, Astrophys. J. (Supplement), 42, 143 (1980)
54. D. Stinebring, Ph.D. Dissertation, Cornell University (1982)
55. J. Rankin, Astrophys. J., in press (1983)
56. N. Bartel, in IAU Symp. No. 95, Pulsars, W. Sieber and R. Wielebinski, eds. (D. Reidel, Dordrecht, 1981), 171
57. N. Bartel, Astron. and Astrophys., 97, 384 (1981)
58. J. Cordes, personal communication (1983)
59. J. Cordes and T. Hankins, Astrophys. J., 218, 484 (1977)
60. B.J. Rickett, Astrophys. J., 79, 8 (1975)
61. J. Cordes, Astrophys. J., 208, 944 (1976)
62. J. Cordes, Astrophys. J., 210, 780 (1976)
63. T.H. Hankins and V. Boriakoff, Nature, 276, 45.
64. V. Radhakrishnan and D.J. Cooke, Astrophys. Lett., 3, 225 (1969)
65. D.C. Ferguson, in IAU Symp. No. 95, Pulsars, W. Sieber and R. Wielebinski, eds. (D. Reidel, Dordrecht, 1981), 141.
66. R. Narayan and M. Vivekanand, Astron. and Astrophys., 113, L3 (1982)
67. J. Arons, in Proc. International Workshop and Summer School on Plasma Astrophysics, T.D. Guyenne, ed., ESA SP-161 (European Space Agency, Paris, 1981), 273
68. J. Barnard and J. Arons, Bull. Amer. Astron. Soc., 13, 851 (1981)
69. R. Buccheri, in Proc. Workshop on Electron-Positron Pairs in Astrophysics, M. Burns, A.K. Harding and R. Ramaty, eds. (American Institute of Physics, New York, 1983), in press.
70. P.N. Bhat, S.K. Gupta, P.V. Ramana Murthy, B.V. Sreekontan, S.C. Tonwar and P.R. Viswonath, in Proc. IAU Symp. No. 95, Pulsars, W. Sieber and R. Wielebinski, eds. (D. Reidel, Dordrecht, 1981), 251
71. Peterson, B.A. Murdin, P.G., Wallace, P.T., Manchester, R.N., Penny, A.J., Jordan, A., Hartley, K.F. and King, D., Nature, 276, 475 (1978)
72. F.D. Seward and F.R. Harnden, Astrophys. J. (Letters), 256, L45 (1982)
73. R.N. Manchester, I.R. Tuohy and N. D'Amico, Astrophys. J. (Letter), 262, L31 (1982)
74. J.H. Taylor, in IAU Symp. No. 95, Pulsars, W. Sieber and R. Wielibinski, eds. (D. Reidel, Dordrecht, 1981), 361
75. D.Q. Lamb and F.K. Lamb, Astrophys. J., 220, 291 (1978)
76. J. Arons and J. Barnard, J. Astrophys. and Astron., in press (1983)
77. M.A. Ruderman and P.G. Sutherland, Astrophys. J., 196, 51 (1975)
78. C.S. Shukre and V. Radhakrishnan, Astrophys. J., 258, 121 (1982)
79. W.M. Fawley, Ph.D. Dissertation, University of California, Berkeley (1978)

80. D.J. Helfand, in Proc. IAU Symp. No. 101, Supernova Remnants and their X-Ray Emission, P.G. Gorenstein and I.J. Danziger, eds. (D. Reidel, Dordrecht, 1982), in press
81. A.J. Deutsch, Ann. d'Astrophys., 18, 1
82. M.A. Ruderman, Ann. Rev. Astron. and Astrophys., 10, 427 (1972) and references therein
83. L. Mestel, Nature Phys. Sci., 233, 149 (1971)
84. M.A. Ruderman, Phys. Rev. Lett., 27, 1306 (1971)
85. E.G. Flowers, J.F. Lee, M.A. Ruderman, P.G. Sutherland, W. Hillebrandt and E. Müller, Astrophys. J., 215, 291 (1977)
86. R.H. Fowler, Statistical Mechanics (Cambridge U. Press, Cambridge, 1966), 356-357
87. P.B. Jones, Mon. Not. Roy. Astron. Soc., 184, 207 (1978)
88. J. Arons, Astrophys. J., 266, 215 (1983)
89. W.M. Fawley, J. Arons and E.T. Scharlemann, Astrophys. J., 217, 227 (1977)
90. D.F. Smith, L. Muth and J. Arons, in Proc. International Summer School and Workshop on Plasma Astrophysics, T.D. Guyenne, ed., ESA SP-161 (European Space Agency, Paris, 1981), 333
91. M.L. Burns and R.V.E. Lovelace, Astrophys. J., 262, 87 (1982)
92. A. Lightman, in Proc. Workshop on Electron-Positron Pairs in Astrophysics, M.L. Burns, A.K. Harding and R. Ramaty, eds. (American Institute of Physics, New York, 1983), in press
93. W. Tsai and T. Erber, Phys. Rev., D10, 492 (1974)
94. J.K. Daugherty and A.K. Harding, Astrophys. J., in press (1983)
95. J. Arons, Astrophys. J., 248, 1099 (1981)
96. J. Arons and E.T. Scharlemann, Astrophys. J., 231, 854 (1979)
97. A.F. Cheng and M.A. Ruderman, Astrophys. J., 214, 598 (1977)
98. A.F. Cheng and M.A. Ruderman, Astrophys. J., 235, 576 (1980)
99. J. Barnard and J. Arons, Astrophys. J., 254, 713 (1982)
100. E. Tademaru, Astrophys. J., 183, 625 (1973)
101. M. Leventhal, C.J. MacCallum, A. Huber and P.S. Stang, Astrophys. J., 240, 338 (1980)
102. G.R. Riegler, J.C. Lang, W.A. Mahoney, W.A. Wheaton, J.B. Willett and A.S. Jacobsen, Astrophys. J. (Letters), 248, L13 (1981)
103. P.A. Sturrock and K. Baker, Astrophys. J., 234, 612 (1979)
104. M.A. Ruderman, in IAU Symp. No. 53, Physics of Dense Matter, C.J. Hanson, ed. (D. Reidel, Dordrecht, 1974), 117
105. P.B. Jones, Astrophys. J., 228, 536 (1979)
106. A.B. Migdal, Phys. Rev., 103, 1811 (1956); Zh. Eksp. Teor. Fiz., 32, 633 (1957), [Sov. Phys.-JETP, 5, 527]. See also T. Stanev, Ch. Varnkov, R.E. Streitmatter, R.W. Ellsworth and T. Bowen, Phys. Rev., D25, 1291 (1982)
107. P.B. Jones, Mon. Not. Roy. Astron. Soc., 197, 1103 (1981)
108. E.A. Jackson, Astrophys. J., 206, 831 (1976)
109. F.C. Michel, Astrophys. J., 227, 579 (1979)
110. J.M. Cornwall and M. Schulz, in Solar System Plasma Physics, L.J. Lanzerotti, C.F. Kennel and E.N. Parker, eds. (North Holland, Amsterdam, 1979), 3, 168-175
111. E.J. Groth, Astrophys. J. (Suppl.), 29, 431 (1975)
112. G.E. Gullahorn and J.M. Rankin, Astron J., 83, 1219 (1978)

113. H. Heintzmann and E. Schrufer, Astron. and Astrophys., 111, L4 (1982)
114. C.E. Max and F.W. Perkins, Phys. Rev. Lett., 29, 1731 (1972)
115. C.E. Max, Phys. Fluids, 16, 1480 (1973)
116. J. Arons, C.A. Norman and C.E. Max, Phys. Fluids, 20, 1302 (1977)
117. F.J. Romeiras, J. Plasma Phys., 22, 201
118. J.N. Leboeuf, M. Ashour-Abdalla, T. Tajima, C.F. Kennel, F.V. Coroniti and J.M. Dawson, Phys. Rev., A25, 1023 (1982)
119. A. Filippenko and V. Radhakrishnan, Astrophys. J., 263, 828 (1982)
120. B. Aschenbach and W. Brinkman, Astron. and Astrophys., 41, 147 (1975)
121. E. Swinbank and G. Pooley, Mon. Not. Roy. Astron. Soc., 186, 775 (1979)
122. F.C. Michel, Comm. Astrophys. and Space Phys., 3, 80 (1971)
123. W.I. Axford, in Proc. International Workshop and Summer School on Plasma Astrophysics, T.D. Guyenne, ed., ESA SP-161 (European Space Agency, Paris, 1981), 425
124. J. Arons and D.F. Smith, Astrophys. J., 229, 728 (1979)
125. A.F. Cheng and M.A. Ruderman, Astrophys. J., 212, 800 (1977)
126. R. Buschauer and G. Benford, Mon. Not. Roy. Astron. Soc., 177, 99 (1977)
127. E. Asseo, R. Pellat and M. Rosado, Astrophys. J., 239, 661 (1981)
128. E. Asseo, R. Pellat and H. Sol, Astrophys. J., 266, 201 (1983)

PAIR PRODUCTION NEAR THRESHOLD IN PULSAR MAGNETIC FIELDS

A. K. Harding
NASA/Goddard Space Flight Center, Greenbelt, MD 20771
and
J. K. Daugherty
University of North Carolina, Asheville, NC 28804

ABSTRACT

In pulsar polar cap models, curvature radiation γ-rays produce e^+e^- pairs in the strong magnetic fields near the surface of the neutron star. While these γ-rays have energies $E_\gamma \gg mc^2$, they also propagate at very small angles to the field, such that the threshold condition, $E_\gamma > 2mc^2/\sin\theta$ is just barely satisfied when they pair produce. Threshold effects on the pair production attenuation coefficient, which are due to the discreteness of the e^+e^- Landau states, must therefore be considered when computing the mean free paths of curvature radiation photons in pulsar magnetic fields. These effects, which are not incorporated in the asymptotic expression for the attenuation coefficient, have some interesting consequences for pulsar models. Since pair production is suppressed near threshold, the photon mean free paths are longer than previously thought. In magnetic fields $\gtrsim 6 \times 10^{12}$ G, the pairs tend to be produced in the ground state Landau level and will not synchrotron radiate. Since synchrotron radiation is an essential ingredient in the electromagnetic cascades which produce low energy pairs above the acceleration region, pulsars with very high magnetic fields may not produce many pairs.

INTRODUCTION

The production of e^+e^- pairs plays a central role in most current pulsar models. Polar cap voltage drops which accelerate particles to ultrarelativistic energies are limited by pair production discharges,[1,2,3] which continue in the form of electromagnetic cascades above the acceleration region.[4] These pair-photon cascades can generate large numbers of e^+e^- pairs which are thought to be essential for the coherent radio emission process. The most important mechanism for producing pairs near the polar cap is pair production by single photons in the intense magnetic field of the neutron star. Recent theoretical study of this process has provided a description of the behavior of the photon attenuation coefficient near threshold and the energy distribution of the pairs.[5,6] In this paper, we discuss the implication of these results for pair production and electromagnetic cascades in pulsar magnetospheres.

PHOTON MEAN FREE PATHS

Magnetic pair production has been studied extensively, and almost exclusively, in the asymptotic limit of low fields, $B \ll B_{cr}$, and high photon energies $\hbar\omega \gg 2mc^2$, where $B_{cr} \equiv 4.414 \times 10^{13}$ G is the critical field strength. If we define the units $\omega' \equiv \omega/2m$ and $B' \equiv B/B_{cr}$ ($\hbar = c = 1$), then this limit takes the form, $\xi \equiv 2\omega'^2/B' \to \infty$. The parameter ξ is related to the number of energetically allowed Landau states available to the e^+e^- pair. In this limit, the polarization-averaged photon attenuation coefficient in the "center of mass" frame, where the photon propagates perpendicular to B, is[7]

$$R_{CM}(\omega',B') \sim 0.23 \frac{\alpha}{\lambdabar} B' \exp\left(\frac{-4}{3\chi}\right) \text{ cm}^{-1}, \quad \chi \ll 1 \quad (1)$$

where $\chi \equiv \omega'_{CM} B'$. In an arbitrary frame, where the photon propagates at an angle θ to B, its energy can be found by Lorentz transforming along B with velocity, $\beta = \cos\theta$, and the result is $\omega' = \omega'_{CM}/\sin\theta$. Similarly, the attenuation coefficient becomes $R = \sin\theta\, R_{CM}$.

To produce a pair, the photon center of mass energy must exceed threshold energy and the attenuation coefficient must be non-negligible. These two conditions are independent and can be expressed in the following simple forms:

(A) $\omega_{CM} = \omega \sin\theta \geq 2m$

(B) $\chi \gtrsim 0.1$

where the latter comes from the exponential dependence of $R(\omega', B')$. Figure 1 illustrates where these conditions are satisfied for different field strengths.

In pulsar magnetospheres, high energy particles produce curvature radiation photons at very small angles to the field such that $\omega\sin\theta < 2m$ even though $\omega \gg 2m$. In order to pair produce, then, these photons must propagate in a straight path until they acquire an angle to the curved field lines which satisfies (A). In figure 1, this is equivalent to moving upward along the diagonal lines of constant field strength. If (B) is satisfied before (A), as it is for $B \gtrsim 0.1\, B_{cr} = 4.4 \times 10^{12}$ G, then the photon will pair produce very near threshold, where Eqn (1) is no longer accurate. It has been

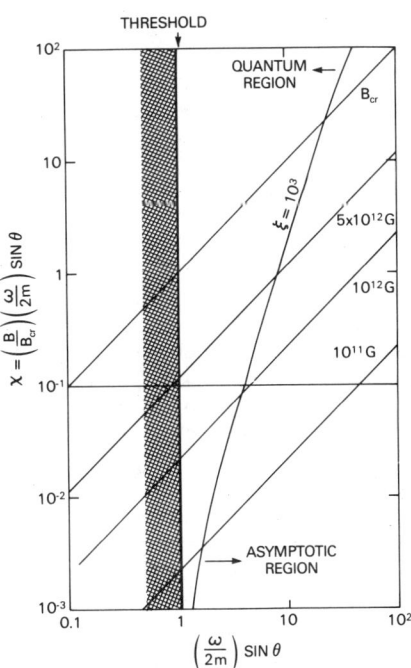

Fig. 1. Pair production parameter χ vs. "center of mass" photon energy for different field strengths(diagonal lines).

found (Ref. 5) that the quantum effects of the discrete pair states suppress the pair production attenuation coefficient below the asymptotic limit, such that a more accurate expression is:

$$R_{CM}(\omega', B') \sim 0.23 \frac{\alpha}{\lambdabar} B' \exp\left[\frac{-4 f (\omega'_{CM}, B')}{3\chi}\right] cm^{-1}, \quad \omega'_{CM} \gtrsim 1$$

$$f(\omega'_{CM}, B') \simeq 1 + .42 \, \omega'_{CM}{}^{-2.7} B'^{-.0038}$$

(2)

The function $f(\omega', B')$ approximates the behavior of the exact attenuation coefficient near threshold, after averaging over the sawtooth pattern of cyclotron resonance spikes (cf. Ref. 6).

Since $\sin\theta$ increases with the photon path length, s, approximately as s/ρ, where ρ is the radius of curvature of the magnetic field, the mean free paths of the photons, from either condition (A) or (B) and Eqn (2) will be

$$\frac{\lambda}{\rho} \simeq \frac{0.1 \, f(\omega'_{CM}, B')}{\omega' B'}, \quad B' \lesssim 0.1 \quad (3a)$$

$$\frac{\lambda}{\rho} \simeq \frac{1}{\omega'}, \quad B' \gtrsim 0.1 \quad (3b)$$

Due to the threshold condition, the mean free path is constant above $\sim 4 \times 10^{12}$ G. Using Eqn (1), the approximate mean free path would be

$$\frac{\lambda_A}{\rho} \simeq \frac{0.1}{\omega' B'}, \quad (4)$$

so that,

$$\frac{\lambda}{\lambda_A} \simeq 1 + 210 \, B'^{2.7}, \quad B' \lesssim 0.1 \quad (5a)$$

$$10 \, B', \quad B' \gtrsim 0.1 \quad (5b)$$

The actual mean free paths of curvature radiation photons are therefore larger than those derived using the well known asymptotic limit and this discrepancy increases with field strength.

The longer mean free paths will cause the voltage drop at the polar cap to be somewhat higher than previously estimated[1,2] although this should not be a large effect. The secondary particle yields of the cascades above the acceleration region will be much more sensitive to changes in the mean free path because the multiplicative effects of large numbers of photons are involved. Longer mean free paths in a magnetic field which falls off with distance from the star should decrease the number of pairs produced per primary particle in these cascades. If proposed effects due to the vacuum index of refraction in a magnetic field are present, λ could be even larger.[8]

DISCRETE e^+e^- PAIR STATES

When the pair is produced, the electron and positron must occupy discrete energy states with well defined energies perpendicular to the magnetic field. The spacing of these states in energy increases with B, reaching $\Delta E \sim mc^2$ at B_{cr}. For a given photon energy and field strength, there is a set of kinematically allowed states into which the pair can be produced. Figure 2 shows the number of these states as a function of photon energy and magnetic field for curvature radiation photons which are emitted parallel to B and travel one mean free path before producing the pair. At each magnetic field, the photon's energy must exceed a certain value before a pair can be produced with at least one member in an excited state. In fields $\gtrsim 5 \times 10^{12}$ G, a significant fraction of photons in the curvature radiation spectrum (for typical pulsar parameters) will produce pairs in the ground state.

Models of pulsar cascades[4] have shown that synchrotron radiation γ-rays from secondary particles are necessary to sustain a cascade with several photon generations and high pair yields. Curvature radiation from the pairs is much less efficient and unless

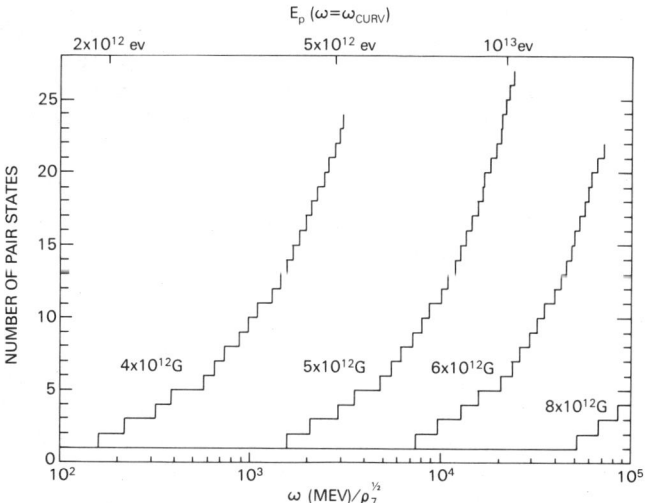

Fig. 2. Number of energetically allowed e^+e^- states vs. curvature radiation photon energy for different magnetic field strengths with constant radius of curvature, $\rho_7 \equiv \rho/10^7$ cm. The top scale is the primary particle energy E_p for which the critical energy $\omega_{curv} = (3/2)(E_p/mc^2)^3 c/\rho = \omega$.

radii of curvature are significantly less than dipole, the cascade terminates after one photon generation. In magnetic fields high enough to suppress production of pairs in excited states, synchrotron radiation cannot take place (the photon energy is fed into particle motion parallel to the field). The efficiency of the cascades should therefore decrease rapidly above some field strength, depending on the critical frequency $\omega_{curv} = (3/2)(E_p/mc^2)^3 c/\rho$ of the primary particle curvature spectrum (Figure 3). Cascade efficiency (ie. the ratio of secondary to primary particles) will increase with field strength below this limit, however. For values of the primary particle energy E_p predicted by various acceleration models [1,2,3], copious pair production by cascades near the polar cap would not be expected to occur for field strengths above $B \sim 6\text{-}8 \times 10^{12}$ G.

CONCLUSIONS

This reevaluation of pair production by curvature radiation photons in pulsar magnetic fields has shown that threshold effects on the attenuation coefficient can have significant consequences for pair-photon cascades. In particular, there should be a maximum in the pair yield per primary particle as a

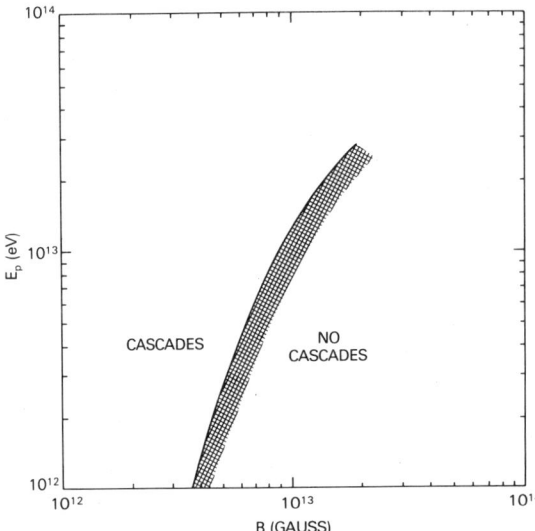

Fig. 3. Primary particle energy (i.e. polar cap acceleration voltage) vs. magnetic field strength above which all curvature radiation photons (with $\omega \leq 3 \omega_{curv}$) produce pairs in the ground state.

function of magnetic field strength with the yields diminishing rapidly in fields $\gtrsim 6 \times 10^{12}$ G where synchrotron radiation is suppressed. Since the copious production of e^+e^- pairs is necessary for coherent radio emission in most pulsar models, this may imply that there is a high magnetic field cutoff for radio pulsars.

REFERENCES

1. J. Arons and E. T. Scharlemann, Ap.J., 231, 854 (1979).
2. J. Arons, this proceeding (1983).
3. M. A. Ruderman and P. G. Sutherland, Ap.J., 196, 51 (1975).
4. J. K. Daugherty and A. K. Harding, Ap.J., 252, 337 (1982).
5. J. K. Daugherty and A. K. Harding, Ap.J., in press (1983).
6. J. K. Daugherty and A. K. Harding, this proceeding (1983).
7. T. Erber, Rev. Mod. Phys., 38, 626 (1966).
8. A. E. Shabad and V. V. Usov, Nature, 295, 215 (1982).

POLARIZATION MODE COUPLING IN RADIO PULSARS

Daniel R. Stinebring
National Radio Astronomy Observatory, Charlottesville, VA 22901

James M. Cordes
Cornell University; National Astronomy and Ionospheric Center
Ithaca, NY 14853

ABSTRACT

Mode coupling (polarization limiting) has been proposed as an explanation for the orthogonally polarized radiation (OPR) observed from pulsars. Using a simplified magnetospheric model we show that it cannot account for the presence of circular polarization in the OPR phenomenon. Mode coupling occurs much closer to the star than the point at which the normal modes of the plasma become significantly elliptically polarized.

INTRODUCTION

Although pulsar intensity and polarization properties fluctuate strongly on a pulse-by-pulse basis, the fluctuations have time-stationary statistics. Early observations showed that the polarization position angle sometimes shows a preference for one of two values (at a given pulse phase), separated by close to 90°. Later study of this orthogonally polarized radiation (OPR) indicated that it is present over some portion of the pulse in almost all pulsars studied[1,2]. Recently it has been shown that OPR is a broadband phenomenon, occurring over the range 430-1404 MHz with little change in statistical properties[3,4].

An example of OPR is shown in Figure 1, where the density of points indicates the number of pulses (at one pulse phase) that had specific position angle and circular polarization values. As first noted by Cordes, Rankin, and Backer[5], there is a correlation between the sign of the circular polarization and the identity of the polarization mode determined from the position angle value. This is shown in Figure 1 along with higher frequency data[3]. This correlation, although currently demonstrated in only a few cases, is highly suggestive since it indicates that circular polarization is integrally tied to the production of OPR. There is a need, then, to find an emission/propagation mechanism that produces orthogonal elliptical polarization.

MODE COUPLING

Most models to explain OPR can be placed into one of two classes, both of which identify the observed polarization modes with the normal propagation modes of the plasma above the polar cap of the star. In the first class of models, separate orthogonally-polarized beams are present which become spatially separated due to refractive effects[6,7]. The other class, which we will consider

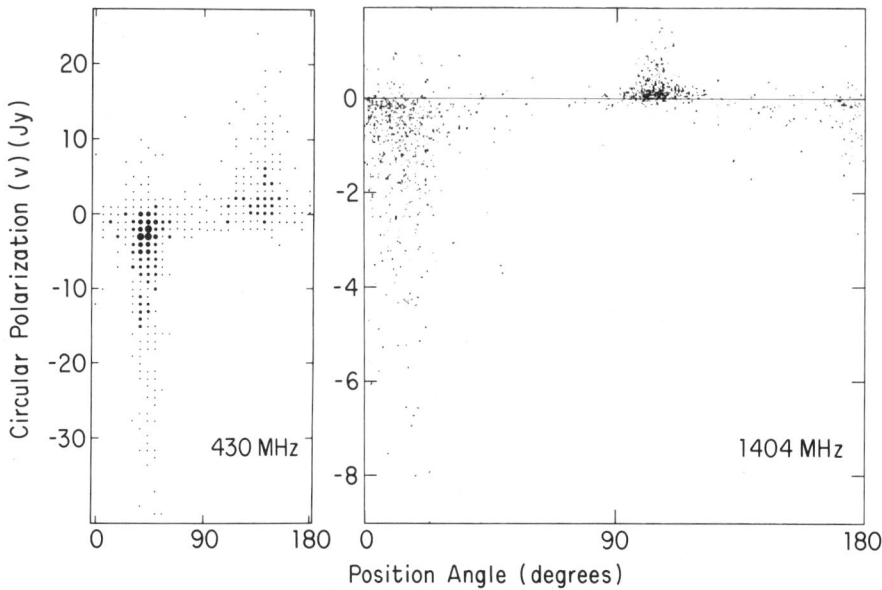

Figure 1. Correlation plot of circular polarization vs. position angle. Data from the second component of PSR 2020+28.

further here, ascribe OPR to mode-coupling (alternatively, adiabatic walking or polarization limiting). In these models the polarization properties of the emergent radiation get determined at the point where normal-mode propagation ceases and the modes become strongly coupled[7,8,9].

Normal mode propagation ceases and strong coupling sets in when the plasma properties change rapidly along the propagation path. When the uncertainty that this inhomogeneity introduces into the wave number ($\sim 1/L$, where $L = |\partial k/\partial s|^{-1}$ is the scale length for change of the plasma parameters) becomes comparable to the difference between the normal mode wave numbers, strong coupling results. The radiation will then propagate with no further change in its polarization properties, since the radiation has decoupled from the normal modes of the plasma.

If mode coupling is to explain orthogonal elliptical polarization, the onset of mode coupling must occur at an altitude where the normal modes of the plasma, which are initially linearly polarized because of the strong magnetic field, have become significantly elliptically polarized. Two parameter boundaries determine the ellipticity of the emergent radiation in this model. The ellipticity boundary separates regions of quasi-transverse (linear modes, near the star) from quasi-longitudinal propagation (circular modes, far from the star). The mode coupling boundary designates the onset of strong coupling and is determined by the condition $Q \equiv |(\Delta kL)^{-1}| \gtrsim 1$, where Q is the mode coupling parameter and Δk is the difference in wave numbers between the plasma modes.

A schematic representation of normal mode development and the onset of mode coupling is displayed in Figure 2. There it can be seen that, due to the extremely large ($\sim 10^{12}$ G) magnetic field near the star, the normal modes of the plasma are initially linearly polarized. As radiation propagates away from the star, the decreasing magnetic field will make the modes more nearly circular. (A gradual circularization of the modes is shown here for clarity; in a realistic magnetosphere the transition from linear to circular modes is abrupt.) The behavior of the mode coupling parameter is shown schematically in the lower part of the figure, with the dashed line indicating the onset of strong coupling.

RESULTS

We have calculated the mode coupling and ellipticity boundaries for a simplified magnetospheric model using the dispersion relation presented by Melrose[7]. The model we have used assumes (1) $n_+ = n_-$ initially, where these are the number densities of positrons and electrons (2) the initial Lorentz factors of the charge species, $\gamma_\pm$, are equal and $\gamma_\pm \sim 200$ (3) $B \propto s^{-3}$, $n_\pm \propto s^{-3}$, where B is the magnetic field strength and s is the distance along the propagation path (4) the co-rotational charge density, $e(n_+ - n_-) = -\underline{\Omega} \cdot \underline{B}/2\pi C$, is provided by a relative streaming of the electron and positron beams in the manner discussed by Cheng and Ruderman[10].

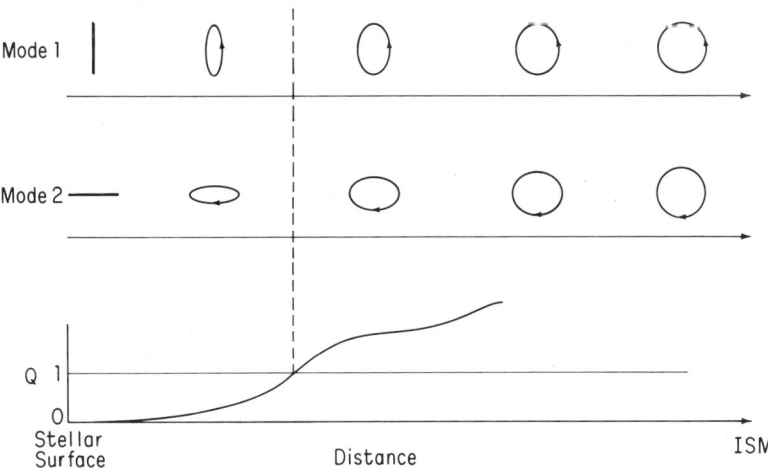

Figure 2. Schematic representation of normal mode circularization and the onset of mode coupling.

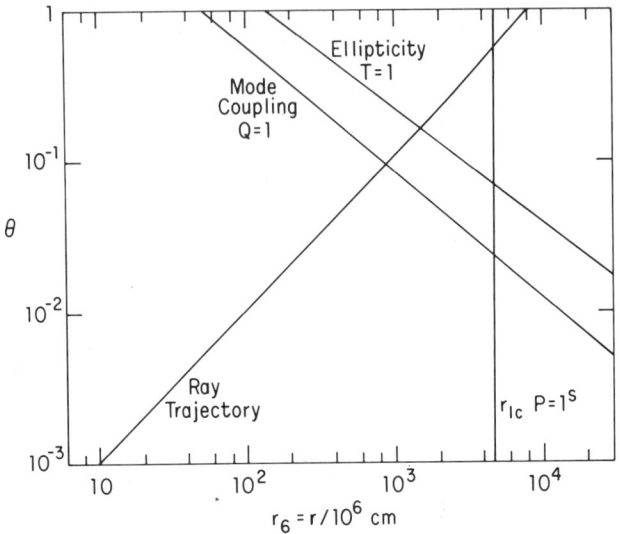

Figure 3. Location of the ellipticity and mode coupling boundaries.

In Figure 3 we present the results of this calculation in a format that does not depend on the detailed geometry of the magnetosphere. The propagation path is plotted as a function of θ (the angle between the ray direction and the local magnetic field) and s. The representative path shown is that for emission along the axis of a rotating dipolar field. A line showing the location of the light-cylinder radius for a pulsar with a period of 1 second is shown for reference.

We find that under the above assumptions the ellipticity boundary is given by

$$\theta_T = \frac{320 \, N^{1/4} \, B_{12}^{1/4}}{\nu_9^{1/4} \, \gamma_<^{3/4} \, r_6^{3/4}} \qquad (1)$$

and the mode coupling boundary by

$$\theta_Q = \frac{440 \, N^{1/3} \, B_{12}^{1/3}}{\nu_9^{1/3} \, \gamma_< \, P^{1/6} \, r_6^{5/6}} \qquad (2)$$

In these expressions $N \equiv |(n_+ - n_-)/2(n_+ + n_-)|$, B_{12} is the surface magnetic field in units of 10^{12} Gauss, ν_9 is the observing frequency in gigahertz, $\gamma_<$ is the Lorentz factor of the less energetic species, r_6 is the distance from the star in units of 10^6 cm, and P is the pulsar period in seconds.

DISCUSSION

Figure 3 shows that the mode coupling and ellipticity boundaries do not occur close to one another under these magnetospheric conditions. Although only the boundaries themselves are displayed, the extremely rapid change of the normal mode ellipticity along the ray path (typically proportional to s^7) guarantees that mode coupling will have occurred under these conditions long before the modes have become significantly circular. We note that the functional form of these two parameter boundaries is similar and that these lines will move toward or away from the star together as the plasma properties are varied. The location of the boundaries is also insensitive to the radiation frequency and the pulsar period.

We are then led to the following conclusion: either the assumptions of this simple model are violated or else mode coupling cannot account for the presence of orthogonal elliptical polarization states in pulsar radiation. Arons[11] has questioned the low-density approximation used in the derivation of the dispersion relation, and some of the model assumptions, such as the s^{-3} dependency of the magnetic field and particle density, may not hold along the entire ray path. But we point out that field line twisting and particle clumping along the ray path can only serve, by moving the mode coupling boundary closer to the star, to move the boundaries further apart.

Mode coupling may be a viable explanation for OPR, but we have shown that it quantitatively fails to produce the circular polarization that is part of the OPR phenomenon.

A more detailed account of this study will be published elsewhere.

REFERENCES

1. R. N. Manchester, J. H. Taylor, and G. R. Huguenin, Ap. J., 196, 83 (1975).
2. D. C. Backer and J. M. Rankin, Ap. J. Suppl., 42, 143 (1980).
3. D. R. Stinebring, J. M. Cordes, J. M. Rankin, J. M. Weisberg, and V. Boriakoff, Ap. J., submitted (1983).
4. D. R. Stinebring, J. M. Cordes, J. M. Weisberg, J. M. Rankin, and V. Boriakoff, Ap. J., submitted (1983).
5. J. M. Cordes, J. M. Rankin, and D. C. Backer, Ap. J., 223, 961 (1978).
6. J. Arons, Proc. Varenna Summer School and Workshop on Plasma Astrophysics, T. D. Guyenne, ed. (European Space Agency, Paris, 1981), p. 273.
7. D. B. Melrose, Aust. J. Phys., 32, 61 (1979).
8. D. B. Melrose and R. J. Stoneham, Proc. Astron. Soc. Aust., 3, 120 (1977).
9. A. F. Cheng and M. A. Ruderman, Ap. J., 229, 348 (1979).
10. A. F. Cheng and M. A. Ruderman, Ap. J., 212, 800 (1977).
11. J. Arons (private communication).

Pulsar Gamma-Ray Lines from Quarkonium Systems

G. Kanbach
Max-Planck-Institut für Physik und Astrophysik
Institut für Extraterrestrische Physik
8046 Garching, W.-Germany

R. Schlickeiser
Max-Planck-Institut für Radioastronomie
Auf dem Hügel 69, 5300 Bonn 1, W.-Germany

ABSTRACT

Excited states of heavy quark-antiquark systems (quarkonium) might be created by colliding e^-e^+ beams in the magnetospheres of gamma-ray pulsars. We investigate the hypothesis that radiative transitions to the ground state, in particular of charmonium, may lead to observable fluxes of gamma-ray lines in the 120 - 450 MeV range and compare the estimates to the continuum flux expected from curvature radiation. If the particle density in the source volumes is high enough the lines could be resolvable with the next generation of high-energy gamma-ray telescopes (Gamma Ray Observatory).

INTRODUCTION

The existence of highly relativistic electron-positron flows in the active parts of a neutron star's magnetosphere is postulated in most theoretical concepts for the emission of high-energy radiation from pulsars. Potential drops with huge accelerating fields (gaps) are thought to develop in regions which are magnetically connected to loss regions beyond the light cylinder and where the resupply of free charges is limited. This can happen either by the inability to release positive ions from the star's surface (polar gap[1], potential $\sim 10^{12}$ V), by the need to conduct and supply charges through neighbouring space charge regions of opposite polarity (outer gaps [2,3], potential $\sim 10^{16}$ V) or in a slot gap model[4]. The growth of these gaps is limited by spark discharges formed of an avalanche of positrons and negatrons which propagates by the production of curvature photons and e^-e^+ pairs in the strong magnetic fields[5]. The charged particles may form oppositely directed beams in the gap regions and since the strong magnetic field does not allow transverse momenta these beams must be highly collimated. The detailed geometrical configuration and temporal structure of such discharge models of the inner magnetospheres of pulsars is far from being understood. Some salient features of these models should however be mentioned: the conversion of gamma-ray photons into a pair plasma due to the transverse electric and magnetic field components leads to plasma densities which are $10^3 - 10^5$ times higher than the corotating charge density would require. Average densities up to 10^{16} cm^{-3} in the relativistic pair plasma above the polar cap of a fast pulsar seem realistic. Furthermore it has been noted[1,4] that the plasma in the pair creation region might be highly inhomogeneous and the densities in the central parts of spark discharges or flow tubes could exceed the average value substantially. The momenta of the positrons and electrons in the relativistic plasma are thought to exhibit a fairly wide distribution: Arons[6] derived a flat spectrum between 10 MeV and 1 GeV while Tademaru[7] suggested a power low distribution with cut-offs at high and low energies and indices in the range -2/3 to -2. Fawley et al.[18] estimated the maximum particle energy above the polar cap to about 3 GeV · P^{-1} but point out that for the production of gamma-ray continuum this energy is insufficient. The overall situation in the discharge region of a gamma-ray pulsar with highly collimated, countermoving electron and positron beams is reminiscent of colliding beam experiments in high energy accelerator laboratories and it is tempting to investigate the application of the fascinating findings in elementary particle physics to a natural environment. Since the particles come with a wide distribution of momenta a wide range of c.m. energies will be realized in the collisions. In particular the range above 1 GeV where the creation of quarkonium systems occurs should be well covered.

PRODUCTION OF QUARKONIUM

The formation of bound systems of heavy quarks in colliding e^-e^+ beam experiments has been extensively investigated since the discovery of the first such system, the ground state of charmonium ψ/J (for review and recent results see [8-13]). We summarize in figure 1 the experimental cross sections schematically: above the lighter quark systems ρ, ω and ϕ the charmonium system made up by a charm quark and its antiparticle is formed most abundantly in its ground state ψ/J and first excited state ψ'. The quasi bound ψ'' state is unimportant. The family of bound bottomonium states at energies between 9.5 and 10.5 GeV is formed with nearly two orders of magnitude less probability than charmonium. Radiative signals from bottomonium are therefore expected at a much reduced rate and we will not consider them further in this paper. The energy levels of charmonium are shown in Figure 2 for the $^3S(\psi)$ and $^3P(\chi)$ states which are most relevant for the emission of gamma-ray lines as indicated by the numbered transitions and a measured photon spectrum[13]. The transition ψ' (3684 MeV) ⟶ χ (triplet 3551, 3503 and 3415 MeV) ⟶ ψ/J (3095 MeV) emits six gamma-ray lines in the energy range 120 - 450 MeV and the branching ratio for this cascade is about 25 percent. The background visible in fig. 2 underlying the lines has not been entirely explained by the experimenters but charged particle leakage and the substantial π^0 fraction from hadronic decays of charmonium could contribute. In the astrophysical application we expect the gamma-ray spectrum from the neutral pions to form a broad continuum around 70 MeV and consider it as part of the pulsar's continuum emission. The production cross section for ψ' is approximated by[14] σ_o = 600 nb for the center of mass energy interval $m_{\psi'} \pm \Gamma/2$, where $m_{\psi'}$ = 3684 MeV and Γ = 224 keV is the width of the resonance, and zero otherwise.

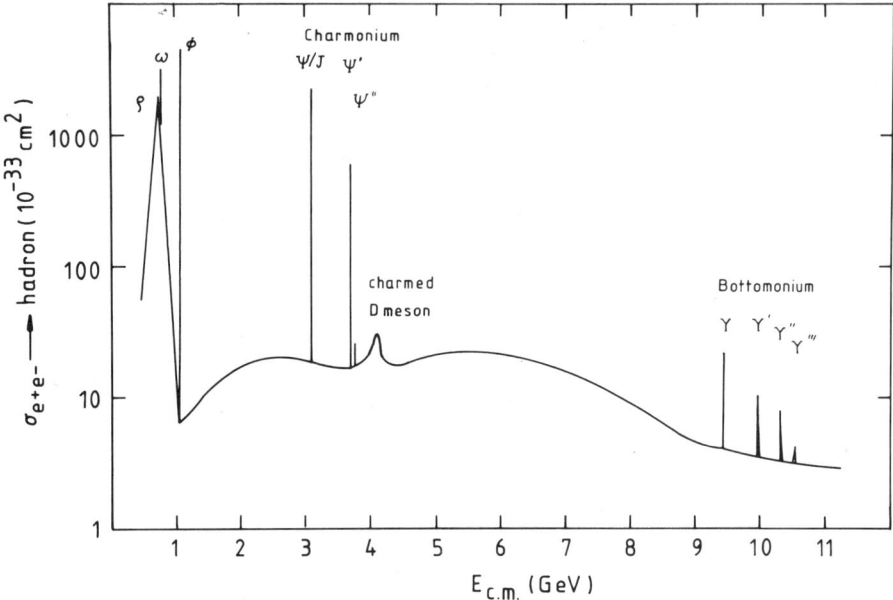

Fig. 1: Schematic cross section for the production of hadrons in colliding e-e+ beam experiments

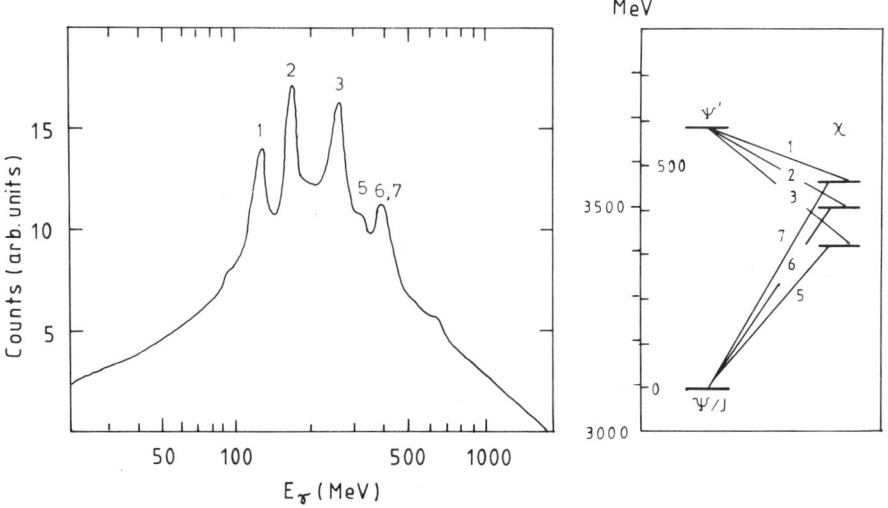

Fig. 2: Inclusive photon spectrum from the decay of ψ' charmonium and corresponding energy levels.

ESTIMATE OF CHARMONIUM LINE INTENSITIES

The source strength of the charmonium gamma-ray line cascade can be written

$$Q = fc \int_{E_n} N_-(E_n)dE_n \int_{E_p} dE_p \, N_+(E_p) \sigma(E_p, E_n) \; photons \; cm^{-3} s^{-1} \quad (1)$$

where $f = 0.25$ is the branching ratio for the cascade and c the speed of light. $N(E_n)$ and $N(E_p)$ are the distribution functions of electrons and positrons respectively (units $cm^{-3} GeV^{-1}$). As was discussed above, theories only provide a rough guidance to the spectral (flat power law with cut-offs) and volume densities ($\gtrsim 10^{16} cm^{-3}$) of a pair plasma in the active region of a pulsar. We assume here that electrons and positrons follow the same distribution function $N(E) = K_e E^{-\beta}$ for $E_{min} < E < E_{max}$ and β in the range 2/3 to 2. Such a distribution function for both species can of course only be strictly valid in the total region if a force-free situation is established as one could envisage for times when the discharges are fully developed. The cut-off energies E_{min} and E_{max} are established in the pair production models[1,4,7] under the influence of accelerating potentials and radiation loss of the relativistic electrons. Estimates are $E_{min} \sim 10^{-1}$ GeV and $E_{max} \sim 10$ GeV, which brackets the necessary energies for the procuction of ψ' without imparting too high a velocity to the ψ'. In fact the Doppler shift of the ψ' gamma-ray lines will be of minor importance if the energy spectra of the colliding beams cut off close to half the rest mass of $\psi' \sim 1.8$ GeV.

In the case of colliding relativistic beams the center of momentum energy available for the creation of particles is approximately given by $E_{cm} \approx 2(E_n E_p)^{1/2}$. If the target electron is at rest, the projectile needs an energy of 1.33×10^4 GeV to create a ψ' system. Such high energies however cannot be achieved in the inner magnetosphere because of radiation reaction limitations.

Inserting the assumed power laws and the resonance cross section into equation 1 one derives:

$$Q = f \cdot c \cdot \sigma_0 \cdot K_e^2 \cdot \Gamma \left(\frac{m_{\psi'}}{2}\right)^{1-2\beta} \ln E \Big|_{max(E_{min}, \frac{m_{\psi'}^2}{4E_{max}})}^{min(E_{max}, \frac{m_{\psi'}^2}{4E_{min}})} \quad (2)$$

The source strength is of course zero if the energy range does not bracket the $m_{\psi'}/2$ value.

ESTIMATE OF ELECTRON DENSITY

An empirical estimate of the relativistic electron density in the active regions of a pulsar can be inferred from gamma-ray continuum observations. Gamma-ray emission from pulsars[15-17] has been definitely detected from PSR0531+21 and PSR0833-45, the Crab and Vela pulsars. The origin of these high energy photons in the 100 MeV to 10 GeV range is thought to be mostly curvature radiation from highly relativistic electrons flowing along the magnetic field lines of the neutron star. The curvature radiation luminosity can be expressed as

$$L(E_\gamma) = \langle E\rangle \cdot \eta \cdot V \cdot N(\underline{E}) \cdot \left|\frac{dE}{dt}\right|_{curvat.radn.} \quad erg\ s^{-1} \qquad (3)$$

with $N(\underline{E})$ the differential electron density, V the source volume, and the mean photon energy $\langle E \rangle$ connected to the typical electron energy $\underline{E}$:

$$\langle E \rangle = \frac{3\hbar c}{2m^3c^6}\frac{\underline{E}^3}{\rho} \approx 30\ MeV\ \gamma_6^3\ \rho_6^{-1} \qquad (4)$$

where ρ_6 is the radius of curvature in units of 10^6 cm.

The energy loss due to curvature radiation in (3) is given by

$$\left|\frac{dE}{dt}\right| = 2.9 \times 10^6\ \gamma_6^4\ \rho_6^{-2} \quad GeV\ s^{-1} \qquad (5)$$

The factor η in (3) allows for the absorption of curvature photons by pair production processes and can therefore also be regarded as the ratio of the number of electrons that produce visible curvature photons to the total number of electrons present in the region. The gamma-ray luminosity of the Vela pulsar has been measured[17] as $L(>100\ MeV) = 3 \times 10^{34}\ erg\ s^{-1}$ and the mean photon energy is around 200 MeV. We then find from (3):

$$N(\underline{E} = 956\ \rho_6^{1/3}\ GeV) = 2 \times 10^{30}\ V^{-1} \cdot \rho_6^{2/3}\ \eta^{-1}\ GeV^{-1} cm^{-3} \qquad (6)$$

Equating $\underline{E} \cdot N(\underline{E})$ with the integral spectrum yields for the spectral constant:

$$K_e = 10^{21}\ \frac{1-\beta}{E_{max}^{1-\beta} - E_{min}^{1-\beta}}\ \rho_6\ r_4^{-2}\ h_4^{-1}\ \eta^{-1}\ \frac{electrons}{cm^3\ GeV^{1-\beta}} \qquad (7)$$

where we wrote for the volume $V = 10^{12} \pi r_4^2 h_4\ cm^3$. The index β is assumed in the range 2/3 to 2. Depending on the exact energy limits the numerical factor in K_e is found in the range 10^{19} to 10^{21}. If we apply $E_{min} = 0.1$ GeV, $E_{max} = 10$ GeV the dependence on β is weak and we find

$$K_e = 10^{20}\ \rho_6\ r_4^{-2}\ h_4^{-1}\eta^{-1}\ electrons\ cm^{-3}\ GeV^{\beta-1} \qquad (8)$$

Using this estimate in (2) yields for the source strength of the charmonium lines:

$$Q = (5 \times 10^{15} - 5 \times 10^{16})\ \rho_6^2\ r_4^{-4}\ h_4^{-2}\eta^{-2}\ photons\ cm^{-3}s^{-1} \qquad (9)$$

The integral flux of charmonium gamma-ray lines from the Vela pulsar (D = 500 pc, beaming factor 0.1, no absorption) is then of the order:

$$F_{lines} = 10^{-14}\ \rho_6^2\ r_4^{-2}\ h_4^{-1}\eta^{-2}\ photons\ cm^{-2}s^{-1} \qquad (10)$$

COMPARISON WITH OBSERVATIONS AND DISCUSSION

The curvature radius is related to the pulsar period[5] as $\rho_6 \simeq 10^{1.9} \cdot p^{1/2}$ which for the Vela pulsar yields $\rho_6 \simeq 10^{1.4}$ and according to (10)

$$F_{lines} \simeq 6 \cdot 10^{-12}\ r_4^{-2} h_4^{-1}\ \eta^{-2}\ photons\ cm^{-2}s^{-1} \qquad (11)$$

which has to be compared with the measured gamma ray flux between 120 and 450 MeV of $7 \cdot 10^{-6}$ ph cm^{-2}s^{-1}. For values of $\eta = r_4 = h_4 = 1$ the line flux is six orders of magnitude smaller than the measured intensity.

In Figure 3a we show the emitted charmonium lines spectrum adjusted to 10 percent of an underlying continuum (in the 120 - 450 MeV band) which is described[17] by a power law $1.4 \cdot 10^{-6}$ E$^{-1.89}$ photons/cm^2 s GeV. According to (11) this corresponds to the case

$$\eta^2 r_4^2 h_4 \simeq 9 \times 10^{-6} \qquad (12)$$

Fig. 3b shows the expected measurement of such an input spectrum in a COS-B like gamma-ray detector with energy resolution $\Delta E/E \approx$ 50% FWHM and the actual COS-B measurement[17]. Fig. 3c demon-

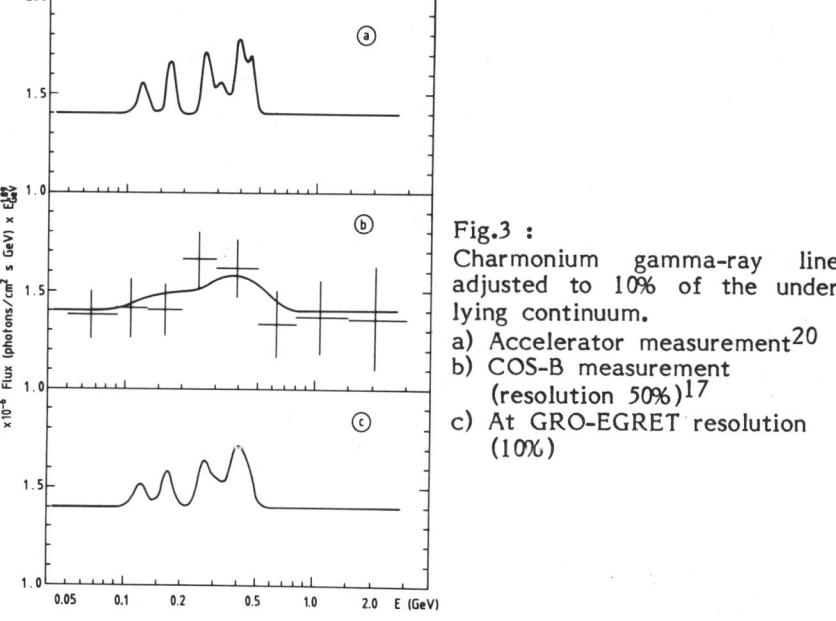

Fig.3 :
Charmonium gamma-ray lines adjusted to 10% of the underlying continuum.
a) Accelerator measurement[20]
b) COS-B measurement (resolution 50%)[17]
c) At GRO-EGRET resolution (10%)

strates the expected spectrum of a GRO-like gamma-ray detector with $\Delta E/E \approx$ 10% FWHM[19].

There seems to be a feature in the COS-B Vela spectrum around 300 MeV (Fig. 3b) the statistical evidence of which, however, is poor. If this feature indeed is due to unresolved charmonium lines the assumption (12) of the combination $\eta^2 r_4^2 h_4$ is justified. Fig. 3c indicates that with the EGRET experiment on the Gamma Ray Observatory it should be possible to identify this flux enhancement with the actual charmonium lines as proposed here.

Physically, a combination $\eta^2 r_4^2 h_4 \approx 9 \times 10^{-6}$ may point to small values of $\eta \approx 10^{-3}$ which means that only 10^{-3} of the total electrons are seen via curvature photon production and that in small volume elements the density of relativistic electrons is as large as 10^{21} cm^{-3} GeV^{-1}. Such a high density seems to be unrealistically large: it is generally thought that the electron-positron number densities can exceed the corotational density by approximately five orders of magnitude through curvature radiation processes[6], limiting it to $\sim 10^{16}$ cm^{-3} GeV^{-1}. However, it was noted[4] that the e$^-$e$^+$-plasma is highly inhomogeneous with density exceeding the average value substantially in central parts of the flow tube. We suggest that these central parts of small volume and high density are the astrophysical sites where charmonium production on a detectable level is possible.

In conclusion we note that gamma-ray lines might be detectable in the spectra of pulsars provided the emission occurs in small volume elements of huge e$^-$e$^+$-densities. The COS-B experiment shows a weak indication of a feature around 300 MeV above the straight power law but their energy resolution is too poor to identify the lines. The GRO mission with its five times better energy resolution should yield a decisive test of this hypothesis.

REFERENCES

1) M.A.Ruderman and P.G.Sutherland, Astrophys.J. 196, 51 (1975)
2) N.J.Holloway, Nature Phys.Sci, 246, 6 (1973)
3) A.Cheng, M.A.Ruderman, P.Sutherland, Astrophys.J., 203, 209 (1976)
4) J.Arons and E.T.Scharlemann, Astrophys.J., 231, 854 (1979)
5) P.A.Sturrock, Astrophys.J., 164, 529 (1971)
6) J.Arons, in Origin of Cosmic Rays, IAU Symp.Nr. 94 ed. by Setti, Spada and Wolfendale, Reidel Publishing Company (1981)
7) E.Tademaru, Astrophys.J., 183, 625 (1973)
8) G.J.Feldmann, M.L.Perl, Phys.Rep., 33, 285 (1977)
9) V.A.Novikov et al., Phys.Rep., 41, 1 (1978)
10) T.Appelquist et al., Ann.Rev.Nucl.Part.Sci., 28, 387 (1978)
11) E.D.Bloom, G.J.Feldmann, Scientific American, May 82, 42 (1982)
12) R.Partridge et al., Phys.Rev.Lett., 45, 1150 (1980)
13) C.Edwards et al., Phys.Rev.Lett., 48, 70 (1982)
14) V.Lüth et al., Phys.Rev.Lett., 35, 1224 (1975)
15) D.J.Thompson et al., Astrophys.J., 200, L79 (1975)
16) K.Bennett et al., Astron.Astrophys., 61, 279 (1977)
17) G.Kanbach et al., Astron.Astrophys. 90, 163 (1980)
18) W.M.Fawley et al., Astrophys.J., 217, 227 (1977)
19) C.E.Fichtel, GRO-EGRET Experiment, priv.comm. (1982)
20) C.J.Biddick et al., Phys.Rev.Lett., 38, 1324 (1977)

Chapter IV   The Galactic Center

OBSERVATIONS OF ANNIHILATION RADIATION FROM THE GALACTIC CENTER REGION

C. J. MacCallum
Sandia National Laboratories
Albuquerque, New Mexico 87185

M. Leventhal
Bell Laboratories
Murray Hill, New Jersey 07974

ABSTRACT

The experiments that have demonstrated variability of the 511 keV positron annihilation line from the Galactic center region are reviewed. The simplest history consistent with all (but one) of the eleven measurements is that of a narrow (<2.5 keV FWHM), unredshifted (dE <.25 keV) line, which was constant in intensity (($1.6+/-0.1)\times10^{-3}$ ph cm^{-2} s^{-1} at earth) from before 1970 until the beginning of 1980, when it dropped to zero in less than a year. The limited evidence for detection of the continuum expected from the three-photon decay of orthopositronium is evaluated, with particular attention to the possibility of simulation by atmospheric scattering. The question of possible correlation between variations in the inverse-power-law Galactic center continuum and the strength of the 511 keV annihilation line is discussed; the evidence is inconclusive.

OBSERVATIONS OF THE 511 KEV ANNIHILATION LINE

The observations of this line, and their astrophysical implications, were reviewed as of a year ago by Jacobson [1], and by Lingenfelter and Ramaty [2]. Since that time, verification that the Galactic center has indeed "turned off" in this line has been published and, as evidenced in these Proceedings, there has been intense international activity in the building of models of possible source regions for this emission. In this section the eleven reported observations spanning the years 1970-1981 are reviewed briefly and chronologically. These observations are summarized in Table I.

The first sighting of the Galactic center annihilation line may have been as early as 23Apr68, on a balloon flight over southern Australia [3] by the pioneering group at Rice University headed by Robert Haymes. The feature was only observed at about the 2-sigma level of significance and was not reported until confirmed by later flights. The original Rice instrument is sketched in Fig. 1; it is the prototype for all subsequent telescopes utilized to date. A massive NaI shield acts in anticoincidence with the enclosed central detector and serves to 1) define the field of view, 2) reject less

TABLE I.

OBSERVATIONS OF THE 511 KEV LINE FROM THE GALACTIC CENTER DIRECTION

DATE	GROUP	RESOLUTION (keV,FWHM)	APERTURE (deg,FWHM)	SHIELD ("NaI)	BACKGROUND MODE	10^{+3}xFLUX (ph cm^{-2} s^{-1})
25/Nov/70	Rice	75	24	3	pointing	1.8 +/- 0.5
20/Nov/71	Rice	66,92	24	3	pointing	1.8 +/- 0.5
2/Apr/74	Rice	60	13	5	pointing	0.80 +/- 0.23
Feb/77	CESR	10,18	50	2	transit	4.2 +/- 1.6
11/Nov/77	B/S	3.2	15	6	pointing	1.22 +/- 0.22
22/Nov/77	UNH	40	100	<4	transit	4.0 +/- 0.6
15/Apr/79	B/S	3.2	15	6	pointing	2.3 +/- 0.7
Oct/79	JPL	2.7	35	*3.5	scanning	1.85 +/- 0.21
Mar/80	JPL	5.7	35	*3.5	scanning	0.65 +/- 0.27
20/Nov/81	GSFC/CENS	2.2	15	6	transit	< 0.5
21/Nov/81	B/S	2.4	15	6	pointing	< 0.4

*(equivalent thickness CsI)

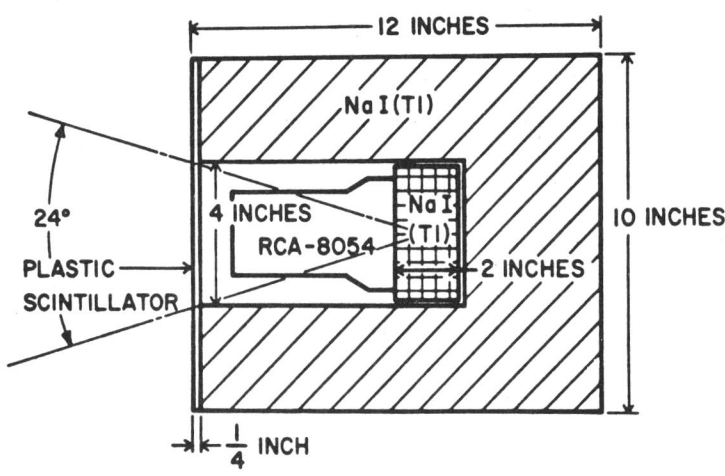

FIGURE 1. Schematic diagram of the original Rice University NaI scintillation gamma-ray telescope.

than full-energy events, and 3) reduce the large atmospheric background. Most later instruments have carried shields significantly thicker than 3 inches of NaI: this seems to be especially important in reducing the instrumental background 511 line. They have also, as large Ge crystals became available, used solid state diodes as the central detector(s) to take advantage of the significantly better energy resolution available. Future instruments currently being built by Bell/Sandia/GSFC and by UCSD/UCB/CESR will have some imaging capability and will employ electronic measures to reduce background induced by radioactivation of the central detectors. In retrospect, however, the importance of the pioneering work at Rice University in this new field of gamma ray astronomy cannot be overemphasized.

On 25Nov70 the Rice group, using a balloon over Argentina, made the first reported measurement [4] of what is now believed to have been positron annihilation radiation from the Galactic center. The experiment was repeated [5] with the same instrument on 20Nov71 with essentially the same result; the combined data are shown in Fig. 2. If the spectrum is analyzed as a power-law continuum plus a Gaussian "feature", then the area under the Gaussian corresponds to a flux at earth of $(1.8 +/- 0.5) \times 10^{-3}$ ph cm^{-2} s^{-1}. The width of the Gaussian is somewhat larger than the instrumental resolution. Its position, definitely below 511 keV, was for some time a puzzlement, generating suggestions of cosmic-ray excitation of interstellar ^7Li and/or ^7Be [6,7] and of large gravitational redshifts of the 511 line [8,9]. The puzzlement seemed to have been dissipated, however, with the realization [10,11] that positron annihilation in a neutral medium of density $<10^{+15}$ H cm^{-3} should occur largely through decay of positronium in a natural mixture of 25% parapositronium (which decays to two photons at 511 keV) and 75% orthopositronium (which decays to three photons with a continuum of energies distributed in essentially a ramp function from 0 to 511 keV). Leventhal [11] demonstrated that this mixture of line plus continuum, folded with the energy resolution function of the Rice instrument, would give a feature whose shape and position was completely consistent with that of Fig. 2.

On 2Apr74 the Rice group flew a new balloon instrument that had been improved in several respects. The shield thickness was increased, the energy resolution was improved, and the field of view was decreased. The telescope was pointed at GX1+4, about 6 degrees away from the Galactic center. A successful flight [12] produced the spectrum shown in Fig. 3. The feature appears again, clearer than ever, but with significantly reduced strength (see Table I) and now shifted to the high-energy side of 511 keV! The change in flux could reasonably be attributed to the change in pointing direction, the decrease in field of view, and/or to a time-varying source. No explanation has yet been put forth for the apparent position of the peak. This result cast a pall of doubt on the reality of the feature, which remained for more than three years following.

On 14Feb77 and 17Feb77 a French/Brazilian consortium flew a

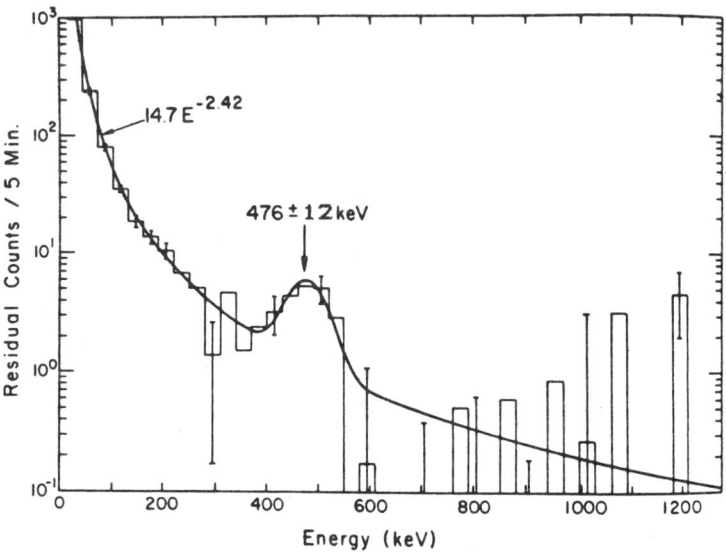

FIGURE 2. Histogram representation and best fit to the measured flux from the Galactic center region; combined data from 1970 and 1971 balloon flights by the Rice group.

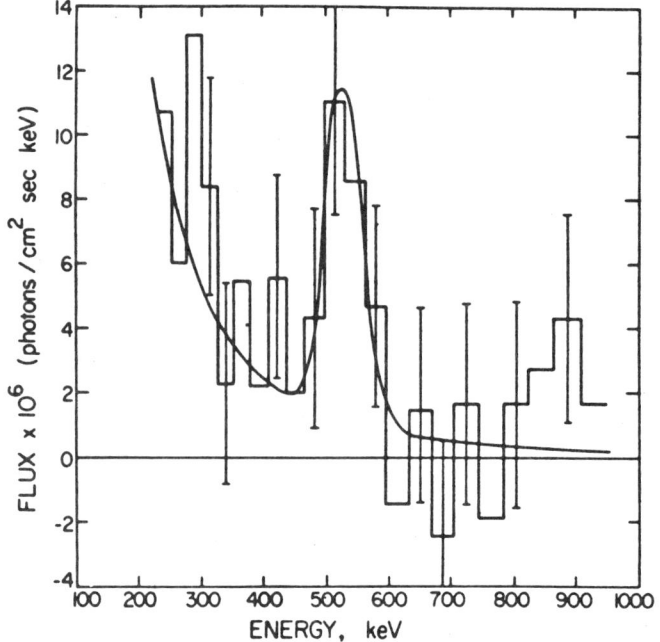

FIGURE 3. Measured photon spectrum of the Galactic center region below 1 Mev; 1974 balloon flight by the Rice group.

balloon-borne instrument over Brazil. The central detector was a lithium-doped germanium crystal. The shield was thin, the field of view large, and the flight short. The result, not communicated until 1981 <13>, was the apparent detection of a line near 511 keV with a flux of either $(4.2 +/- 1.6) \times 10^{-3}$ or $(3.4 +/- 2.1) \times 10^{-3}$ ph cm^{-2} s^{-1} depending on the mode of analysis, the former being preferred by the authors.

The first completely convincing observation was made on 11Nov77 by the Bell/Sandia group <14>. With a thick shield, a large Ge detector, and a full two day balloon flight, the spectrum shown in Fig. 4 was obtained. Figure 5 shows the region around 511 keV in expanded scale. The center of the line is at $(510.7 +/- 0.5)$ keV and its FWHM is 3.2 keV, equal to the instrumental resolution. If the radiation is assumed to originate in the decay of positronium at a distance of 10 kpc, then about $3 \times 10^{+43}$ annihilations/s must be taking place. Lingenfelter and Ramaty <2> point out that the corresponding luminosity is comparable to the hard x- and gamma-ray continuum luminosity and is perhaps 10^{-4} of the estimated bolometric luminosity of the Galactic core! The observed narrowness of the line (since refined by the JPL satellite measurements) has been shown by Bussard et al <15> to require an annihilating medium of temperature $<\sim 10^{+5}$ deg K and degree of ionization $>\sim 10\%$. Figure 4 also shows an apparent excess just below 511 keV, which is tentatively ascribed to the orthopositronium continuum as discussed below.

Just ten days later, on 21Nov77, the University of New Hampshire group flew a quite different instrument <16> and obtained a significantly different result. With a lightly shielded NaI central detector and a very wide field of view, the spectrum in Fig. 6a was obtained. The data do not constrain the underlying inverse-power-law continuum, but can be fit to a line at 511 keV plus an orthopositronium continuum, folded with the known instrument resolution. The fit is marginally poor (chi-square probability ~8%). Figure 6b shows the 99% confidence contours obtained in line-intensity,positronium-fraction space. Taken at face value, this analysis indicates a much larger line intensity than was observed by the Bell/Sandia group ten days earlier.

On 15Apr79 the Bell/Sandia instrument was flown again <17>. An electronic failure spoiled the energy resolution a few hours into the flight. Significant information was obtained regarding the inverse-power-law continuum (see below) but only limited information on the line intensity could be extracted.

During Oct79 the JPL satellite experiment HEAO-3 (described in detail by Mahoney et al <18>) observed the Galactic plane. This telescope was similar in principle to those previously mentioned but operated in a scanning mode, rotating with twenty minute period about a fixed axis perpendicular to the viewing direction. For two weeks

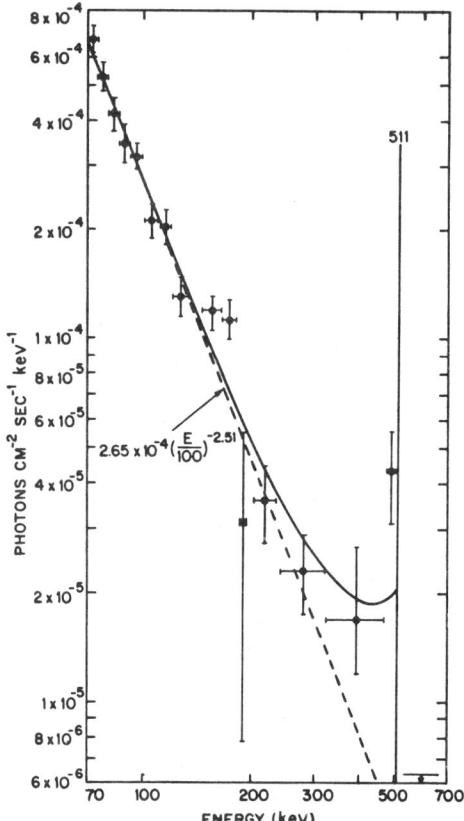

FIGURE 4. Differential photon spectrum from the Galactic center region for the energy range 70-700 keV as measured by the Sandia/Bell 1977 balloon experiment. Best least-squares fit is shown.

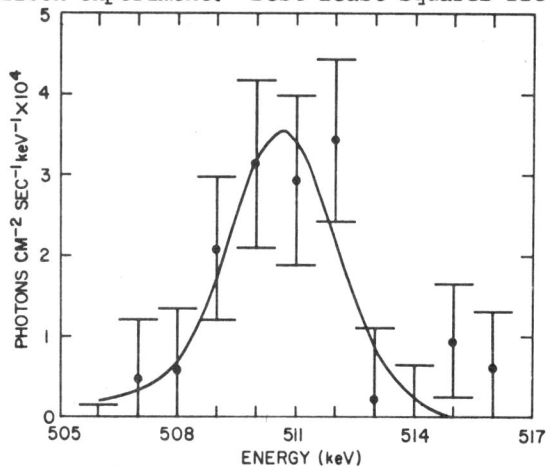

FIGURE 5. Differential photon spectrum, expanded scale near 511 keV.

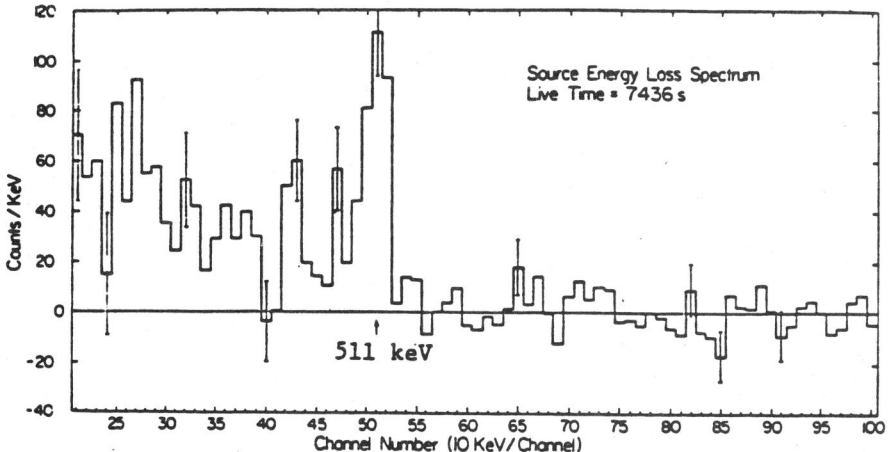

FIGURE 6a. Photon spectrum obtained by the New Hampshire group.

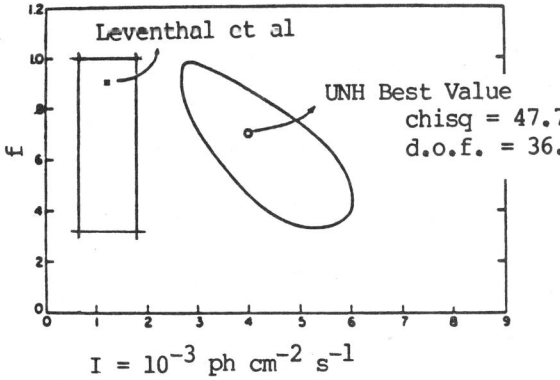

FIGURE 6b. 99% confidence contours of line-intensity versus positronium-fraction, determined from data in Fig. 6a (right curve) and from Sandia/Bell data taken ten days earlier (left rectangle).

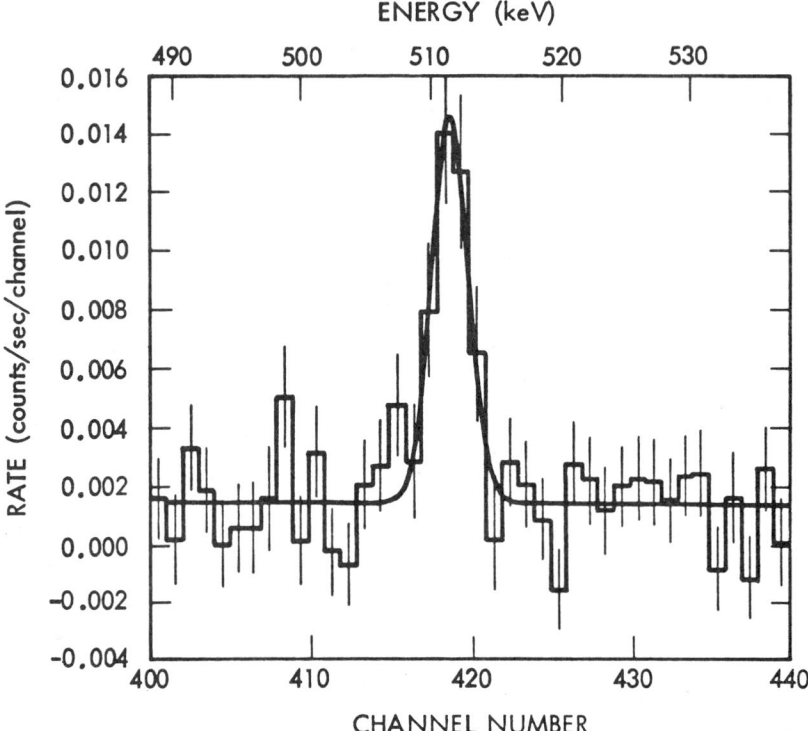

FIGURE 7. Galactic center spectrum near 511 keV obtained from HEAO-3 satellite experiment, October 1979.

in Oct79, and again in Mar80, this axis was aligned with the Galactic poles. Limited information was obtained <19> for the first time regarding the spatial extent of the source. Figure 8a shows the measured intensity of the 511 line as a function of Galactic longitude; the solid line is the expected result for a constant background plus a point source convolved with the instrument resolution. Although the data are completely consistent with a point source, it is not possible on a statistical basis to exclude a line source up to ~25 deg wide. Figure 7 shows the spectrum obtained near 511 keV. Refined values are obtained for the position (510.9 +/- 0.25 keV) and the width (<2.5 keV FWHM) of the line.

The remarkable result from HEAO-3 came five months later, in Mar80 <19>. For a steady source, Fig. 8b should be entirely comparable with Fig. 8a. The only instrumental change was a radiation-induced degradation in energy resolution from 2.7 to 5.7 keV FWHM. The 511 line is still present, but its intensity has decreased by a factor of three (a decrease of 3.5 sigma, probability of no source change ~5×10^{-4}) in five months! The implication

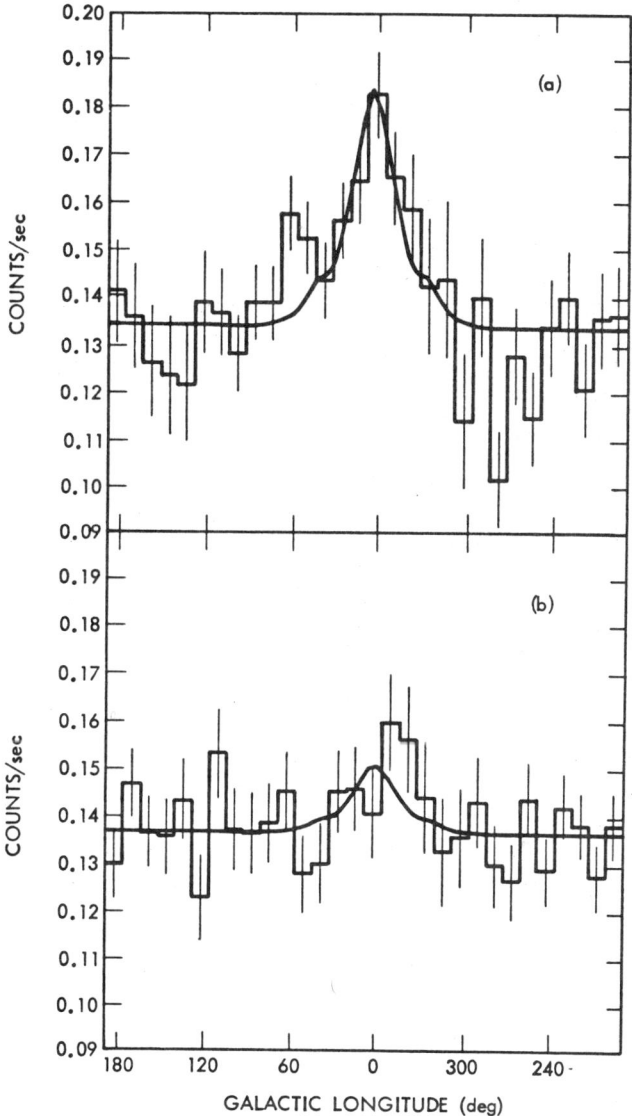

FIGURE 8. Galactic scans from HEAO-3 experiment in energy bin containing the 511 keV annihilation line. (a):Oct79, (b):Mar80.

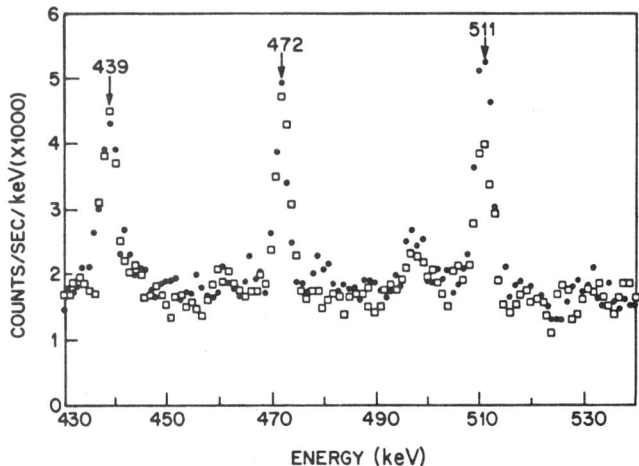

FIGURE 9. Energy spectra from Bell/Sandia experiment of Nov77 (raw data corrected only for drift in energy calibration). Circles represent data taken from Galactic center direction, squares are background measurements from other directions.

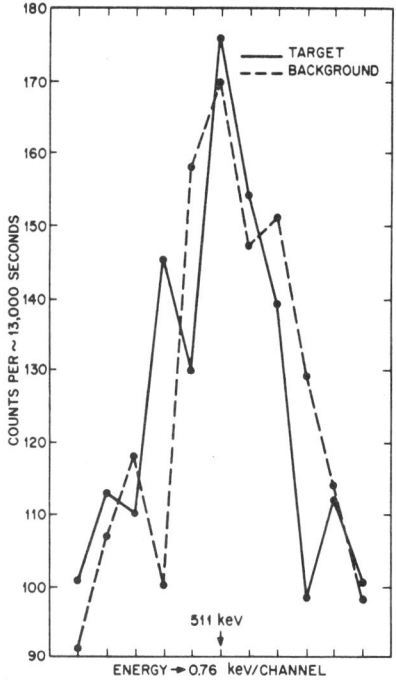

FIGURE 10. Target and background data from Bell/Sandia flight Nov81.

regarding the size of the source region is obvious. It probably also becomes necessary <2> to assume that the annihilation region has a density $>\sim 10^{+5}$ H cm^{-3}, in order for fast positrons to slow down and annihilate in less than half a year.

The final points in this history were provided by the GSFC/CENS group <20> and the Sandia/Bell group <21> with flights on consecutive days in Nov81. Neither group saw any evidence whatsoever of an astrophysical line at 511 keV. Figures 9 and 10 compare the Sandia/Bell raw data for Nov77 and Nov81: in the first case the on-target spectrum shows a clear doubling of the 511 line strength above the continuum compared to the off-target spectrum; in the second case there is no discernible difference. "One-sigma" upper limits for the line flux are listed in Table I.

Figure 11 displays the "light curve" representing the eleven measurements cited. What is the simplest time-dependence compatible with the data? The last four observations certainly require a sudden drop in 1979-80 to a low value. Do the first eight, taken at face value, pass the chi-square test for a constant flux? The answer is no. The entire excess, however, is due to two data: the New Hampshire result and the third Rice result. The latter datum can be treated in one of two ways. It can be rejected as incompatible with the other data on the basis of the inexplicable peak position. Or, one can surmise that, whatever happened, something was measured that by hindsight must have come from the Galactic center. Then, as pointed out by Matteson <22>, the listed flux should be divided by about 0.62 to account for the angular response at a six degree pointing offset from the Galactic center. In either case, the chi-square test for a constant flux then yields about 23 for seven degrees of freedom, with the New Hampshire experiment contributing 15. If the New Hampshire experiment is eliminated, the remaining data provide an acceptable fit to a constant flux of $(1.60 +/- 0.13) \times 10^{-3}$ ph cm^{-2} s^{-1} from the Galactic center prior to 1980.

A caveat must be entered here. Once one begins reanalyzing old data, multiple avenues beckon. First, Table I strongly hints at a positive correlation between measured flux and detector aperture. (There is also a strong correlation between measured flux and shield thickness!) The time variability required by the last four entries demands, of course, that the bulk of the emission come from a "point" source. But various authors have suggested models that combine a variable point source with a constant extended line source in the Galactic plane. Most recently Jacobson <1>, and Dunphy (these Proceedings), have proposed that a better fit to all the data (including the New Hampshire anomaly) would be provided by a point source of strength about 1.0×10^{-3} ph cm^{-2} s^{-1} that turned off at the end of 1979, plus a constant distributed line source of strength about 3.0×10^{-5} ph cm^{-2} s^{-1} deg^{-1} lying in the Galactic plane. A

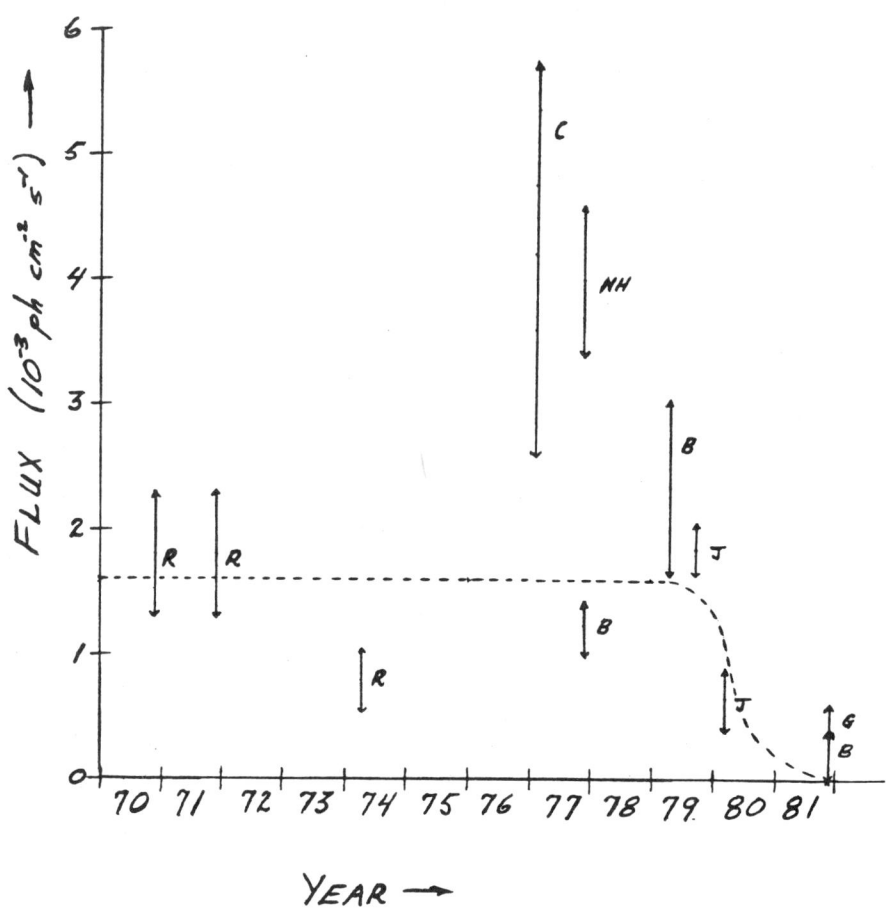

FIGURE 11. Measured intensity of 511 keV annihilation line from Galactic center region versus time. See Table I for details. If the third Rice experiment is omitted (see text) or corrected for pointing offset, then the dashed line is an acceptable fit to all data except the New Hampshire result.

distributed source of this magnitude is quite in line <23> with ideas that the late time light curves of Type I supernovae are powered by the kinetic energy of positrons emitted in the decay of ^{56}Co, with some small fraction of these positrons escaping into the interstellar medium. Analysis of the old data in terms of this model would be a nontrivial undertaking for the scanning and transit experiments.

Second, with the benefit of hindsight it is clear that the Rice spectra should be reanalyzed in terms of a line at 511 keV plus the orthopositronium continuum, as was done with the New Hampshire data, instead of in terms of a single Gaussian. (Of course, it is not rational to analyze the results of their third experiment in this fashion.) One can estimate that roughly half of the wide Gaussian feature centered on 476 keV would then be ascribed to the orthopositronium continuum and only half to a line centered on 511 keV. If this is correct and the reanalysis is carried out without using the distributed line source model just described, then the chi-square test for a constant source prior to 1980 fails.

OBSERVATIONS OF THE ORTHOPOSITRONIUM CONTINUUM

In Figures 4 and 5 the solid line represents the best least squares fit to the data in terms of a power law (2 free parameters), an annihilation line (3 free parameters), and the orthopositronium continuum (1 free parameter: the fraction f of annihilations which proceed through the decay of positronium). The best value found for f is about 0.9. The overall fit passes the chi-square test even for f equal to zero, but the improvement in the value of chi-square shown in Fig. 12 for finite values of f is significant according to the statistical F test at the 99% level of confidence for $f > 0.3$, and at the 95% level for $f > 0.6$. It seems clear that something is being measured on the low-energy side of the 511 line, but it cannot be identified with certainty as orthopositronium. Some other possibilities are considered below.

Compton scattering in the instrument and imperfect functioning of the anticoincidence shield produce an observed spectral "tail" to the low energy side of any line. This "tail" is rather flat in the Sandia/Bell instrument with a value ~0.5% of the line peak intensity; i.e. it is off scale below Fig. 4.

Compton scattering in the residual air above a balloon-borne instrument can produce a continuum to the low energy side of a line, particularly in a telescope with a wide field of view. Figure 13 shows a calculation of this continuum (in the single-scattering approximation) at a typical atmospheric depth (3.4 gm cm^{-2}) and zenith angle (30 deg), for several aperture sizes. The units of the abscissa are flux/keV scaled to an incident line flux of 10^{+4}. None of the curves would be visible on Fig. 4 (line flux 1.2×10^{-3}). The

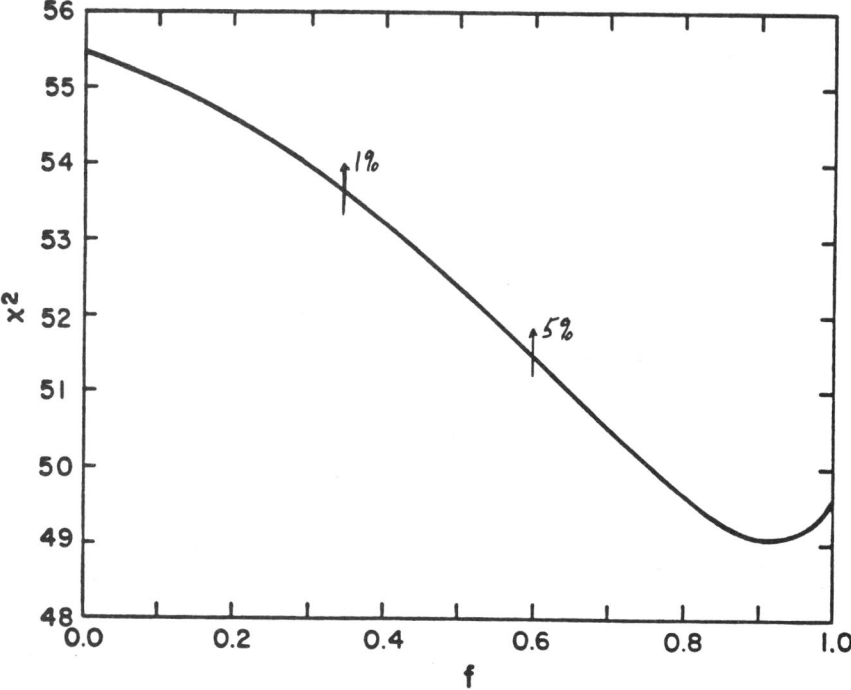

FIGURE 12. Chi-square versus f, the fraction of positrons annihilating via positronium. 57 degrees of freedom.

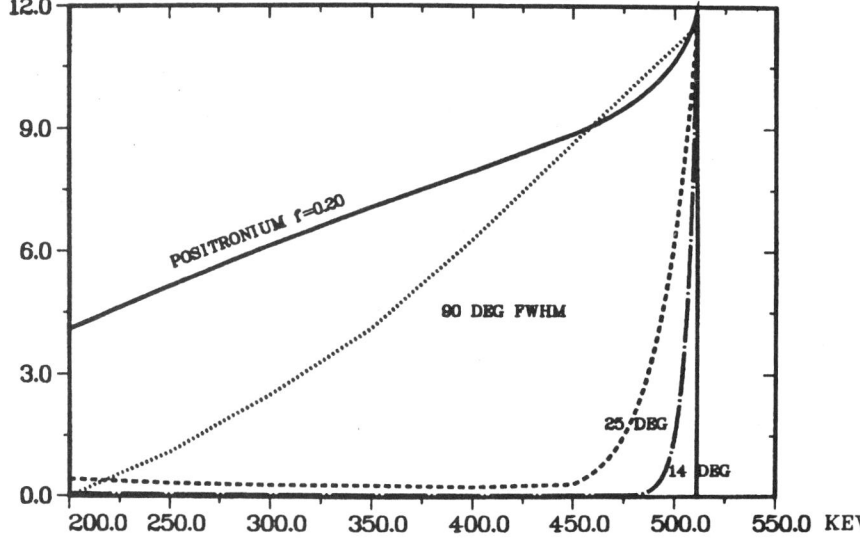

FIGURE 13. Continuum (in units keV^{-1} for incident line flux 10^{+4}) produced by scattering of the 511 keV line in the residual atmosphere above a typical balloon position. Parameter is telescope aperture.

appropriately scaled orthopositronium continuum is included, however, for the case f=0.2, to show that a wide-aperture, poor-resolution balloon instrument could mistake atmospheric scattering for roughly this much orthopositronium.

Aside from instrumental effects, there is the possibility that the continuum is astrophysical but not orthopositronium. Forrest <24> has shown that Compton scattering in an atmosphere of order 5 gm cm^{-2} thick surrounding the source could mimic orthopositronium if the spectral shape is not well measured. Riegler and others have pointed out that a distribution of gravitational redshifts could do the same.

The only other reported measurement of the orthopositronium continuum, by the New Hampshire group <16>, is already shown in Fig. 6. They find that a large positronium fraction f best fits their measured spectral shape. Of course, Leventhal's analysis <11> of the early Rice results points clearly to the same conclusion concerning that data.

JPL is currently analyzing results from HEAO-3 that for the first time will provide good-sensitivity, high-resolution continuum data at energies above the 511 keV line. Riegler (these Proceedings) reports that these data lie well above the inverse-power-law continuum as determined by data below 511 keV. This is in agreement with the HEAO-1 data of Sept77 <22> but may be in marginal disagreement with the high-energy Sandia/Bell data of Nov77. He finds that the best two-power-law fits to all the data require quite a low value for the orthopositronium continuum. Unless the spectral form is very well characterized either by data or theory, however, it is difficult to apportion the flux below 511 keV uniquely between orthopositronium and the underlying continuum. The question is interesting because the absence of the expected orthopositronium continuum would presumably be attributed to an ionizing ultraviolet flux in the annihilating medium of at least 10^{+25} ph cm^{-2} s^{-1}. Even the total ultraviolet luminosity inferred by Lacey et al <25> for the Galactic center region (~10^{+51} positronium-ionizing photons per second) could not produce such a flux at >~10^{+3} Schwarzschild radii (required by the JPL upper limits on the gravitational redshift of the 511 keV annihilation line) from a presumed central black hole, unless the black hole mass were less than $3\times10^{+4}$ solar masses. McKinley and Ramaty (private communication) are currently calculating self-Comptonized bremsstrahlung spectra, however, whose shape (for thick sources) would seem to allow for considerable orthopositronium continuum in fitting the experimental data.

OBSERVATIONS OF THE POWER-LAW CONTINUUM: CORRELATED VARIABILITY?

Continuum measurements of x- and gamma-radiation from the Galactic center region to date have been summarized by Matteson <22>.

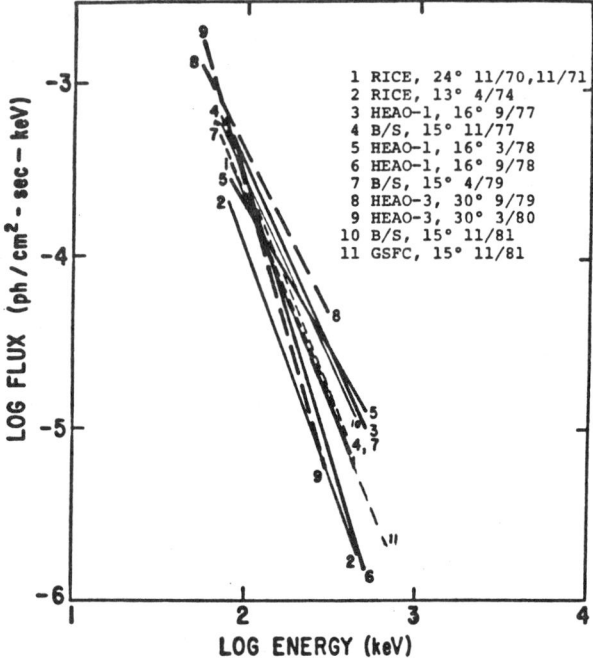

FIGURE 14. Best power-law fits to measured spectra of the Galactic center region which contain data above 200 keV.

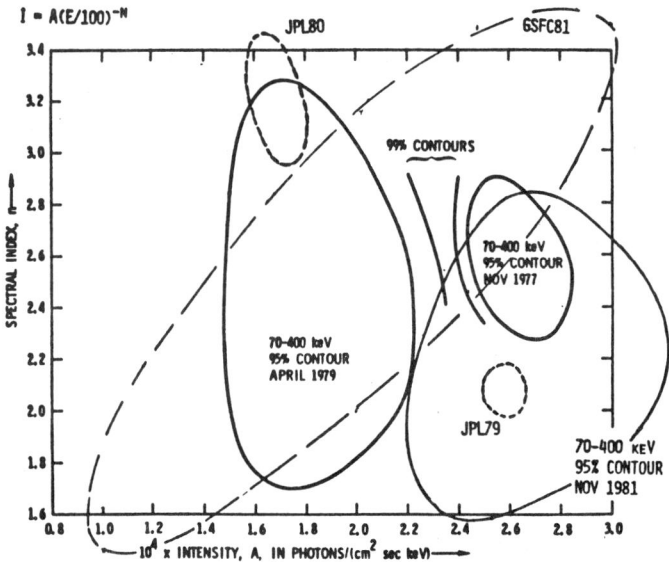

FIGURE 15. 95% confidence contours, spectral index versus intensity, for selected Galactic center spectra. Solid curves are calculated without allowing for non-diagonal instrument response; A values should be reduced about 40%, n values are little affected.

To oversimplify, it seems highly probable that a compact, variable source exists exactly at the Galactic center that is distinguished from the other dozen or so sources in the region by an extremely hard spectrum. Above 200 keV it is the dominant source. Observed spectra that include data above 200 keV are shown in Fig. 14. (As already mentioned, the Rice observation in Apr74 should be divided by an average number something like 0.62 to account for the angular response at a 6 degree offset.) Obvious variability is indicated, especially at higher energies. However, the angular resolutions indicated are large enough to include half a dozen or more of the multiple variable sources that HEAO-1 resolved in the 80-180 keV band. Only the power law fits to the data are shown, and these are, of course, strongly influenced by data below 200 keV where the statistical precision is greater.

The most extreme variability is shown by fits 8 and 9 to data taken by HEAO-3 just six months apart, precisely the data that caught the Galactic center in the act of "turning off" in the 511 keV annihilation line. This raises the question <26> of whether variations in the continuum and in the line strength are generally correlated. Figure 15 shows confidence contours in the intensity, spectral index plane for some of the fits shown in Fig. 14. The solid curves were calculated without correcting for non-vetoed response to partial-energy-loss events and should be reduced in intensity by a factor of about 1.4; the spectral index is not much affected. After this is done, the Sandia/Bell contour of Nov77 and the GSFC and Sandia/Bell contours of Nov81 all overlap (the 511 line being at those two times "on" and "off"), the Sandia/Bell contour of Apr79 and the JPL contour of Oct79 are disjoint (the 511 line being "on" at both times), and the JPL contours of Oct79 and Mar80 are disjoint (the 511 line being at those two times "on" and "turning-off"). Thus there is no obvious and consistent correlation. But certainly any conclusion is rendered suspect by the presence of unresolved and variable sources (e. g., Nova Ophiuchi 77). The question is a most important one and calls strongly for better data.

FURTHER OBSERVATIONS NEEDED

The mainstream of model building as reported in these Proceedings divides into two branches. Lingenfelter, Phinney, and others propose thermal pair production by photon-photon collisions near a small ($<10^{+3}$ solar mass) black hole, with the positrons escaping isotropically to distances much greater than the Schwarzschild radius. Burns, Lovelace, Novikov, and many others propose massive ($>10^{+6}$ solar masses) rotating black hole dynamos surrounded by rotating accretion disks and emitting huge jets of gamma rays (or e+e- pairs) which then interact with orbiting gas clouds. Measurements will help prune the proliferation of models:

1) Further measurements of the Galactic center continuum above 511 keV and up to several MeV will provide information about the central e+e- production region, which may be characterized by a temperature of some 10^{+10} K <22>. New balloon-borne instruments being built by GSFC/Bell/Sandia and by UCSD/UCB/CESR should begin providing such data in 1987. The Oriented Scintillator Spectrometer Experiment (OSSE) aboard the Gamma Ray Observatory (GRO) to be launched in 1988 will have an excellent sensitivity in this energy range.

2) Imaged measurements of the Galactic center continuum with about one degree resolution are sorely needed below 511 keV to reduce source confusion. The new balloon instruments will be the first to have this capability. OSSE's five degree resolution will be marginal in this respect but, with high sensitivity and sophisticated data analysis, should prove extremely useful.

3) Measurements of the time variability of the 511 line, correlated with the continuum measurements 1) and 2), will help distinguish between the two theoretical branches mentioned above. Cooperation will be required between OSSE and the balloon instruments, which alone will have adequate resolution to measure line shape, position, etc.

4) Firm determination of the (time dependent?) fraction of annihilations proceeding via positronium will provide constraints on the nature of the annihilating medium. This should follow from 1) and 2) above.

REFERENCES

<1> A. S. Jacobson, THE GALACTIC CENTER, editors G. R. Riegler and R. D. Blandford (New York: A. I. P., 1982), p. 123.

<2> R. E. Lingenfelter and R. Ramaty, THE GALACTIC CENTER, editors G. R. Riegler and R. D. Blandford (New York: A. I. P., 1982), p. 148.

<3> R. C. Haymes et al, Ap. J. 157, 1455 (1969).

<4> W. N. Johnson, III, F. R. Harnden, Jr. and R. C. Haymes, Ap. J. 172, L1 (1972).

<5> W. N. Johnson, III and R. C. Haymes, Ap.J. 184, 103 (1973).

<6> G. J. Fishman and D. D. Clayton, Ap. J. 178, 337 (1972).

<7> B. Kozlovsky and R. Ramaty, Ap.J. 191, L43 (1974).

<8> P. Guthrie and E. Tademaru, Nature Phys. Sci. 241, 77 (1973).

<9> R. Ramaty, G. Borner and J. M. Cohen, Ap.J. 181, 891 (1973).

<10> F. W. Stecker, Ap. and Space Sci. 3, 579 (1969)

<11> M. Leventhal, Ap. J. 183, L147 (1973).

<12> R. C. Haymes et al, Ap. J. 201, 593 (1975).

<13> F. Albernhe et al, Astron. Astrophys. 94, 214 (1981).

<14> M. Leventhal, C. J. MacCallum and P. Stang, Ap. J. 225, L11 (1978).

<15> R. W. Bussard, R. Ramaty and R. J. Drachman, Ap. J. 228, 928 (1979).

<16> B. M. Gardner et al, THE GALACTIC CENTER, editors G. R. Riegler and R. D. Blandford (New York: A. I. P., 1982), p.144.

<17> M. Leventhal et al, Ap. J. 240, 338 (1980).

<18> W. A. Mahoney et al, Nucl. Instrum. Methods 178, 363 (1980).

<19> G. R. Riegler et al, Ap.J. 248, L13 (1981).

<20> W. S. Paciesas et al, Ap.J. 260, L7 (1982).

<21> M. Leventhal et al, Ap. J. 260, L1 (1982).

<22> J. L. Matteson, THE GALACTIC CENTER, editors G. R. Riegler and R. D. Blandford (New York: A. I. P., 1982), p. 109.

<23> R. Ramaty and R. E. Lingenfelter, Phil. Trans. Roy. Soc. London A301, 671 (1981).

<24> D. Forrest, THE GALACTIC CENTER, editors G. R. Riegler and R. D. Blandford (New York: A. I. P., 1982), p. 160.

<25> J. H. Lacey, C.H. Townes and D. J. Hollenbach, Ap. J. 262, 120 (1982).

<26> G. R. Riegler et al, Proceedings of the 17th International Cosmic Ray Conference, Paris, July 1981.

THE GAMMA-RAY SPECTRUM OF THE GALACTIC-CENTER REGION

G. R. Riegler, J. C. Ling, W. A. Mahoney,
W. A. Wheaton, and A. S. Jacobson
Jet Propulsion Laboratory,
California Institute of Technology,
Pasadena, CA 91109

ABSTRACT

The HEAO-3 High Resolution Gamma-Ray Spectrometer observed the galactic center region in fall 1979 and in the spring of 1980. Variation of the positron annihilation line at 511 keV has been reported previously. The fall 1979 observations show a significant high-energy continuum at energies above 511 keV. The intensity of a possible positronium triplet-state continuum is found to be less than that expected for direct positron annihilation and positronium decay in an ionized, warm (T $\lesssim 10^5$ K) plasma; depending on assumption for the shape of the high-energy continuum spectrum, positronium fractions between 0.0 and 0.75 (at 90% statistical confidence level) are consistent with the observations.

INTRODUCTION

Measurements of the spectrum of the galactic center region in the 50 keV to 3 MeV range have been carried out with scintillators[1,2] and with solid-state detectors[3-8,12]. We have previously reported[6] position information for the observed 511 keV positron annihilation radiation, and that it is consistent with emission from a single point source. In the discussions and conclusions below, we assume that the observed gamma-radiation is emitted by a single source which is taken to be the "galactic center" defined by radio[9] and infrared[10] measurements. We point out, however, that the available gamma-ray data are also consistent with emission from one or more sources at locations near the galactic center which have been observed at hard x-ray energies[11]. Therefore, we will refer to emission from the "galactic center region".

In this paper we describe results of a detailed spectrum analysis for the first observations of the galactic center region with the HEAO C-1 High-Resolution Gamma-Ray Spectrometer between September 24, 1979 and October 13, 1979 (the "fall 1979" observation). The results of the second observation during spring 1980 will be reported elsewhere. We present here the first observation of an astronomical gamma-ray source with a high-resolution detector over the energy range from 50 keV to ~ 3 MeV, and the interpretation in terms of a complex spectrum with line and continuum emission.

The gamma-ray spectrum of the galactic center region shows four distinct features: (a) a low-energy continuum up to ~ 170 keV; (b) a medium-energy continuum from ~ 170 keV to 511 keV; (c) the positron annihilation line at 511 keV; and (d) the high-energy continuum above 511 keV.

The low-energy continuum has a powerlaw shape, and was previously reported to vary in intensity and slope[3,12]. The positron annihilation line was first unambiguously observed by Leventhal, MacCallum and Stang[3], and was later found to be variable[5-8]. In the present analysis we have fitted the high-energy continuum from 511 keV to ~ 3 MeV with either powerlaw or blackbody functions.

Our model for the observed gamma-ray flux in the medium-energy band is the sum of extrapolations of the low-energy and high-energy continua. As an additional component of some of the models we have also added an independently variable triplet-state positronium continuum. It will be shown below that the amount of triplet-state positronium continuum required for a spectrum fit depends strongly on our assumption for the shape of the high-energy continuum.

Positrons annihilate with electrons either directly or by first forming positronium. When triplet-state positronium decays, it yields three photons of variable energy E < 511 keV which are part of the "triplet-state continuum"; when singlet-state positronium decays, it produces two photons of energy E = 511 keV. Because of the threefold degeneracy of the azimuthal spin quantum numbers, the triplet-ortho state of positronium is formed three times as often as the singlet-para state. Therefore, for normal positronium decay there are $(0.75 \times 3) / (0.25 \times 2) = 4.5$ times as many photons in the full triplet-state continuum I_t as there are in the singlet-state annihilation line I_s at 511 keV. For each spectrum model we can use the observed 511 keV line intensity I_{511} and the best-fit triplet-state continuum intensity I_t to define a positronium fraction $f = I_s/I_{511} = I_t/(4.5 \times I_{511})$. In other words, the positronium fraction is that the fraction of the observed 511 keV line which would be associated with normal positronium decay as indicated by the triplet-state continuum.

A positronium fraction of f=0.9 was observed by Leventhal, MacCallum and Stang[3] for the galactic center region. This value is, in fact, consistent with the relative likelihood of direct positron annihilation and positronium formation in warm, mildly ionized plasmas which are consistent with the observed annihilation line width[13].

DATA ANALYSIS METHOD

We have previously[6] described analysis methods which are useful for the signal-to-background ratios and background variability encountered in gamma-ray astronomy. After extensive study we find that the most reliable results are obtained from the following method: for each scan of the candidate source (the HEAO-3 spectrometer executed a great-circle scan of the sky once every approximately 20 minutes), we determine the background level and source intensity in each of several energy bands. The intensity valves from many scans are then averaged to obtain the final estimates.

The line flux values presented here are somewhat different from those quoted earlier[6] for three reasons: (1) in the present analysis we have used a larger amount of data, (2) the present analysis method outlined above, is more mature than the method used previously, and (3) the dependence of the line flux calculation on the shape of the underlying continuum (see below) has been included.

OBSERVATIONS

The spectrum of the galactic center region in 18 bands from 50 keV to 3.12 MeV is shown in the figure. The major spectrum parameters for five spectrum fits are given in the table. All spectrum models contain at least one powerlaw (to fit the low-energy region of the spectrum) and the 511 keV annihilation line.

Best Spectrum Fits for the Fall 1979 Observations
of the Galactic Center Region*

Model	1	2	3	4	5
Continuum Shape	1 PWR	1 PWR+ triplet	2 PWRs	2 PWRs+ triplet	1 PWR + triplet + blackbody
Powerlaw Index	2.68 ± 0.03	2.93 ± 0.03	2.93 ± 0.03	2.92 ± 0.03	2.95 ± 0.03
Second Index α or kT(keV)	--	--	1.22 ± 0.04	1.14 ± 0.05	242 ± 4
Positronium Fraction f	0.0	0.81 ± 0.12	0.0	0.26 ± 0.10	0.60 ± 0.08
Annihilation Line Flux 10^{-3} photons cm^{-2} s^{-1}	2.12 ± 0.22	1.70 ± 0.22	1.77 ± 0.22	1.70 ± 0.22	1.57 ± 0.22
χ^2/d.o.f.**	18.4	10.2	2.6	2.5	2.8

*All errors at 68% confidence level for a single parameter[14].
PWR = powerlaw.
**The figure shows that the main contributions to the relatively poor χ^2-values come from the 170-500 keV region. This suggests that the spectrum is more complex than the models used here.

The χ^2/d.o.f. (degree of freedom) values in the table show that models 1 and 2, i.e., a single powerlaw without or with a triplet-state continuum, do not fit the data adequately and must be rejected. Inspection of the figure shows that the observed high-energy flux values require the addition of a second continuum component to the spectrum models.

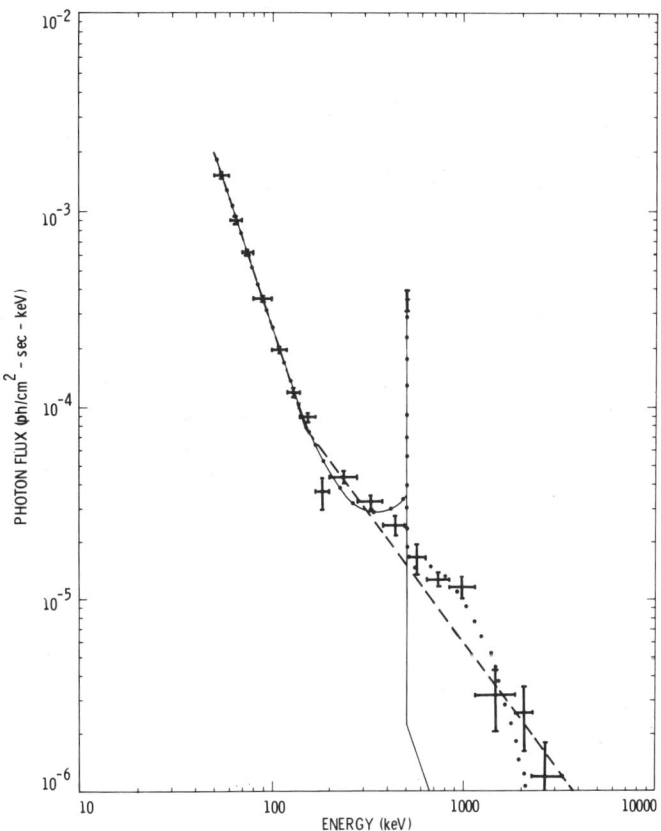

Spectrum of the galactic center region observed during a 20-day period between September 24 and October 13, 1979. The 511 keV data point is shown for a band width of 6.0 keV (see Reference 6 for line width and center energy). In addition to the annihilation line at 511 keV, the model spectra consist of: a single powerlaw plus triplet-state continuum (thin solid line; Model 2 in table); two powerlaws (broken curve, Model 3); or a powerlaw plus blackbody distribution plus triplet-state continuum (dotted curve, Model 5).

Models 3, 4 and 5 in the table include a second continuum component, and all have comparable χ^2/d.o.f. values. Model 3 uses two powerlaws without any triplet-state continuum. In Model 4, a triplet-state continuum was added; for a best-fit positronium fraction of $f = 0.26 \pm 0.10$, no significant improvement in the goodness-of-fit was obtained. In Model 5, we include a blackbody (Planck-function) spectrum and a triplet-state continuum in addition to the low-energy powerlaw. This model results in a maximum value for the positronium fraction since the blackbody distribution shows a positive slope at energies below the Wien peak at $E \sim 2.8kT$ and therefore requires addition of a larger amount of triplet-state continuum than, say, a powerlaw distribution with negative slope. The best-fit positronium fraction for this case is $f = 0.60 \pm 0.08$ (where the quoted error is calculated at 68% confidence for a single parameter[14]). At the 90% confidence level, this fit is consistent with positronium fractions from $f = 0.41$ to $f = 0.75$.

INTERPRETATION

If the observed source is at a distance $d = 10$ kpc and radiates isotropically, then its high-energy luminosity at $E = (511-3120)$ keV is $L_h = (2.9 \pm 0.4) \times 10^{38}$ erg/sec. This luminosity is higher than the value of $L \sim 2 \times 10^{38}$ erg/sec observed during measurements in 1974[1] and 1977[2]. However, the shape of the continuum spectrum presented here is qualitatively similar to that obtained previously with a scintillator instrument[2].

As was pointed out by Lingenfelter and Ramaty[15], the ratio of the positron annihilation line flux to the high-energy continuum flux is an indication of the efficiency of the positron production process.

For the fall 1979 measurement and spectrum model 3 we deduce a 511 keV luminosity of $L_{511} = (1.6 \pm 0.2) \times 10^{37}$ erg/s and a resulting efficiency of $e = L_{511}/L_h = 0.05$. For the spectrum-fit models 4 and 5, the added triplet-state continua yield $e = 0.12$ and $e = 0.20$, respectively. These high efficiencies place stringent requirements on models for the production of positrons; however, Lingenfelter and Ramaty[15,16] have pointed out that photon-photon collisions of high-energy (> 511 kev) photons may be consistent with the range of values required here.

We point out that the observed positrons and positronium may have been generated at an earlier epoch, perhaps at a lower efficiency from a more intense flux of high-energy continuum radiation. As mentioned above, an observation of the galactic center region two years prior to that reported here yielded a lower luminosity for the high-energy continuum.

The spectrum fits presented above demonstrate that the choice of the high-energy continuum model shape has a dramatic effect on the question of whether or not a triplet-state continuum must be added to the spectrum model, and influences conclusions regarding the positronium fraction?

Depending on the model, we have observed positronium fractions of 0.0, 0.26 or 0.60; at the 90% confidence levels, all positronium fractions from 0.0 to 0.75 are consistent with models 3, 4 and 5 in the table. In view of the theoretical predictions for a positronium fraction $f \approx 0.9$[13], the important conclusion from the fall 1979 observations is that the observed positronium fraction is lower than expected. If we assume that most of the positrons, once they have been generated, form positronium rather than annihilate directly[13], then either spin-flip or ionization of positronium must alter the continuum-to-triplet intensity ratio. Collisions[17] could produce an apparent change in the positronium fraction; however, this process is not likely since it requires ambient plasma densities of at least 10^{14} atoms/cm^3. Photoionization by UV photons[18] may be more efficient efficient, and would require UV photon densities which are consistent with current models for emission from the galactic center region[16].

ACKNOWLEDGMENT

The research described in this paper was carried out at the Jet Propulsion Laboratory, California Institute of Technology, under contract with the National Aeronautics and Space Administration.

REFERENCES

1. R. C. Haymes, G. D. Walraven, C. A. Meegan, R. D. Hall, F. T. Djuth, and D. H. Shelton, Ap. J., 210, 593 (1975).
2. J. L. Matteson, in "The Galactic Center", AIP Conference Proceedings 83, G. R. Riegler and R. D. Blandford Editors, American Institute of Physics, p. 109 (1982).
3. M. Leventhal, C. J. MacCallum, and P. D. Stang, Ap. J. (Letters) 225, L11 (1978).
4. M. Leventhal, C. J. McCallum, A. F. Huters, and P. D. Stang, Ap. J., 240, 338 (1980).
5. M. Leventhal, C. J. MacCallum, A. F. Huters, and P. D. Stang, Ap. J. (Letters), 260, L1 (1982).
6. G. R. Riegler, J. C. Ling, W. A. Mahoney, W. A. Wheaton, and T. A. Prince, Ap. J. (Letters) 248, L13 (1981).
7. W. S. Paciesas, T. L. Cline, B. J. Teegarden, J. Tueller, P. Durouchoux, and J. M. Hamejry, Ap. J. (Letters), 260, L7 (1982).
8. F. Albernhe, J. F. Leborgne, G. Vedrenne, D. Boclet, P. Durouchoux, and J. M. DaCosta, Astron. Astrophys., 94, 214 (1981).
9. K. Y. Lo, in "The Galactic Center", AIP Conference Proceedings 83, G. R. Riegler and R. D. Blandford Editors, American Institute of Physics, p. 1 (1982).
10. J. H. Lacy, in "The Galactic Center", AIP Conference Procedings 83, G. R. Riegler and R. D. Blandford Editors, American Institute of Physics, p. 53 (1982).
11. A. M. Levine et. al, submitted to Ap. J. Suppl. (1983).
12. G. R. Riegler, J. C. Ling, W. A. Mahoney, W. A. Wheaton, and A. S. Jacobson, in Proceedings of 17th Int. Cosmic Ray Conf., Paris, 9, 85 (1981).
13. R. W. Bussard, R. Ramaty, and R. J. Drachman, Ap. J., 228, 928 (1979).
14. Y. Avni, Ap. J., 210, 642 (1976).
15. R. E. Lingenfelter and R. Ramaty, in "The Galactic Center", AIP Conference Proceedings 83, G. R. Riegler and R. D. Blandford Editors, American Institute of Physics, p. 148 (1982).
16. R. E. Lingenfelter and R. Ramaty, in "Positron-Electron Pairs in Astrophysics", AIP Conference Proceedings, M. L. Burns, A. K. Harding and R. Ramaty Editors, American Institute of Physics (1983).
17. C. J. Crannell, G. Joyce, R. Ramaty and C. Werntz, Ap. J., 210, 582 (1976).
18. R. C. McCray, private communication (1983).

EFFECTS OF LINE WIDTH AND SPATIAL EXTENT ON MEASUREMENTS OF THE 0.51 MEV GALACTIC CENTER LINE

P. P. Dunphy, E. L. Chupp, and D. J. Forrest
University of New Hampshire, Durham, NH 03824

ABSTRACT

Over the past 15 years, a number of measurements have been made of positron-electron annihilation radiation from the Galactic Center region. The results after 1979 show a significant decrease in measured flux intensity from that previously observed with the same instruments. This is probably due to time variations; however, the contributions of a spatially extended and/or energy-broadened component should also be considered. The entire data set is consistent with a time variable point source plus a distribution along the Galactic Disk. There is no strong evidence for a component broadened in energy.

INTRODUCTION

A number of observations of the Galactic Center region in positron-electron annihilation radiation have been carried out since the first indication of a measurable flux in 1968 [1]. Several reviews of these observations have been published recently [2,3,4]. When plotted versus the epoch of observation, the flux measurements show a clear variation which is not consistent with a constant value. The question is, what are the causes of the variations? In the following, we will discuss some possible explanations.

POSSIBLE CAUSES OF MEASUREMENT VARIATIONS

A possible (but the least interesting) reason for variations in the flux measurements is simply the systematic errors in the individual measurements. It can be assumed, however, that every effort was made by the experimenters to minimize such errors. A second class of systematic effects includes differences between detector systems and observational techniques which lead to differences in the apparent flux values.

Before considering these problems, however, we look into the question of intrinsic source variability. The most effective way to evaluate the data for source variability is to compare measurements made at different times with the same detector. This bypasses the problem of differences among detectors. The strongest evidence for variability is shown in the HEAO-3 JPL γ-ray spectrometer scans of the Galactic Center during the fall of 1979 and the spring of 1980 [5]. The annihilation line flux from a point-like source near the Galactic Center decreased by a factor of 3 between 1979 and 1980. This interpretation is supported by a series of 3 balloon-based observations by the Bell-Sandia collaboration [3]. The fact that all measurements before 1980 show a positive flux, while all measurements afterward show a lower flux, lends further support to the reality of this change. On the other hand, there is no such clear-cut evidence for time variations before 1980. The Rice measurements [1] carried out with the same detector in 1970 and 1971 show no variation, and the Bell-Sandia measurements [3] of 1977 and 1979 are marginally consistent ($\sim 10\%$ probability) with a constant source.

We now consider various detector properties and source characteristics (other than time variations) which might explain the differences in the measured fluxes. One possibility is the presence of both narrow ($\lesssim 1$ keV) and broad ($\gtrsim 10$ keV) line components at the source.

The positron annihilation line width can be affected by the temperature and ionization fraction in the annihilation region[6] as well as macroscopic kinematic and gravitational effects. If the Galactic Center radiation consists of both a narrow and an energy-broadened line component, this would lead to higher measured fluxes in the broad resolution scintillation detectors compared to the narrow resolution solid state detectors. The energy resolution of the various detectors have been tabulated by Leventhal and MacCallum [3]. In Fig. 1 we have plotted the measured flux intensity against detector energy resolution. Since there is no apparent correlation between these parameters, we can conclude that the differences are dominated by some other effect.

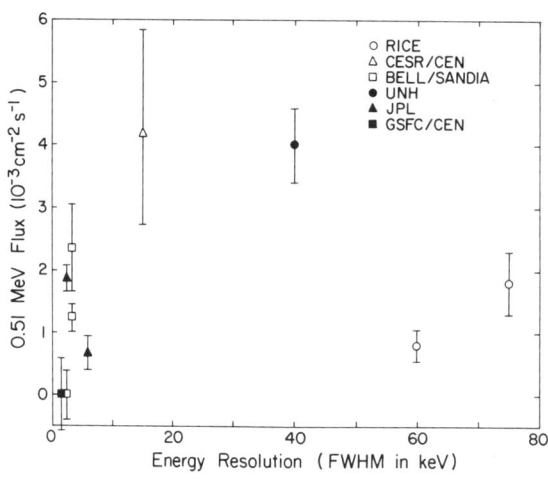

Fig. 1. Measurements of the positron annihilation flux plotted versus detector energy resolution. References can be found in the review of Leventhal and MacCallum [3].

Measurements of the 0.51 MeV line could also be influenced by a combination of detector field-of-view (angular aperture) and a source extent that is large compared to the aperture. Here, detectors with larger apertures would effectively see a larger flux. Fig. 2 shows the measurements plotted versus the aperture size of the detectors.

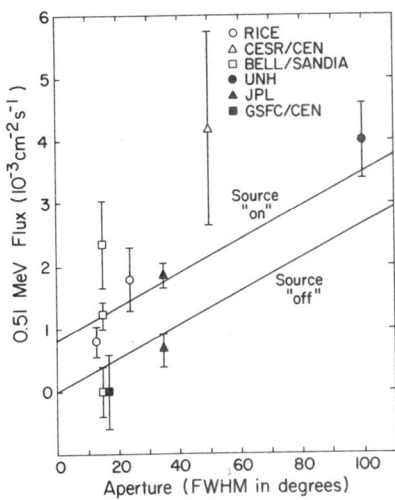

Fig. 2. Measurements of the positron annihilation flux plotted versus detector aperture angle. The sloping lines represent a model consisting of a linear source distribution along the Galactic Disk, plus a time-varying point source.

Unlike the plot in Fig. 1, there is an ordering of the data which suggests a correlation between the measured flux and detector aperture. To fit the data, we use a model discussed by Jacobson [2], but we also include more recent data points. Consider a linear distribution of radiation along the Galactic Plane that extends $\gtrsim 45°$ on both sides of the Galactic Center. This would explain the linear relationship between flux and aperture size. Furthermore, we postulate a point source at the Galactic Center which remained approximately constant in intensity until the end of 1979, when it decreased to near or below the level of detectability. We have performed a weighted least-squares fit to the data using the reported flux values and standard deviations. The fit for this model, with the 2 parameters of point source and Disk source intensity, is shown in Fig. 2. The flux from the point source (when "on") is $8.2(\pm 1.9) \times 10^{-4}$ cm^{-2}s^{-1} and that of the Disk source is $1.5(\pm 0.3) \times 10^{-3}$ cm^{-2}s^{-1}rad^{-1}. It is important to remember that the original data points were obtained assuming a point source model only, and so this quantitative approach is by no means exact. The important point is the general behavior of the data.

DISCUSSION AND CONCLUSIONS

The above analysis indicates that source extent may be an important factor in comparing measurements of the Galactic Center annihilation line. This is not to say that other systematic effects, such as line width and time variations before 1980, do not come into play, but that source distribution appears to be the dominant one. One factor which has not been considered above, although it would act like a broad-line component, is the effect of a positronium continuum in the spectrum. This does not affect the measurements with solid state detectors, but would increase the apparent line flux measured by scintillation detectors. Best-fit values for the positronium fraction in several measurements range from 0.5 to 0.9[7,8,9]. Taking the positronium fraction to be $\sim$0.7, a reduction of $\sim$40% should be applied to the Rice 1970-71 flux measurements [1] and perhaps to the Rice 1974 flux measurement[10] for consistency.

Another factor which can affect the comparability of the results is the technique for determining background. In the case of the balloon-borne experiments, background was found by varying the detector azimuth or by using drift scans. Since drift scans move across the Galatic Disk rather than along it, any linear distribution of the source along the Galactic Disk should not affect the results appreciably. The HEAO-3 satellite measurements, however, were obtained by scans along the Galactic Disk. This would tend to underestimate the flux from the Galactic Center region if there is a non-negligible flux extending all along the Disk. The Rice observation in 1974, furthermore, was centered on GX 1 + 4, so this data point should be corrected upward by a factor of $\sim$1.3 if the source is exactly at the Galactic Center. Note that this factor, together with the correction for a hypothetical positronium contribution in the Rice data, does not significantly affect the model in Fig. 2.

In summary, we can say that: 1. There is conclusive evidence for a decrease in the 0.51 MeV Galactic Center flux between the boundaries of the HEAO-3 observations which are supported by all measurements made to date, and 2. There is strong evidence for the existence of a Disk distribution in addition to a time-varying point source. A number of models have been proposed which could explain a distributed source; these include low energy cosmic rays, supernovae, novae, and pulsars[11]. Confirmation of the existence of such a source could come from further observations with collimated detectors of varying aperture, scans perpendicular to the Galactic Plane, or observations with imaging detectors having large fields-of-view.

ACKNOWLEDGEMENTS

The authors thank Mary M. Chupp for editing and Robert

Hoffman for typing this paper. This work was supported by NASA Grant NGL 30-002-021.

REFERENCES

1. W. N. Johnson and R. C. Haymes, Ap. J. 184, 103 (1973).
2. A. S. Jacobson, The Galactic Center, AIP Conference Proceedings No. 83 (American Institute of Physics, New York, 1982), p. 123.
3. M. Leventhal and C. J. MacCallum, The Galactic Center, AIP Conference Proceedings No. 83 (American Institute of Physics, New York, 1982), p. 132.
4. C. J. MacCallum (this volume).
5. G. R. Riegler, J. C. Ling, W. A. Mahoney, W. A. Wheaton, J. B. Willet, and A. S. Jacobson, Ap. J. (Letters) 248, L13 (1981).
6. R. W. Bussard, R. Ramaty, and R. J. Drachman, Ap. J. 228, 928 (1979).
7. M. Leventhal, C. J. MacCallum, and P. D. Stang, Ap. J. (Letters) 225, L11 (1978).
8. M. Leventhal, C. J. MacCallum, A. F. Huters, and P. D. Stang, Ap. J. 240, 338 (1980).
9. B. M. Gardner, D. J. Forrest, P. P. Dunphy, and E. L. Chupp, The Galactic Center, AIP Conference Proceedings No. 83 (American Institute of Physics, New York, 1982), p. 144.
10. R. C. Haymes, G. D. Walraven, C. A. Meegan, R. D. Hall, F. T. Djuth, and D. H. Shelton, Ap. J. 201, 593 (1975).
11. R. Ramaty and R. E. Lingenfelter, Nature 278, 127 (1979).

LOW-TEMPERATURE POSITRON ANNIHILATION

Richard J. Drachman
Laboratory for Astronomy and Solar Physics
NASA/Goddard Space Flight Center, Greenbelt, MD 20771

ABSTRACT

For a satisfactory understanding of astrophysical annihilation radiation, expecially that observed from the galactic center direction, the interaction of positrons with the ambient medium must be carefully investigated. Although hot, ionized regions may be important sources of annihilation radiation, in this report I will examine mainly the simpler processes occurring in low-temperature neutral hydrogen gas. The goal is to set limits on conditions in the annihilation region by using the predictions of atomic theory compared with the observed γ-ray line width, continuum strength and time dependence.

INTRODUCTION

Suppose that positrons of energy high compared to the ionization energy of atomic hydrogen (13.6 eV) enter a region containing cold, non-ionized atomic hydrogen and nothing else. Assume further that no magnetic field or radiation is present in the region and that the hydrogen density is much less than 10^{12} cm^{-3}. We are thus considering a very simple, idealized model for the galactic center annihilation source and will try to predict the properties of the annihilation radiation emerging from this source. In spite of the simplicity of the model, it will be seen that a good deal of interesting physics is involved in its analysis. At the present time, there are no experimental data on the collision of positrons with atomic hydrogen, as there are for molecular hydrogen[1] and atomic helium.[2] On the other hand, the theoretical situation, although still imperfect, is relatively good, as I will show.

The observations[3-5] of the galactic center annihilation radiation are easily summarized:
1. A narrow line is seen; the best measurement[5] of its width (FWHM) gives 3.13 ± 0.57 keV. With a detector resolution of 2.72 keV, the source line width is 1.6(+0.9,-1.6) keV, consistent with zero.
2. The energy of the line is very accurately that expected for $e^+ - e^-$ annihilation at rest, 510.90 ± 0.25 keV compared with 511.00 keV in the laboratory.
3. There is some evidence[4] for the existence of a continuum component on the low-energy side of the line in addition to a power-law background.
4. The intensity of the line radiation is definitely time-variable,[5] with a time constant less than about 1/2 year.

In the next section the physical processes leading to annihilation of positrons will be described, and the resulting line width will be shown to be too large to agree with observations. The standard modification, allowing the annihilation region to be partly ionized will also be discussed. In the final section a possible loophole will be described, perhaps allowing retention of the original picture of a cold, non-ionized region and predicting some observational consequences. Much of the discussion below is based on the work of Bussard et al.[6] and Crannell et al.[7]

POSITRON ANNIHILATION

Five different reactions can occur in positron-hydrogen collisions at moderate energies (13.6 eV < E < 1 keV):

$e^+ + H \rightarrow e^+ + H$ (Elastic Scattering) (1a)
$\rightarrow 2\gamma + H^+$ (Direct Annihilation) (1b)
$\rightarrow e^+ + H^*$ (Excitation) (1c)
$\rightarrow e^+ + e^- + H^+$ (Ionization) (1d)
$\rightarrow Ps + H^+$ (Positronium Formation) (1e)

Stecker[8] pointed out very early that in this energy range only the last process is effective in annihilating positrons, since Ps formation has an atomic-sized cross-section of order a_0^2 (~10^{-16} cm^2) while process 1b is radiative, of order r_0^2 (~10^{-26} cm^2). Essential to this argument is the low density of scatterers; once a Ps atom is formed it has no further collisions in the short time before it annihilates into 2 γ-rays (1.25×10^{-10} sec) or 3 γ-rays (1.4×10^{-7} sec). (The opposite occurs in most laboratory experiments where high densities are maintained; Ps formation in this energy range is merely an inelastic process, since the Ps atom is almost always collisionally ionized before it annihilates.)

With these considerations in mind, I can describe the life story of a positron very simply. It enters the cold atomic hydrogen region at some high energy and loses energy by processes (1c) and (1d). At about 100 eV process (1e) begins to become non-negligible, and for the rest of its lifetime every collision carries a certain increasing risk of Ps formation and immediate annihilation. It is the velocity distribution of the Ps atoms at the instant of their annihilation into 2 γ-rays that determines the width of the observed annihilation line. Before examining this question more quantitatively let us first note that positrons of energy E produce positronium atoms of energy $E - E_0$ (where $E_0 = 6.8$ eV, the threshold for process (1e)) giving a rectangular distribution in the line-of-sight component of velocity v_z, ranging between $v_z = \pm \sqrt{(E-E_0)/m_e}$. Since the Doppler shift of one of the annihilation γ-rays is $\Delta = (v_z/c)m_ec^2$, one gets a rectangular line profile of width

$$\Gamma = 2\Delta = 1.430\sqrt{E(ev)-6.80} \text{ keV}. \tag{2}$$

For this monoenergetic positron distribution to satisfy the

experimental constraint $\Gamma \leq 2.5$ keV Eq. (2) requires $E \leq 9.86$ eV; we will see shortly that this is an unreasonably low energy.

In Ref. 6, a detailed Monte Carlo simulation is carried out, in which an ensemble of positrons is followed downward in energy until annihilation takes place. The resulting γ-ray line histogram corresponds to a width of 6.5 keV, in clear disagreement with observation. Rather than repeating the details here, I will use a simple continuous slowing-down model to describe the thermalization of the positrons. First, however, it is necessary to review the status of positron-hydrogen scattering theory, beginning with the Ps-formation cross-section.

This cross-section is difficult to compute accurately, in part because of the unsymmetrical relationship between initial and final states in Eq. (1e); the center of mass coordinate of the Ps atom is not a natural one for the initial e^+ - H state. Nevertheless, a number of two-channel calculations have been carried out both for s-waves[9] and p-waves,[10] and they agree in predicting a great reduction below the Born cross-section[11] for Ps formation, but they do not extend up to energies of interest here. For that reason, a phenomenological extension to higher energies was carried out,[12] which used a modified Born approximation whose L=0 and L=1 partial cross-sections were reduced to agree with Refs. 9 and 10, and whose L > 1 terms were unchanged. The only real test of this approximation comes from a comparison with the total inelastic cross-section of Winick and Reinhardt,[13] obtained by a sophisticated analytic technique. Although no distinction is made between Ps formation, excitation and ionization, only the first of these is energetically allowed for energies between 6.8 eV and 10.2 eV. There is reasonably good agreement in this energy range between the results of Refs. 12 and 13, encouraging me to use the results of Ref. 12 in the rest of the analysis. (Note that Ps formation in excited states has been neglected here; an increase of less than 20 percent might be expected in the cross-section at higher energy.)

I do not know of any positron-hydrogen ionization calculations (except for the Born approximation which does not distinguish e^+ from e^-.) A very simple analytic form has been devised by Lotz[14] for the e^- - H ionization cross-section:

$$\frac{\sigma_I}{\pi a_0^2} = \frac{2.47 \ln E [1 - 0.6 \, e^{-0.56(E-1)}]}{E}, \qquad (3)$$

where E is the energy in Rydbergs. I will use Eq. (3) in the analysis, although there is no estimate of error in the e^+ case.

On the other hand there is a recent close-coupling calculation of the positron impact cross-section for excitation of the n=2 levels of hydrogen.[15] Up to at least $E = 7$ Ry (≈ 100 eV) these are well fitted by a formula like that of Eq. 3:

$$\frac{\sigma_{exc}}{\pi a_0^2} = \frac{2.73 \ln (4/3 \, E)(1 - 0.63 \, e^{-0.531E})}{E}. \qquad (4)$$

The n=2 excitation should dominate the total excitation process, so I will use Eq. (4) in the further analysis.

In Fig. 1 the Ps-formation cross-section and the inelastic scattering cross-section ($\sigma_I + \sigma_{exc}$) are plotted as functions of energy. To give an idea of the uncertainty in these quantities,

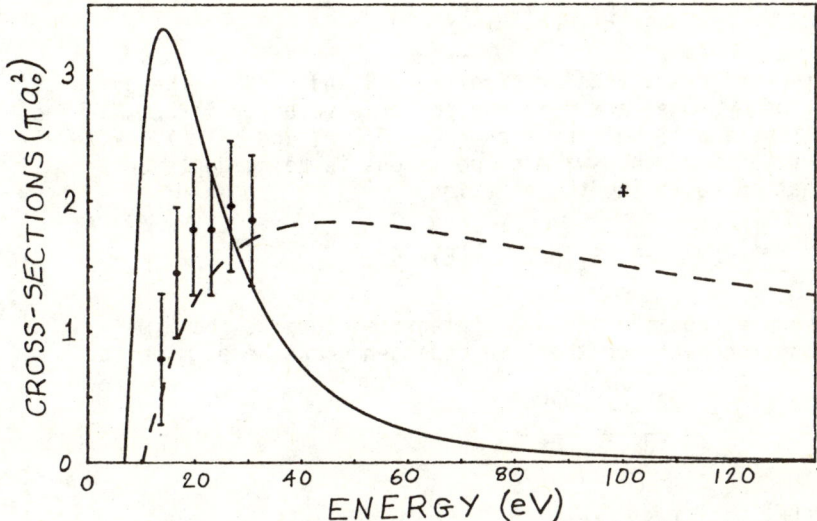

Fig. 1. Positron-hydrogen cross-sections as functions of energy. The solid line is for Ps formation (Ref. 12) and the dashed line is $\sigma_I + \sigma_{exc}$ (Eqs. 3 and 4). The dots and the cross are estimates of $\sigma_I + \sigma_{exc}$ obtained by combining the results of Ref. 12 with those of Ref. 13 and Ref. 16, respectively.

I have added a few additional points: the six low-energy points represent the difference between total inelastic from Ref. 13 and Ps-formation from Ref. 12; the point at 100 eV is a similar result from an eikonal approximation.[16] These are both higher than the cross-section used here, but not, I think, in too serious disagreement; for atomic research more accuracy is desirable, but for astrophysics the present results are quite satisfactory.

Before carrying out a more detailed calculation of positron slowing down and γ-ray line shape I can draw some semi-quantitative conclusions from Figure 1 itself. Notice that the two competing processes, inelastic scattering and Ps formation, are equally probable at a positron energy of 28 eV. Of the positrons surviving to reach this energy one-half will form Ps at their next collision, and from Eq. (2) they will give a line width of 6.58 keV. It is thus unlikely that the observed width of $\Gamma \leq 2.5$ keV can be achieved.

To account correctly for the large energy losses occurring at each inelastic collision a Monte Carlo simulation is needed.[6] But an approximate, qualitatively correct treatment involving a continuous slowing-down approximation is easy to formulate and yields results much like the more exact ones. Assume that all positrons enter with the same high initial energy E_1 and lose energy according to the equation

$$\frac{dE}{dt} = -[\sigma_{exc}(E) \Delta_E + \sigma_I(E) \Delta_I] Nv(E) \tag{5}$$

where the cross-sections are from Eqs. 3 and 4, N is the number density of hydrogen and v is the positron velocity, $\sqrt{2E/m_e}$. The energy losses will be taken[6] as $\Delta_E = 10.2$ eV and $\Delta_I = 17$ eV. At the same time, the swarm of positrons is being depleted by Ps-formation following the equation

$$\frac{dn(E)}{dt} = -\sigma_{Ps}(E) n(E) v(E) N \tag{6}$$

These coupled equations can be integrated to give the time history of a positron swarm or the time dependence can be eliminated:

$$\frac{dn(E)}{dt} = \frac{dn(E)}{dE} \frac{dE}{dt}, \tag{7a}$$

$$\frac{dn(E)}{dE} = g(E) n(E), \tag{7b}$$

where $g(E) = \sigma_{Ps}(E)/[\sigma_{exc}(E) \Delta_E + \sigma_I(E) \Delta_I]$. Eq. (7b) has the trivial solution

$$n(E) = n(E_1) \exp \int_{E_1}^{E} dE' g(E'), \tag{8}$$

and I have plotted $n(E)/n(E_1)$ in Fig. 2, where $E_1 = 20Ry = 272eV$ is a high enough energy to be considered asymptotic, since $g(E_1)$ is very small. It is, of course, unrealistic to carry the solution below $E \approx \Delta_I$, since there the continuous slowing-down approximation is grossly incorrect.

The principal conclusion to be drawn from Fig. 2 is that the half-value energy, where only 1/2 of the original positrons still survive, is at a high energy, $E_{1/2} = 39$ eV. This is a further indication that most of the Ps atoms formed are moving too fast to give the required narrow annihilation line. Furthermore, the exact line profile corresponding to this form of n(E) can be derived by integrating the rectangular line shapes of Eq. (2) normalized to an area proportional to dn(E). That is,

$$P(\Delta) \propto \int_{E_0+\Delta^2/m_ec^2}^{E_1} dE' g(E')n(E') / \sqrt{E'-E_0}. \tag{9}$$

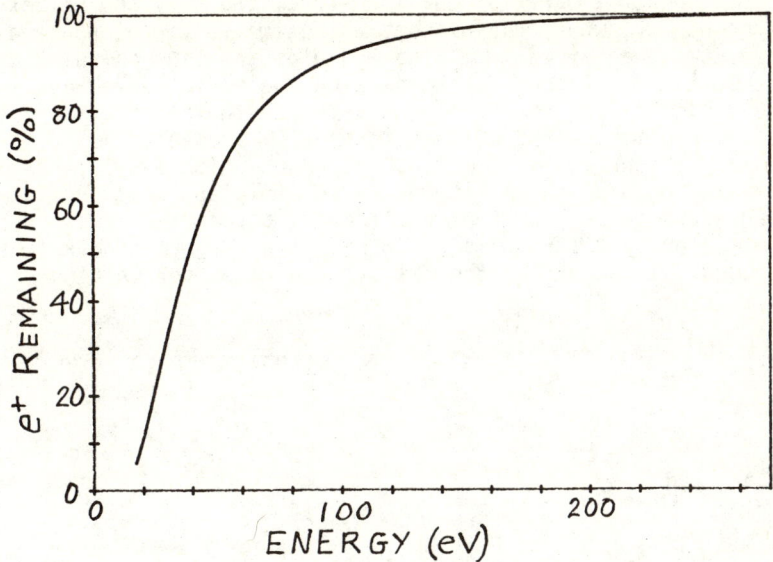

Fig. 2. Percent of positrons originally at 272 eV surviving to reach a given lower energy while slowing down in neutral atomic hydrogen.

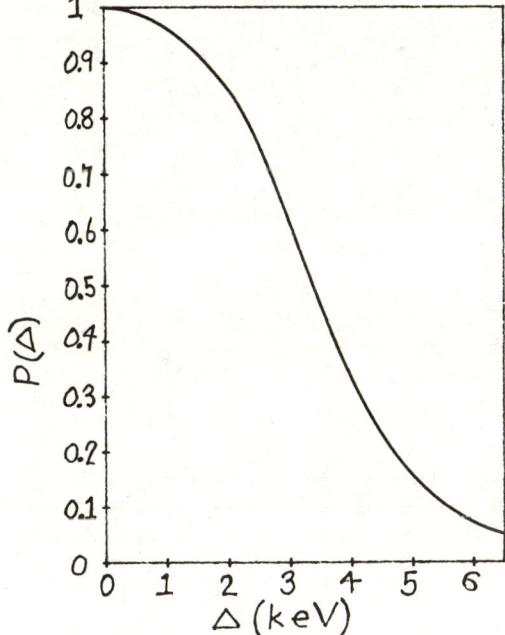

Fig. 3. Gamma-ray line intensity as a function of distance from the line center Δ. The line is symmetric about $\Delta = 0$.

In Fig. 3 I have plotted the annihilation line profile against Δ in keV. (The central parts of the line involving very low values of positron energy, where the present approximation is poor, have been extrapolated quadratically.) The width of the line calculated this way is 6.75 keV, in good agreement with the Monte Carlo results of Ref. 6, and once and for all inconsistent with the observations.

The solution to the dilemma is usually presented as follows[6]: If one allows some 5-10 percent ionization in the gas the slowing down of the positrons is so efficient that they do not form an appreciable amount of positronium before they take up a thermal velocity distribution. In Fig. 4 a diagram from Ref. 6 is reproduced. It shows the thermal-average rates for the four

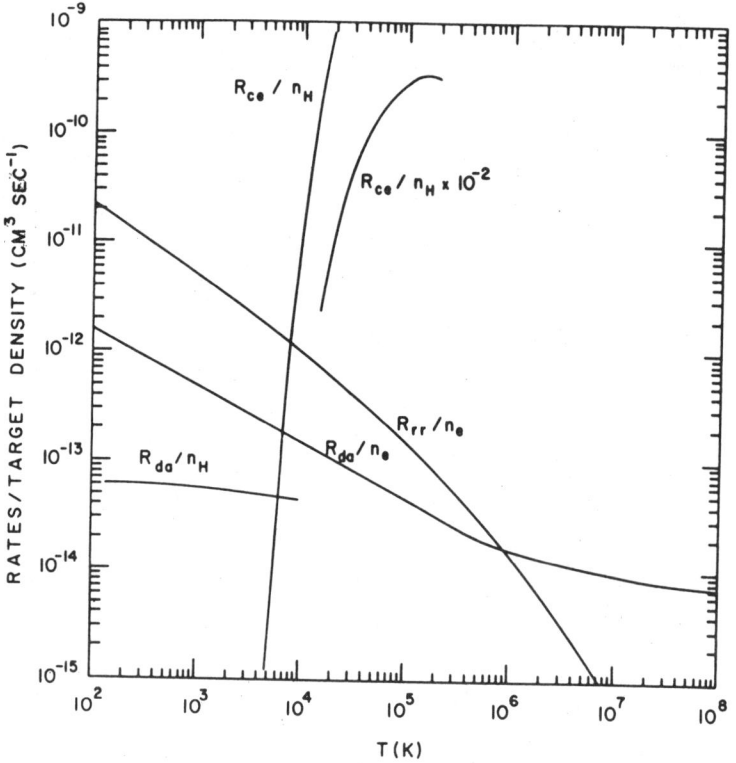

Fig. 4. Rates (per unit target density) at which thermal positrons form Ps by charge exchange with H (R_{ce}/n_H) or by radiative recombination with free electrons (R_{rr}/n_e), and annihilate directly with free electrons (R_{da}/n_e) or with bound electrons (R_{da}/n_H), as functions of temperature. (From Ref. 6)

processes occurring in partially ionized hydrogen. At temperatures below about 5×10^3 K the dominant process is radiative recombination with an electron to form Ps followed by annihilation. The shape of the resulting line is

$$P(\Delta) = e^{-\Delta^2/kTm_ec^2}, \qquad (10)$$

its width is[7] $1.105\,(T_4)^{1/2}$ keV where T_4 is the temperature in 10^4K, and a satisfactory line width can be obtained. At higher temperatures charge exchange (Ps formation) with neutral hydrogen dominates; it rises very rapidly since only the tail of the thermal distribution above $E_0 = 6.8$ eV can contribute.

Once convinced that annihilation from a thermal swarm is occurring, we can, in principle determine the ionization fraction and the temperature if we know the line width and the Ps fraction. At present, of course, we have only an upper limit on the line width, while there is only rough and discordant data on the Ps fraction, which is obtained by a delicate process of curve fitting in the presence of a large γ-ray background. In Ref. 4 the most probable Ps fraction was given as 92 percent (but consistent with zero). At this Workshop, however, Riegler[17] suggested a value of about 20 ± 20 percent. Clearly no conclusion is yet possible. If the lower value proves to be correct, however, severe constraints will be placed on the scenario described above. In particular, it is not possible to keep the Ps fraction below about 60 percent while keeping the ionization fraction >5 percent. Several possible ways out of this dilemma would be to allow the annihilation to occur at high density[17] (to pick off the triplet Ps) or to flood the annihilation region with ultraviolet radiation (to ionize the triplet Ps before decay.) I will propose another possibility in the next section, by going back to the completely neutral case.

A TIME-DEPENDENT SOLUTION

In discussing Fig. 2, I did not mention the fact that about 5 percent of the original high-energy positrons avoid being annihilated as Ps during the slowing down process. This conclusion was originally drawn from the Monte Carlo calculation of Ref. 6, but a similar estimate can be obtained from Fig. 2. Note that the continuous slowing-down model should not be carried to an energy below 23.8 eV, since below that point a single ionization event will bring the positron below the Ps-formation threshold. But 19.0 percent of all the positrons reach 23.8 eV, and the probability that the next collision is an ionization event is $P_I(23.8) = \sigma_I(23.8) / [\sigma_I + \sigma_{exc} + \sigma_{Ps}] = 0.1335$. Similarly, the probability that the next collision is an excitation event is 0.2776, lowering the positron energy to 13.6 eV. At 13.6 eV, the excitation probability is 0.1303; in effect we are doing a very simple Monte Carlo calculation for the last few collisions before the Ps threshold is reached. The result is

$$n(E<5.8) = n(23.8) [P_I(23.8) + P_{exc}(23.8) \times P_{exc}(13.6)]$$

$$= 3.2 \text{ percent,}$$

in qualitative agreement with Ref. 6. Three questions may be asked about this positron residue: how wide is the γ-ray line it produces, what is the resulting Ps fraction, and what is its time dependence

In this region (E < 6.8 eV) where energy loss is by elastic scattering only ($\Delta E/E = 2m_e/m_H$ per collision) and the annihilation rate is very low (due to collisions with H atoms), the continuous approximation is reliable. Using the almost exact e^+-H annihilation and momentum-transfer cross-sections[18] in the following equations

$$\frac{dE}{dt} = \frac{-2m_e \sigma_{MT}(E) Ev(E)N}{m_H} \qquad (11)$$

$$\frac{dn(E)}{dt} = -\sigma_A(E) n(E) v(E) N \qquad (12)$$

it is very easy to show that almost all the positrons cool rapidly and annihilate at very low evergy. (For example, 94 percent of all the positrons initially at 6.8 eV reach an energy of 0.05 eV, corresponding to about T = 600 K.) They will then annihilate with the bound electrons in the atomic hydrogen ground state; the quantum-mechanical momentum distribution is the cause of the annihilation line width, since the energy of the positrons is nearly zero.

If one has an accurate e^+-H zero-energy scattering wave function $\Psi(\vec{x}, \vec{r})$, where $\vec{x}, \vec{r}$ are the co-ordinates of incident e^+ and atomic e^- (in atomic units) respectively, the annihilation line profile is given by[19]

$$P(\Delta) \propto \int_{\frac{2\Delta}{\alpha m_e c^2}}^{\infty} dq\, q \left| \int d\vec{x}\, j_0(qx) \Psi(\vec{x}, \vec{x}) \right|^2. \qquad (13)$$

Humberston[20] has carried out such a calculation, although with a different situation in mind, and from it an annihilation width of 1.3 keV can be derived. (Several references[6,21,22] have misquoted this as 2.6 keV.) This positron component gives a narrow annihilation line with a vanishingly small Ps component. (If some Ps is observed, a very small ionized fraction will account for it easily; at T = 10^3K, n_e/n_H = 0.0025 gives 20 percent Ps fraction.)

In Ref. 6 we assumed that the broad line due to Ps formation and this narrow component were properly weighted and superposed, since we assumed a time-independent source. This combined annihilation line is not narrow enough to agree with the observations. The new suggestion I would like to make is based on the observed time dependence of the radiation.

Suppose a short burst of positrons is injected somehow into our cold atomic hydrogen region. We would observe two successive annihilation phases: The positrons would slow down in a relatively short time and, as described, would form Ps in flight giving a wide line at first. Soon, however, the only remaining positrons would be in the component below 6.8 eV, annihilating slowly with a narrow line width. From Eq. (12) it follows that if this second component is being observed $N > 10^6$ cm^{-3} for a mean life of 1/2 year. At this density the first phase should take less than 10 percent as long,[23] about one or two weeks. The conclusion is that randomly timed observations of the galactic center are much more likely to see narrow lines than wide ones, and this may be the explanation of the present observational situation.

It is important, then, to try to observe the galactic center source on a continuous basis, for the general purpose of charting its time dependence and specifically to look for the unique time dependence of the line width discussed above.

REFERENCES

1. K.R. Hoffman et al., Phys. Rev. A **25**, 1393 (1982); M. Charlton et al., J. Phys. B **16**, 323 (1983).
2. T.S. Stein et al., Phys. Rev. A **17**, 1600 (1978); K.F. Canter et al., J. Phys. B **6**, L 201 (1973); B. Jaduszliwer and D.A.L. Paul, Can. J. Phys. **52**, 1047 (1974); G. Sinapius, W. Raith, and W.G. Wilson, J. Phys. B **13**, 4079 (1980).
3. W.N. Johnson and R.C. Haymes, Astrophys. J. **184**, 103 (1973); R.C. Haymes et al., Astrophys. J. **201**, 593 (1975); F. Albernhe et al., Astron. Astrophys. **94**, 214 (1981); M. Leventhal et al., Astrophys. J. **240**, 338 (1980); M. Leventhal et al., Astrophys. J. **260**, L1 (1982); W.S. Paciesas et al., Astrophys. J. **260**, L7 (1982).
4. M. Leventhal, C.J. MacCallum, and P.D. Stang, Astrophys. J. **225**, L 11 (1978).
5. G.R. Riegler et al., Astrophys. J. **248**, L 13 (1981).
6. R.W. Bussard, R. Ramaty, and R.J. Drachman, Astrophys. J. **228**, 928 (1979).
7. C.J. Crannell et al., Astrophys. J. **210**, 582 (1976).
8. F.W. Stecker, Cosmic Gamma Rays (NASA, Washington, 1971). "Ps" is the chemical symbol for positronium, the hydrogen-like atom composed of one positron and one electron. Its two possible spin states annihilate into 2 γ-rays (singlet) or 3 γ-rays (triplet).

9. J. Stein and R. Sternlicht, Phys. Rev. A 6, 2165 (1972); Y.F. Chan and P.A. Fraser, J. Phys. B 6, 2504 (1973); J.W. Humberston, Can. J. Phys. 60, 591 (1982).
10. Y.F. Chan and R.P. McEachran, J. Phys. B 9, 2869 (1976).
11. H.S. W. Massey and C.B.O. Mohr, Proc. Phys. Soc. (London) A 67, 695 (1954); I.M. Cheshire, Proc. Phys. Soc. 83, 227 (1964).
12. R.J. Drachman, K. Omidvar, and J.H. McGuire, Phys. Rev. A 14, 100 (1976).
13. J.R. Winick and W.P. Reinhardt, Phys. Rev. A 18, 925 (1978).
14. W. Lotz, Z. fur Physik 206, 205 (1967).
15. L.A. Morgan, J. Phys. B 15, L 25 (1982).
16. F.W. Byron, Jr., C.J. Joachain, and R.M. Potvliege, J. Phys. B 15, 3915 (1982).
17. G.R. Riegler, this volume.
18. A.K. Bhatia, R.J. Drachman and A. Temkin, Phys. Rev. A 16, 1719 (1977).
19. R.J. Drachman, in VII ICPEAC, Invited Talks and Progress Reports edited by T.R. Govers and F.J. de Heer (North-Holland, Amsterdam, 1972), p. 277.
20. J.W. Humberston and J.B.G. Wallace, J. Phys. B 5, 1138 (1972).
21. R.J. Drachman, Can. J. Phys. 60, 494 (1982).
22. R.J. Drachman, in Positron Annihilation edited by P.G. Coleman, S.C. Sharma, and L.M. Diana (North-Holland, Amsterdam, 1982), p. 37.
23. R.E. Lingenfelter and R. Ramaty, in The Galactic Center edited by G.R. Riegler and R.D. Blandford (Am. Inst. Phys., NY, 1982).

"GAMMA-GUN" IN THE CENTER OF THE GALAXY

N. S. Kardashev, I. D. Novikov, A. G. Polnarev and B. E. Stern
Academy of Sciences of the USSR, Space Research Institute
Moscow, USSR

ABSTRACT

This paper discusses a model of physical processes that occur in the center of the Galaxy and cause the emission of the 511 keV annihilation γ - line and of a continuous γ - spectrum. The hypothesis of the paper is that electron positron pairs whose annihilation produces the observed γ - line are formed when a gas-cloud target is irradiated by a directed beam of hard (~ 100 MeV) γ-quanta. A model is developed that shows how a γ - beam (gamma-gun) forms near a supermassive rotating black-hole surrounded by a gaseous accretion disk with a magnetic field. The results of numerical simulations are compared with observational data.

INTRODUCTION

In recent years the hypothesis of a massive black hole being a source of activity of galactic nuclei and quasars has become more and more attractive[1-4].

The center of the Galaxy is observable in many spectral ranges from radio to γ - rays (for reviews see ref. 5). Of interest is the fact that the luminosity in all of these ranges is comparatively low. There is, however, indirect evidence that in the UV-range, which is not observable due to absorption, the luminosity is ~10^{41} erg s^{-1}. This follows from infrared observations of gas clouds at a distance of about 3 pc from the center. The state of ionisation in these clouds, estimated from the observed IR lines[6], should be maintained by thermal radiation with T~3.1×10^4K and power ~10^{41} erg s^{-1}.

Most valuable for revealing the physical character of the processes in the center of the Galaxy are peculiarities observed in the γ - spectral region. This 10 keV to 2 MeV radiation, when observed[7,8,9], has the following typical features:

1. A narrow line with E = 511 keV, width <2.5 keV (the respective velocity dispersion <560 km.s^{-1}) and luminosity ~ 2×10^{37} erg.s^{-1}. The center of the line corresponds to the laboratory frequency to an accuracy of 0.25 keV (the respective Doppler velocity is <150 km.s^{-1}). The line varies[9-11] appreciably in intensity, probably in less than 100 days.

2. There is a gamma and X-ray continuum, with variable intensity and spectral index, and this variability seems to correlate with the variability of the line[12].

3. There is an intensity rise in the continuum, just below the line at E = 511 keV, while the continuum intensity decreases sharply above about 2 MeV (ref. 12).

4. There is an upper limit of the luminosity in the region >1 MeV, which is $\lesssim 10^{38}$ erg s^{-1} (ref. 12).

We show that the annihilation line and the continuous spectrum over the range of energies greater than 100 keV may result from the interaction of a directed beam of γ - quanta (with a characteristic energy of $\sim$ 100 MeV), formed near a massive black hole (some kind of a "gamma-gun"), and the gaseous matter of the cloud target. The present paper gives the results of calculations of such interactions and shows how the various typical features of the spectrum can be explained. We regard the hard radiation as the most probable carrier of energy from the neighborhood of a black hole to the target where $e^{\pm}$ pairs are born and an annihilation line forms. The present paper also offers a possible model of the gamma-gun. Comparison with other hypotheses is given below.

GENERAL SCENARIO

We begin with the description of a possible γ - gun model where the rotational energy of a black-hole is converted into a directed beam of γ - quanta. It should always be kept in mind that we regard the model of such a gun only as a possibility.

Assume that there is a black-hole of $\sim 10^6$ M_θ in the center of the Galaxy. There may even possibly be a binary system of black holes[13,14] with about 10^6 M_θ each[15]. An accretion disk forms around the black hole, from the matter of stars destroyed by tidal forces in the gravitational field[16-19]. If a disk near the black hole is geometrically thick[20-25], then a sort of a funnel forms, its diameter being of the order of several gravitational radii (see Figure 1). The following features of disk accretion are essential for our paper and should be emphasized.

1. Accretion of the magnetized matter leads to the appearance of a poloidal magnetic field around the black hole, i.e. a field aligned along the axis of black hole or disk rotation[26-32].

2. The funnel is filled by the thermal emission from the disk at a temperature (in our model) of the order of 10^5 K, with an almost equilibrium density. This value of temperature is controlled by the temperature of the funnel walls near the black hole. This temperature may be different in different models of accretion disks[25,33]. There could be some other sources of non-thermal radiation. For the consequences (in terms of our model) of the radiation present in the funnel to best match the available observations, the radiation temperature in the funnel should be (as we see below) within 3×10^4 K $< T < 10^6$ K.

3. It has been shown[28,34] that if a black hole of mass M, being the object of disk accretion, has an angular momentum J (Kerr black hole), then it operates as an equivalent electric battery with a power:

$$L_H \simeq 10^{40} \, (M/10^6 M_\odot)^2 \, (J/J_m)^2 \, (H/10^4 \text{ gauss})^2 \text{ erg s}^{-1},$$

where J is the rotational moment, J_m is the maximum rotational moment, and H is the magnetic field in the vicinity of the black hole.

We show that for such conditions, a directed γ - flux with a characteristic photon energy of $\sim 10^2$ MeV (gamma-gun) may be ejected along the disk axis. If at a large distance from the black hole this γ - quanta flux meets a target (say a gas cloud), the interaction of the γ - radiation with the matter of the cloud causes the formation of an annihilation line and of hard-radiation continuum. This is the general picture. Now we come to the details.

GAMMA-BEAM FORMATION

Accordings to refs. (26-32), an almost force-free electromagnetic field forms near a rotating black hole. In this field, a charged particle is accelerated by $E_\parallel$, the small component of an electric field aligned along the magnetic field, and simultaneously it interacts with the thermal radiation in the funnel of the disk. This field cannot be very strong in our model. Indeed, if the field were too strong, the particle would gain an energy which would be sufficient for effective production of electron-positron pairs, either in direct collisions with thermal photons or by losing energy to these photons via the inverse Compton effect. The resultant energetic γ - quanta, in turn, interact with thermal photons leading to $e^\pm$ pair production. The process is stationary only if the number of the pairs thus born is of the order of unity per one seed particle. In the opposite case, the number of charge carriers would grow till the $E_\parallel$ component achieved the required value $E_{\parallel crit}$. If the situation is such, then the pattern is similar to a glow discharge in a gas.

We consider the case when the energy of particles accelerated by the $E_\parallel$ field is not yet sufficient for the direct pair production with the thermal photons to be the dominant process ($kTE_e < 30m^2$, $c = 1$), so that a particle loses its energy via the inverse Compton effect. The electron mean-free-path in the photon field is of the order of

$$\ell_e \sim 10^8 \text{ cm } (T/10^5 K)^{-3}.$$

At $T \sim 10^5$ K this value is much less than the characteristic scales of the magnetic and electric fields ℓ_E near the black hole, the latter being larger or of the order of the gravitational radius $\ell_E = r_g \simeq 3 \times 10^{11}$ cm. It is essential for what follows that $\ell_E \gg \ell_e$.

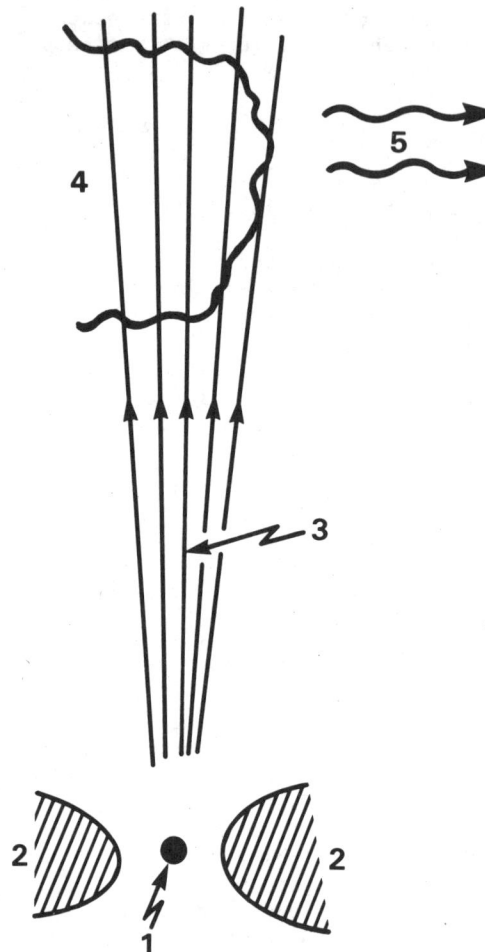

A computer - aided simulation of the whole process was conducted by the Monte-Carlo method. The results of numerical calculations for a uniform radiation field with kT = 20 eV when a particle passes a distance of $\ell = 3 \times 10^{11}$ cm are summarized in Table 1. (The ℓ - dependence of the spectrum and characteristic energy of the resultant γ - quanta is only logarithmic, hence the specific value of ℓ is of no importance in the simulation, if $\ell \gg \ell_e$.).

Thus, for the critical strength $E_{\parallel crit} \simeq 17 V/cm$ a steady-state process is set up when electrons and positrons (we shall refer to them as e_1) accelerated by $E_\parallel$ lose their energy converting thermal photons into γ_2 - quanta (see caption to Figure 1 for definitions of the particle and photon subscripts). The

Figure 1. Schematics of processes that lead to the formation of the γ-radiation from the center of the Galaxy: (1) Black Hole; (2) Accretion disk; (3) Directed γ-quanta flux, (4) Cloud target; (5) Annihilation line, scattered and bremsstrahlung gamma rays. The following notation is used in the text: e_1 - electrons and positrons accelerated near the black hole; γ_2 - gamma rays produced by Compton scattering of e_1 on thermal photons, e_3 - positrons and electrons produced in the cloud by γ_2 photons on ambient matter; γ_{4_b} and γ_4 - hard photons produced by e_3 and γ_2 in the cloud by bremsstrahlung and Compton scattering, respectively; γ_{4_s} - scattered γ_{4_b} and γ_4 photons in the cloud.

steady-state spectra of these γ_2 - quanta and e_1, estimated by the numerical Monte Carlo method, are shown in Figure 2. For the dashed curves, the input energy of the e_1 pairs is $E_1 \sim 1$ GeV, their mean free path is about 10^7 cm, and the mean energy of the resultant γ_2 - quanta is ~ 170 MeV. The energy distribution of the γ_2 - quanta, however, stretches to energies of the order of several GeV. Most γ_2 - quanta do not interact with thermal photons, or with each other, or with the electromagnetic field. It is only the hardest γ_2 - quanta, which, interacting with thermal photons give pair production, thus sustaining the stationary current. One pair is born per about 10^4 γ_2 - quanta. Note that the plasma in the funnel is also transparent for γ_2 quanta.

TABLE 1

Number of pairs produced by a charged particle over the distance 10^{11} cm versus the accelerating field.

$E_\parallel$	33 V/cm	27 V/cm	20 V/cm	17 V/cm
Number of Pairs	140	30	10	1

Table 1 shows that the dependence of $n_\pm$ on $E_\parallel$ is rather steep and the critical strength of the electric field corresponding to the glow - discharge mode is $E_{\parallel \text{crit}} \simeq 17$ V/cm.

It is of interest that if there is some fraction of high-energy non-thermal radiation near a black hole, then $E_{\parallel \text{crit}}$ will be lower than the above mentioned value.

Collimation of a γ - beam from the vicinity of a black hole surrounded by a disk is determined by the geometry of a disk funnel and by the pinch effect.

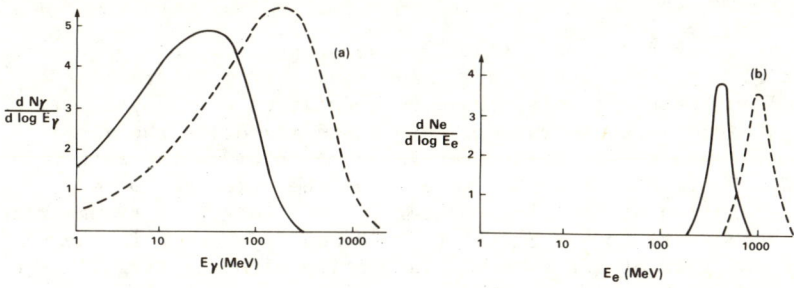

Figure 2. (a) Spectra of γ-gun photons, N_γ. (b) Spectra of electrons and positrons N_e in the gamma-gun: Solid curves—mean energy of photons 35 MeV, dashed curves—mean photon energy 170 MeV.

Note that the essential feature of the model required to explain the observed γ-line is the tremendously strong current, $I = 10^{40}$ particles per second, that flows in the vacuum near the black hole with $M = 10^6 \, M_\odot$. A strong toroidal magnetic field is related to that current. The existence of a γ - gun of the given type depends on whether such a current exists. The closure of this current may occur through the disk and the horizon between the equator and a pole of the black hole. The required I is proportional to $M^{1/2}$, while the net energy related to the current varies as $\sim M^2$. Hence all problems connected with the current become less complicated the smaller the black hole mass is.

THE GAMMA-QUANTA BEAM INTERACTION WITH THE TARGET AND THE FORMATION OF THE ANNIHILATION γ-LINE

Assume that at a distance $R \sim 10^{16}$ to 10^{17} cm (see ref. 15) the γ_2 - beam enters a gas cloud. All the processes in the cloud have been calculated by the Monte-Carlo method and the fate of particles have been followed until they stop, annihilate or leave the cloud. The program of electromagnetic cascade simulation SIMEX has been used in the calculation[35]; the calculations also assumed a hydrogen cloud.

Processes in the cloud may be pictured as follows. The photon γ_2 hitting the cloud produces $e^\pm$ pairs in the field of nuclei and electrons of the target. This photon also produces high-energy recoil electrons due to the Compton effect. Then the relativistic charged particles (e^- and e^+) are decelerated mainly due to ionization losses. At the simulation stage it was supposed that as soon as a positron is decelerated down to the energy ~ 1 MeV it is rapidly thermalized, with no interesting effects whatsoever, and then it is annihilated producing two or three photons.

As is shown in ref. (36) 70% to 90% of the annihilation events should involve positronium formation and therefore only 50% to 30% of the positrons will annihilate producing two 511 keV photons. We assumed in our calculations that positronium forms in 70% of the cases.

A star destroyed by the tidal forces of a black hole or by a collision with another normal star, and moving at a distance of $\sim 10^{16}$ cm from the black hole, could be considered as a physically reasonable target. While moving, this cloud may cross the γ - beam. Gamma-quanta lose their energy and momentum to the matter in the cloud. The energy transfer rate is of the order of 10^{40} erg s^{-1}. Acceleration of the cloud matter (if the target is transparent or semitransparent) is ~ 1 cm.s^{-2} $(r/10^{14}$ cm$)^{-2}$, where r is the radius of the beam at the time of its meeting with the target. For the velocity of the matter accelerated over half a year not to exceed 100 km.s^{-1} (or it would contradict the observations), the beam radius should be larger than 10^{14} cm, i.e. the divergence of the beam should be more than 10^{-2}. The expansion rate of the heated cloud should not be very high either.

The most preferable situation for the formation of the observed annihilation γ - line is when the matter stretches along the beam, so that the cloud is thick enough longitudinally to effectively absorb the beam and not very thick across so that line photons can easily reach the observer.

Such a situation may, for instance, occur if the beam hits the edge of the cloud facing the observer. Parameters of the cloud-target are approximately: mass $M \sim M_\odot$, radius $R \sim 10^{15}$ cm, density $\rho \sim 10^{-13}$ g/cm^3; the temperature of the cloud region illuminated by the γ - quanta beam is about $T \sim 1$ to 3×10^4K, the cross-section of the γ_2 quanta beam near the cloud-target is about 10^{14} cm. Gamma-line variability may be attributed to the cloud movement (along with changes in the physical conditions in the "gamma-gun"). Indeed, we know too little about the central regions of the Galaxy ($R \lesssim 10^{17}$ cm) to be sure about the target for the γ - beam.

Figure 3a shows the efficiency of annihilation line formation versus the energy of the photons hitting the target. To be more specific, the target is assumed to be 30 g cm^{-2} thick along the direction of the beam. If the target along the beam is infinitely thick, ~ 3 times as many positrons are born. For a thickness $\lesssim 10$ g cm^{-2} the efficiency starts to decrease rapidly, since

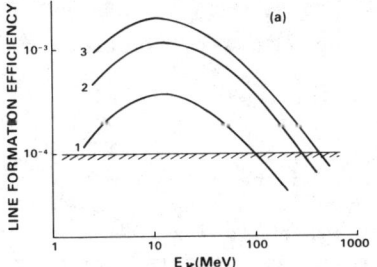 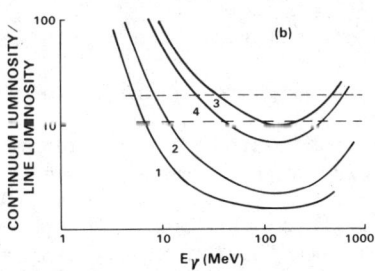

Figure 3. Line formation efficiency (a) and the ratio of luminosities in the continuum and in the line (b) versus the energy of photons hitting the target. An assumption is made that 70% of the annihilation events occur via positronium production. (a) Curve 1 - hydrogen spherical target 30 g/cm^2 in diameter; Curve 2 - hydrogen target stretched along the γ-beam, its length 30 g/cm^2, cross-beam thickness 4 g/cm^2; Curve 3 - the same geometry as curve 2 but for a helium target. Hatched in the Figure is the least possible efficiency which may be derived from observational data. (b) Ratio of luminosities in the continuum (for $E_\gamma > 300$ keV) and in the line for a target 30 x 4 g/cm^2 for several observation angles: (1) $\cos\theta = -0.7$, (2) $\cos\theta = 0$, (3) $\cos\theta = 0.7$, (4) $\cos\theta = 0.7$ and hydrogen is replaced by helium. Horizonal dashed lines in the Figure show the ratios obtained in 1974 and 1977 observations.

newly born positrons do not have enough time to be decelerated within the target. As has been mentioned above, the total energy release in the center of the Galaxy does not exceed 3×10^{41} erg s^{-1}. It is reasonable to restrict the intensity of the γ - beam to the same value. Then, the luminosity in the line will be 10^{37} erg. s^{-1} if the efficiency is 10^{-4}. Hence the characteristic energies of the beam γ - quanta should not exceed 100-200 MeV (see Figure 3a). The width of the annihilation line depends in the given conditions only on the temperature in the cloud and it can be as narrow as possible.

FORMATION OF THE CONTINUUM

The following processes contribute to the formation of a continuous spectrum (Figure 1).

1. Bremsstrahlung of relativistic electrons e_3^- and positrons e_3^+ that gained energy from photons γ_2. Bremsstrahlung quanta come to the observer after they have been scattered by electrons in the cloud.

2. Photons γ_2 scattered toward observer via the Compton effect.

3. Inverse Compton-effect of e_3 via the thermal radiation of the cloud. This thermal radiation should exist in the cloud since the latter is heated by the beam γ_2. These quanta reach the observer after scattering by electrons in the cloud.

4. Annihilation-line quanta scattering on plasma electrons.

5. Three-photon annihilation of electron-positron pairs.

If the motion of the relativistic particles in the cloud were isotropic, their bremsstrahlung would exceed the upper limit on the flux of 1 MeV γ-quanta which is 10^{38} erg.s^{-1}, and what is more, the upper limit on the luminosity at 10 MeV, which is 5×10^{36} erg s^{-1}. But the motion of these particles in the cloud is extremely anisotropic, with particle velocities predominantly along the γ - beam direction. Their deflection angles in case they are multiply scattered by ions is expected to be very small. Furthermore, not much scattering is expected due to the chaotic magnetic field in the cloud, since an overall expansion of the original star by more than a factor of 3×10^4 leads to a field decrease by a factor of 10^9. Furthermore, the energy density of relativistic particles in the cloud is high: the computed value is $\varepsilon_{rel} \sim 10$ erg cm^{-3} for the parameters considered above, which corresponds to the energy density of a magnetic field of ~ 10 Gauss. Hence, the bremsstrahlung is collimated along the direction of the γ - beam and is not directly observable.

The second process is interesting since it may introduce another, very prominent feature (in addition to the 511 keV line).

After a photon with the initial energy E_0 is scattered through the angle θ it has the energy $E'=m_e/(m_e/E_0+1-\cos\theta)$, that is for $E_0 \gg m_e$, $E'=m_e/(1-\cos\theta)$. Thus, in the observed γ - spectrum there should be a maximum, the position of which is determined by the beam direction, and its width by the beam divergence. The ratio between the luminosity in the 511 keV line and in this Compton maximum is determined both by the beam direction and the energy of photons in the beam. To the accuracy of a logarithmic correction, we have that $N/N_{511} \simeq 0.2(E_\gamma/100\text{MeV})^{-1}(1-\cos\theta)^{-1}$. Thus, this maximum could well be observed.

The third process, according to the calculations, strongly affects the observed spectrum if the ratio of the optical thermal-photon density to that of protons in the cloud $n_\gamma/n_p \gtrsim 1$. If the beam is not very wide, $d \lesssim 10^{15}$ cm, and the energy release in the target is $\sim 10^{40}$ erg.s^{-1}, then this condition is fulfilled even if the cloud is almost transparent for optical photons.

Figure 3b shows the ratio of the luminosity in the continuum in the range $E_\gamma > 300$ keV to that in the 511 keV line depending on the photon energy in the γ - beam. With decreasing energy the ratio grows since the ratio of the Compton scattering cross section to the pair generation cross section grows. The Figure shows that the observed continuum - to - 511 keV line ratio agrees with the calculated ratio in a rather narrow energy range near 100 MeV. Therefore, the γ - beam spectrum must be sufficiently narrow. In particular this requirement is met by the spectrum obtained with the mechanism given in Section 2, but not by a wide power spectrum with any index.

Figure 4 gives observational data on the continuous spectrum and its computed analogs derived under various assumptions. Most interesting is the version with a beam making an angle θ to the line of sight, such that $\cos\theta \sim 0.7$, and the mean energy of the γ - beam is ~ 100 MeV. In this case, a typical spectral feature in the range 1-2 MeV mentioned[12] in the HEAO-1 data and in greater detail in ref. 37 can be naturally explained by a partially blurred Compton maximum, discussed above.

We may also try to explain the power spectrum in the 60-300 keV range with the index 2; Figure 4d shows the respective version. In this case the energy of the γ - beam photons must be much lower while the angle θ should be equal to or more than 90°, since otherwise the intensity above 0.5 MeV would exceed the observed value.

Thus it is possible to match either the slope of the power spectrum (Figure 4d) or the behavior of the spectrum at $E_\gamma > 300$ keV (Figure 4c), but not these two together.

The confirmation of a maximum in the spectrum should be a strong argument in favor of the γ - gun model. In this case we may say that we know for sure the direction of the beam relative to the

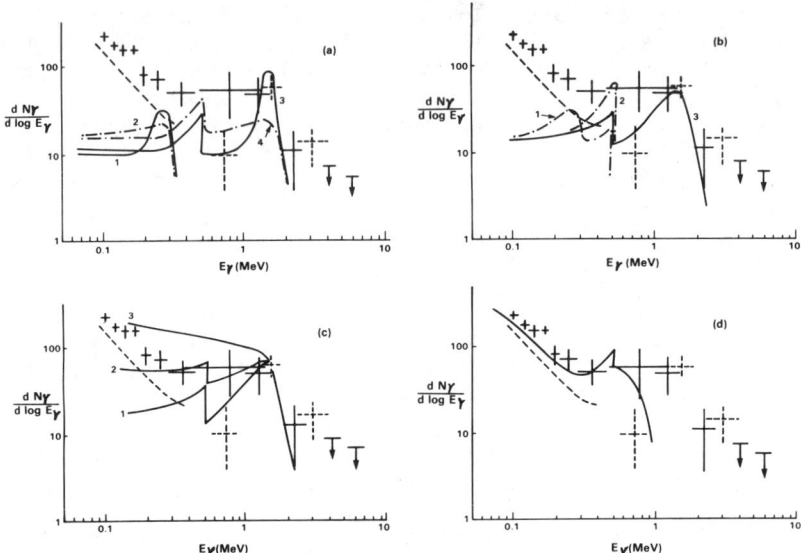

Figure 4. Comparison of calculated spectra of observed photons with the measured spectra. Solid crosses are HEAO-1 data, the dashed crosses and curve show the data from ref. (37). All spectra are normalized (relative to each other) to the γ-line intensity. The target was assumed stretched along the beam (30 x 4 g/cm^2), the fraction of photon annihilation ~50%, and the annihilation line is not shown. (a) Results for the "monochromatic" γ-gun for different energies and observation angles: 1. $\cos\theta = -0.7$, $E_\gamma = 40$ MeV; 2. $\cos\theta = -0.7$, $E_\gamma = 200$ MeV; 3. $\cos\theta = 0.7$, $E_\gamma = 40$ MeV; 4. $\cos\theta = 0.7$, $E_\gamma = 200$ MeV. (b) γ-gun with a spectrum similar to the one given in Figure 2 (mean energy $\bar{E}_\gamma = 100$ MeV). 1. $\cos\theta = -0.7$; 2. $\cos\theta = 0$; 3. $\cos\theta = 0.7$. (c) Variations of the spectrum with varying density of thermal photons in the cloud. (The spectra mentioned above were obtained for zero photon density). $E_\gamma = 100$ MeV, $\cos\theta = 0.7$, $T' = 10^4$ K); 1. $n_\gamma/n_p = 0$; 2. $n_\gamma/n_p = 3$; 3. $n_\gamma/n_p = 9$. (d) Mean energy of the beam 35 MeV, the spectrum is averaged over the observation angle range $|\cos\theta| < 0.5$; photon density in the cloud corresponds to $n_\gamma/n_p = 50$ at $T = 10^4$ K.

observer, while the power spectrum may be attributed not to Comptonized photons of the cloud, but to some other mechanism or maybe to some other sources not resolved in the range >100 keV (ref. 12) and not directly associated with the source discussed here. If the observed intensity of the continuum for energies $\lesssim 300$ keV is attributed to those other sources, the observed strong variations of intensity in the range 300-500 keV are fairly well described by the rise and fall of the continuum (see curves 1 to 3 in Figure 4c) with varying density of thermal photons in the cloud.

Thus the γ - gun model may explain the observational data if the following restrictions are imposed.

1. The γ - gun spectrum should be sufficiently narrow. It excludes versions of the model in which electrons are accelerated to high energies with the subsequent cascade degradation of energy. Electrons which produce the γ - beam should have almost constant energy over the whole path through the region of dense radiation.

2. The mean energy of photons in the γ - gun should be ~ 100 MeV; variations by more than a factor of 3 would obviously contradict the observations.

3. To form an annihilation line the γ - beam should meet along its path $\gtrsim$ 10 to 20 g cm^{-2} of matter.

4. The target should be preferably stretched along the beam (at least in the periods when the line is most intense), otherwise it would be difficult to match the intensity in the range 300-500 keV with respect to the intensity of the line.

These restrictions are somewhat less stringent if the target consists of He, as indicated indirectly in ref. 38. If the target consists of even heavier elements (which could be possible if the target results from the destruction of a highly evolved star), the restrictions imposed on the model become even less stringent.

ALTERNATIVE VERSIONS OF THE γ-GUN

We have considered a version in which an accelerating electric field exists near a black hole, the scale of this field being much larger than the mean-free-path of electrons before they are scattered by thermal photons. In that case a strict limitation is imposed on the accelerating field value and the current of relativistic electrons and positrons becomes similar to the glow discharge.

Probably, another situation is possible when the spatial charge distribution of electron-positron plasma is adjusted so that the voltage drop occurs over their mean-free-path length (e.g. near the horizon) and may be arbitrarily large. Then the cascade multiplication of particles may occur via thermal radiation, but this time without an accelerating electric field, due to which newly born charge carriers do not contribute to the current. This situation obviously happens in pulsars[39], the only difference being that multiplication there takes place in the strong magnetic field and the probability of pair production at a certain distance to the neutron star abruptly grows from zero up to large values. As a result, the acceleration and multiplication regions are separated by a steep front. However, the spectrum of photons in this case is a wide power spectrum and this contradicts the observational data, as has already been mentioned.

Synchrotron radiation of relativistic particles, if they are born at large angles to the magnetic fields, may act as an optically thick medium for accelerating particles, the role assigned to the thermal radiation of the disk.

The alternative to the generation of positrons by the γ - beam is the photon-photon (γ-γ) generation of positrons considered in references (19), (34), (36). Fluxes of MeV photons interacting with each other could be a more effective source of positrons than the flux of ~ 100 MeV photons interacting with matter. However, in this case it is important that MeV photons contribute most to the γ - flux power, since the cross-section of pair production decreases with increasing energy as $1/E_{cm}^2$ and the width of the energy range where γ-γ generation is effective is ~ 1 MeV wide.

Versions with the γ-γ generation of positrons were considered by Rees[19] (their GeV - Comptonized photons interacted with keV synchrotron photons), by Lingenfelter and Ramaty[36] (positrons were assumed to be produced by thermal MeV - photons in the hot accretion disk of the black hole, M~ 100 M_θ), and by Blandford[34], the latter paper was already mentioned above.

CONCLUSION

Thus, the model of a gamma-gun, in the context of reasonable assumptions, yields both a sufficiently high efficiency of line formation (10^{-4} to 10^{-3}) and a required intensity ratio of the γ - line and the continuous γ - spectrum in its vicinity. The observed abrupt bending in the spectrum (or even the maximum) near 2 MeV is easily explained within this model. The presence of another maximum or plateau at about 1.8 to 2 MeV in addition to the 511 keV line in the γ-spectrum, if confirmed, may be a serious argument in favor of the γ-beam model, while the absence of such a feature, deduced from better limits than the current values, may question the validity of the model.

The variability of the γ - line and continuum may be explained by changes both in the target and in the source. Figure 4 illustrates how the intensity of the continuum changes with respect to the line (and this is indeed observed) with changing density of the thermal radiation in the target. The line intensity may be changed by both the increase in the target thickness across the beam (here the line intensity falls with respect to the intensity of the continuum) and the change of the γ-gun intensity (here the line and the continuum vary simultaneously).

A γ-gun seems very promising for describing the best collimated jets among those observed. In fact, an intense γ-beam is an ideal means to control over large distances the direction pre-set on the scale of the order of a gravitational radius of a black hole.

If, near rotating black holes, the directional acceleration of electrons and positrons occurs along the rotation axis, then it is sufficient to have thermal (or synchrotron) radiation with intensity $L \sim 10^{38}$ erg s^{-1} $(M/10^6 M_\odot)$ $(T/10^5 K) \ll L_{Edd}$ within the volume restricted to several gravitational radii, for most of the energy of accelerated particles to be converted into a beam of γ-quanta. Thus a γ-gun may be a wide-spread phenomenon. Unfortunately the scattered γ-radiation from extragalactic objects is below the sensitivity-threshold of current gamma-ray detection. We may hope, however, that directed beams will be detected from certain bright objects. Then, with a γ-beam divergence of $\sim 10^{-2}$ and an intensity 10^{44} erg s^{-1}, the object may be observed from any distance. The criterion is quite evident in this case: the luminosity in the γ-range is very high as compared with energy values which seem reasonable for such an object. In this case the γ-spectrum may be a wide power spectrum or a narrow one similar to that in Figure 2, depending on the steady state current mode. Most probable candidate objects here may be extragalactic sources showing the effects of high superluminal velocities[40].

The authors thank S. S. Herstein, A. F. Illarionov, V. S. Imshennik, V. A. Usov and I. S. Shklovskii for discussions and assistance.

REFERENCES

1. Robinson, I., Schild, A., Schucking, E. L., Quasi-Stellar Sources and Gravitational Collapse, (University of Chicago Press, Chicago, 1965).
2. Zeldovich, Ya. B., Dok. Akad. Nauk. SSSR, 9, 195 (1964).
3. Lynden-Bell, D., Nature, 223, 690 (1969).
4. Lynden-Bell, D. and Rees M. J., Monthly Notices Roy. Astron. Soc. 152, 46 (1971).
5. Reigler, G. R., Blanford, R. D., The Galactic Center, (Am. Inst. Physics, New York, 1982).
6. Lacy, J. H., The Galactic Center (Am. Inst. Physics, New York, 1982) P. 453.
7. Johnson, W. N. and Haymes, R. C., Astrophys. J., 184, 103 (1973).
8. Leventhal, M., MacCallum, C. J. and Stang, P. D., Astrophys. J. Lett. 225, L11 (1978).
9. Riegler, G. R. et al. Astrophys. J. Lett., 248, L13 (1981).
10. Leventhal, M. et al., Astrophys. J. Lett., 260, L1 (1982).
11. Paciesas, W. S. et al., Astrophys. J. Lett., 260, L7 (1982).
12. Matteson, J. L., The Galactic Center (Am. Inst. Physics, New York, 1982), p. 109.
13. Komberg, B. V., Astr. Z., 44, 906 (1967).
14. Begelman, M. C., Blanford, R. D., Rees, M. J., Nature, 287, 307 (1980).
15. Kardashev, N. S., Preprint IKI ANSSSR, 1982, N 728.
16. Hills, J. G., Nature, 254, 295 (1975).
17. Frank, J. and Rees, M. J., Monthly Notices Roy. Astron. Soc., 176, 633 (1976).

18. Frank, J., Monthly Notices Roy. Astron. Soc., 187, 833 (1979).
19. Rees, M. J., The Galactic Center (Am. Inst. Physics, New York, 1982), p. 166.
20. Lightman, A. P. and Eardley, D. M., Astrophys. J. Lett., 187, L1 (1974).
21. Kozlowski, M., Jaroszynski, M. and Abramowicz, M., Astron. Astrophys., 63, 209 (1978).
22. Abramowicz, M., Jaroszynski, M. and Sikora, M., Astron. Astrophys., 63, 221 (1978).
23. Paczynski, B. and Witta, P. J., Astron. Astrophys., 88, 23 (1980).
24. Rees, M. J. Physica Scripta, 17, 193 (1978).
25. Rees, M. J., Proceedings of International School and Workshop on Plasma Astrophysics (Varrena, Italy, 1981), p. 267.
26. Bisnovatyi-Kogan, G. S. and Ruzmaikin, A. A., Astrophys. and Space Sci., 28, 45 (1974).
27. Bisnovatyi-Kogan, G. S. and Rusmaikin, A. A., Astrophys. and Space Sci., 42, 401 (1976).
28. Blandford, R. D., Znajek, R. L., Monthly Notices Roy. Astron. Soc., 179, 433 (1977).
29. Znajek, R. L., Monthly Notices Roy. Astron. Soc., 179, 457 (1977).
30. Znajek, R. L., Monthly Notices Roy. Astron. Soc., 185, 833 (1978).
31. Thorne, K. S. and Macdonald, D., Monthly Notices Roy. Astron. Soc., 198, 339 (1982).
32. Macdonald, D. and Thorne, K. S., Monthly Notices Roy. Astron. Soc., 198, 2 345 (1982).
33. Novikov, I. D. and Thorne, K. S., Black Holes, (Gordon and Breach, New York, 1973), p. 343.
34. Blandford, R. D., The Galactic Center (Am. Inst. Physics, New York, 1982), p. 177.
35. Stern, B. E., Preprint of Nuclear Research Institute Academy of Science USSR, N 81-82 (1978).
36. Lingenfelter, R. E., and Ramaty, R., The Galactic Center, (Am. Inst. Physics, New York, 1982), p. 148.
37. Haymes, R. C. et al., Astrophys. J., 201, 593 (1975).
38. Hall, D. N., Kleinmann, S. G. and Scoville, N. Z., Astrophys. L. Lett., 26, L53 (1982).
39. Arons, J., Space Sci., Rev., 24, 437 (1979).
40. Blandford, R. D., Rees, M. J., Physica Scripta, 17, 265 (1978).

THE ORIGIN OF THE GALACTIC CENTER ANNIHILATION RADIATION

R.E. Lingenfelter
Center for Astrophysics & Space Sciences, C-011
University of California, La Jolla, CA 92093 USA

R. Ramaty
Laboratory for High Energy Astrophysics
NASA/Goddard Space Flight Center, Greenbelt, MD 20771 USA

ABSTRACT

Observations of the e^+-e^- annihilation radiation from the Galatic Center suggest that something truly extraordinary is occurring there. We review the observations of this intense, time-varying, 0.511 MeV emission and discuss the implications of these and other recent observations on the positron production process, the annihilation region and the fundamental nature of the Galactic Center source.

INTRODUCTION

A year ago we reviewed[1] the possible origins of the e^+-e^- annihilation radiation from the Galactic Center and concluded that the most likely process for producing the annihilating positrons is photon-photon pair production in the vicinity of a massive black hole at the center of our Galaxy. We briefly summarize here the evidence and arguments leading to this conclusion, and at the same time discuss the implications of new observations and calculations that shed further light on the problem.

We briefly summarize the observations[2] and then discuss in turn their implications on the nature of the annihilation region, the positron production process and the Galactic Center source itself.

OBSERVATIONS

Intense positron annihilation radiation at 0.511 MeV has been observed from the direction of the Galactic Center for over a decade. This emission was first seen in a series of balloon observations[3-5] with low-resolution NaI detectors starting in 1970. But it was not until 1977 that the annihilation line energy of 0.511 MeV was clearly identified with high-resolution Ge detectors flown by Leventhal, MacCallum and Stang[6]. The latter observation also revealed that the line is very narrow (FWHM $\leq$ 3.2 keV) and suggested that the continuum below 0.511 MeV may include a significant contribution from three-photon positronium annihilation, consistent with $\sim$ 90% of the positron annihilation taking place through positronium formation.

The existence of this very narrow line was confirmed by Riegler et al.[7] with Ge detectors on HEAO-3 in the fall of 1979. These observations set an even more stringent limit on the line width (FWHM < 2.5 KeV) and determined the line center energy as 510.90 $\pm$ 0.25 keV.

The HEAO-3 observations[7] also provided new information on the location and spacial extent of the emission region and most important showed that the line intensity varies significantly in time. In particular, these observations showed that the line emitting region is smaller than the angular resolution of the detector (35° FWHM) and that the direction of the source coincides with that of the Galactic Center, within the observational uncertainty of $\pm$ 4°. Moreover the observations showed that the 0.511 MeV line intensity decreased by a factor of three in six months, from $(1.85 \pm .21) \times 10^{-3}$ photons/cm² sec in the fall of 1979 to $(0.65 \pm .27) \times 10^{-3}$ photons/cm² sec in the spring of 1980. This variability has been confirmed by balloon-borne Ge detector observations[8-9]. Observations[10] with a NaI detector in the fall of 1977 could also indicate a variation on a time scale as short as 10 days, but it seems much more likely that the higher 0.511 MeV intensity observed with this detector results from a larger diffuse galactic component seen in its much greater (100° FWHM) field of view.

Observations of continuum emission in the hard X-ray and gamma-ray bands have recently been reviewed by Matteson[11]. The hard X-ray emission is also time variable and is weakly correlated with the variability of the 0.511 MeV line (e.g. Ref. 9). These observations set an upper bound of 2×10^{38} erg/sec on the Galactic Center continuum luminosity at photon energies $> m_e c^2$, since only part of this emission may come from the same source as the annihilation radiation.

The luminosity of the Galactic Center region at various photon energies, implied by these and other observations are summarized in Figure 1 from Reference 1.

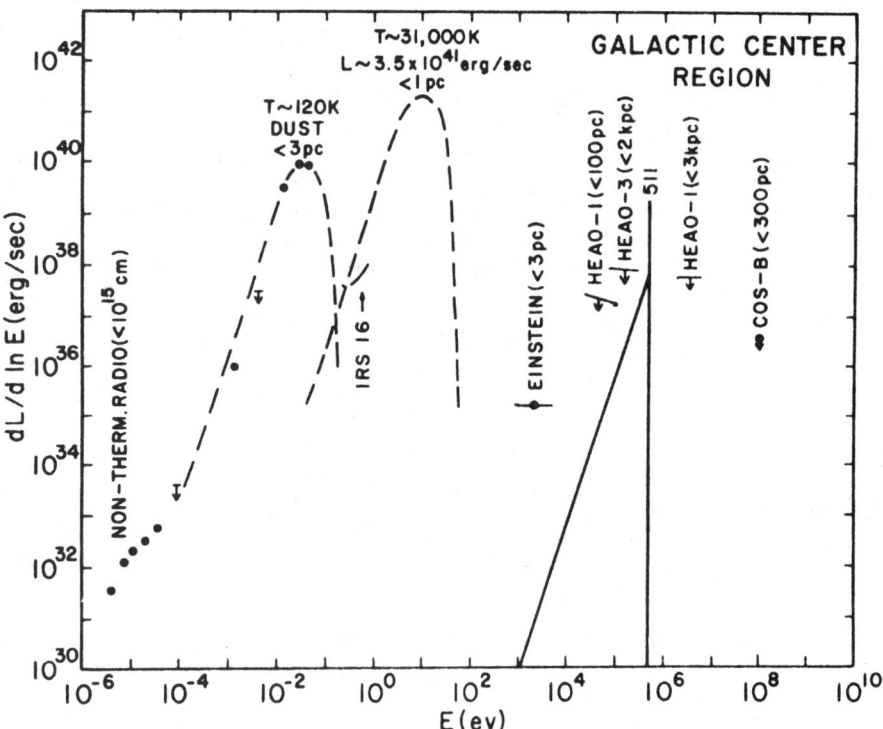

Figure 1. The luminosity per unit lnE as a function of photon energy, E, from the region around the Galactic Center. Data are shown for the compact ($< 10^{15}$ cm) nonthermal radio source[12], the $\sim$ 3pc dust ring[13], the nonthermal infrared source IRS 16 (Ref. 14), the soft X-ray emission (< 3pc) from the EINSTEIN satellite measurements[15], the hard X-ray emission from HEAO-1 (Ref. 11), the 511 KeV line and positronium continuum[6,7], and the gamma ray emission from HEAO-1 (Ref. 11) and COS-B (Ref. 16). Also shown as dashed curves are the blackbody luminosity at $\sim$ 31,000 K required[17] to account for ionization in the warm IR clouds within < 1pc of the Galactic Center and a blackbody luminosity at 120 K as inferred[13] from the far infrared observations of the $\sim$ 3pc dust ring.

THE ANNIHILATION REGION

The nature of the positron annihilation region is constrained by the intensity variations, the line width and the line center energy. The size of the region should not exceed about 10^{18} cm, the distance traveled by relativistic positrons in 1/2 year. The density of the gas in which the positrons annihilate should be larger than 10^5 H/cm^3, the minimum density required to slow them down in 1/2 year, but less than 10^{15} H/cm^3 in order not to break up the triplet positronium before it annihilates. Since triplet positronium could also be broken up (R. McCray, private communication) by photons of energy > 6.8 eV, the energy density of such photons should not exceed ~ 10^3 erg/cm^3 in the annihilation region, or if the 31,000 K emission[17] (Fig. 1) comes from a single source the annihilation region must be > 2×10^{13} cm away from it, assuming a positronium ionization cross section of 3×10^{-18} cm^2.

The observed line width requires[18] that this gas also be at least partially ionized ($n_e < 0.1n$). If the gas were neutral, the line width would be larger than observed because it would be Doppler broadened, not by the thermal motion of the gas, but by the velocity of energetic positrons forming positronium in flight by charge exchange with neutral hydrogen. In a partially ionized gas, however, positrons lose energy to the plasma fast enough that the positrons thermalize before they annihilate or form positronium. The line width thus reflects the temperature of the medium, so that the observations require a temperature $\lesssim 5 \times 10^4$ K. The line width further limits any velocities of rotation, expansion or random motion to < 700 km/sec, while the line center energy implies a bulk velocity along with line of sight $-90 < v < +200$ km/sec and a gravitational redshift $z < 7 \times 10^{-4}$.

The strongest of these constraints are summarized in Table I.

Table I

CONSTRAINTS ON THE e$^+$-e$^-$ ANNIHILATION REGION AT THE GALACTIC CENTER

Physical Parameter	Constraint	Observation
Size	< 10^{18} cm	variability
Gas density	> 10^5 H/cm^3	variability
Ionization state	$n_e/n > 0.1$	line width
Temperature	$\lesssim 5 \times 10^4$ K	line width
Rotation, expansion or random motion	< 700 km/sec	line width
Bulk motion along line of sight	$-90 < v < +200$ km/sec	line center energy
Gravitational redshift	$z < 7 \times 10^{-4}$	line center energy

As we previously suggested[19] possible annihilation sites which satisfy these constraints are the warm clouds[17] and the compact source IRS 16 (Ref. 12), observed within the central parsec of the Galaxy.

THE POSITRON SOURCE

The nature of the positron source is also strongly constrained by the observed variation of the 0.511 MeV intensity and by observations at other wavelengths. The decrease of a factor of three in the line intensity in six months clearly excludes any of the multiple, extended sources, such as cosmic rays, pulsars[20], supernovae[21], or primordial black holes[22], previous proposed. Instead, it essentially requires a single, compact (< 10^{18} cm) source which is apparently located either at, or close to, the Galactic Center and which is inherently variable on time scales of six months or less.

The observed 0.511 MeV line intensity of $\sim 2 \times 10^{-3}$ photons/cm^2 sec requires at the distance of the Galactic Center (~ 10 kpc) a positron annihilation rate of 4×10^{43} e$^+$/sec, if $\sim 90\%$ of the positrons annihilate via positronium. This rate corresponds to minimum luminosity of $\sim 6 \times 10^{37}$ erg/sec in both the line and three-photon continuum. With such a luminosity the Galactic Center is the most luminous gamma-ray source in the galaxy. The uniqueness of this source makes it unlikely that it results from the chance occurrence of the youngest supernova or pulsar along with line of sight to the center of the galaxy.

The strongest constraints on the various positron production processes are set[1] by observations of continuum emission at energies $> m_e c^2$ from the direction of the Galactic Center[11]. When compared with the annihilation radiation luminosity, the continuum gamma ray luminosity implies a very efficient positron production process, one in which more than 30% of the total radiated energy $> m_e c^2$ goes into electron-positron pairs. If the positron production occurs on time scales comparable to that of the observed variation and in an essentially optical thin region which emits isotropically, only photon-photon pair production can provide the required high efficiency.

We considered[1] two geometries for the positron production region: a spherical volume in which e$^+$–e$^-$ pairs are produced by photons interacting isotropically and a beam in which the pairs are produced by photon interactions only at small angles.

The most efficient pair production occurs in isotropic interactions of photons at energies close to $m_e c^2$. The pair production rate Q in a spherical source of radius r may be approximated by

$$Q \sim \tfrac{1}{2} n_\gamma^2 <\sigma c> \frac{4\pi}{3} r^3$$

where $<\sigma c>$ is the average pair production cross section times the velocity of light, equal to $\sim 2 \times 10^{-15}$ cm^3/sec for a Wien spectrum of temperature $\sim m_e c^2/2$, and n_γ is the photon number density. The photon density can also be related to the continuum luminosity at energies $\gtrsim m_e c^2$ by

$$L \sim \frac{\epsilon n_\gamma c 4\pi r^2}{3},$$

where ϵ is the average photon energy and r/c is the photon residence time. Combining these two equations and setting $\epsilon \sim 3 m_e c^2/2$, we see that for a given continuum luminosity the positron production rate depends only on the source size, such that the radius,

$$r \sim \frac{3 <\sigma c>}{8\pi c^2 \epsilon^2} \frac{L^2}{Q} \sim 2 \times 10^{-25} \frac{L^2}{Q} \text{ (cm)}.$$

From the observed luminosity limit of $L \leq 2 \times 10^{38}$ erg/sec and a production rate Q equal to the annihilation rate of 4×10^{43} e$^+$/sec, the radius of the positron source must be $\leq 2 \times 10^8$ cm.

Pair production by isotropic photon-photon interactions thus requires an exceedingly compact course, but with a high luminosity. The most obvious candidate is a blackhole. But if this source is a blackhole releasing gravitational energy of accreting matter close to its Schwarzschild radius, then it must have a mass $\leq 10^2$ M$_\odot$, which is much smaller than the masses of 10^6 to 10^7 M$_\odot$ blackholes that have been suggested[17,23] at the Galactic Center. Yet such a small size would be consistent with arguments by Ozernoy[24] that the Galactic Center cannot contain a blackhole larger than about 10^2 M$_\odot$, if tidal disruption of stars is the principal source of the accreting matter on which it grows.

The photons needed to produce the pairs could themselves be produced in a hot accretion disk around the blackhole[25]. A luminosity of $\sim 2 \times 10^{38}$ erg/sec requires an accretion rate of $\sim 3 \times 10^{-8}$ M$_\odot$/yr which could form a $\sim 10^2$ M$_\odot$ hole in the age of the Galaxy. A major fraction of the e$^\pm$ pairs produced by photon-photon collisions above the disk could then escape from the source region before they annihilate, a constraint (Table 1) set by the absence of any measurable redshift in the energy of the annihilation line.

We turn now to the alternative geometry of pair production by small angle photon interactions in a beam, which may be produced[26,27] by dynamo action in a magnetic field accreting onto a blackhole. We previously showed[1] that with a beam the constraint on the size of the production region could be greatly relaxed, but at the expense of a much higher beam luminosity in gamma rays of energy $\gg m_e c^2$. This possibility for producing positrons in the Galactic Center was first suggested by M. Burns (private communication 1982) and differs from Novikov's[28] model (in this volume), which relies on the relatively less efficient production by beam photons interacting with gas in a cloud.

The pair production rate Q for small angle ($\theta \sim r_b/l$) photon interactions in a beam of radius r_b and length l may be approximated by

$$Q \sim \tfrac{1}{2} n_\gamma^2 <\sigma v_\perp> \pi r_b^2 l,$$

where $v_\perp \sim (r_b/l)c$ is the mean transverse velocity of the interacting photons, and n_γ is the density of those photons with energies greater than the small angle pair production threshold $E_{th} \sim (l/r_b) m_e c^2$. This density can be related to the beam luminosity of such photons by

$$L_b \sim (l/r_b) m_e c^2 n_\gamma c \pi r_b^2.$$

Combining these two equations, we see that the beam radius is

$$r_b \sim \frac{<\sigma c>}{2\pi c^2 (m_e c^2)^2} \frac{L_b^2}{Q} (\frac{r_b}{l})^2 \sim 8 \times 10^{-25} \frac{L_b^2}{Q} (\frac{r_b}{l})^2 \text{ cm}.$$

Thus for a pair production rate Q of 4×10^{43} e$^+$/sec the beam radius could be as big as $\sim 10^{12}$ cm, or equal to the Schwarzschild radius of a $3 \times 10^6 M_\odot$ blackhole, if the beam luminosity at photon energies greater than 25 MeV were as high as half the Galactic Center bolometric luminosity limit of $\sim 3.5 \times 10^{41}$ erg/sec, and the aspect ratio of the beam were 0.02, corresponding to angle of 1°.

The resulting pairs would also have energies of ~ 25 MeV, comparable to those of the photons which produced them. But they could be stopped and annihilate to give narrow 0.511 MeV line emission, if the beam hit a gas cloud. The bulk of the pair energy, amounting to $\sim 10^{40}$ erg/sec, would be dissipated in heating the gas which could in turn reradiate it isotropically as thermal radiation consistent with the constraints on the $\lesssim 30,000$ K luminosity. Since the radiation yield of ~ 25 MeV electrons and positrons is $\lesssim 3\%$, their bremsstrahlung could also be consistent with the hard X-ray and gamma-ray luminosity limit of $\lesssim 2 \times 10^{38}$ erg/sec.

The detailed energetics of both of these geometries, however, are still under study.

SUMMARY

The observed time variations and line width of the e$^+$–e$^-$ annihilation radiation from the Galactic Center require that the positrons be produced essentially by a single source and that they annihilate in an ambient gas of density $> 10^5$ H/cm^3, ionization fraction $> 10\%$, temperature $< 5 \times 10^4$ K, and confined to a region of size $< 10^{18}$ cm. Such conditions may exist in the warm clouds and the compact source IRS 16 within the central parsec of the galaxy.

The limits on the accompanying continuum emission at energies $> m_e c^2$ set strong constraints on the positron production process, requiring an exceedingly high efficiency, such that $> 30\%$ of the total radiated energy $> m_e c^2$ goes into e$^+$–e$^-$ pairs. The most likely mechanism appears to be pair production in photon-photon collisions in the close vicinity of a massive blackhole, either near the hot (kT $\sim m_e c^2$) inner part of an accretion disk around a $\sim 10^2 M_\odot$ blackhole, or in a beam of ~ 25 MeV photons produced by a

$\sim 10^6 M_\odot$ hole. In either case the absence of any measurable redshift in the line center energy requires that a large fraction of the positrons escape from the central source and annihilate at great distances from the hole ($> 10^3$ times the Schwarzschild radius).

ACKNOWLEDGEMENTS

The work of R.E.L. was supported by NASA grant NSG-7541.

REFERENCES

1. R.E. Lingenfelter and R. Ramaty, The Galactic Center (Am. Inst. Physics, New York, 1982), p. 148.
2. C.J. MacCallum and M. Leventhal, these proceedings (1983).
3. W.N. Johnson, F.R. Harnden and R.C. Haymes, Ap. J., *172* L1 (1972).
4. W.N. Johnson and R.C. Haymes, Ap. J., *184*, 103 (1973).
5. R.C. Haymes et al., Ap. J., *201*, 593 (1975).
6. M. Leventhal, C.J. MacCallum and P.D. Stang, Ap. J., *225*, L11 (1978).
7. G.R. Riegler et al., Ap. J., *248*, L13 (1981).
8. M. Leventhal et al., Ap. J., *260*, L1 (1982).
9. W.S. Paciesas et al., Ap. J., *260*, L7 (1982).
10. B.M. Gardner et al., The Galactic Center (Am. Inst. Physics, New York, 1982), p. 144.
11. J.L. Matteson, The Galactic Center (Am. Inst. Physics, New York, 1982), p. 109.
12. K.Y. Lo et al., Ap. J., *249*, 504 (1981).
13. I. Gatley, The Galactic Center (Am. Inst. Physics, New York, 1982), p. 25.
14. E.E. Becklin et al., Ap. J., *219*, 121 (1978).
15. M.G. Watson et al., Ap. J., *250*, 142 (1981).
16. B.N. Swanenburg et al., Ap. J., *243*, L69 (1981).
17. J.H. Lacy et al., Ap. J., *241*, 132 (1980).
18. R.W. Bussard, R. Ramaty and R.J. Drachman, Ap. J., *228*, 928 (1979).
19. R. Ramaty and R.E. Lingenfelter, Phil. Trans. R. Soc. Lond., *A301*, 671 (1981).
20. P.A. Sturrock and K.B. Baker, Ap. J., *234*, 612 (1979).
21. R. Ramaty and R.E. Lingenfelter, Nature, *278*, 127 (1979).
22. P.N. Okeke and M.J. Rees, A. Ap., *81*, 263 (1980).
23. D. Lynden-Bell and M.J. Rees, M.N.R.A.S., *152*, 461 (1971).
24. L.M. Ozernoy, Large Scale Characteristics of the Galaxy (Reidel, Dordrecht, 1979), p. 395.
25. D.M. Eardley et al., Ap. J., *224*, 53 (1978).
26. R.D. Blandford, Active Galactic Nuclei (Cambridge Univ. Press, London, 1979), p. 241.
27. R.V.E. Lovelace, J. McAuslan and M. Burns, Particle Acceleration Mechanisms in Astrophysics (Am. Inst. Physics, New York, 1979), p. 399.
28. N.S. Kardashev, I.D. Novikov, A.G. Polnarev and B.E. Stern, these proceedings (1983).

POSITRONS FROM SUPERNOVA AND THE ORIGIN OF THE GALACTIC CENTER POSITRON ANNIHILATION RADIATION

Stirling A. Colgate

Los Alamos National Laboratory, Los Alamos, NM 87545

ABSTRACT

The emission of positrons from supernova ejecta is discussed in terms of the galactic center annihilation radiation. The positrons from the radioactive sequences $^{56}Ni \rightarrow {}^{56}Co \rightarrow {}^{56}Fe$ are the most numerous source from supernova. Only type I supernova will allow a significant fraction to escape the expanding ejecta. For a neutron star model of a type I SN a fraction 4×10^{-3} of the escaped positron is enough to create the observed several year fluctuation of the annihilation radiation. The likelihood of this model is discussed in terms of other astrophysical evidence as well as the type I SN light curve.

INTRODUCTION

It is now generally agreed that the origin of the energy that powers the entire light curve of type I SN[1-5] and possibly part of the late time light curve of type II SN[6] is due to the radioactive sequence $^{56}Ni \xrightarrow{6.1 \text{ d}} {}^{56}Co \xrightarrow{77 \text{ d}} {}^{56}Fe$. The nuclear physics that leads to this unique isotope is the thermonuclear synthesis of alpha particle nuclei C, O, and Si. The isotope ^{56}Ni is the minimum in the packing fraction curve of alpha particle nuclei, or the end point of nuclear synthesis starting from lighter alpha particle nuclei. As a consequence, the preponderance of ^{56}Ni causes its decay energy to dominate the heating of ejected matter during most of the subsequent expansion. Without a persistent energy source, adiabatic cooling during expansion would lead to much cooler matter. In this decay chain, positrons are produced in the last decay $^{56}Co \xrightarrow[20\% \beta^+]{77 \text{ d}} {}^{56}Fe$. It is primarily these positrons that are candidates for the production of the annihilation radiation from the galactic center from a previous supernova event, although other radioactive species particularly $^{22}Na \xrightarrow[90\% \beta^+]{3.8 \text{ y}} {}^{22}Ne$ could contribute,[7] to the positron annihilation flux.

The problems with each of these radioactive sources for an explanation of the galactic center radiation are:
1. number of positrons at the source
2. escape from the source (matter thickness and magnetic field)
3. lack of other gamma rays
4. special annihilation conditions.

REQUIRED SOURCE SIZE

If the usual assumption is made that the annihilation gamma ray flux of 1 to 2×10^{-3} photons cm^{-2} s^{-1} observed at the earth indeed comes from the galactic center, then the recent several year fluctuation[7,8] of roughly 2×10^{-3} photons cm^{-2} s^{-1} for a year at the distance of the galactic center requires a single source of 3×10^{50} positrons, or for example, 10^{-4} M$_\odot$ of ^{56}Ni, or 10^{-5} M$_\odot$ of ^{22}Na. The likely production of ^{22}Na per supernova is less than 10^{-6} M$_\odot$ for type I[9] and so it is not further considered here. This is a small mass of ^{56}Ni but isotope production, decay, and positron escape fraction significantly restrict the possibilities for even these small requirements. The escape fraction of ^{56}Ni positrons from supernova ejecta depend critically on the supernova model.

ESCAPE OF POSITRONS

Electrons or positrons lose energy by ionization in the medium. The spectrum of beta particles results in a distribution in range and hence an escape fraction dependent on thickness for a source medium such as supernova ejecta. In ref. 3 the effective range for the ^{56}Co beta particles and the escape fraction for various supernova models of the radial distribution of the source function were considered. Surprisingly, the range distribution to first order behaves like an exponential absorption similar to the exponential absorption of gamma rays, or a range of equivalent constant cross section. This result also applies whether or not a "combed" dipole magnetic field as would occur in the radial expansion of the electrically conducting ejecta is present. Hence the Monte Carlo deposition and escape calculations which were applied to determine the gamma ray heating and escape fraction can be carried over to the positron case with an effective range or absorption mean free path of $\lambda_+ = 0.1$ g cm^{-2}. (This range assumed high atomic number matter, i.e., ^{56}Fe ionized no more than several times. Ionized hydrogen would result in approximately 1/10 this effective range.) The escape fraction can then be calculated using $(1 - D)$, where D is the deposition fraction of Table 1 of ref. 3. In essence, 50% escape occurs when $\tau \equiv \rho R/\lambda_+ = 1$, or when $\rho R = 0.1$ g cm^{-2}. At early times when $\tau > 1$, the escape fraction is increasing rapidly as τ decreases. Later it approaches unity for $\tau \ll 1$. Hence if the time, t_1, when $\tau = 1$ is large compared to the lifetime of the radioactive positron decay nucleus, t_+, then the attenuation acts as a window opening near $\tau = 1$, or $M/[4\pi R^2 \lambda_+] = 1$. This model would apply to a type II SN where a large envelope of hydrogen blankets a small core of radioactive ^{56}Co. The number of positrons escaping through such a blanket is proportional to $e^{-(\rho R/\lambda_+)}$. The actual escape is more complicated than this, but the ratio, t_1/t_+ gives some indication of the escape probability for various supernova models. Generally supernova all seem to have the same total kinetic energy of roughly 10^{51} ergs or less, independent of type as interpreted from an integration of supernova remnant emission,[10]. For a uniformly expanding mass M whose velocity is v at the outer radius R, then the kinetic energy is $(3/5) M v^2/2$, so that the time t_1 to reach a thickness λ_+ becomes:

$$t_1 = 5.4 \times 10^6 \, M(\lambda_+ E_{51})^{-\frac{1}{2}} \quad \text{s.}$$

$$= 620 \, M \, (\lambda_+ E_{51})^{-\frac{1}{2}} \quad \text{d.} \tag{1}$$

Here M is in solar masses λ_+ in 0.10 g cm^{-2} and E_{51} in 10^{51} ergs. Thus for a type II supernova where the ^{56}Ni would be a small central region surrounded by 10 M_θ of expanding hydrogen and where $\lambda_+ \cong$ 0.01 g cm^{-2}, $t_1 = 2 \times 10^4$ days, or $t_1/t_+ = 177$ for $t_+ = 111$ days of ^{56}Co. The escape fraction is then a number far too small to reasonably account for the positrons of the galactic center even for a solar mass of ^{56}Ni. Thus positron escape from models of SN I is more likely.

Three models of SN I are considered:
1. A bare mass of ^{56}Ni, 1.41 M_θ, and $E^{51} = 1$ resulting from a thermonuclear explosion.
2. A sphere of 0.7 M_θ of ^{56}Ni surrounded by 0.7 M_θ of lighter elements again from a thermonuclear explosion $E_{51} \cong 1$.
3. A neutron star model for which 0.25 M_θ of ^{56}Ni is surrounded by 0.25 M_θ of lighter elements, (ref. 3) $E_{51} = 0.67$.

The bare ^{56}Ni mass allows significant escape from a surface layer λ_+ thick. However, such a model places too much iron into the interstellar medium,[11] does not have the lighter element mantle necessary to give the absorption spectra on top of a photosphere characteristic of SN I[12] and results in too high a peak luminosity to be consistent with $H_0 > 50$.[5] Hence the second thermonuclear model is favored over the first. Here positron escape is more complicated due to details of the model, but in general positron escape is more attenuated than the bare source model because of the interior source and finite range of positrons. The positrons from models 2 and 3 will be roughly the same, but I prefer the third model of a neutron star because we have already calculated the positron escape fraction as $\cong 10\%$, ref. 3. In this case 0.025 M_θ of ^{56}Ni would contribute to escaped positrons, and only a fraction 4×10^{-3} of these escaped positrons are needed to supply the galactic center source.

The reasons for considering the collapse to a neutron star for type I supernova are:
1. The high doppler velocity of the ejecta (1.5 to 2×10^9 cm s^{-1}) and limited kinetic energy give better agreement with low ejected mass;[13]
2. The observation of binary neutron stars (pulsars and rapid pulsating x-ray sources) demands neutron star formation with a low mass of ejecta (less than the total mass, to remain bound) and

3. The exponential, late-time light decay of 56 day half life is explained only by the escape of positrons from a small mass of ejecta.
4. The peak luminosity of type I SN can be interpreted in terms of the Hubble constant and mass of ^{56}Ni,[5] as $H \cong 40 \, M_{Ni}^{-\frac{1}{2}}$, an uncomfortably small value for a white dwarf explosion of $1.4 \, M_\theta$ of ^{56}Ni.
5. The rate of iron production in the solar neighborhood as well as the galaxy is too large by 5 to 10 fold for $1.4 \, M_\theta$ of ^{56}Ni per type I SN.

The reasons for believing that a thermonuclear explosion disrupts the whole star are:
1. Stellar evolution calculations predict the formation of a degenerate carbon oxygen core whose early ignition would explode the star.
2. No soft x-ray source (cooling neutron star) or pulsar is seen at the site of Tycho[13] or Kepler's SN remnants.[14]

We examine some of these reasons for low mass ejecta (1-5) in greater detail because they determine the likelihood that positrons from supernova can contribute* to the galactic center annihilation radiation.

(1) Doppler velocities of the type I SN ejecta are inferred from the line shifts. To obtain these requires an interpretation of the spectra. The most convincing interpretation is that of Branch[12] and colleagues who superimpose absorption lines on a continuum emission from a photosphere. They estimate an expansion velocity of 10^9 cm s^{-1} for the photosphere at its maximum radius at 30 days after maximum light or 45 days after the supernova origin. We can estimate the thickness and hence outer radius of the matter distribution on the basis of assuming a uniform expansion and an opacity as great outside as inside the photosphere and the Eddington surface approximation. The result is that the velocity of the surface of the uniform region becomes 1.5×10^9 cm s^{-1}. Using the above and the fact that the maximum occurs at roughly 45 days results in an ejected mass $0.14/\kappa$ solar masses. The opacity, κ, unfortunately, is most uncertain because of the line scattering and doppler shifts. It could be as low as the Compton cross section $\times \, 1/<A>$, A the average atomic weight, and hence $\leq 10^{-2}$ cm^2 g. On the other hand Karp et al.[15] have estimated opacities significantly greater than Compton for smaller doppler shifts. Here very small changes in composition, temperature and doppler velocities make large changes in opacity.

(2) Several low mass, binary pulsars and one low mass fast pulsating x-ray source are known. It is highly unlikely that these binaries can be formed in a type II supernova because the mass ejected is greater than the mass of the sum of the residual stars. An alternate mechanism to produce neutron stars unobservably seems unlikely when collapse inside a $10 \, M_\theta$ envelope (SN II) manages to eject the envelope rather than collapsing the envelope to a black hole. In other words if collapse to a neutron star ejects matter in one case with a massive envelope, then collapse with a much less massive envelope should eject some matter more easily. A very small ejected mass will make an observable supernova.

(3) The exponential light curve of SN I is well known to give a constant exponential decrement from roughly 100 days to ≅ 700 days after maximum luminosity with a half-life of 56 ± (a few) days. The uniqueness of the exponential behavior is open to some question, especially when one attempts to average the very large data base of SN I light curves assembled by Barbon et al.[16] Similarly a scatter of decay times is presented by Rust et al.[17] However, as is even evident from their Table I whenever the late time light curve is measured with high accuracy, the value of the decay half-life approaches 56 days with significantly greater accuracy. One surmises that the very brightest supernova are measured most carefully because of the potential for a more meaningful data set. The 56-day half life is observed with better than 5% accuracy for IC4182 (1937c), NGC4214 (1954a), NGC5253 (1972e). These are the only cases in which a data point beyond 600 days (after maximum) was measured. These data points are shown in Figs. 1 and 2 from ref. 3 with the 700-day data point of 5253 emphasized as well as the corresponding ones for 4182. These points were obtained after great effort; the first photometric by Kirshner and Oke[18] and the second photographic by Deutsch as reported by Van Hise[19] and Baade and Zwicky.[20] The late time exponential decay, when measured accurately extends over a range of luminosity of 10^3. There are examples of fainter type I supernova light curves less accurately measured that do not demonstrate this striking feature, but where there is the potential for observation, the exponential feature occurs.

We have interpreted this feature as being due to the progressive escape of positrons[2,3] from the expanding debris. Figures. 1 and 2 are from ref. 3 where the deposition function for both gamma rays and positrons were calculated with a Monte Carlo numerical calculation. This deposition calculation shows that in a region of expansion appropriate to the problem, an exponential radioactive decay was transformed into a half life roughly 3/4 of the nuclear decay half life (77 d for ^{56}Co). The factor 3/4 is a fortuitous circumstance previously noted but unexplained.[19]

The alternate explanation of this phenomena has been put forth for the thermonuclear explosion models.[4] Here, because the escape fraction is small or an assumed closed magnetic field within the uniform density region of the ejecta, leads to the assumption of fully trapped positrons.

If there were no other losses, then one obtains the dashed curve in Fig. 2 for the luminosity corresponding to the 77-day half life of ^{56}Co positrons. One observes that something like a factor of 10 loss must be introduced to reduce the positron luminosity source to the observed luminosity decay.

Axelrod[4] has proposed that the reduction in late time luminosity occurs due to infrared emission. This produces the curve with the dotted cutoff. The reason for the sudden transition is derived in Meyerott[21] where he shows that the infrared emission states in Fe II and Fe III are strongly saturated and hence result in an emission relatively independent of density whereas the optical emissivity is density dependent, his Figs. 6 and 7. Hence the transition from optical to infrared emission will occur suddenly in time as shown in

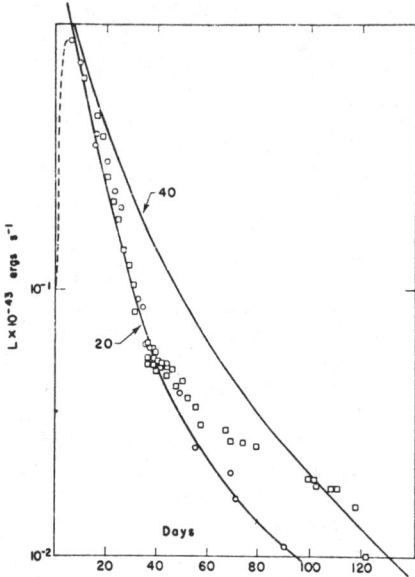

Fig. 1.—The calculated luminosity at early and intermediate times for $M_{Ni} = 0.25$ solar masses and the corresponding deposition functions for $\tau = 1$ at 20 days and 40 days. Gamma-ray deposition and the Ni → Co → Fe decay determine the solid curves. The dashed curve is the modification of the deposition function due to diffusion and expansion (Colgate and McKee 1969). The extrapolation of the deposition curves reaches 2×10^{43} ergs s^{-1} at $t = 0$, and the difference between this and the dashed curve is due to heat energy converted to kinetic by expansion. The circles give NGC 5253 data (Kirshner and Oke 1975), and the squares give NGC 4182 data (Baade and Zwicky 1938; Van Hise 1974).

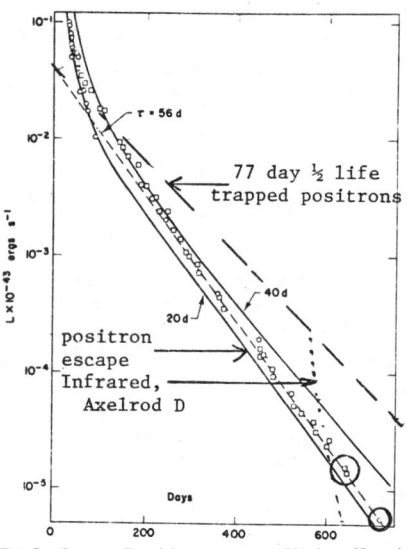

Fig. 2.—Same as Fig. 1 for times out to 700 days. Here the curves are primarily determined by the deposition of positrons from the Co → Fe decay. The dashed curve is a fit to the data with a slope corresponding to a 56 day half-life.

Fig. 2. The infrared emission depends upon the low lying levels in Fe II and Fe III at wave lengths of roughly 23 and 33 microns. The excitation of these levels can take place by relatively low energy electrons, $\cong 0.1$ eV, and so a large infrared radiation flux saturated in the lines will take place at relatively low temperature. However, if the density at late times is such that the emissivity in the optical is low, then the temperature will increase. This higher temperature will then ionize Fe III to Fe IV and no further infrared emission will take place. I estimate that for the expected conditions that the temperature maximizes a second time at roughly one year (15,000°) and is 3 × higher than necessary to ionize Fe III to Fe IV. I believe this is what happens with the low mass high velocity expansion conditions of the neutron star SN I model.

(4) The peak luminosity of SN I has been interpreted by Branch in terms of distance, red shift and ejected mass of ^{56}Ni. The value he derives is $H \cong (M_{Ni}/M_\theta)^{-\frac{1}{2}}$ km s^{-1} Mpc^{-1}. For a full white dwarf mass

of 1.4 M_θ of ^{56}Ni this corresponds to H = 34 and 0.7 M_θ ^{56}Ni to ≅ 50. The ¼ M_θ of ^{56}Ni, neutron star model results in H ≅ 80. The continuing debate on the correct value of H may yet resolve supernova or be resolved by them.

(5) The iron synthesized by SN I is bounded by the likely amount of iron in the solar neighborhood[11] and places an upper bound of 0.3 M_θ of iron per all supernova. Since SN II should produce some iron, this restricts SN I models to low mass ejecta.

LACK OF OTHER GAMMA RAYS

The spectra of radiation from the galactic center shows no other gamma ray lines to roughly 10% of the annihilation peak. The greatest sensitivity to the gamma ray lines of ^{56}Co→^{56}Fe decay is the gamma ray of 0.84 MeV that is emitted each transition. Thus there will be 2/5 as many annihilation quanta per decay as the 0.84 gamma rays. The lower limit background instantaneous rate of 1×10^{-4} photon cm^{-2} s^{-1} from the galactic center requires that the neutron star model of 0.25 M_θ of ^{56}Ni has decayed by 2×10^{-6}, or observed after a time of 13 life times or 4 years. This 4-year delay must precede the earliest observations of roughly 10 years ago. Therefore we are observing at least 14 years after such a possible event. On the other hand only a fraction 4×10^{-3} of the emitted positrons need to annihilate during the one year peak so that the escape positrons may run into a local molecular cloud whose density graident is large and solid angle small to result in the observed peak. The possibility of such a SN I event occurring 15 to 20 years ago in the galactic center and not being observed in some wave length would seem possible but unlikely.

ACKNOWLEDGEMENT

I am deeply indebted to Albert Petschek for extensive help in preparing this paper and to Reuven Ramaty and Richard Lingenfelter for discussion.

REFERENCES

1. Colgate, S. A. and McKee, C. R., 1969, Ap. J., 157, 623.
2. Arnett, W. D. 1979, Ap. J. (Letters) 230, L37.
3. Colgate, S. A., Petchek, A. G., and Kriese, J. T., 1980, Ap. J. 237, L81.
4. Weaver, T. A., Axelrod, T. S., and Woosley, S. E., 1980, Proceedings, "Type I Supernovae," ed. J. C. Wheeler, University of Texas, Austin, TX, p. 113.
5. Branch, David, "Models of Type I Supernovae and Observations of SN 1981b," presented at the Eleventh Texas Symposium on Relativistic Astrophysics, Austin, TX, December 13-17, 1982.
6. Arnett, W. D. and Falk, S. W., 1976, Ap. J. 210, 733.
7. Lingenfelter, R. E. and Ramaty, R., 1982, The Galactic Center, Am. Inst. Phys., ed. Riegler and Blandford, p. 148.
8. Mahoney, W. A. Wheaton, W. A., Willett, J. B., Jacobson, A. S. 1981, Ap. J. (Letters) 248, L13.

9. Wallace, R. K., private communication.
10. Chevalier, R. A., 1977, Ann. Rev. Astron. Astrophys. 15, 175.
11. Twarog, B. A. and Wheeler, J. C., 1982, Ap. J. 261, 636.
12. Branch, D., Lacy, D. H. McCall, J. L., Sutherland, P. G., Uomoto, A., Wheeler, J. C., and Wills, B. J., 1983, "The Type I Supernova 1981b, in NGC 4536: The First Hundred Days," preprint, University of Oklahoma, Norman, Oklahoma.
13. Chevalier, R. A. Kirshner, R. P., and Raymond, J. C. (1980) Ap. J. 235, 186.
14. Nomoto, K. and Tsuruta, S., 1981, Astrophys. J. Letts. 250, L19.
15. Karp, A. H., Lasher, G., Chan, K. L., and Salpeter, E. E. (1977) Ap. J. 214, 161.
16. Barbon, R. Ciatti, F., and Rosino, L. (1973) Astron. & Astrophys. 25, 241.
17. Rust, B. W., Leventhal, M., McCall, S. L. (1976) Nature 262, 118.
18. Kirshner, R. P., and Oke, J. B. (1975) Ap. J. 200, 574.
19. Van Hise, J. R. (1974) Ap. J. 186, 1007.
20. Baade and Zwicky (1938) Ap. J. 88, 411.
21. Meyerott, R. E. (1980) Ap. J. 239, 253.

AN ELECTRON-POSITRON JET MODEL FOR THE GALACTIC CENTER

M. L. Burns
NASA/Goddard Space Flight Center
Greenbelt, MD 20771
and
Department of Physics and Astronomy
University of Maryland
College Park, MD 20742

ABSTRACT

High energy observations of the galactic center on the subparsec scale seem to be consistent with electron-positron production in the form of relativistic jets. These jets could be produced by a $\sim 10^6$ $M_\odot$ black hole dynamo transporting pairs away from the massive core. An electromagnetic cascade shower would develop first from ambient soft photons and then non-linearly; the shower using itself as a scattering medium. This is suited to producing, cooling and transporting pairs to the observed annihilation region. It is possible the center of our galaxy is a miniature version of more powerful active galactic nuclei that exhibit jet activity.

OBSERVATIONS

Interesting activity from the heart of our galaxy has been observed over the past few years. Specifically, a luminous line at 0.511 MeV has been detected which is suprisingly narrow and variable.[1] In addition, the region from which this line emanates coincides with a small and less luminous radio source.[2] A short list of these observations follows:

1. A 10^6 $M_\odot$ object at the galactic center could be inferred from the widths of Neon lines in HII clouds orbiting the central region.[3]

2. Luminosities are: 10^{38} erg/s in continuum gamma rays[4]; 10^{37} erg/s in 0.511 MeV line[1]; 10^{37} erg/s in hard X-rays[4]; 10^{35} erg/s in soft X-rays[4]; 10^{40} erg/s in UV[3]; 10^{41} erg/s in IR[5] and 10^{33} erg/s in radio.[2]

3. The 0.511 MeV line is variable on approximately a six month time scale.[1]

4. The line was measured at (510.9 ± 0.25) keV.[6]

5. A compact radio source is resolved to less than 10^{15} cm in size.[2]

CONSTRAINTS

Two important constraints come from the line. First, the six month variability constrains the size of the region producing the line to be less than or equal to six light months in extent. It is possible the radio source is an even more stringent constraint on size. Secondly, the annihilation region must be separated from any massive object by at least $\sim 10^3 [2GM/c^2]$ ($\sim 10^{15}$ cm for a 10^6 $M_\odot$ black hole) so that the line will not be gravitationally redshifted out of the observed narrow energy range. Since the compact radio object exhibits spectral properties similiar to the cores of active galaxies, this coupled with the observed higher energy phenomena leads one to consider scaled down versions of power sources for active galaxies.

MODEL

A black hole dynamo producing relativistic e^+-e^- jets of the type proposed by Lovelace[7,8] for double radio sources will be considered here. Other beam models have been considered by Brown[9], Blandford[10] and Novikov.[11] Relativistic electron-positron jets seem necessary because:

1. A line is observed at 0.511 MeV implying annihilation of e^+-e^- pairs.

2. A jet would beam pairs into an annihilation region far enough from the massive engine to be consistent with the width of the line and bypasses the problem of producing pairs too close to the engine.

3. The natural propagation of a relativistic jet would convert kinetic energy into rest mass of pairs thereby simultaneously producing them and transporting them out.

4. Relativistic beaming maintains a confined interaction region in which high energy particles can interact with one another.

A dynamo, possibly powering a quasar, produces a luminosity, $L \sim 10^{44} (M/10^8 M_\odot)^2 (B/10^3 G)^2$ erg/s.[8] For the galactic center a $\sim 10^6$ $M_\odot$ black hole and line luminosity of $\sim 10^{37}$ erg/s results in a dynamo magnetic field of $\sim$ 30 Gauss. Using this field, an accelerating potential can be calculated. It is $V \sim 10^{19} (M/10^8 M_\odot)(B/10^3 G)$ V $\sim 10^{15}$ V. The resulting electron flow from the engine is $I \sim 10^{37} (M/10^8 M_\odot)(B/10^3 G)$ e/s $\sim 10^{34}$ e/s. These collimated fast electrons will then interact with the dominant scattering background. This background initially appears to be the infrared radiation; however the majority of interstellar dust that could be correlated with IR is believed to be outside of the central parsec[5] thereby making it impossible for electrons to cool against it. However $\sim 10^{40}$ erg/s in ionizing UV must be produced from a 3.2×10^{4} °K region within the central parsec to be consistent with HII

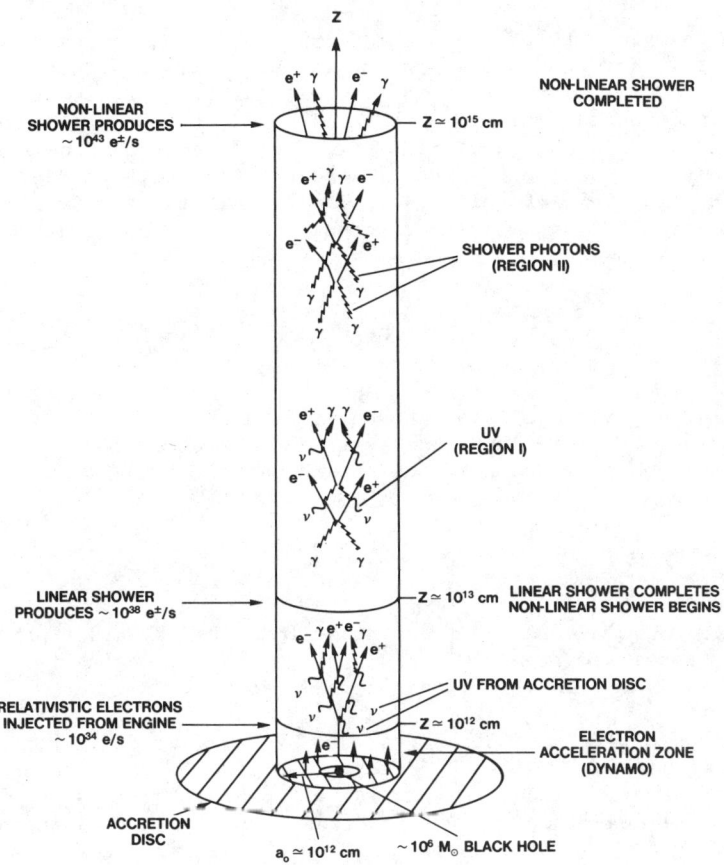

Figure 1. Schematic of Electron-Positron Jet at Galactic Center

observations.[3] An accretion disc emitting UV at a density of ~10^{15} cm^{-3} from ~10^{26} cm^2 could supply a cooling medium for fast electrons, this setting a cross-sectional radius of a_o~10^{12} cm for the jet (Figure 1). Since the compact radio source is smaller than ~10^{15} cm, it would seem the bulk of fast particles are not propagated past that distance. The radio therefore sets a length of the jet at ℓ~10^{15} cm. Knowledge of the observed 0.511 MeV line flux and necessary photon reaction rate quite independently enables one to constrain the dimensions of the source.[12] For beam geometry it is found the length and cross-sectional radius must be related by a_o^2/ℓ~10^9 cm for photon-photon interactions to occur. The UV-radio estimates of the jet dimensions are then consistent with constraints from photon-photon reactions.

CASCADE SHOWER

In the vinicity of the dynamo (~10^{12} cm) an electromagnetic cascade will develop by the interaction of collimated fast electrons with UV. Pairs will be produced, in a linear fashion, as kinetic energy is degraded from ~10^{15} eV to ~10^{11} eV, the pair production threshold. At a distance of ~10^{13} cm, pair production from the UV will cease because the UV density is decreasing with increasing distance. The yield of pairs, $f_\pm$, can be calculated from coupled relativistic Boltzmann transport equations.[13] The spectrum behaves roughly as a power law,

$$\text{Linear:} \quad f_\pm(z,E) \sim e^{\tau(z)} \left(\frac{E_o}{E}\right)^{s+1} \qquad (1)$$

Here z is the distance into the shower from the injection point at ~10^{12} cm, E_o is the primary energy from the engine and E is the energy. The optical depth of the shower particles to UV is

$$\tau(z) \simeq 10 \left[1-\left(\frac{a_o}{z}\right)^3\right] \qquad (2)$$

where $a_o \simeq 10^{12}$ cm is the engine radius. The flux of shower particles has then increased from ~10^{34} e/s injected by the engine to ~10^{38} $e^\pm$/s as a result of linear cascade action.

Although pair production from the UV has ceased, continued cooling of pairs from the UV (Figure 1, Region I) will produce hard photons energetic enough to sustain pair production in a non-linear fashion. The non-linear shower proceeds as follows. Hard photons produced by inverse Compton of the UV will pair produce. Those pairs then cool by inverse Compton, producing more hard photons which further pair produce and so on. However, since the UV density is dropping with distance and the shower population is growing, at some point the soft photons in the shower will dominate the UV (Figure 1, Region II). The cascade then proceeds using soft photons in the shower as a medium until pair production threshold is reached. The end products are then cooled pairs and hard X-rays, below threshold, of comparable luminosities (~10^{37} erg/s). The yield of pairs, $f_\pm$, can be calculated for the non-linear shower from coupled transport equations.[14] The simpler solutions yield power laws of the form

$$\text{Non-linear:} \quad f_\pm(z,E) \sim \left[1 - \lambda \left(\frac{z}{z_o} - 1\right)\right]^{-1} \left(\frac{\varepsilon_o}{E}\right)^{\bar{s}+1} \qquad (3)$$

Here z is the distance into the shower, $z_o \simeq 10^{13}$ cm is the starting position of the non-linear shower and ε_o, the primary energy, is the typical energy emerging from the linear shower. The spatial dependence exhibits the large non-linear growth of the pairs possible at $z/z_o \simeq 1+\lambda^{-1}$. Depending on the value of λ, the growth may be explosive ($\lambda \gg 1$) or adiabatic ($\lambda \ll 1$). This parameter estimated for the galactic center would imply a slow buildup of pairs. An interesting aside is the possible application of the explosive solutions to gamma ray bursts. In any case, the completed non-linear shower has now increased the particle flux from $\sim 10^{38}$ e±/s to $\sim 10^{43}$ e±/s at a distance of $\sim 10^{15}$ cm.

Electrons and positrons will now drift into a region, possibly an HII cloud[12], and annihilate to form the 0.511 MeV line. Pairs are now past the point where the black hole could gravitationally redshift the line out of the observed range and are also cool enough not to significantly contribute to the compact radio source past $\sim 10^{15}$ cm. It can now been seen that variability in the 0.511 line could be caused by variability in the dynamo itself. For instance pinching of the beam could cause the observed changes of line intensity and would predict correlated hard X-ray and 0.511 MeV line variability.

The density of pairs leaving the beam is $\sim 10^7$ cm^{-3}. If an HII cloud is to act as a beam bag and subsequent annihilation region, the energy density of the beam should be comparable to the energy density of the cloud. If the pairs have energies of 1-10 MeV, a calculated cloud density is $n_{cloud} \simeq \gamma m_e n_+/m_p \simeq 10^5$ cm^{-3}. This is consistent with H_2CO(formaldehyde) and NH_3(ammonia) observations.[15]

SUMMARY

The model developed here supplies an annihilation region with the correct flux of relatively cool pairs at the proper distance from the engine. Hard X-rays would be produced with comparable luminosity to the 0.511 MeV line and their variabilities would be directly correlated. An HII cloud appears to be consistent with stopping cooled pairs from the jet. The proposed jet at the galactic center deposits more of its energy in the 0.511 MeV line as opposed to the radio, unlike its more powerful counterpart in an active galaxy. This could be telling us the position of the acceleration point with respect to the central material. For an active galaxy acceleration occurs further out, the kinetic energy feeding a radio source. For an object like the galactic center, acceleration occurs closer in, the kinetic energy converting to rest mass which feeds an annihilation line.

ACKNOWLEDGEMENTS

The author would like to thank R. Lovelace, D. Leiter, R. Ramaty, R. Lingenfelter and the High Energy Astrophysics Theory Group at Goddard for useful discussions.

REFERENCES

1. M. Leventhal and C. J. MacCallum, The Galactic Center, AIP Conf. Proc. **83**,132(1982).
2. K. Y. Lo, The Galactic Center, AIP Conf. Proc. **83**,1(1982).
3. J. H. Lacy, The Galactic Center, AIP Conf. Proc. **83**,53(1982).
4. J. L. Matteson, The Galactic Center, AIP Conf. Proc. **83**,109(1982).
5. I. Gatley, The Galactic Center, AIP Conf. Proc. **83**,25(1982).
6. A. S. Jacobson, The Galactic Center, AIP Conf. Proc. **83**,123(1982).
7. R. V. E. Lovelace,Nature,**262**,649(1976).
8. R. V. E. Lovelace, J. MacAuslan and M. Burns, Particle Acceleration Mechanisms in Astrophysics, AIP Conf. Proc. **56**,399(1979).
9. Robert L. Brown, Ap. J.,**262**,110(1982).
10. R. D. Blandford, The Galactic Center, AIP Conf. Proc. **83**,177(1982).
11. I. D. Novikov, these proceedings.
12. R. E. Lingenfelter and R. Ramaty, The Galactic Center, AIP Conf. Proc. **83**,148(1982).
13. M. L. Burns and R. V. E. Lovelace,Ap. J.,**262**,87(1982).
14. M. L. Burns in preparation.
15. R. Gusten, The Galactic Center, AIP Conf. Proc. **83**,9(1982).

POSITRON PRODUCTION BY A HOT, YOUNG PULSAR

Kenneth Brecher and Apostolos Mastichiadis
Department of Astronomy, Boston University, Boston, MA 02215

ABSTRACT

We consider a way of positron production in pulsars: pair production from electron – photon collisions. We show that this process must be the dominant one for a young pulsar when its surface temperature is around 10^7 K. We apply this model to the observed positron flux coming from the Galactic Center and we show that/a pulsar with parameters close to those of the Crab at birth can explain, in principle, the flux.

I. INTRODUCTION

Positron production in pulsars was first proposed by Sturrock[1]. The basic process considered was the interaction between curvature photons and the pulsar's intense magnetic field. However in the early stages of the life of a pulsar the star can produce pairs in a more efficient way: the dominant pair production mechanism will be the interactions between the fast electrons accelerated in the magnetic field of the pulsar with the thermal soft X-ray photons that are radiated from the surface of the star. We find that this process can dominate for the first 10-100 years of the life of the star. After that time the star's surface temperature will drop substantially and the Sturrock mechanism will prevail, producing positrons at a reduced rate. The star will then become, presumably, a regular radio pulsar.

We apply these considerations to the problem of the source of the positrons producing the 511 keV annihilation line observed from the Galactic Center region and we argue that such a model can in principle explain the observed flux if it has parameters close to those suggested for the Crab pulsar at birth ($\omega = 350$ s^{-1}, $B = 5 \times 10^{12}$ G, $T = 10^7$ K). Under these assumptions it is argued that the Galactic Center should not be a scaled down version of active galactic nuclei[2]. The model also gives a natural explanation for the disappearance of the source[3].

II. PHYSICAL PROCESSES FOR THE HOT, YOUNG PULSAR

A pulsar at birth has a temperature $T \sim 10^7$ K. This results in an almost Planckian photon field of average energy $\varepsilon_0 \simeq 2$ keV. This temperature drops by a factor of 3 in the first 100 years of the pulsar's life as a result of the cooling of the neutron star[4]. On the other hand pulsars are able to accelerate particles to relativistic energies (for example, the Crab nebula is filled with relativistic electrons that must be supplied from the central pulsar[5]). Detailed acceleration mechanisms are unavailable, but most authors favor an electric field parallel to the magnetic field lines. Here we will assume that electrons are accelerated to high energies but the actual mechanism is of secondary importance.

Assuming a strong electric field, the electrons are accelerated essentially instantaneously to high energies and they do not suffer significant radiative losses since $\tau_{acc} \ll \tau_{comp}$. However as the energy of the electron increases the Compton scattering cross section decreases while the photon-electron pair production cross section increases. So, for the assumed young pulsar's photon distribution (with $T \simeq 10^7$ K), the two processes become equal for electron energies around 20 GeV. Above that energy, photon-electron pair production dominates.

By photon-electron pair production we mean the triplet pair production in the Coulomb field of an electron studied first by Borsellino[6], Votruba[7] and others. In this case the soft X-ray photon of energy ε_0 appears Lorentz boosted in the rest frame of the electron and it has an energy of $\gamma\varepsilon_0(1-\cos\theta)$. When its energy is above the threshold for pair production ($\varepsilon_{th} = 4m_ec^2$) then the photon is absorbed and an electron-positron pair is produced. For cross-sections and some of the kinematics involved the reader is referred to a review paper by Motz et al[8].

According to this picture, a cascade will occur since each of the produced pairs will be reaccelerated in the field and, in that way, new pairs will be formed. Moreover, possible synchrotron radiation (if the pairs are produced with some angle $\theta \neq 0$ with respect to the local magnetic field) will produce more pairs either by γ-γ or by γ-B interactions. Once the acceleration stops, the particles cascade down to energies around the threshold and the process eventually stops.

We should also mention how the proposed mechanism compares with the Sturrock mechanism. The latter is based on the assumption that accelerated electrons emit curvature photons which pair produce on the magnetic field. This mechanism has a threshold (since $\varepsilon_{ph} \propto E_e^3$ and curvature photons should have $\varepsilon_{ph} > 500$ GeV in order to start a cascade). By balancing the accelerating force and the radiation reaction force one can find that in the case of a radiation field of high T the energy of the accelerating electrons can never exceed the threshold for curvature pair production so the photon-electron pair production remains dominant throughout.

III. THE GALACTIC CENTER

As an application of the above we consider the case of the Galactic Center. Lingenfelter and Ramaty in their review paper[9] estimate that the 511-keV annihilation line observed from the Galactic Center direction corresponds to a positron flux of 10^{43} e$^+$/sec. Existing models [1,10] of pair production in pulsars have rates not exceeding 10^{41} e$^+$/sec, so a radio pulsar seems inadequate to explain the positron flux.

In order to see whether a young pulsar can explain the positron flux we have applied a simple model considering two extreme cases for the processes in the cascade. We have assumed that the pulsar has a surface temperature $T = 10^7$ K that corresponds to a monochromatic photon spectrum of energy ~ 2.3 keV and that the only process that occurs is the triplet pair production. The first type (A) of cascade considered is when the electron behaves like a heavy particle, i.e. there is no recoil momentum. In that case the high energy electron does not suffer any losses and the produced pairs share essentially the energy of the photon. The overall picture is then highly relativistic electrons moving through a photon gas and leaving behind ∼ 1 keV electrons and positrons.

The second type (B) is the other extreme case where the recoil momentum of the particle is equal to the momenta of the produced particles. In that case 2/3 of the initial energy of the particle goes to the pairs while the other 1/3 remains with the primary. A cascade then occurs that goes down to energies around the threshold for the process (∼ 450 MeV) and all particles share the same energy.

Table 1 shows the estimated values (for case A and case B) for the optical depth (τ), the number of secondaries per primary ($N_{sec}/_{pri}$), the final energy of the produced pairs (E_f), the distance (d) from the star's surface at which 95% of the pairs have already formed and the pulsar's non-thermal luminosity(*) (L_{nth}) required to explain the observed positron flux. The above parameters have been calculated for two energies of the primaries, E_{pr} = 50 GeV and 500 GeV. The parameters have been calculated using a step-by-step numerical method.

Table 1

	E_{pr} = 50 GeV		E_{pr} = 500 GeV	
	Case A	Case B	Case A	Case B
τ	25	25	70	70
$N_{sec}/_{pri}$	90	250	150	2200
E_{fin}	1 keV	200 MeV	1 keV	230 MeV
d(cm)	5×10^7	5×10^6	5×10^7	5×10^6
L_{nth}(erg/sec)	$\sim 10^{41}$	$\sim 10^{40}$	$\sim 10^{41}$	$\sim 10^{40}$

This first order of magnitude calculation gives us a required non-thermal luminosity for the young pulsar ∼ 10^{40}-10^{41} erg/sec. This is an upper limit since we have assumed that there is no reacceleration of the secondaries. If we take a typical value L_{nth} ∼ 10^{40} erg/sec this corresponds to ω ∼ 350 s^{-1}, R = 10^6 cm and B = 5×10^{12} G. These parameters are not unreasonable since they are very close to the ones estimated for the Crab pulsar at birth.

(*) In evaluating L_{nth} we have assumed that the primaries pair produce <u>after</u> their acceleration is over. We have taken also η=1 (where η is the fraction of initial energy going to pair production).

Thereafter, when the temperature falls to $\sim 3\times 10^6$ K, the optical depth for the γ-e pair production process becomes $\tau \ll 1$ and so pairs will continue to form at a lower rate by the Sturrock mechanism. This gives an explanation why the line has disappeared and consequently we can make the prediction that the line will not reappear.

This model supports the conclusion[2] that the Galactic Center positron source is not a scaled down version of an active galactic nucleus since it attributes its activity to a young pulsar that happens to lie in the region and which formed less than 100 years ago.

IV. CONCLUSION

The mathematical problem, as it was described in § II is not an analytical one and is highly non-linear. The approach attempted in § III can give us only an order of magnitude estimate. We are now undertaking the following program: First, to calculate the distribution of the produced particles and of the target particle for the triplet production (that includes complications such as exchange effects between the two electrons and the γ-e interaction (in the terminology of the Feynman diagrams)); next by a numerical simulation to compute the emerging photon and particle spectrum; and, finally, to place the young pulsar in an astrophysical environment (such as the Galactic Center) and find how the produced positrons annihilate and what is the emergent spectum.

This research was supported in part by NSF Grant AST-8020756 and NASA contract number NAG8-435.

REFERENCES

1. P.A. Sturrock, Ap.J. 164, 529 (1971).
2. A.P. Marscher, K. Brecher, W.A. Wheaton, J.C. Ling, W.A. Mahoney and A.S. Jacobson, this volume.
3. M. Leventhal, C.J. MacCallum, in The Galactic Center, ed. G.R. Riegler and R.D. Blandford (New York: Amer. Inst. Physics), p. 132 (1982).
4. S. Tsuruta, in IAU Symposium 53, Physics of Dense Matter, ed. C.J. Hansen (Dordrecht: Reidel) p. 209 (1974).
5. R.N. Manchester and J.H. Taylor, in Pulsars, W.H. Freeman and Company, p. 64 (1977).
6. A. Borsellino, Nuovo Cimento, 4, 112 (1947).
7. V. Votruba, Phys. Rev. 73, 1468 (1948).
8. J.W. Motz, H.A. Olsen and H.W. Koch, Rev. Mod. Phys. 41, 581 (1969).
9. R.E. Lingenfelter and R. Ramaty, in The Galactic Center, ed. G.R. Riegler and R.D. Blandford (New York: Amer. Inst. Physics) p. 148 (1982).
10. A.F. Cheng and M.A. Ruderman, Ap.J. 216, 865 (1977).

PHENOMENA IN THE CENTER OF THE GALAXY AS A CONSEQUENCE OF A RECENT SUPERNOVA OUTBURST

I. Shklovskii
Academy of Sciences of the USSR
Space Research Institute, Moscow, USSR

ABSTRACT

The paper gives arguments supporting the idea that the whole range of phenomena in the center of the Galaxy can be attributed to a supernova outburst in a binary system about 100 years ago. Such an object should be quite similar to SS433, only much younger, more compact and more active. In particular, the anomalously high abundance of helium, as compared to hydrogen, recently detected in the IRS-16 source that is emitting a "broad" ($\Delta\nu\sim1500$ km·s^{-1}) helium line at $\nu = 4857$ cm^{-1} is explained by the fact that the emitting plasma is an expanding remnant of the exploded massive He star. Gas ejection from the central region ($\dot{M}\sim10^{-3}M_\theta$/year) is due to supercritical accretion in the close binary system whose compact component is the remnant of the exploded helium star. Rapid overflowing of gas via the Lagrangian point, which started such accretion, began several thousand years ago, long before the explosion of the He component. Emission of the E = 511 keV annihilation line is related to the powerful generation of electron-positron pairs in the magnetosphere of the neutron star, the remnant of a supernova outburst. IRS-16 is most probably a plerion. If this is the case, its infrared and X-ray radiation should be synchrotron emission and linearly polarized. This radiation may also be variable. Thus the activity observed in the center of the Galaxy is naturally explained by the evolution of stars in that region, the character of which qualitatively differs from that of phenomena in quasars and radio galaxies. In particular it becomes clear why the nucleus of M31, similar to the nucleus of our Galaxy, shows no activity whatsoever: there are no young remnants of supernovae in M31.

X-RAYS AND GAMMA-RAYS FROM ACTIVE GALAXIES

James L. Matteson
Center for Astrophysics and Space Sciences, C-011
University of California, San Diego
La Jolla, CA 92093

ABSTRACT

Photon-photon pair production in active galaxies is considered and the concept of the annihilation efficiency, the efficiency for the conversion of > 511 keV continuum luminosity into positron annihilation luminosity, is introduced. Equations that give a source's annihilation luminosity and 511 keV flux as a function of its size, continuum luminosity and distance are developed. These are applied to the available X-ray and gamma-ray data on active galaxies in order to make specific predictions. Efficiencies as high as > 6 percent and fluxes up to 8×10^{-4} ph/cm^2-sec result. While the latter are below present limits, they are within the reach of advanced instruments now in development.

INTRODUCTION

The large luminosities, small source volumes and hard spectra which are characteristic of X-ray and gamma-ray emission from active galaxies imply that pair production is likely to be an important phenomenon in these objects. Although 511 keV positron annihilation radiation, the major signature of pair production, has not yet been detected from any active galaxy, its eventual detection should provide a great deal of information on the conditions in active galaxies. This has already occurred for our Galaxy as the observations of the 511 keV radiation from the galactic center [1,2,3] have been interpreted [4,5] in terms of the temperature, density, velocity, size and central mass of the annihilation region. When combined with other data, the observations constrain the positron production process to be most likely photon-photon pair production.

In this paper physical considerations with regard to pair production are presented and photon-photon pair production is considered in some detail. Equations are developed that predict pair production and annihilation radiation in terms of observed properties of active galaxies. Then X-ray and gamma-ray observations are discussed and their implications for pair production are developed. The boundary between X-rays and gamma-rays is assumed to be ~ 30 keV, a typical upper limit to the region of good spectral measurement with proportional counter experiments. Active galaxies are taken to be characterized by a compact, highly luminous and often variable source of X-ray, and sometimes gamma-ray, emission and include Seyfert galaxies, BL Lac objects, quasars and Cen A.

PAIR PRODUCTION AND ANNIHILATION IN ACTIVE GALAXIES

Positrons and electron-positron pairs may be produced by a variety of processes as discussed by Lingenfelter and Ramaty [6]. The large photon densities in active galaxies which follows from their large luminosities and small sizes suggests that photon-photon pair production is the dominant positron production process in these objects. (If the ratio of particles to photons is greater than $1/\alpha$ (=137), then particle-photon or particle-particle pair production would be dominant.) Several authors have discussed photon-photon pair production in the context of astronomical sources [4,6,7,8]. For photons with energies E_1 and E_2 which collide at angle Θ, the pair production threshold is given by

$$E_1 E_2 \geq \frac{2(mc^2)^2}{1-\cos \Theta} \qquad (1)$$

and the maximum value of the cross section, 1.7×10^{-25} cm^2, occurs when $E_1 E_2$ equals twice the threshold value. In the case of an isotropic region, i.e. $\Theta = 90°$, some (E_1, E_2) values at the maximum cross section are: (1 keV, 1 GeV), (10 keV, 100 MeV), (100 keV, 10 MeV) and (1 MeV, 1 MeV). Pair production will remove photons of both energies, but, since the photon density is normally much lower at the higher energy, the greatest effect on the spectrum will occur there and a break or a steepening of the spectrum will result. Calculation of the effects of pair production generally requires information on the spectrum of photons from the same source region over a wide energy range, i.e. E_1 to E_2, information which is not yet available for any active galaxy.

Bassani and Dean dealt with this problem in their calculations of photon-photon opacities in active galaxies [8] by assuming that the spectra measured in the 2-10 keV range extend as an $E^{-1.6}$ power law to higher energies. However, at $\sim$ 1 MeV the problem is simplified since $E_1 \simeq E_2$ at the maximum cross section and data on the spectra and sizes of active galaxies at these energies are beginning to be available. Therefore, this discussion is generally restricted to the effects of pair production by $\sim$ 1 MeV photons and X-ray and gamma-ray observations which are relevant to the emission from active galaxies up to several MeV.

Using a similar restriction, Lingenfelter and Ramaty [4] have developed the useful relationship,

$$d \simeq 3 \times 10^{-25} \frac{L^2(>511)}{Q}, \qquad (2)$$

between a source region's diameter, d (cm), photon luminosity above 511 keV, L(>511) (erg/sec), and the pair production rate, Q (sec^{-1}). L(>511) actually refers to energies up to a few MeV. Since the annihilation energy is 2 mc^2 per pair, the annhilation efficiency, η, i.e. the efficiency for the conversion of > 511 keV luminosity into pairs and ultimately annihilation luminosity, is

$$\eta \stackrel{\sim}{-} 5 \times 10^{-31} \frac{L(>511)}{d} . \qquad (3)$$

Of course, the annihilation luminosity is originally in the form of pairs and only becomes observable radiation after the positrons slow and annihilate. The flux of 511 keV annihilation radiation at the earth can be calculated given the fraction of positrons that annihilate in the active galaxy and the fraction that form positronium, both of which are assumed to be 100 percent. Since only the singlet state of positronium produces 511 keV photons, only 25 percent of the annihilation energy goes into 511 keV radiation. Thus the 511 keV flux at the earth is

$$F(511) \text{ (ph/cm}^2\text{-sec)} = \frac{\eta L(>511)}{16\pi D^2 mc^2} \qquad (4)$$

where D(cm) is the distance to the active galaxy. Combining equations (3) and (4) and using $d = c\Delta t$, where Δt is the time scale for variability, gives

$$F(511) = 4.2 \times 10^{-85} \frac{L^2(>511)}{D^2(\text{mpc}) \Delta t(\text{sec})} . \qquad (5)$$

Since the continuum brightens, B, is proportional to $L(>511)/D^2$,

$$F(511) \propto \frac{B \, L(>511)}{\Delta t} . \qquad (6)$$

Thus for objects of equal continuum brightness and variability, those with the greatest continuum luminosity, i.e. at the greatest distance, should produce the greatest 511 keV flux.

A consequence of pair production will be a steepening of the spectrum above ~ 1 MeV and the conversion of the photons being removed by pair production into annihilation luminosity. Since the spectrum of the continuum and annihilation radiation as well as limits on the source size, via time variations, can all, in principle, be measured, equations (3) and (5) can provide a useful, model-independent constraints on the physical conditions in active galaxies.

X-RAY OBSERVATIONS OF ACTIVE GALAXIES

X-ray emission from active galaxies was discovered during the early days of X-ray astronomy. However, it was not until the extensive observations at high sensitivity with the Uhuru, OSO-8, Ariel V, HEAO-1 and HEAO-2 that the nature of their spectra and variability became clear. These have been reviewed by several authors [9,10]. Here the relevant results are simply summarized. The spectra of 24 Seyfert 1 and 2 galaxies [11,12] are well fitted by photon power laws with indices of ~ -1.7 and exhibit a dispersion of ~ 0.1 in the index. The spectra of Cen A, the quasars 3C273 and 0241+622 and the hard component of BL Lac objects are similar, having indices of ~ -1.7 [13], -1.5 [14], -1.9 [15] and -1.5 [16], respectively. Variability on time scales from several hours to years is a property of these objects. The shortest time scales are of interest here and a tabulation of many of them [8] indicates values of $\sim 1/2$ day in Seyfert 1 galaxies and BL Lac objects, a few days in Seyfert 2 galaxies and a few hours in quasars. In addition, the Seyfert 2 galaxy NGC 7582 has been observed [12] to flare on a $\lesssim 1$ day time scale, and variability with a characteristic time of ~ 100 sec has been observed in the Seyfert 1 galaxy NGC 6814 [17] and the quasar 1525+227 [18]. Recent work [19] has shown that variability on time scales from minutes to a few hours is not a feature of most active galaxies, with limits on the RMS variability on these time scales typically being $\sigma_{I/I} \lesssim 10$ percent.

If one assumes that the power law spectra of active galaxies measured in the X-ray range extend to ~ 1 MeV and that the emitting region at ~ 1 MeV has a size given by $d = c\Delta t$, where Δt is the time scale of the observed X-ray variations, then the efficiency, η, for the production of pairs may be determined from equation (3). The results of this calculation are indicated in Figure 1 which is based on a tabulation of X-ray luminosities and time variations [8]. BL Lac objects and quasars have η values $> 10^{-2}$ and several exceed 1, while most Seyfert galaxies have η values from 10^{-3} to 10^{-2}. NGC 6814 has $\eta \sim 1$ due to its short time scale of variation. Of course, η values > 1 are unphysical and indicate that the assumptions of the size, luminosity, and isotropic nature of the sources at ~ 1 MeV cannot all be satisfied. Beamed emission [8] and a gamma-ray production region which does not coincide with the X-ray production region [20] have been invoked to deal with this problem.

GAMMA-RAY OBSERVATIONS

Although the X-ray results indicate that active galaxies are likely to have significant pair production, there have not yet been the necessary confirming observations of 511 keV annihilation radiation. As an alternative, observations above 511 keV may be used to estimate the amount of pair production and annihilation flux through the use of equations (3) and (5). The key questions to be addressed by these data are: to how high an energy does the $\sim E^{-1.7}$ power law spectrum extend and what are the shortest time variations at the highest energies? The extensive gamma-ray observations performed by the HEAO-1 and the recent, high-sensitivity balloon observations at ~ 1 MeV have provided important progress on these questions.

The HEAO-1 measured the mean spectrum of 12 active galaxies, mostly Seyfert 1's, in the 2 to 165 keV range [21]. The results show that the power law spectrum measured at X-ray energies does, indeed, extend to higher energies, with a mean power law index of -1.62 and a dispersion of 0.10. Further data available on several more active galaxies are discussed below.

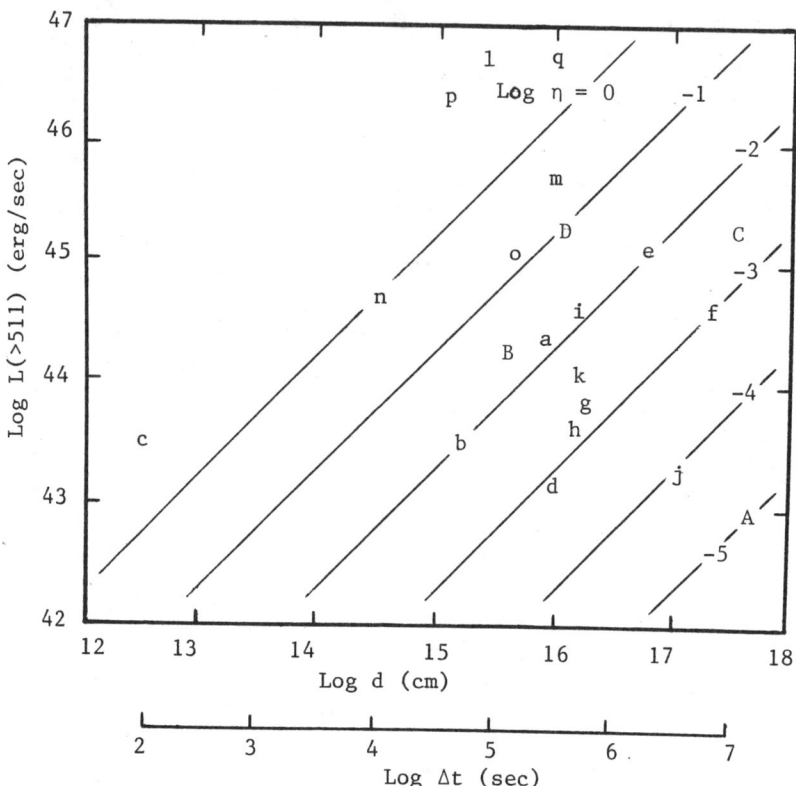

Figure 1. The annihilation efficiency calculated from equation (3) using the observed variability and luminosity of active galaxies. Lower case letters indicate the use of X-ray data [8,19] which were extrapolated to ~ 1 MeV assuming an $E^{-1.7}$ spectrum. Upper case letters indicate the use of gamma-ray data that are given in Table 1.

Objects Observed in X-rays*

Seyfert 1	Seyfert 2	Quasar	BL Lac
a-NGC7469	g-NGC2992	l-3C273	o-MKN421
b-NGC4151	h-NGC5506	m-0241+622	p-PKS2155-304
c-NGC6814	i-NGC526A	n-OX169	q-3C66A
d-NGC3783	j-NGC7582		
e-MKN509	k-MCG 5-23-16		
f-MCG 8-11-11			

* all are from reference 8 except for c which is from reference 19.

Objects Observed in Gamma-Rays**

A - Cen A   B - NGC4151   C - NGC4151   D - MKN509

** Refer to Table 1

a) Cen A

This nearby object is the brightest X-ray emitting active galaxy. Its $\sim$ 1 MeV radiation was discovered in a 1974 balloon observation [21] in which the spectrum above $\sim$ 30 keV was fit by an $E^{-1.9}$ power law, and a > 1 MeV excess above this was interpreted as due to nuclear lines. The presence of nuclear lines in the spectrum has not been confirmed. Histories of the $\sim$ 100 keV flux and spectral slope [22,23] show that both are variable on $\sim$ 6 month time scales. The total range of intensity variation is a factor of 10 and the power law slope varies from -1.3 to -1.9. The HEAO-1 placed a 35 percent limit on intensity variations on time scales from 8 hours to 10 days and showed that the flux at 1 MeV varied over a 6 month period [23]. It also measured the 2 keV to 2 MeV spectrum, which was well described by a power law with a slope of -1.60 up to $\sim$ 140 keV where it steepened to a slope of -2.0. A spectral break is expected since the spectrum above $\sim$ 100 keV must have an average slope that is < -2.3 in order to satisfy the upper limits at $\sim$ 100 MeV [24,25]. Other observers have not detected the break [21,26] which could easily be variable in view of the overall variability of Cen A.

b) NGC 4151

This Seyfert 1 galaxy is the second brightest active galaxy. Histories of its flux and spectral slope at $\sim$ 100 keV [27] also indicate variability over $\sim$ 6 months. The total range of intensity variation is a factor of $\sim$ 4 and the spectral slope varies from -0.9 to -1.6. A correlation of the intensity and slope [27] shows that they tend to be coupled with the higher flux corresponding to the harder spectrum, implying greater variability at higher energies. This is supported by results at $\sim$ 1 MeV where a factor of $\sim$ 20 variability and $\sim$ 6 month time scales are indicated by a comparison of the results of various observations [27,28,29,30,31]. Recent HEAO-1 results [27] have shown a $\sim$ 2 day X-ray/low energy gamma-ray correlated variation, showing for the first time that the $\sim$ 1 day variations observed in X-rays extend up to $\sim$ 200 keV. The HEAO-1 results also show that the spectrum breaks from a slope of -1.6 to $\sim$ -2.0 at $\sim$ 100 keV. The extrapolation of the high energy spectrum is just consistent with the $\sim$100 MeV upper limits [24,25]. The break must be variable in order to be reconciled with the high flux observations at $\sim$ 1 MeV [29,30].

c) Mkn 509

Since this Seyfert 1 is typically $\sim$ one-third as bright as NGC 4151, it has required extensive observations with the HEAO-1 [32] to measure its spectrum up to $\sim$ 100 keV. These were a series of four pointed observations, each separated by an average of 3 days, which found a power law slope of -1.8 and discovered a factor of > 2 decrease in flux, at 2 σ confidence, over a 6 day period. This decrease was correlated with a decrease measured at the same time in X-rays and is further evidence that the $\sim$ 1 day variability of active galaxies extends to gamma-ray energies.

d) Others

Flux from several active galaxies has been observed at gamma-ray energies, but there is not yet data on their variability. The X-ray emitting Seyfert 1 galaxy MCG 8-11-11 has been detected from balloons in the 20 to 300 keV range [33] and the 0.1 to 20 MeV range [34]. The latter result gives a power law spectrum with slope -1.0 ± 0.7 up to ~ 2 MeV where it steepens to < -3.8 and thus comes into agreement with limits at ~ 100 MeV [24]. 13 to 120 keV observations of 3C273 by the HEAO-1 [35] have shown that its spectrum extends up to 120 keV with a power law slope of -1.7. When these results are combined with the X-ray [14] and > 50 MeV γ-ray [36,20] spectra, with indices of -1.4 and -2.7, respectively, there is a clear indication of a steepening in the spectrum over several decades in energy, with a maximum in the luminosity per logarithmic energy band at ~ 1 MeV.

It has been noted [37] that the ~ 100 MeV gamma-ray source CG 135+1 may be associated with the quasar 0241+622. However, a refined position for CG 135+1 [25] shows that it is not associated with the quasar. BL Lac objects have not yet been detected above a few tens of keV. However, the hard spectra sometimes observed at ~ 10 keV [16] imply that they may also produce significant gamma-ray flux.

DISCUSSION

The gamma-ray results discussed above show that the hard power law spectra of Seyfert 1 galaxies extend up to ~ 200 keV and that in 2 cases, NGC4151 and MCG 8-11-11, they extend up to several MeV, as does the spectrum of Cen A. Correlated X-ray/gamma-ray variability from a few keV up to ~ 100 keV on a time scale of days has been observed in NGC 4151 and Mkn509, also a Seyfert 1. This is compelling evidence that the X-ray and low energy gamma-ray, i.e. ~ 100 keV, emission arise from the same region and makes plausible the assumption that the ~ 1 MeV emission arises there also. Thus the predicted annihilation efficiency shown in Figure 1 for Seyfert 1 galaxies is supported by the gamma-ray observations. However, this will not be certain until ~ 1 day variability is observed at ~ 1 MeV.

3C273 is the only active galaxy that has been detected at ~ 100 MeV where its spectrum and flux indicate that the total spectrum apparently smoothly steepens from a few tens of keV up to a few tens of MeV, suggesting a common emission region from a few keV up to $\gtrsim 100$ MeV. However, several authors [8,20] have pointed out that the assumption that the X-ray source volume is coincident with the 100 MeV source volume leads to severe difficulties. This follows from the large optical depth at 100 MeV against pair production, calculated to be ~ 250 [8] and ~ 20 [20]. Such large values would attenuate the ~ 100 MeV flux to far below the observed level. This difficulty can be removed by invoking (a) jet models which result in small angles of incidence between the pair producing photons, thereby raising threshold energies or (b) models in which the ~ 100 MeV source region is simply separate from the X-ray source region. Thus the annihilation efficiency for quasars, as well as Seyfert 2 galaxies and BL Lac objects, predicted in Figure 1 will remain highly speculative until information on their spectra and variability at ~ 1 MeV is available.

Limits on the $\sim$ 100 MeV flux from active galaxies [24,25] are typically a factor of $\sim$ 10 to $\sim$ 50 below the extrapolation of the X-ray spectra of the brighter active galaxies, i.e. those with a flux of $\sim$ 1 Uhuru count/sec. Therefore, the spectra of active galaxies must break by at least $\sim$ 0.3 to $\sim$ 0.6 in the power law slope over the $\sim$ 100 keV to $\sim$ 100 MeV energy range. The breaks in the gamma-ray spectra of Seyfert 1's and Cen A are consistent with this, but whether or not they are due to pair production is difficult to assess at this time. Calculations of pair production optical depths[8] give values of $\sim$ 0.01 to 0.08 at 1 to 10 MeV, too small to account for the observed breaks. However, sources that are more compact than indicated by their time variations would result in greater optical depths. Specific model calculations which incorporate fundamental emission processes and photon reprocessing by pair production and annihilation are necessary to interpret the spectral breaks and predict the associated 511 keV fluxes from active galaxies.

The gamma-ray results on Cen A, NGC4151 and Mkn509 can be used to obtain predictions of pair production and annihilation using equations (3) and (5). Here it is assumed that the variability observed in Cen A, NGC4151 and Mkn509 at $\sim$ 100 keV also occurs at $\sim$ 1 MeV and the $\sim$ 1 MeV luminosity of Mkn509 is given by an extrapolation of the power law spectrum measured up to $\sim$ 100 keV. The data and results are summarized in Table I and the efficiencies are indicated in Figure 1. Table I also gives some limits on the 511 keV flux. Additional limits have been presented at this workshop [26,38].

Table I Pair Production and Annihilation in Active Galaxies

Object	Ref to Fig 1	Δt	Luminosity @$\sim$1 MeV (erg/sec)	D (mpc)	η^*	$F(511)^{**}$ (ph/cm^2-sec)	2σ limits on 511 keV line† (ph/cm^2-sec)
Cen A	A	6 mo @ .1 MeV	10^{43}	4.4	10^{-5}	1.4×10^{-7}	6.5×10^{-4}
NGC 4151	B	2 days @ .1 MeV	2×10^{44}	11	2×10^{-2}	8×10^{-4}	8×10^{-4}
NGC 4151	C	5 mo†† @ 1 MeV	2×10^{45}	11	2×10^{-3}	1×10^{-3}	--
Mkn 509	D	6 days @ .1 MeV	2×10^{45}	210	6×10^{-2}	7×10^{-5}	--

  * Predicted annihilation efficiency, from equation (3)
 ** Predicted 511 keV flux, from equation (5)
  † From the HEAO-1 at the time of the Δt and luminosity measurements
 †† From a comparison of 2 balloon observations [28,29]

The pair production efficiency for Cen A is quite low and the predicted 511 keV flux is four orders of magnitude below the current limit. This is a consequence of its relatively low luminosity and the lack of observed short time scale variability. The parameters are more favorable for NGC4151 where the efficiency is several percent and the predicted 511 keV flux, $\sim 10^{-3}$ ph/cm^2-sec, is at the level of the present limit. The values in Table I, B are from the HEAO-1 observation when the flux was near the low end of the range of variability, while the values in Table I, C result, in part, from the balloon observation [29] of the maximum luminosity at ~ 1 MeV. That nearly the same 511 keV flux is predicted in the two cases is a result of their different Δt values. In the latter case this is simply the time between two balloon observations. However, if, in reality, it were the 2 day value observed by the HEAO-1, then the annihilation efficiency would be ~ 20 percent and a transient 511 keV flux of $\sim 10^{-1}$ ph/cm^2-sec would result if the positrons could annihilate in a few days. The large luminosity and short time scale of variation of Mkn 509 results in a large efficiency and therefore a large predicted 511 keV flux, $\sim 7 \times 10^{-5}$ ph/cm^2-sec, for such a distant object.

Several uncertainties affect the predictions of equations (3) and (5) and Table I. The assumption that all the positrons annihilate is too optimistic and causes the 511 keV flux to be overestimated. However, the assumption of 100 percent positronium formation leads to an underestimate of the 511 keV flux; annihilation of free positrons would result in a flux that is 4 times greater. The assumption that $d = c\Delta t$ must overestimate the source diameter, so that the annihilation efficiency and thus the 511 keV flux are underestimated. Further underestimates of them result from the Δt values themselves, which are usually overestimated because the infrequent observations do not resolve the variations. Table I, B is an exception since the continuous observation of the HEAO-1 resolved the variation of NGC 4151.

The next generation of gamma-ray spectroscopy instruments will have the potential to detect 511 keV radiation from active galaxies. The advanced germanium spectrometers now under development for balloon observations will have a sensitivity of $\sim 10^{-4}$ ph/cm^2-sec in an ~ 8 hour observation and the Oriented Scintillation Spectrometer Experiment of the Gamma-Ray Observatory will have a sensitivity of $\sim 3 \times 10^{-5}$ ph/cm^2-sec in a 30-day observation from space. The latter will be able to perform nearly continuous observation of a single object for several weeks and thus will be very sensitive to ~ 1 day variability in the ~ 1 MeV flux from the brighter active galaxies.

CONCLUSION

The X-ray and gamma-ray observations of the spectra and variability of active galaxies imply that photon-photon pair production and annihilation are likely to be significant phenomena in these objects. Annihilation efficiencies up to ~ 10 percent and 511 keV fluxes up to $\sim 10^{-3}$ ph/cm^2-sec are predicted from the gamma-ray data and, in some cases, greater values are predicted from the X-ray data. The predicted 511 keV fluxes are below the present upper limits. The gamma-ray data, especially at $\sim$ 1 MeV, are crucial to the predictions but are very limited at this time. This situation should improve in the mid to late 1980's when advanced spectrometers carried on balloons and the Gamma-Ray Observatory conduct observations of active galaxies.

ACKNOWLEDGEMENTS

This work was supported by NASA grants NGL-05-005-003 and NAGW-449.

REFERENCES

1. M. Leventhal, C.J. MacCallum and P.D. Stang, Ap.J. (Letters), 225, L11 (1978).
2. G. R. Riegler, J.C. Ling, W.A. Mahoney, W.A. Wheaton, J.B. Willet and A.S. Jacobson, Ap.J. (Letters), 248, L13 (1981).
3. C.J. MacCallum, this volume.
4. R.E. Lingenfelter and R. Ramaty, in: The Galactic Center, G.R. Riegler and R.D. Blandford, eds., AIP, New York, 1982, p. 148
5. R.E. Lingenfelter, this volume.
6. R.J. Gould and G.P. Schreder, Phys. Rev., 155, 1408 (1967).
7. J.B. Pollack, P.D. Guthrie and B.S.P. Shen, Ap.J. (Letters), 169, L113 (1971).
8. L. Bassani and A.J. Dean, Nature, 294, 332 (1981).
9. K.A. Pounds, in: X-ray Astronomy, R. Giacconi and G. Setti, eds., D. Reidel, Dordrecht, 1980, p. 273.
10. H. Tananbaum, op.cit., p. 291.
11. R.F. Mushotzky, F.E. Marshall, E.A. Boldt, S.S. Holt and P.J. Serlemitsos, Ap.J., 235, 377 (1980).
12. R.F. Mushotzky, Ap.J., 256, 92 (1982).
13. R.F. Mushotzky, P.J. Serlemitsos, R.H. Becker, E.A. Boldt and S.S. Holt, Ap.J., 220, 790 (1978).
14. D.M. Worrall, R.F. Mushotzky, E.A. Boldt, S.S. Holt and P.J. Serlemitsos, Ap.J., 232, 683 (1979).
15. D.M. Worrall, E.A. Boldt, S.S. Holt and P.J. Serlemitsos, Ap.J., 240, 421 (1980).
16. D.M. Worrall, E.A. Boldt, S.S. Holt, R.F. Mushotzky and P.J. Serlemitsos, Ap.J., 243, 53 (1981).
17. A.F. Tennant, R.F. Mushotzky, E.A. Boldt and J.H. Swank, Ap.J., 251, 15 (1981).
18. T. Matilsky, C. Schrader and H. Tannanbaum, Ap.J. (Letters), 258, L1 (1982).
19. A.F. Tennant and R.F. Mushotzky, Ap.J., 264, 92 (1983).
20. G.F. Bignami, K. Bennett, R. Buccheri, P.A. Caraveo, W. Hermsen, G. Kanbach, G.G. Lichti, J.L. Masnou, H.A. Mayer-Hasselwander, J.A. Paul, B. Sacco, L. Scarsi, B.N. Swanenburn and R.D. Wills, Astron. Astrophys., 93, 71 (1981).
21. R.D. Hall, C.A. Meegan, G.D. Wallraven, F.T. Djuth and R.C. Haymes, Ap.J., 210, 631 (1976).

22. J.H. Beall, W.K. Rose, W. Graf, K.M. Price, W.A. Dent, R.W. Hobbs, E.K. Conklin, B.L. Ulich, B.R. Dennis, C.J. Crannell, J.F. Dolan, K.J. Frost and L.E. Orwig, Ap.J., 219, 836 (1978).
23. W.A. Baity, R.E. Rothschild, R.E. Lingenfelter, W.A. Stein, P.L. Nolan, D.E. Gruber, F.K. Knight, J.L. Matteson, L.E. Peterson, F.A. Primini, A.M. Levine, W.H.G. Lewin, R.F. Mushotzky and A.F. Tennant, Ap.J., 244, 429 (1981).
24. G.F. Bignami, C.E. Fichtel, R.C. Hartman and D.J. Thompson, Ap.J. 232, 649 (1979).
25. A.M.T. Pollack, G.F. Bignami, W. Hermsen, G. Kanback, G.G. Lichti, J.L. Masnou, B.N. Swanenburg and R.D. Wills, Astron. Astrophys., 94, 116 (1981).
26. N. Gehrels, T.L. Cline, W.S. Paciesas, B.J. Teegarden and J.Tueller, this volume.
27. W.A. Baity, R.F. Mushotzky, D.M. Worrall, R.E. Rothschild, A.F. Tennant and F.A. Primini, UCSD preprint SP-83-10, submitted to Ap.J. (1983).
28. C.A. Meegan and R.C. Haymes, Ap.J., 233, 510 (1979).
29. F. Perotti, A. Della Ventura, G. Sechi, G. Villa, G. Di Cocco, R.E. Baker, R.C. Butler, A.J. Dean, S.J. Martin and D. Ramsden, Nature, 282, 484 (1979).
30. F. Perotti, A. Della Ventura, G. Villa, G. Di Cocco, L. Bassani, R.C. Butler, J.N. Carter and A.J. Dean, Ap.J. (Letters), 247, L63 (1981).
31. R.S. White, B. Dayton, R. Gibbons, J.L. Long, E.M. Zanrosso and A.D. Zych, Nature, 284, 608 (1980).
32. S. Dil, F.A. Primini, E. Basinska, M. Bautz, S.K. Howe, F. Lang, A.M. Levine, W.H.G. Lewin, D.M. Worrall, P.L. Nolan and J.L. Matteson, Ap.J., 250, 513 (1981).
33. F. Frontera, F. Fuligni, F. Morelli and G. Ventura, Ap.J., 234, 477 (1979).
34. F. Perotti, A. Della Ventura, G. Villa, G. Di Cocco, R.C. Butler, J.N. Carter and A.J. Dean, Nature, 292, 133 (1981).
35. F.A. Primini, B.A. Cooke, C.A. Dobson, S.K. Howe, A. Scheepmaker, W.A. Wheaton, W.H.G. Lewin, W.A. Baity, D.E. Gruber, J.L. Matteson and L.E. Peterson, Nature, 278, 234 (1979).
36. B.N. Swanenburg, K. Bennett, G.F. Bignami, P. Caraveo, W. Hermsen, G. Kanbach, J.L. Masnou, H.A. Mayer-Hasselwander, J.A. Paul, B. Sacco, L. Scarsi and R.D. Wills, Nature, 275, 298 (1978).
37. K.M.V. Apparao, G.F. Bignami, L. Maraschi, H. Helmken, B. Margon, R. Hjellming, H.V. Bradt and R.G. Dower, Nature, 273, 450 (1978).
38. A.P. Marscher, K. Brecher, J.C. Ling, W.A. Mahoney, W.A. Wheaton and A.S. Jacobson, this volume.

ARE MILDLY ACTIVE GALAXIES SOURCES OF e± ANNIHILATION RADIATION?

Alan P. Marscher* and Kenneth Brecher*
Department of Astronomy, Boston University
725 Commonwealth Ave., Boston, MA 02215

William A. Wheaton, James C. Ling,
William A. Mahoney, and Allan S. Jacobson
Jet Propulsion Laboratory, California Institute of Technology
4800 Oak Grove Dr., Pasadena, CA 91103

INTRODUCTION

The Galactic Center has been established as a source of 511 keV line radiation resulting from electron-positron annihilations[1,2,3,4]. Using the HEAO-3 Gamma-Ray Spectrometer, Riegler et al.[5] have found that the Galactic Center 511 keV line is variable over a timescale of less than six months. The variability is supported by the failure of balloon experiments to detect the line in November 1981 [6,7]. This implies that the source size is less than about one light-year. Thus, a relatively powerful, compact source of positrons exists near (or in the direction of) the Galactic Center.

The Galactic Center is also the site of a compact, nonthermal radio source similar to a scaled-down version of those found in the nuclei of active galaxies and quasars[8]. This suggests, by analogy, that there is a mildly active "central engine" at the Galactic Center, which is capable of producing the high output of positrons ($\sim 10^{43}$ positrons s^{-1}) required to explain the strength of the 511 keV line [9,10]. If this is the case, then one would expect other active galaxies to also contain positron sources in their nuclei. Then the predicted strength of the consequent 511 keV line emission would depend on how the positron production and annihilation rates scale with other indicators of activity, such as nonthermal radio emission, optical emission-line luminosity, and nuclear X-ray emission.

The most straightforward assumption is that the positron production rate increases linearly with compact nonthermal radio luminosity. Then, if the mean annihilation time of a positron in another galactic nucleus is similar to that in the Galactic Center, the observed 511 keV flux should scale roughly as the observed compact nonthermal radio flux, which is a readily measured quantity. This simple relation can be justified if one assumes that the number of positrons in a source scales directly as the number of relativistic electrons required to account for the observed synchrotron radiation. However, it is clear that any breakdown in the chain of arguments which we have constructed (e.g. due to variations in ambient gas density from one galactic nucleus to another) would destroy the validity of the proposed relation.

*HEAO-3 Guest Investigator

Using the above criterion, one would at first glance expect the many compact radio sources with flux densities exceeding that of the Galactic Center to be detectable 511 keV line emitters. This is known not to be the case: for example, the relatively nearby radio galaxy, Centaurus A (NGC 5128), has compact radio emission with flux density 5 times that of the Galactic Center[11], yet the 511 keV line has not been detected in this source[12]. This result can be explained if the mean time for a positron to annihilate is longer in such sources than in the Galactic Center. This could occur if the positrons are produced at such high energies (thus making it more difficult to form positronium), or traverse regions of such low densities, that they cannot annihilate as rapidly as the Galactic Center positrons apparently do[9]. Centaurus A and most of the other high luminosity objects in fact contain jets (of relatively low density) via which most of the nonthermal material is channeled into intergalactic space (see, e.g., ref. 13).

Because of these possible "escape routes" for positrons, we decided that a promising candidate for 511 keV line detection would be any active galaxy whose nonthermal radio activity is totally contained within the galaxy. We have chosen 7 galaxies of various types using the criterion that they have compact radio flux densities within an order of magnitude of that of the Galactic Center compact nonthermal source, and that their radio components do not extend beyond the optical boundaries of the galaxies. This restricted the sample to galaxies with nonthermal activity intermediate between that of our Galaxy and very active galaxies and quasars. These galaxies represent those suitable objects whose compact radio structures were best studied at the time of selection.

OBSERVATIONS

The HEAO-3 Gamma Ray Spectrometer was operable from 20 September 1979 until 1 June 1980, during which nearly 1.5 scans of the sky were completed. The instrument is described by Mahoney et al.[14], and a discussion of the method used for measuring the 511 keV line radiation from the Galactic Center is given by Riegler et al.[5]. The details of the data reduction will be published separately by these authors.

None of the galaxies were detected in the 511 keV line. Table 1 gives the nominally measured values of and 2.3 σ upper limits (98% confidence level) to the 511 keV line flux for each source (listed according to increasing redshift) and the Galactic Center. These are preliminary results; refined values will be published at a later date. Other relevant information, taken from Jones, Sramek, and Terzian [15] and references cited therein, is also listed. The Galactic Center radio fluxes are detailed in Lo [16] and the 511 keV line flux history is summarized by Jacobson [17]. In the next section we discuss the significance of our results.

DISCUSSION

The significance of our negative results is limited somewhat by the sensitivity of the instrument. Nevertheless, if our proposed scaling law were valid, we should have detected the 511 keV line from 3 of our 6 source positions. This is shown in Table 1, which compares the expected values of and observed upper limits to the 511 keV line. Average radio flux densities and our proposed scaling law were used; the flat spectra and lack of extreme variability in the radio allow us to compare the radio parameters of the galaxies with those of the Galactic Center without concern over which frequency or epoch to choose.

Our results take on added significance when we consider that the weighted mean flux in the 511 keV line from all our sources taken together is $0 \pm 0.76 \times 10^{-4}$ photons cm^{-2} s^{-1}. This can be compared with the average of 4.6×10^{-4} photons cm^{-2} s^{-1} expected according to our scaling arguments, a difference which is significant at the 6 σ level.

Thus our proposed scaling law for 511 keV line emission from galactic nuclei is not valid. From our results, we can state that our scaling overestimates the 511 keV line flux by a factor of 3 or more. One possible cause of this discrepancy could be that the approximations (rather than the assumptions) leading to our scaling conspire to predict stronger scaling of 511 keV flux with radio flux than is actually the case. Perhaps further very-high-resolution radio studies of the Galactic Center and other galactic nuclei could yield more accurate comparisons of the physical conditions (particularly the relativistic electron contents) inside these objects, in order to determine whether our assumptions hold true.

Another possible reason for the discrepancy between observed and predicted fluxes could involve the variablilty of the Galactic Center 511 keV line. If the Galactic Center was (fortuitously) in an infrequent "high" state for the 10 years during which the 511 keV line was detected, then the 10-year mean 511 keV flux which we adopted (1.29×10^{-3} photons cm^{-2} s^{-1}) for comparison is too high for our purposes. This would again lead to an overestimation of the predicted 511 keV fluxes from the galaxies in our sample.

Barring these two possibilities, we can conclude that one of our assumptions is invalid. Thus it could be that the production of positrons is decoupled from the generation of relativistic electrons. Or perhaps higher luminosity sources produce positrons at such a high energy that their cooling (and hence annihilation) times are longer than in the Galactic Center. The cooling and annihilation times could also be lengthened if higher luminosity objects expel any dense gas from their central regions, thus forming an evacuated volume in which the positrons spend a large fraction of their time. The presence of emission lines in most of these galaxies (see ref. 18 for a summary), however, indicates that dense, ionized gas is, in fact, present. Nevertheless, the degree of mixing between any positrons produced and the thermal gas is unknown, and could conceivably be higher in the Galactic Center.

TABLE 1

Results of Search for 511 keV Line from Galaxies (Preliminary Results)

Galaxy	Type	S_ν(compact radio) (Jy)	$F_{511}(\pm 1\sigma)$ (10^{-4} photons cm^{-2} s^{-1})	F_{511}(expected)[a]	2.3σ upper limit
M81	Sb, Sey 1.5	.08 to .15	$-.7 \pm 1.5$	5.3	3.5
M82	Irr	.1 to .25			
NGC 4278	E1	.4 to .6	2.0 ± 2.1	9.2	4.9
M104	Sa	.05 to .1	0 ± 2.4	1.4	5.5
NGC 6500	Sa	~.2	-3.3 ± 2.0	3.7	4.7
NGC 2911	S0	~.2	$.6 \pm 1.7$	3.7	3.9
NGC 262 (Mkn 348)	S0/Sa Sey II	.2 to .3	1.5 ± 1.8	4.6	4.1
Galactic Center	Sb(?)	.4 to 1	~10 to 20 (mean)		

[a]Calculated from expression (2) using average radio flux densities and a mean Galactic Center 511 keV line flux of 1.29×10^{-3} photons cm^{-2} s^{-1}.

Another possibility is that the Galactic Center radio emission is relativistically beamed at a large angle to the line of sight[19]. The actual radio luminosity could then be higher than that inferred from the observed radio flux. The ratio of 511 keV line flux to angle-averaged radio flux in mildly active galaxies would then be correspondingly lower. Other galaxies with unbeamed radio flux would not then be detected at 511 keV even if our assumed scaling law is correct.

Finally, it could be that the Galactic Center positron source is not a scaled-down version of an active galactic nucleus, but rather a highly energetic stellar system or pulsar (for example) which happens to lie in the Galactic Center region owing to the high density of stars there. In this case, comparisons with other galactic nuclei would be inappropriate.

A.P.M. and K.B. gratefully acknowledge support from HEAO-3 Guest Investigator grant number NAG8-435. K.B. also acknowledges partial support from NSF Grant AST-8020756.

REFERENCES

1. Johnson, W.N., and Haymes, R.C., Ap. J. <u>184</u>, 103 (1973).
2. Haymes, R.C., Walraven, G.D., Meegan, C.A., Hall, R.D., Djuth, F.T., and Shelton, D.M., Ap. J. <u>201</u>, 593 (1975).
3. Leventhal, M., MacCallum, C.J., and Stang, P.D., Ap. J. (Letters) <u>225</u>, L11 (1980).
4. Leventhal, M., MacCallum, C.J., Huters, A.F., and Stang, P.D., Ap. J. <u>240</u>, 338 (1980).
5. Riegler, G.R., Ling, J.C., Mahoney, W.A., Wheaton, W.A., Willett, J.B., Jacobson, A.S., and Prince, T.A., Ap. J. (Letters) <u>248</u>, L13 (1981).
6. Paciesas, W.S., Cline, T.L., Teegarden, B.J., Tueller, J., Durouchoux, P., and Hameury, J.M., Ap. J. (Letters) <u>260</u>, L7 (1982).
7. Leventhal, M., and MacCallum, C.J., in The Galactic Center, ed. G.R. Riegler and R.D. Blandford (Amer. Inst. Physics, New York, 1982), p. 132.
8. Kellermann, K.I., Shaffer, D.B., Clark, B.G., and Geldzahler, B.J., Ap. J. (Letters) <u>214</u>, L61 (1977).
9. Lingenfelter, R.E., and Ramaty, R., in The Galactic Center. ed. G.R. Riegler and R.D. Blandford (Amer. Inst. Physics, New York, 1982), p. 148.
10. Rees, M.J., in The Galactic Center, ed. G.R. Reigler and R.D. Blandford (Amer. Inst. Physics, New York, 1982), p. 166.
11. Preston, R.A., Wehrle, A.E., Morabito, D.D., Jauncey, D.L., Batty, M., Haynes, R.F., Wright, A.E., and Nicholson, G.D., in IAU Symposium 97, Extragalactic Radio Sources, ed. D.S. Heeschen and C.M. Wade (D. Reidel, Boston, 1982), p. 119.
12. Hall, R.D., Meegan, C.A., Walraven, G.D., Djuth, F.T., and Haymes, R.C., Ap. J. <u>210</u>, 631 (1976).

13. Feigelson, E.D., in IAU Symposium 97, Extragalactic Radio Sources, ed. D.S. Heeschen and C.M. Wade (D. Reidel, Boston, 1982), p. 107.
14. Mahoney, W.A., Ling, J.C., Jacobson, A.S., and Tapphorn, R.M. Nucl. Instr. Methods $\underline{178}$, 363 (1980).
15. Jones, D.L., Sramek, R.A., and Terzian, Y., Ap. J. $\underline{261}$, 422 (1982).
16. Lo, K.Y., in The Galactic Center, ed. G.R. Riegler and R.D. Blandford (Amer. Inst. Physics, New York, 1982), p. 1.
17. Jacobson, A.S., in The Galactic Center, ed. G.R. Riegler and R.D. Blandford (Amer. Inst. Physics, New York, 1982), p. 123.
18. Jones, D.L., Sramek, R.A., and Terzian, Y., Ap. J. $\underline{246}$, 28 (1981).
19. Reynolds, S.P., and McKee, C.F., Ap. J. $\underline{239}$, 893 (1980).

UPPER LIMITS TO THE ANNIHILATION RADIATION LUMINOSITY OF CENTAURUS A

N. Gehrels*, T. L. Cline, W. S. Paciesas[+]**,
B. J. Teegarden, and J. Tueller**
NASA/Goddard Space Flight Center, Greenbelt, MD   20771

P. Durouchoux and J. M. Hameury
Centre d'Etudes Nucléaires de Saclay, France

ABSTRACT

A high-resolution observation of the active nucleus galaxy Centaurus A (NGC 5128) was made by the GSFC Low Energy Gamma-ray Spectrometer (LEGS) during a balloon flight on 1981 November 19. The measured spectrum between 70 and 500 keV is well represented by a power law of the form 1.05×10^{-4} $(E/100 \text{ keV})^{-1.59}$ ph cm^{-2} s^{-1} with no breaks or line features observed. The 98% confidence (2σ) flux upper limit for a narrow (< 3 keV) 511-keV positron annihilation line is 9.9×10^{-4} ph cm^{-2} s^{-1}. Using this upper limit, the ratio of the narrow-line annihilation radiation luminosity to the integral ≥ 511 keV luminosity is estimated to be < 0.09 (2σ upper limit). This is compared with the measured value for our galactic center in the Fall of 1979 of 0.10-0.13, indicating a difference in the emission regions in the nuclei of the two galaxies.

INTRODUCTION

Centaurus A (NGC 5128) is a nearby galaxy (~ 5 Mpc) with an active nucleus that is an intense source of X-rays and gamma-rays. The nuclear source is variable on time scales from days to years and is spatially unresolved at X-ray energies; the upper limit to the size of the nuclear component of the X-ray emission is 0.3 arcsec[1] (8 pc). Einstein observations indicate that above ~ 2 keV the nuclear component dominates the X-ray emission from the galaxy[1].
There are three previously reported observations of Cen A by instruments capable of measuring a positron annihilation feature in the spectrum: the Rice University instrument in 1968[2] and 1974[3], and HEAO A-4 in 1978[4]. No lines were seen at 511 keV in any of the three observations, with 2σ flux upper limits for an unresolved line being 1.8×10^{-3}, 8×10^{-4}, and 6.5×10^{-4} ph cm^{-2} s^{-1}, respectively. These instruments employed scintillation detectors with energy resolutions at 511 keV in the 40-70 keV FWHM range. We report here the results of the first high-resolution gamma-ray observation of Cen A, made by an instrument employing Germanium detectors with a resolution of 2.2 keV FWHM at 511 keV.

---

*   NAS/NRC Resident Research Associate
[+]   Present Address: Department of Physics, University of Alabama, Huntsville, AL   35899
** Also the University of Maryland

OBSERVATIONS

The GSFC Low Energy Gamma-ray Spectrometer (LEGS) performs high-resolution spectroscopy between $\sim$ 70 keV and 8 MeV using an array of three cooled high-purity Ge detectors surrounded by an active NaI scintillation shield. The detectors have an active volume of 230 cm^3 and a peak effective area of 35.5 cm^2 at 130 keV. The average in-flight energy resolution rises from 1.8 keV FWHM at 70 keV to 3.5 keV at 2.6 MeV. At 511 keV the effective area is 13.3 cm^2, and the resolution is 2.2 keV FWHM. The active NaI shield is $\sim$ 12 cm thick and collimates the field-of-view of the detectors to $\sim$ 16° FWHM. The instrument is balloon-borne and is mounted in a servo-controlled gondola that uses an altazimuth pointing system under microcomputer control; the pointing precision is $\sim 0.5°$. A detailed description of LEGS is given by Paciesas et al.[5].

The instrument was launched from Alice Springs, Australia on 1981 November 19 and observed Cen A for 3 hours at an average line-of-sight atmospheric depth of 3.5 g cm^{-2}. The observation was divided into 20-minute intervals during which the telescope was alternately pointed at the source and away from the source for background determination. The source flux was calculated by subtracting the average background level from each source interval, correcting for detector efficiency and atmospheric attenuation, and summing the resulting residual fluxes.

Figure 1 shows the observed spectrum of Cen A, calculated as described above. The best-fitting power law of the form $A(E/100 \text{ keV})^{-\alpha}$ has $A = 1.05 \times 10^{-4}$ ph cm^{-2} s^{-1} keV^{-1} and $\alpha = 1.59$. The joint 90% confidence error limits (χ^2_{min} + 4.6; reference 6) for A and α are shown in the inset. The data are consistent over the entire measured energy range with a power law of photon index -1.59, showing no evidence for a spectral break. The observed values of A and α are both intermediate in the range of values measured previously for Cen A (reference 4, and references therein).

The data were searched for features in the spectrum with the result that no statistically significant narrow or broad lines were seen. The 2σ flux upper limit for a narrow (< 3 keV FWHM) 511-keV positron annihilation line is 9.9×10^{-4} ph cm^{-2} s^{-1}, calculated from the source flux in a 4-keV wide bin centered on 511 keV. This limit is a factor of $\sim$ 1.9 larger than the upper limits at nearby energies due to the intense instrumental background line at 511 keV[5].

DISCUSSION

The lack of a 511-keV line in the Cen A spectrum is of particular interest in light of the 511-keV emission that has been observed from the central region of our own galaxy (see, e.g., reference 8). In order to compare the nucleus of Cen A with the galactic center source, we calculate here the ratio of the 511-keV line luminosity (or upper limit to the luminosity) to the luminosity of photons with energies ≥ 511 keV. This second quantity is not as accurately determined for either source as lower-energy integrals of

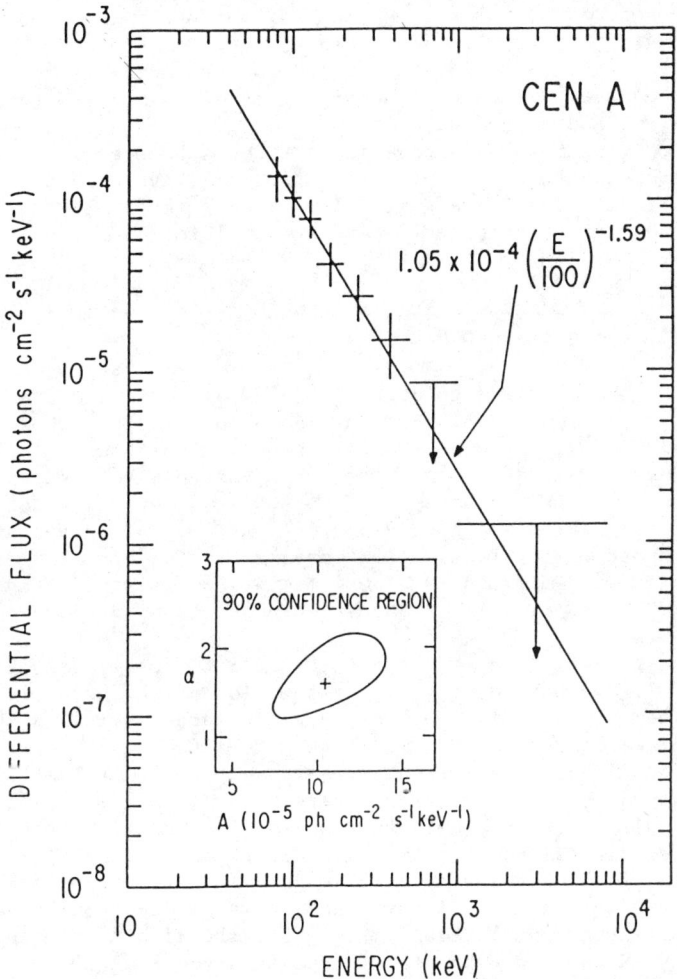

Figure 1. Cen A spectrum and best-fitting power law. The points between 70 and 500 keV are shown with 1σ statistical error bars, while the upper limits between 500 keV and 8 MeV are 2σ (98 percent confidence) upper limits. Inset--90 percent confidence contour for power-law parameters A and α defined in text.

the spectra but is more relevant for the comparison; for instance, the positrons responsible for the galactic center 511-keV emission are likely produced by γ-γ interactions of photons with $\gtrsim m_e c^2$ energies[9]. Also, for the galactic center region there is considerable source confusion in the low-energy gamma-ray measurements made to date, whereas at energies greater than a few hundred keV the nuclear source dominates[10].

Since no flux measurements exist for Cen A between 1 MeV and 1 GeV, an extrapolation of our measured spectrum (Figure 1) was required to obtain the Cen A luminosity at energies ≳ 511 keV. We performed the extrapolation using the spectral form for the differential flux

$$\frac{dF}{dE} = \frac{0.159}{E^{1.59}[1 + (E/2000 \text{ keV})^{1.3}]} \text{ photons cm}^{-2} \text{ s}^{-1} \text{ keV}^{-1} \quad (1)$$

in analogy to the observed spectrum for 3C 273[11,12]. The spectral form in equation (1) tends toward the measured spectrum below 1 MeV and is consistent with the SAS-2 flux upper limits[12] between 35 and 200 MeV. Based on equation (1), the Cen A luminosity at energies ≳ 511 keV is 2.7×10^{43} ergs s^{-1}. The LEGS 2σ upper limit for narrow-line 511-keV emission of 9.9×10^{-4} ph cm^{-2} s^{-1} gives a line luminosity upper limit of 2.4×10^{42} ergs s^{-1}. The upper limit for the line-to-continuum ratio is, therefore, 0.090; a similar analysis using the HEAO A-4[4] line limit and spectrum gives a ratio upper limit of 0.073.

For the galactic center, we use the HEAO C-1 observations in the Fall of 1979. Depending on the assumption of the shape of the spectrum, the ≳ 511 keV luminosity is in the range $1.4-1.7 \times 10^{38}$ ergs s^{-1}, based on data extending above 1 MeV (G. Riegler, private communication, 1983). The measured narrow-line 511-keV luminosity at that time was $(1.8 \pm 0.2) \times 10^{37}$ ergs s^{-1} [13], yielding a line-to-continuum ratio in the approximate range 0.10-0.13. Both the line and continuum fluxes from the galactic center are variable[13,14] with the line flux falling below detectable levels in recent measurements[15,16]. However, there is evidence for a correlation betwen the line and continuum flux levels[14,16] so that the line-to-continuum ratio of ~ 0.12 may apply to other times than the Fall 1979 measurement. Both the LEGS and HEAO A-4 2σ upper limits for the Cen A line-to-continuum ratio fall below the measured galactic center ratio. Their combined weight gives evidence that the emission regions in the nuclei of the two galaxies are different. This result is not surprising given the factor of ~ 10^5 difference in gamma-ray luminosity of the two sources. The higher temperature and activity level that one might expect in the nucleus of Cen A could produce a broadening of any 511-keV line emission. None of the observations to date have detected a broadened annihilation-radiation feature in the spectrum, but the sensitivity for line detection decreases as the line width increases above the instrumental resolution width.

SUMMARY

Cen A was observed on 1981 November 19 with a high-resolution gamma-ray spectrometer. The photon spectrum between 70 and 500 keV was found to be well fit by a power law of the form 1.05×10^{-4} (E/100 keV)$^{-1.59}$, with no evidence of a break. No lines or features were seen in the spectrum; the 2σ flux upper limit for a narrow (< 3 keV) 511-keV positron-annihilation line is 9.9×10^{-4} ph cm^{-2} s^{-1}. A comparison with the observed 511-keV line flux from the galactic center in the Fall of 1979 indicates that Cen A is less luminous than the galactic center in 511-keV narrow-line emission relative to the $\geq$ 511 keV continuum emission.

REFERENCES

1. E. D. Feigelson, E. J. Schreier, J. P. Delvaille, R. Giacconi, J. E. Grindlay, and A. P. Lightman, Ap. J. **251**, 31 (1981).
2. R. C. Haymes, D. V. Ellis, G. J. Fishman, S. W. Glenn, and J. D. Kurfess, Ap. J. **155**, L31 (1969).
3. R. D. Hall, C. A. Meegan, G. D. Walraven, F. T. Djuth, and R. C. Haymes, Ap. J. **210**, 631 (1976).
4. W. A. Baity, R. E. Rothschild, R. E. Lingenfelter, W. A. Stein, P. L. Nolan, D. E. Gruber, F. K. Knight, J. L. Matteson, L. E. Peterson, F. A. Primini, A. M. Levine, W. H. G. Lewin, R. F. Mushotzky, and A. F. Tennant, Ap. J. **244**, 429 (1981).
5. W. Paciesas, et al., submitted to Nucl. Inst. Meth. (1983).
6. M. Lampton, B. Margon, and S. Bowyer, Ap. J. **208**, 177 (1976).
7. A. S. Jacobson, R. J. Bishop, G. W. Culp, L. Jung, W. A. Mahoney, and J. B. Willett, Nucl. Inst. Meth. **127**, 115 (1975).
8. M. Leventhal, C. J. MacCallum, and P. D. Stang, Ap. J. **225**, L11 (1978).
9. R. E. Lingenfelter and R. Ramaty, in The Galactic Center, ed. G. R. Riegler and R. D. Blandford (New York: Am. Inst. Phys.), p. 148.
10. J. L. Matteson, in The Galactic Center, ed. G. R. Riegler and R. D. Blandford (New York: Am. Inst. Phys.), p. 109.
11. B. N. Swanenburg, K. Bennett, G. F. Bignami, P. Caraveo, W. Hermsen, G. Kanback, J. L. Masnou, H. A. Mayer-Hasselwander, J. A. Paul, B. Sacco, L. Scarsi, and R. D. Wills, Nature **275**, 298 (1978).
12. G. F. Bignami, C. E. Fichtel, R. C. Hartman, and D. J. Thompson, Ap. J. **232**, 649 (1979).
13. G. R. Riegler, J. C. Ling, W. A. Mahoney, W. A. Wheaton, J. B. Willett, A. S. Jacobson, and T. A. Prince, Ap. J. **248**, L13 (1981).
14. G. R. Riegler, J. C. Ling, W. A. Mahoney, W. A. Wheaton, and A. S. Jacobson, 17th Int. Cosmic Ray Conf. **9**, 85 (1981).
15. M. Leventhal, C. J. MacCallum, A. F. Huters, and P. D. Stang, Ap. J. **260**, L1 (1982).
16. W. S. Paciesas, T. L. Cline, B. J. Teegarden, J. Tueller, P. Durouchoux, and J. M. Hameury, Ap. J. **260**, L7 (1982).

ELECTRON/POSITRON/GAMMA RAY BEAMS IN COSMIC RADIO SOURCES

R. V. E. Lovelace and C. B. Ruchti
Cornell University, Ithaca, N.Y. 14853

ABSTRACT

Further study has been made of an electrodynamic model for the origin of the beams which are thought to power extra-galactic double radio sources. In this model the rotating magnetized accretion disk of a massive black hole acts as a unipolar induction dynamo. Energy and angular momentum of the infalling matter of the disk is extracted electromagnetically and propagated outward along the ±z axes initially as the Poynting flux of a relativistic space-charge flow. Such a non-neutral flow is possible in the otherwise empty vortex funnels which exist in accretion flows of matter with angular momentum. An analysis of the space charge flow is made utilizing the analogy with laboratory high-voltage, magnetically insulated transmission lines. The magnetic insulation is predicted to undergo electrical breakdown at a distance $\sim z_*$ providing a termination for the transmission line. At the termination the Poynting flux is transformed into particle kinetic energy. In general, the termination will not be matched and some fraction of this energy will be directed back towards the central object where it heats the disk. The remaining fraction of the energy is directed outward, and under appropriate conditions it gives rise to electromagnetic cascade showers. The showers result in collimated, ±z directed beams of electrons, positrons, and gamma rays. The relevant shower equations of Burns and Lovelace have been integrated numerically to determine the ratio of the energy in the gamma rays to the energy in the electrons and positrons for different physical conditions. The electron-positron beams are eventually stopped and scattered in pitch angle by the ram pressure of an external medium. Synchrotron radiation of the scattered electrons and positrons produces the observed radio lobes.

I. INTRODUCTION AND SUMMARY

A massive black hole is assumed to have formed in the nucleus of an active galaxy or quasar with a power output due to the accretion of gaseous matter through an accretion disk.[1-3] In the presence of an axial magnetic field $\sim 10^3$ Gauss, an electric field $\sim 10^5$ V/cm and a potential drop of $\sim 10^{19}$ V are generated by the unipolar induction of the rotating disk.[4-6] A potential drop of a similar order of magnitude is generated across a rotating black hole in the presence of a $\sim 10^3$ G axial magnetic field (maintained by the accretion disk).[7,8] The electromagnetic field configuration acts to generate oppositely ±z directed space charge flows[6,9] analogous to those thought to exist in the axi-symmetric pulsar model of Goldreich and Julian.[10] In contrast with the pulsar case, the accretion disk is naturally axi-symmetric. The power output of these flows, and the associated angular momentum output, are extracted electromagnetically from the infalling matter of the disk. The space charge flows are initially confined within the otherwise empty vortex funnels which are predicted to exist in accretion flows of matter with angular momentum.[11-14]

Near to the black hole, the power output is carried predominantly by the Poynting flux of the space charge flow. The flow behaves as a DC electromagnetic wave. The analysis of such flows is facilitated by the analogy which exists with laboratory high-voltage magnetically insulated transmission lines.[15-18] The magnetic insulation is predicted[6] to undergo electrical breakdown at a distance $\approx z_* \sim 10^{16}$ cm from the central object. In the breakdown region, which acts as a termination for the transmission line, the Poynting flux is converted into particle kinetic energy.

A self-consistent plasma model for the termination does not exist for the unipolar dynamo model (or for the axi-symmetric pulsar model). Nevertheless, the different possible physical processes can be assessed. Consideration has been given to the acceleration of ultra high-energy ($\sim 10^{19}$ eV) cosmic rays in the breakdown region for the case where the axial magnetic field is anti-parallel to the angular velocity vector of the disk.[19,4] A general discussion and estimates of the relevant high-energy scattering processes are given in Ref. 6. For the estimated physical conditions at the termination, the scattering processes can readily trigger electromagnetic cascade showers.[6] The showers give rise to collimated beams of relativistic electrons, positrons, and gamma rays.[6,20,21]

The electron-positron beams, which are electrically neutral and propagate ballistically, provide a possible mechanism for energizing the double radio components associated with some active galaxies and quasars. The interpretation that the radio lobes (~ 1 pc to $\sim$Mpc) are powered by beams originating in a central object is supported by a large body of observations and is exemplified by the radio map of NGC 6251.[22] An alternative model for the collimated transport of energy to the lobes involves collimated hydrodynamic flows.[23] Considerations of electron-positron plasmas[24] and of electron-positron beams[6,25] have been stimulated in part by the small upper limits on the Faraday rotation obtained for many sources.[26,27] Recently, a simple dynamical model[28] has been put forward for the generation of the apparent super-luminal motion seen in many strong compact (VLBI) sources.[29]

The gamma ray beams generated in the electromagnetic cascades may have an energy content comparable or larger than that in the electrons and positrons.[21] Under appropriate conditions, electron-positron pairs can be created at large distances from the central source via gamma-ray/synchrotron-photon collisions.[30] At present the observations of high energy gamma rays is limited to the detection of variable (~ 1 yr), high energy ($\sim$ TeV) gamma ray emission ($\sim 10^{41}$ erg/s) from Cen A.[31,32]

Section II of the paper discusses the relativistic space-charge flows using the analogy with the laboratory magnetically insulated transmission lines. The electrical breakdown of the insulation at z_* is interpreted as the termination of a transmission line. A treatment is given of the possibility of a time-independent reflected space-charge flow arising from an unmatched termination. A brief discussion is given of a case involving time dependent space charge reflections between $z = 0$ and z_* (cf. Ref. 19).

Section III of the paper presents new results on the nature of electron/positron/gamma-ray showers which may occur in the vicinity of the termination. Numerical solutions of the relevant transport equations[21] are discussed for the case of showers arising from scattering of energetic particles off of an optical to X-ray photon background having a flux density proportional to $\nu^{-\alpha}$, where ν is the frequency and α the spectral index. The ratio of the energy in the gamma rays to that in the electrons and positrons is evaluated as a function of α.

II. ACCRETION FLOW ELECTRODYNAMICS

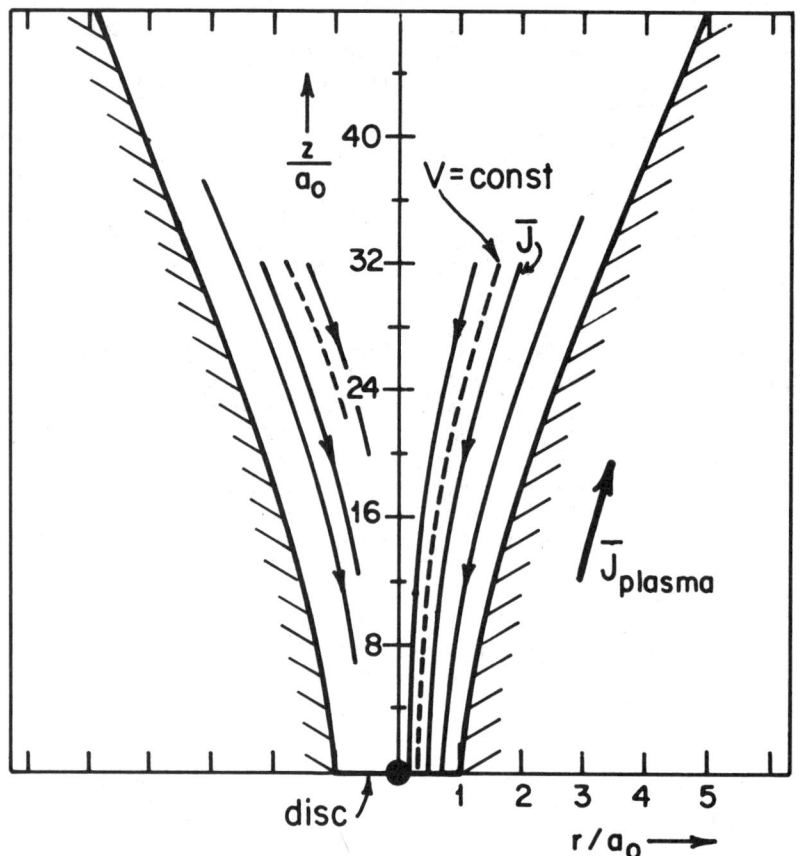

Fig. 1. Schematic drawing of unipolar induction dynamo.

The geometry of the unipolar induction dynamo is indicated in Fig. 1.[6] The inflow of ionized matter in the disk carries in and "amplifies" any ambient magnetic field owing to flux conservation. In general, the field has poloidal $\underset{\sim}{B}_p = (B_r, 0, B_z)$ and toroidal $(0, B_\phi, 0)$ components in cylindrical (r, ϕ, z) coordinates. The toroidal field within the disk may be important for short time-scale, flare-like activity[33,34] which acts to limit the magnitude of B_ϕ. The buildup of the poloidal field is limited by the Rayleigh-Taylor instability to values $|\underset{\sim}{B}_p| \lesssim 10^4$ G $(\dot{M}/M_\odot/yr)^{1/2}(10^8 \, M_\odot/M)$, where M is the black hole mass and $\dot{M}$ is the mass accretion rate.[35] Above and below the accretion disk the poloidal field, $\sim B_z \hat{z}$, is trapped[6] in the vortex funnels which are a consequence of the inflow of matter with angular momentum.[11-14]

Owing to the rotation of the accretion disk, a non-rotating observer at $z = 0$ sees an electric field

$$E_r = - [v_\phi(r)/c] B_z(r) ,$$
$$\sim 3 \times 10^5 (V/cm)(B_z/10^3 G) , \qquad (1)$$

where v_ϕ is the azimuthal velocity of the disk material. For numerical estimates, v_ϕ and B_z are taken at $r = r_1 \equiv 2 \, GM/c^2 \sim 3 \times 10^{13}$ cm $\times$ $(M/10^8 M_\odot)$. The potential drop across the disk at $z = 0$ is

$$V_{12} = - \int_{r_1}^{r_2} dr \, (v_\phi/c) B_z ,$$
$$\sim 2 \times 10^{19} \, V \, (B_z/10^3 G)(M/10^8 M_\odot) . \qquad (2)$$

The potential V_{12} acts to accelerate electrons in the $\pm z$ directions if $\underset{\sim}{\Omega} \cdot \underset{\sim}{B} > 0$, where $\underset{\sim}{\Omega}$ denotes the angular velocity vector of the disk material. If $\underset{\sim}{\Omega} \cdot \underset{\sim}{B} < 0$, V_{12} acts to accelerate protons. The plasma constituting the wall of the vortex funnel is rotating relatively slowly so that the electrical potential along it can be taken to be zero.

The electromagnetic field configuration at $z = 0$ is analogous to that in laboratory relativistic foil-less diodes.[36] In the absence of a reflected space charge wave (which we discuss subsequently), the voltage drop V_{12} gives rise to a space charge limited current flow.[36,37] In the $+z$ direction,

$$I \approx c \, V_{12} ,$$
$$\sim 10^{18} \, A \, (B_z/10^3 G)(M/10^8 M_\odot) , \qquad (3)$$

and there is an equal current in the $-z$ direction. The current flow of Eq. (3) corresponds to an impedance

$$Z_o = c^{-1} \, (cgs) = (4\pi)^{-1}(\mu_o/\varepsilon_o)^{1/2} \, (MKS) ,$$
$$\approx 30 \, \Omega. \qquad (4)$$

This impedance and the current flow (3) is assumed in the Goldreich-Julian axi-symmetric pulsar model.[10,38,39]

Looking from z = 0 one has in effect a transmission line with a characteristic impedance Z_0. Thus the current flow of Eq. (3) is relevant only for the case where the transmission line has a matched termination. A laboratory analogue of the astrophysical situation exists in the form of magnetically insulated high-voltage transmission lines. Such lines have been investigated experimentally[16] and theoretically.[17,18]

Consider first the case of a matched termination. The non-zero particle rest mass, m, results in a modification of Eq. (3). A small fraction, $\sim \varepsilon = (mc^2/eV_{12})^{1/2} \ll 1$, of the total potential drop V_{12} is "used up" in a thin Langmuir sheath ($\Delta z \sim \varepsilon r_2$) in which the particles are accelerated up to relativistic energies ($\sim \varepsilon eV_{12} \gg mc^2$).[36,37] As a result, the current flow is smaller than Eq. (3) by a fractional amount of order ε. Aside from the influence of a radiation field, discussed below, the particle motion is force-free, $\underline{E} + (\underline{v}/c) \times \underline{B} = 0$.

Thus for a matched termination, the power output in the space charge flow (in the +z direction) is

$$L_{B+} \approx 1/2 \, V_{12} I \, ,$$
$$\sim 10^{44} \text{ erg/s } (B_z/10^3 G)^2 (M/10^8 M_\odot)^2 \, . \quad (5)$$

L_{B+} is initially carried by the Poynting flux, $S_z = (c/4\pi) E_r \times B_\phi$. Hence, the space charge flow behaves as a DC electromagnetic wave. Only a small part of the power output is carried in the form of particle kinetic energy. This power is of the order of εL_{B+}.[36,37]

For $\underline{\Omega} \cdot \underline{B} > 0$, the space charge current of electrons may initially flow through an intense continuum radiation field. The geometry of the radiation field can be roughly characterized by a radius r_{op}, with the photon number density within r_{op} given by $n_{ph} = L_{op}(\pi r_{op}^2 \bar{\varepsilon} c)^{-1}$, where L_{op} is the total photon luminosity, and $\bar{\varepsilon} \ll mc^2$ is the mean photon energy. The electron motion is highly relativistic, with Lorentz factors $\gamma \gg 1$, and the motion is approximately in the z direction. Thus,

$$m_e c \frac{d\gamma}{dz} = -eE_z - F_d \, , \quad (6)$$

where $-e$ is the electron charge, E_z is a non-force-free electric field contribution, and F_d is the drag force due to Compton scattering off of photons. For $\gamma < \gamma_* \equiv (m_e c^2/\bar{\varepsilon})$, $F_d = m_e c^2 \sigma_T n_{ph} (\gamma)^2/\gamma_*$, whereas for $\gamma > \gamma_*$, $F_d = (3/8) m_e c^2 \sigma_T n_{ph} \gamma_* \cdot \ell n(2\gamma/\gamma_*)$, where $\sigma_T \equiv (8\pi/3)(e^2/mc^2)^2$ is the Thomson cross section.

One has quasi-steady beam propagation if $|E_z| \approx F_d/e \equiv E_{zo} \gg (m_e c^2/e)(d\gamma/dz)$. In the full distance r_{op} through the radiation field, the potential drop arising from the E_{zo} field is

$$eE_{zo}r_{op}/(m_e c^2) = \sigma_T n_{ph} r_{op} \begin{cases} \gamma^2/\gamma_* & \text{for } \gamma < \gamma_* \\ (3/8)\gamma_* \ln(2\gamma/\gamma_*) & \text{for } \gamma > \gamma_*. \end{cases} \quad (7)$$

Here,

$$\sigma_T n_{ph} r_{op} \sim 4 \times 10^7 \left(\frac{L_{op}}{10^{46} \text{ erg/s}}\right)\left(\frac{10^{15} \text{ cm}}{r_{op}}\right)\left(\frac{1 \text{ eV}}{\bar{\varepsilon}}\right).$$

The potential drop of Eq. (7) can readily be balanced by a small axial gradient in the space charge density if

$$\gamma < [\Gamma \gamma_*/(n_{ph} \sigma_T r_{op})]^{1/4} \quad \text{and} \quad \gamma < \Gamma^{1/3}, \quad (8)$$

where $\Gamma \equiv eV_{12}/(m_e c^2)$. For the previous reference values, we have $\Gamma = 4 \times 10^{13}$, and the two inequalities become $\gamma < 10^3$ and $\gamma < 3.4 \times 10^4$, respectively. In the following, the electron space charge flow ($\underline{\Omega} \cdot \underline{B} > 0$) is assumed to obey Eq. (8) for $z \lesssim r_{op}$.

For a range of distances, z, extending well beyond r_{op}, the particle motion is to an excellent approximation force-free because of the smallness of

$$\beta \approx \frac{2mc^2\gamma}{eV_{12}}[2 + (B_z/B_\phi)^2]^{-1}, \quad (9)$$

which is the ratio of the energy density of the particles to that of the electric and magnetic fields. Therefore, the particle motion can be described by simplified drift equations. Using these equations, a self-consistent model for the electric and magnetic fields can be constructed. For this purpose we introduce a new, slightly stretched cylindrical coordinate system (η,ϕ,z), where ϕ and z are identical to the prior cylindrical coordinates, but $\eta = r/a(z)$ with $a(z)$ the beam channel radius. We assume $da/dz \ll 1$. The particle drift velocity is $\underline{v} = v_\parallel (\underline{B}/B) + \underline{E} \times \underline{B}/B^2$ with $v_\parallel$ the speed parallel to the magnetic field, and $(\underline{v})^2 = 1 - (\gamma)^{-2}$. To lowest order in da/dz we have

$$v_\phi = (v_\parallel B_\phi B - E_\eta B_z)/B^2,$$
$$v_z = (v_\parallel B_z B + E_\eta B_\phi)/B^2, \quad (10)$$

where $B \equiv (B_\phi^2 + B_z^2)^{1/2}$ and $v_\parallel = [1 - (E_\eta/B)^2 - (1/\gamma)^2]^{1/2}$. To lowest order in da/dz, $\underline{v} \cdot \underline{E} = 0$ and γ is a constant.

To lowest order in da/dz, the field equations are

$$\frac{1}{\eta}\frac{\partial}{\partial \eta}(\eta e_\eta) = \tilde{\rho} , \qquad (11a)$$

$$\frac{1}{\eta}\frac{\partial}{\partial \eta}(\eta b_\phi) = \tilde{\rho} v_z , \qquad (11b)$$

$$\frac{\partial}{\partial \eta}(b_z) = -\tilde{\rho} v_\phi . \qquad (11c)$$

Here, $e_\eta \equiv E_\eta/B_0$, $b_\phi \equiv B_\phi/B_0$, and $b_z \equiv B_z/B_0$ are the dimensionless fields; $\tilde{\rho} \equiv 4\pi\rho\, a/B_0$ is the dimensionless charge density; and $B_0 \equiv |B_\phi(r=a)|$.

The normalization of Eq. (10) utilizes the charge-conservation condition $\int_0^a r\, dr\, j_z(r) = \text{const.}$, and it implies $b_\phi(\eta=1) = S$, where $S \equiv -\text{sign}(\underline{\Omega}\cdot\underline{B})$. Thus, we have the constraint $\int_0^1 \eta\, d\eta\, \tilde{\rho} v_z = S$. The conservation of magnetic flux in the z-direction gives a second constraint,

$$\langle b_z \rangle = 2\int_0^1 \eta\, d\eta\, b_z = 2k[a_0/a(z)] , \qquad (12)$$

where $a_0 \equiv a(z=0)$ and $k = |B_{zv}/B_0(a_0)| = \text{const.}$, with B_{zv} the axial component of the vacuum magnetic field at $z = 0$. The parameter k is roughly the speed of light divided by an average azimuthal velocity of the disk material. Hence, $k > 1$.

The power carried by the flow is, to lowest order in β, the integral of the Poynting flux,

$$L_{B+} = (1/2)\int_0^1 \eta\, d\eta\, e_\eta b_\phi + \mathcal{O}(\beta) .$$

The angular momentum carried by the flow is

$$J_+ = (1/2)\int_0^1 \eta\, d\eta\, \eta\, b_\phi b_z [a(z)/a_0] + \mathcal{O}(\beta) .$$

To lowest order in da/dz, L_{B+} and J_+ are independent of z.

Appropriate boundary conditions on Eqs. (11) at $z = 0$ are required to uniquely specify the field solution. The disk rotation curve gives $e_\eta(\eta,z=0)$, which in turn gives $\tilde{\rho}(\eta,z=0)$. Equations (10) and (11) can then be used to derive $\tilde{\rho} v_z$ as a function of η at $z = 0$. To lowest order in da/dz, the continuity equation implies $\partial(\tilde{\rho} v_z)/\partial z = 0$ because $v_\eta = 0$. The z-dependence of the solutions of Eqs. (10) and (11) enters through the second constraint, Eq. (12). Thus, the solutions depend on two parameters: γ which is a constant, and $\langle b_z \rangle = k[a_0/a(z)]$, which decreases as the beam radius increases. In the drift approximation of Eqs. (10) and (11), the β ratio of Eq. (9) has been neglected relative to unity. However, the solutions do depend on γ in contrast with the case of the "pulsar equation"[38,39] where $\beta = 0$ and $\gamma = \infty$.

For a simple illustrative case, we consider electron flow with $\tilde{\rho} v_z = -2$. Equation (11b) then gives $b_\phi = -\eta$ so that the first constraint, $b_\phi(1) = -1$, is satisfied identically. Figure 2 shows sample field profiles for $\langle b_z \rangle = 0.5$ and 0.013, both with $\gamma = 100$, obtained by numerical integration of Eqs. (10) and (11). We have been able to find numerical solutions only for

$$\langle b_z \rangle = 2[a^2 B_o]^{-1} \int_0^a r \, dr \, B_z \quad > \sim 1.3/\gamma \, . \tag{13}$$

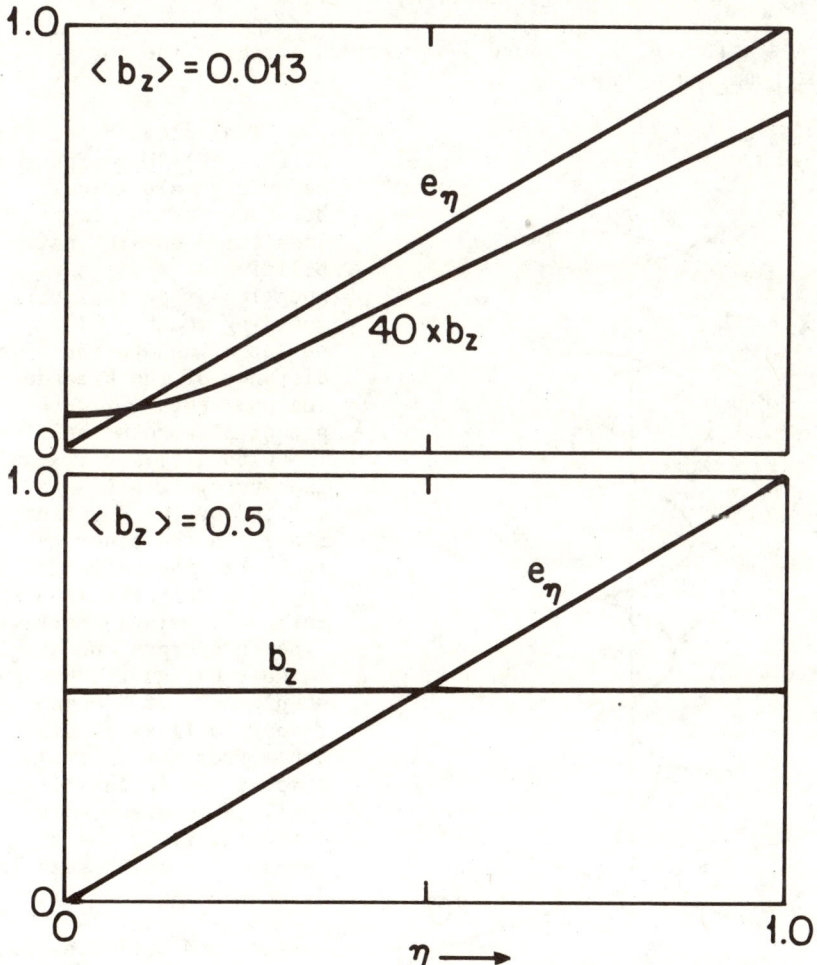

Fig. 2. Electric and magnetic fields of self-consistent space charge flows from Eqs. (10)-(12).

A qualitative explanation of this inequality follows from the fact that the confining pressure of the B_z field, $B_z^2/8\pi$, is needed to balance the outward (+η) pressure of the B_ϕ-E_η fields which is $\sim B_\phi^2/(8\pi\gamma^2)$.

The condition $<b_z> > 1.3/\gamma$ for the equilibrium of the space charge flow allows for a very large axial propagation distance. For example, for $k = 2$ and $\gamma = 10^3$ the expansion of the beam radius, $a(z)/a_0$, is limited to values $\lesssim 3\times10^3$. For most of this expansion, $B_\phi \approx B_{\phi 0}(a_0/a)$ and $B_z \approx B_{z0}(a_0/a)^2$. Hence the equation for a field line lying on the beam edge is

$$x,y = a(z)[\cos(z/A), \sin(z/A)],$$

where $A \equiv a_0(B_{z0}/B_{\phi 0})$. Figure 3 shows such a field line for a conical beam channel.

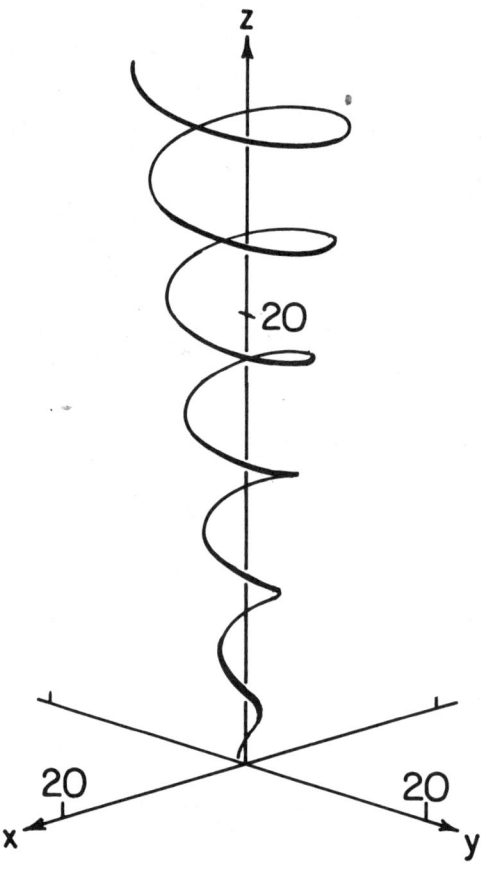

Fig. 3. Projection of field line on the beam edge for the case $a = a_0 + 0.3z$ and $\phi = 1.33z$.

The space charge flow of Eqs. (9)-(11) is expected to undergo electrical breakdown of the magnetic insulation due to instabilities at a distance much less than that corresponding to $<b_z> = 1.3/\gamma$. We let z_* denote the axial distance of the breakdown region. For $z < z_*$, the potential across the flow (r = 0 to a(z)) is V_{12} and the energy flow L_{B+} is carried by the Poynting flux. On the other hand, for $z > z_*$ the potential is approximately zero and the energy flow is in particle kinetic energy. We argue in Ref. 6 that the breakdown occurs at a comparatively small axial distance from the central object. An instability of possible importance in determing z_* is the kink instability which sets in at distances $\gtrsim 2\pi(B_{z0}/B_{\phi 0})a_0$. A discussion and estimates of the different scattering processes in the breakdown region have been given previously.[6] For the estimated physical conditions one is led to investigate the nature of electromagnetic cascade showers.[20,21] Further results on the showers is presented in Sec. III.

The breakdown of the magnetic insulation at $z = z_*$ is in effect the termination of the magnetically insulated transmission line. As mentioned earlier, Eqs. (3)-(5) correspond to the special case of a matched termination (or load) where all of the incident (+z) electromagnetic power is absorbed. Although a self-consistent model of the termination does not exist, some general statements can be made: A time-independent termination can have a "reflected" space charge flow traveling in the -z direction from $z = z_*$ to 0 as shown in Fig. 4. We let the voltage reflection coefficient be denoted by

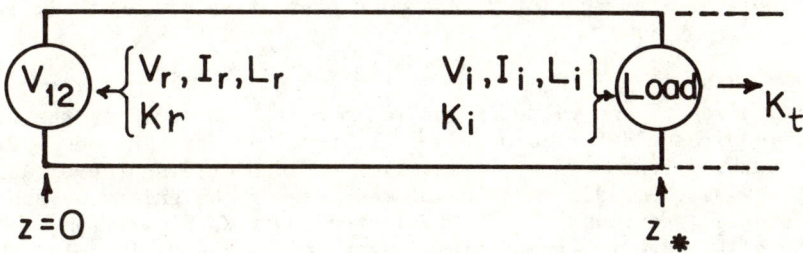

Fig. 4. Transmission line parameters.

$\alpha = -V_r/V_i$, where V_i is the voltage drop [r = 0 to a(z)] of the incident flow, and V_r is that of the reflected flow. For specificity, consider $\underline{\Omega} \cdot \underline{B} > 0$. For $\alpha > 0$, the reflected flow may consist of counter accelerated protons from an ambient plasma, or positrons created in an $e_+/e_-/\gamma$ shower at $z = z_*$. For $\alpha < 0$, it is possible that the reflected flow consists of incident electrons reflected by a negative space charge accumulation at $z = z_*$.

Evidently, we have $V_{12} = V_i + V_r = (1 - \alpha)V_i$, with V_{12} of Eq. (2) fixed by the rotation of the disk. Similarly, for the current carried by the incident (I_i) and reflected (I_r) flows, $I = I_i + I_r = (1 + \alpha)I_i$. Thus the effective impedance seen by the disk is

$$Z_{eff} = V_{12}/I = \frac{1-\alpha}{1+\alpha} Z_o ,$$

where $Z_o = V_i/I_i = (1/4\pi)(\mu_o/\varepsilon_o)^{\frac{1}{2}} = 30\Omega$. The total electromagnetic power output of the +z side of the disk is

$$L_{B+} = \frac{V_{12}^2}{2Z_{eff}} = \frac{1+\alpha}{1-\alpha} \frac{V_{12}^2}{2Z_o} . \qquad (14a)$$

L_{B+} is larger than Eq. (5) for $0 < \alpha < 1$. The electromagnetic power carried by the incident (+z) flow is

$$L_i = (1/2)V_i I_i = \frac{V_{12}^2}{2(1-\alpha)^2} \frac{1}{Z_o} \qquad (14b)$$

The EM power carried by the reflected (-z) flow is

$$L_r = (1/2)V_r I_r = (1/2)\left(\frac{\alpha}{1-\alpha}\right)^2 V_{12}^2 / Z_o \qquad (14c)$$

Thus, $L_{B+} = L_i - L_r$.

We let K_i and K_r denote the power carried in particle kinetic energy by the incident and reflected flows, respectively. Further, K_t denotes the power carried by particle kinetic energy beyond z_*. As in the earlier discussion, K_i is assumed negligible compared with L_i. Therefore, energy (or momentum) conservation at $z = z_*$ implies

$$K_r + K_i = L_{B+} . \qquad (15)$$

Determination of the value of the ratio K_r/L_{B+}, as well as the value of α, requires a detailed model of the termination. The power K_t flows outward from z_* and ultimately goes into driving the expansion and powering the radio emission of the jets of the galaxy or quasar as discussed previously.[6] The reflected power K_r flows from z_* back to the disk where it irreversibly heats the disk. Ultimately, this power could appear as a non-thermal optical to X-ray continuum. In contrast, the electromagnetic power L_r flows from z_* to the disk where it is reversibly absorbed in "torquing up" the disk.

The transmission line termination does not necessarily give rise to a time-independent configuration of the type described by Eqs. (13)-(15). Figure 5 shows a time dependent case corresponding to a short circuit load where $\alpha = 1$. From Eq. (14a) the steady power output would be infinite which is impossible. The time evolution shown in Fig. 5 starts at $t = 0$ with $I = 0$. The current magnitude increases in steps, whereas the voltage drop switches between V_{12} and 0. The characteristic time is $T_o = z_*/c$, where, for simplicity, we consider z_* to be a constant. Clearly, the increase of I will cease at some point due to instabilities. One possibility is the kink instability.[40] The non-linear development of the instability acts to increase the circuit impedance which returns the transmission line to its $t = 0$ configuration. An independent discussion of possible time dependent current flow in the dynamo model is given by Lake and Pudritz.[19] (See also Ref. 41.)

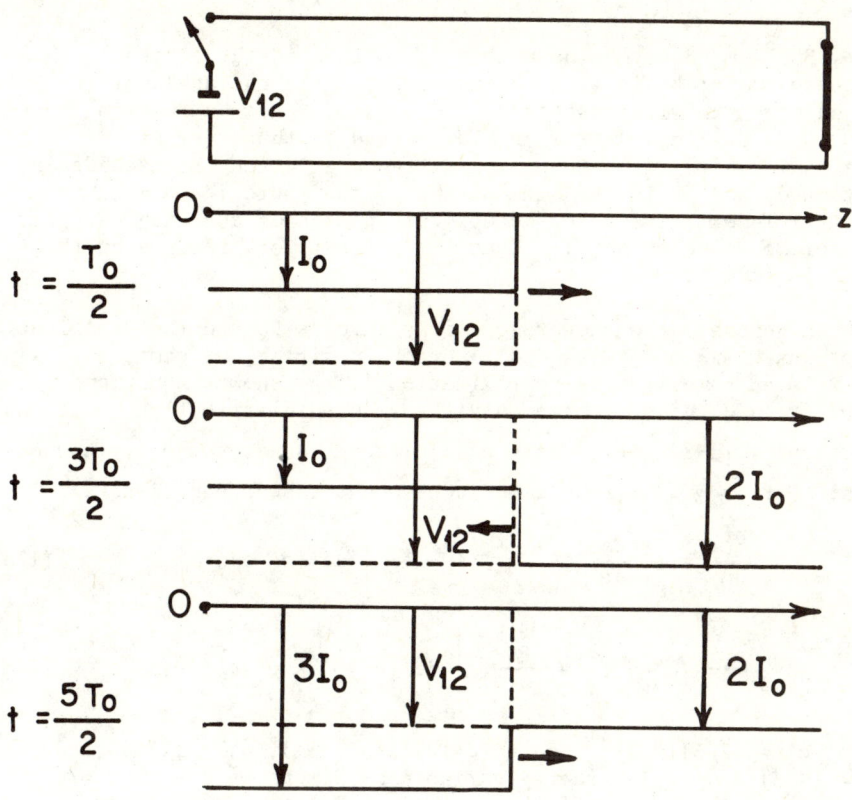

Fig. 5. Current and voltage on a short circuited line.

III. ELECTRON/POSITRON/GAMMA RAY CASCADES

In earlier work, coupled relativistic Boltzmann-type transport equations were derived for the electrons, positrons, and gamma rays in a cascade shower.[20,21] The scattering medium for the shower was considered to be a spectrum of "low-energy" photons ($\sim 10^{-3}$ to 10^3 eV) originating from the inner region or the accretion disk of the black hole. The scattering processes involved (a) the production of an energetic electron-positron pair by the scattering of a gamma ray and a low energy photon, and (b) the production of a gamma ray by the Compton scattering of an energetic electron off a low energy photon. The possibly important higher order scattering process involving pair production in the field of an electron has recently been pointed out to us,[42] but this process[43] has not yet been incorporated into the transport equations.

The distribution function for the low energy photons is written as

$$F_\nu(z, k_\nu, \theta_\nu) = \mathcal{F}_\nu(z, \theta_\nu)(1/k_\nu)^{3+\alpha} . \qquad (16)$$

Here, $F_\nu d^3x\, d^3p$ denotes the number of photons in the element $d^3x\, d^3p$ of phase space; k_ν is the photon energy; θ_ν is the angle of the photon momentum with respect to the z axis; $\mathcal{F}_\nu$ is a function peaked about $\theta_\nu = 0$ with a narrow angular spread of the order of $a_0/z_* \ll 1$, where z_* is the axial distance at which the cascade is initiated, and a_0 is the size of the photon source; and α is the spectral index. The observed optical spectra of quasars[44] suggest an average value of $\alpha \sim 1$. Thus the present work focuses on the limit $1 - \alpha \ll 1$.

In general, the shower equations involve F_ν and the distribution functions $F_n(x_\mu, p_\mu)$, with n = electron, positron, or gamma ray. In the relevant small angle approximation[21], the shower equations simplify, and these involve only the reduced distribution functions, $f_n(z,t,p) = \int d^2x_\perp d^2p_\perp F_n$, where $p = |\underline{p}|$ and $\perp$ denotes the x,y components. For $1 - \alpha \ll 1$, the time independent shower equations are[21]

$$\frac{\partial f_+}{\partial \eta} = \frac{c_1}{e^2} \frac{\partial}{\partial e}(e^{3+\alpha} f_+) + \frac{c_2}{e^{2-\alpha}} \int_e^1 de'(e'-e) f_\gamma' , \qquad (17a)$$

$$\frac{\partial f_\gamma}{\partial \eta} = -c_3 e^\alpha f_\gamma + \frac{c_4}{e^{3+\alpha}} \int_e^1 de'(e')^{1+2\alpha}(e'-e) f_+' . \qquad (17b)$$

Here, $c_1 = (2/3)(1-\alpha)^{-1}$, $c_2 = (c_3/2)(2+\alpha)(3+\alpha)$, $c_3 = 11/180$, and $c_4 = (4/3)(2-\alpha)$. The energy variable is dimensionless, $e \equiv \varepsilon/\varepsilon_0 \leq 1$, where ε_0 is the energy of the particle initiating the cascade. The distance or optical depth variable for the shower is

$$\eta = (\varepsilon_0)^\alpha \int_{z_*}^z dz'/X(z'), \text{ where } X^{-1} = (\pi r_e/m_e^\alpha)^2 \int_0^\infty d\theta_\nu^2 (\theta_\nu)^{2+2\alpha} \mathcal{F}_\nu(z, \theta_\nu).$$

A finite limiting value of η is approached as $z \to \infty$. The influence of an electric field is not included so that $f_+ = f_-$.

The first term on the right-hand side of Eq. (17a) gives the loss of electrons (or positrons) due to Compton scattering, whereas the second term gives the gain due to pair creation. The first term on the right-hand side of Eq. (17b) gives the loss of gamma rays due to pair production and the second term gives the gain due to Compton scattering. Energy is conserved in the cascade,

$$E = \int_0^1 de\, e^3 (2f_+ + f_\gamma) = \text{const.}$$

For the numerical integration of Eqs. (17), we found it useful to work with the scaled distribution functions $g_+ = e^{3+\alpha} f_+$ and $g_\gamma = e^{3+\alpha} f_\gamma$, and a logarithmic energy variable $x = \log(1/e) \geq 0$. The functions g_+ and g_γ are represented by their values at a set of equally spaced x values, x(i), for i=1,...,501, with x(1) = 6 to 10

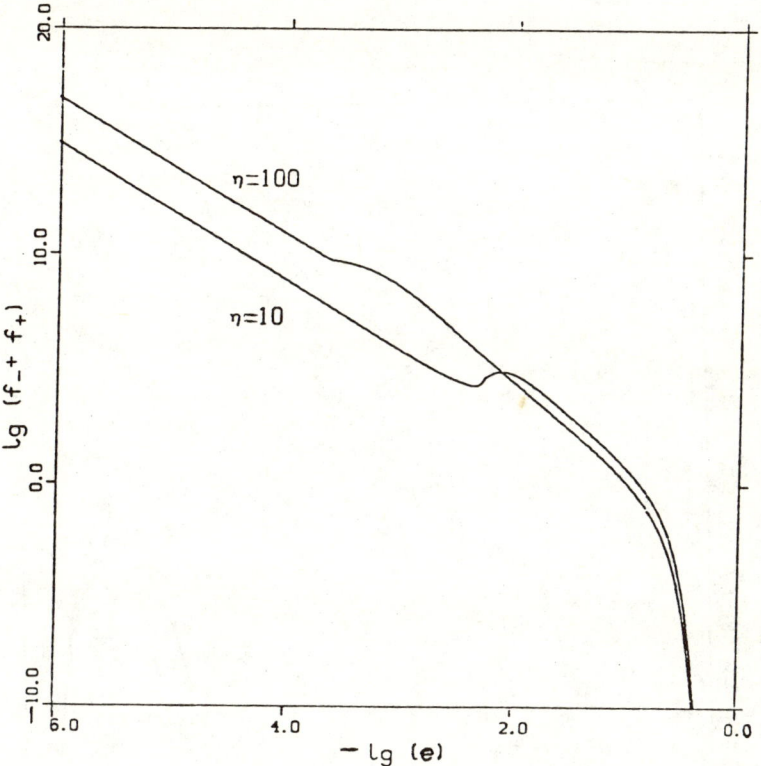

Fig. 6. Electron-positron energy distribution functions.

and $x(501) = 0$. Thus Eqs. (17) are reduced to a set of 1002 coupled differential equations. Basically, these equations have been solved numerically using the package DE described in Ref. 45. A detailed discussion of the techniques developed to suppress numerical instabilities will be given elsewhere.[46]

Figures 6 and 7 show the energy dependence of the electron and gamma ray distribution functions for the case of a shower initiated by an energetic gamma ray in a background photon distribution with an index $\alpha = 0.9$. Power-law dependences of f_+ and f_γ are seen to develop rapidly with increasing η. The power laws correspond approximately to $f_+ \propto e^{-3.9}$ and $f_\gamma \propto e^{-3}$ for $e \leq 0.1$. Alternatively, $dN_+/de \propto e^{-1.9}$ and $dN_\gamma/de \propto e^{-1}$, where dN_n denotes the number of particles in the energy interval de. Qualitatively similar power-law forms were found earlier for the case $\alpha = 0$.[20,21]

Figure 8 shows the α dependence of the ratio of the energy in the gamma rays to that in the electrons and positrons for two optical depths. This ratio is large compared with unity and it diverges as $1 - \alpha \to 0$, as predicted.[21] In contrast, for $\alpha = 0$ this ratio is of

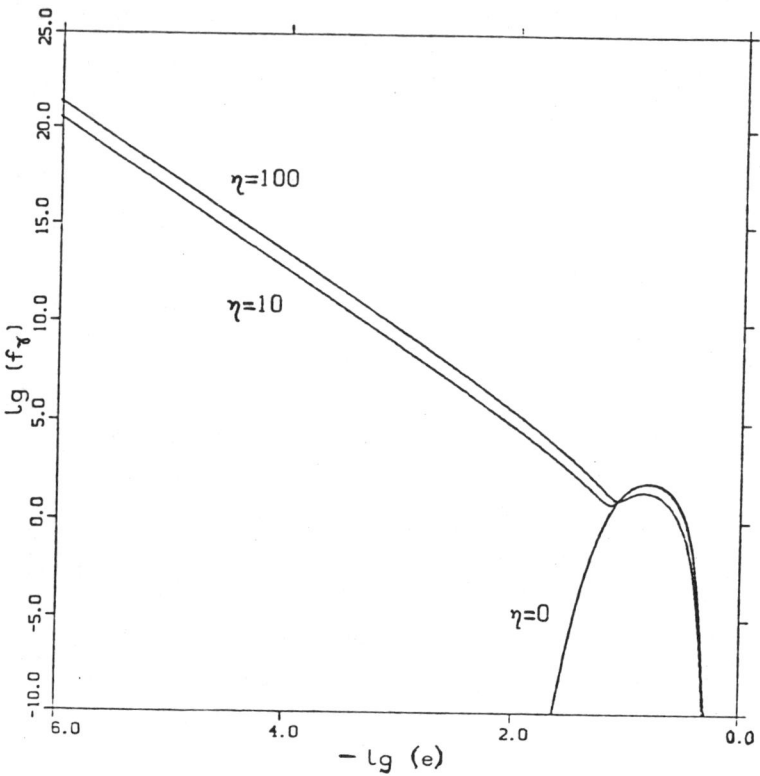

Fig. 7. Gamma ray energy distribution function.

the order of unity.

At completion of the shower, a collimated beam of relativistic electrons, positrons, and gamma rays emerges. Initially, this beam propagates ballistically in the +z direction.[6] The ballistic propagation of the electron-positron component is possible in the absence of strong magnetic field inhomogeneities. Eventually, a region of inhomogeneous field between the beam and an external medium will stop the beam by pitch angle scattering.[28] The electron-positron distribution from an $\alpha = 0.9$ cascade then gives rise to radio emission, $dL_R/df \propto f^{-0.45}$, in the optically thin limit, where dL_R is the radio power in the frequency interval df. Of course, for a detailed comparison with observed radio spectra[47] the influence of diffusion of particles in space and energy must be included as well as the possible influence of self-absorption. The electrons and positrons in the cascades considered here are highly

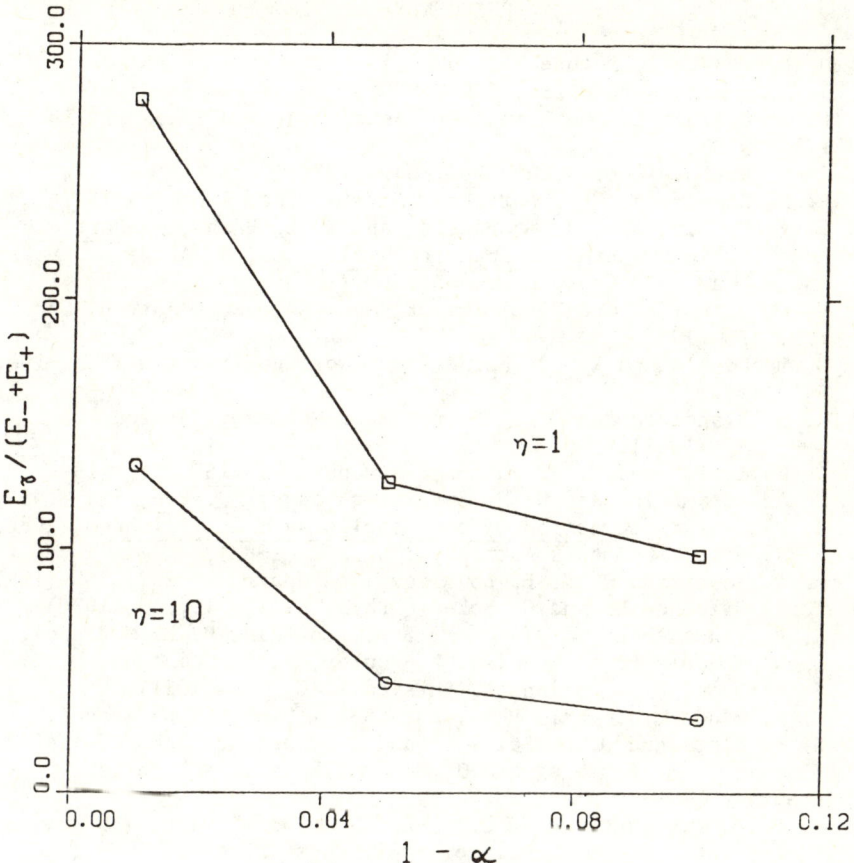

Fig. 8. Ratio of energy carried by gamma rays to that carried by electrons and positrons.

relativistic so that only a small fraction of the e_+/e_- energy appears in the form of annihilation radiation. An alternative, moderately relativistic cascade model[48], may underlie the e_+/e_- generation which gives the intense variable annihilation line source seen in the direction of the galactic center.[49,50]

The gamma ray component of the shower is expected to propagate ballistically under most conditions. However, it is possible to have pair creation triggered by gamma-ray/synchrotron photon collisions.[30] Observations and limits on the high energy gamma ray emission of extra galactic sources are clearly of interest.[31,32]

REFERENCES

1. D. Lynden-Bell, Nature 223, 690 (1969).
2. J. G. Hills, Nature 254, 295 (1975).
3. L. M. Ozernoy and M. Reinhardt, Astrophys. and Spa. Sci. 59, 171 (1978).
4. R. V. E. Lovelace, Nature 262, 649 (1976).
5. R. D. Blandford, Mon. Not. Roy. Astron. Soc. 176, 465 (1976).
6. R. V. E. Lovelace, J. MacAuslan, and M. L. Burns, in Particle Acceleration Mechanisms in Astrophysics, ed. J. Arons, C. McKee, and C. Max, AIP Conf. Proc. 56, 399 (1979).
7. R. D. Blandford and R. L. Zrajek, Mon. Not. Roy. Astron. Soc. 179, 433 (1977).
8. D. MacDonald and K. S. Thorne, Mon. Not. Roy. Astron. Soc. 198, 345 (1982).
9. R. D. Blandford and D. G. Payne, Mon. Not. Roy. Astron. Soc. 199, 883 (1982).
10. P. Goldreich and W. J. Julian, Astrophys. J. 157, 869 (1969).
11. M. A. Abramowicz and W. H. Zurek, Astrophys. J. 246, 314 (1981).
12. B. Paczynski, in Astronomische Gesellschaft Mitterlungen, 1982.
13. P. J. Wiita, Comments Astrophys. 9, 251 (1982).
14. H. A. Scott and R. V. E. Lovelace, Astrophys. J. 252, 765 (1981).
15. T. J. Orzechowski and G. Bekefi, Phys. Fluids 19, 43 (1976).
16. J. P. VanDevender, J. T. Crow, B. G. Epstein, D. H. McDaniel, C. W. Mendel, E. L. Neau, J. W. Poukey, J. P. Quintenz, D. B. Seidel, and R. W. Stinnett, Physica 104C, 167 (1981).
17. C. W. Mendel, J. Appl. Phys. 50, 3830 (1979); C. W. Mendel, S. A. Slutz and D. B. Seidel, Sandia Report No. SAND 82-1740 (1982); D. E. Pershing and J. Golden, Naval Research Lab. preprint, 1982.
18. R. V. E. Lovelace and E. Ott, Phys. Fluids 17, 1263 (1974); E. Ott and R. V. E. Lovelace, Appl. Phys. Lett. 27, 378 (1975).
19. G. Lake and R. E. Pudritz, Bell Labs Report, 1983.
20. M. L. Burns, Ph.D. thesis, Cornell University (1980).
21. M. L. Burns and R. V. E. Lovelace, Astrophys. J. 262, 87 (1982).
22. A. C. S. Readhead, M. H. Cohen and R. D. Blandford, Nature 272, 131 (1978).
23. R. D. Blandford and M. J. Rees, Mon. Not. Roy. Astron. Soc. 169, 395 (1974).
24. P. D. Noerdlinger, Phys. Rev. Lett. 41, 135 (1978).
25. W. Kundt and Gopal-Krishna, Nature 288, 149 (1980).
26. T. W. Jones and S. L. O'Dell, Astr. and Astrophys. 61, 291 (1977).
27. J. F. C. Wardle, Nature 269, 563 (1977).
28. H. A. Scott and R. V. E. Lovelace, Astrophys. and Spa. Sci., in press (1983).
29. M. H. Cohen, T. J. Pearson, A. C. S. Readhead, G. A. Serelstad, R. S. Simon, and R. C. Walker, Astrophys. J. 231, 293 (1979).
30. M. L. Burns, in preparation, 1983.
31. J. E. Grindlay, H. F. Helmken, R. Hanbury-Brown, J. Davis, and L. R. Allen, Astrophys. J. Lett. 197, L9 (1975).
32. J. E. Grindlay, in Proc. of International Workshop on Very High Gamma Ray Astronomy, Ootacamund, India (1982).

33. G. A. Shields and J. C. Wheeler, Astrophys. Lett. 17, 69 (1976).
34. A. A. Galeev, R. Rosner and G. A. Vaiana, Astrophys. J. 229, 318 (1979).
35. R. V. E. Lovelace and H. A. Scott, in Proc. of Internat. School on Plasma Astrophys, ESA SP-161, 215 (1981).
36. J. Chen and R. V. E. Lovelace, Phys. Fluids 21, 1623 (1978).
37. W. M. Fawley, J. Arons, and E. T. Scharlemann, Astrophys. J. 217, 227 (1977).
38. E. T. Scharlemann and R. V. Wagoner, Astrophys. J. 182, 951 (1973).
39. F. C. Michel, Rev. Mod. Phys. 54, 1 (1982).
40. R. V. E. Lovelace, Phys. Fluids 19, 723 (1976); Phys. Rev. Lett. 41, 1801 (1978).
41. R. Rudenberg, Electrical Shock Waves in Power Systems (Harvard, Cambridge, 1968), Ch. 4.
42. J. Saba, in preparation (1983).
43. E. Haug. Z. Naturforsch 30a, 1099 (1975).
44. K. Davidson and H. Netzer, Rev. Mod. Phys. 51, 715 (1979).
45. L. F. Shampine and M. K. Gordon, Computer Solution of Ordinary Differential Equations (Freeman, San Francisco, 1974).
46. C. B. Ruchti, in preparation (1983).
47. A. G. Pachalczyk, Radio Galaxies (Pergamon, New York, 1977).
48. M. Burns, in preparation (1983).
49. M. Leventhal and C. J. MacCallum, Ann. N.Y. Acad. Sci. 336, 248 (1980).
50. R. Ramaty, D. Leiter, and R. E. Lingenfelter, Ann. N.Y. Acad. Sci. 375, 338 (1981).

COMPACTNESS AND PAIR PRODUCTION IN ACTIVE GALACTIC NUCLEI

M. Salvati
European Southern Observatory, Garching, F.R.G. and
Istituto di Astrofisica Spaziale, Frascati, Italy

A. Cavaliere
Istituto di Astronomia, Università degli Studi di Padova
Padova, Italy

E. Costa
Istituto di Astrofisica Spaziale, Frascati, Italy

E. Massaro
Istituto Astronomico, Università di Roma "La Sapienza",
Roma, Italy

ABSTRACT

Barring relativistic bulk motions, the continuum emitted by AGN's implies very compact sources. We find that the compactness parameter $k = Lc/R$ cannot, especially in the X-ray band, exceed a level interestingly close to the current data. In sources driven by high energy electrons, the main limitation is associated with pair production by heavily absorbed, inverse-Compton gamma-rays.

INTRODUCTION

The strong, highly variable continuum radiated by active galactic nuclei (AGN's) must come either from relativistic jets, or from very compact sources.

In the present context, a compact source is one where the photon density is high, and Thomson-cross-section electromagnetic reactions drastically interfere with the radiation transfer. A convenient measure of the compactness is the quantity $k = Lc/R$, where L is the source luminosity in a given spectral range, and R a typical size of the emitting region; this quantity is also accessible to observation, under the canonical assumption on the variability time scale $\Delta t \simeq R/c$.

Many authors have pointed out the relevance of k to the physics of AGN's, and have discussed limits to it deducible from the photon escape time[1], or the photon escape probability[2]. Also governed by k is the ratio of the competing time scales which describe the emitters' behavior both in thermal and non-thermal scenarios[3]; these are the collisional and bremsstrahlung times, and, respectively, the confinement and inverse-Compton times, but in any case obvious consistency requirements translate into a limit on the source compactness.

The viewpoint which we take in the present work is suggested by the general consensus that the optical continuum is of a synchrotron origin in at least a subclass of AGN's (OVV's in the first place[4]); then a canonical line of reasoning (see, e.g., the discussion by Burbidge[5]) shows how the relativistic electrons cannot live long enough to travel at speed c the source size R_o, and a reacceleration mechanism has to be postulated throughout the source volume.

Much more debated is the origin of the X-rays, for which no compelling evidence such as polarization is yet available. However, rather than on models, our argument mostly depends on observable properties of the X-ray emission; of particular importance is the source size R_x, which from variability data appears to obey $R_x \lesssim R_o$.

Assume now that the X-ray volume is not only smaller than, but also included within the optical source. The X-rays will be Compton scattered into the gamma-ray range by the optically emitting electrons; the emerging gamma-ray luminosity might be negligible because of the pair-producing collisions with the X-ray photons. This however would result in the injection of new particles into the acceleration volume, and a positive feedback would be established.

We will indicate a simplified way of computing the effect of the feedback, and find the range of k for which a steady solution exists.

A SIMPLE LIMITING CASE

The different kinds of particles are distinguished by means of the labels x (for X-ray photons), γ (for gamma-ray photons), and e (for electrons and positrons). Since outside R_x the target photon density is falling off as r^{-2}, we neglect there the inverse-Compton production of gamma-rays and their partial conversion into electron-positron pairs. Such processes are taking place, though, and our assuming $R_\gamma \simeq R_x$ will result in a lower limit to the magnitude of the effect.

We further neglect gradients and anisotropies of the various species within the respective radii, which amounts to a trivially simplified transfer problem, and to a straightforward relation between densities and luminosities. For instance, the specific energy density of the X-ray photons becomes

$$w_x = n_x h \nu_x \simeq \frac{L_x}{\pi R_x^2 c} \frac{1-\alpha}{\nu_2} \left(\frac{\nu_2}{\nu_x}\right)^\alpha \qquad \nu_x < \nu_2 \qquad (1)$$

As suggested by the observations, we have taken a power-law spectrum with energy index $\alpha \simeq 0.5 - 0.7 < 1$; the cutoff frequency ν_2 is generally unknown, and has to be approximated with the highest observed frequency, $\nu_2 \simeq 5$ KeV $- 50$ KeV $\ll (mc^2/h)$; note that, by using a lower limit on ν_2, one is again underestimating the effect of pair production.

If all electrons and positrons within $R_e = R_o$ are kept at relativistic energies by the distributed reacceleration mechanism, and if the power-law optical continuum is indicative of this relativistic population, then

$$\frac{dn_e}{d\gamma} = n_e \frac{p-1}{\gamma_1} \left(\frac{\gamma_1}{\gamma}\right)^p \qquad \gamma_1 < \gamma < \gamma_2, \; p \simeq 2 - 3 \qquad (2)$$

The total density n_e is obtained by balancing the particle outflow at R_o with the source terms; these in turn can be distinguished into a primary injection term, I, due to the central machine, and the secondary injection S connected with pair production

$$n_e = (I + S) / (\pi R_o^2 c) \qquad (3)$$

The energy density of the gamma-rays, w_γ, is proportional to the net volume emissivity, i.e. the difference between the inverse-Compton production, $\varepsilon_{\gamma+}$, and the photon-photon absorption, $\varepsilon_{\gamma-}$

$$w_\gamma = \frac{4}{3} \frac{R_\gamma}{c} (\varepsilon_{\gamma+} - \varepsilon_{\gamma-}) \qquad R_\gamma \simeq R_x \qquad (4)$$

The volume production rate is easily computed from Eqs. 1, 2, and 3 in the monochromatic approximation

$$\varepsilon_{\gamma+} = \frac{4}{3} c \, \sigma_T \frac{L_x}{\pi R_x^2 c} \frac{1-\alpha}{\nu_2} \left(\frac{\nu_2 \gamma_1^2}{\nu_\gamma}\right)^\alpha n_e (p-1) \left(\frac{mc^2 \gamma_1}{h\nu_\gamma}\right)^{p-1-2\alpha} *$$

$$* \left[\frac{1 - \left(\frac{h\nu_\gamma}{mc^2 \gamma_2}\right)^{p-1-2\alpha}}{p-1-2\alpha}\right] \; ; \quad \frac{h\nu_\gamma}{mc^2} \frac{h\nu_2}{mc^2} > 1, \; \gamma_2 > \frac{h\nu_\gamma}{mc^2} > \gamma_1 \qquad (5)$$

Here we restrict ourselves to gamma-rays capable of pair producing at least with the highest-frequency X-rays; this requirement implies that the gamma-ray spectrum is determined by the Klein-Nishina threshold, rather than the primaries' cutoffs. Note also that the square-bracket term is of order unity for $p > 1+2\alpha$, and needs not be retained in the following.

As for the volume absorption rate, we include the frequency dependence of the photon-photon cross-section in a form factor η, of order unity, and obtain

$$\varepsilon_{\gamma-} = \eta \, \sigma_T c \frac{L_x}{\pi R_x^2 c} \frac{1-\alpha}{h\nu_2} \left(\frac{h^2 \nu_2 \nu_\gamma}{m^2 c^4}\right)^\alpha w_\gamma \qquad (6)$$

After solving Eqs. 4, 5 and 6 for w_γ, we insert it in the right-hand side of Eq. 6, multiply by $(2/h \, \nu_\gamma)(4\pi/3)R_x^3$, and integrate with respect to ν_γ so as to obtain S; the integral is dominated by the contribution of the lower limit, where $\nu_\gamma = (mc^2/h)^2/\nu_2$, so that, within a form factor δ, we have

$$S = \frac{8\pi}{3} R_x^2 \frac{I+S}{\pi R_o^2} \frac{\delta}{n} (p-1) \left(\frac{mc^2}{h\nu_2 \gamma_1}\right)^{1-p} \frac{\tau_x^2}{1+\tau_x}$$

and

$$\frac{S}{I} = \left[\frac{3n}{8\delta(p-1)} \left(\frac{mc^2}{h\nu_2 \gamma_1}\right)^{p-1} \left(\frac{R_o}{R_x}\right)^2 \frac{1+\tau_x}{t_x^2} - 1\right]^{-1}$$

$$\tau_x = \frac{4}{3} n \sigma_T R_x \frac{L_x}{\pi R_x^2 c} \frac{1-\alpha}{h\nu_2} \qquad (7)$$

Physically meaningful solutions for S are only obtained if

$$\frac{\tau_x^2}{1+\tau_x} < \frac{3n}{8\delta(p-1)} \left(\frac{R_o}{R_x}\right)^2 \left(\frac{mc^2}{h\nu_2 \gamma_1}\right)^{p-1} \qquad (8)$$

The right-hand side is typically larger than 1, and the interesting cases will also have $\tau_x \gtrsim 1$; then $\delta \simeq 1/(p-1-\alpha)$, and Eq. 8 can be transformed into the following condition on k_x

$$k_x \lesssim \frac{2 \cdot 10^{39}}{\gamma_1} \text{ erg s}^{-1} \left(\frac{R_o}{R_x}\right)^2 \left(\frac{mc^2}{h\nu_2 \gamma_1}\right)^{p-2} \qquad (9)$$

DISCUSSION

While the dimensionless terms of Eq. 9 contain information on the geometry and physics of the source, the dimensional part is determined solely by the electromagnetic nature of the relevant interactions, and equals mc^4/σ_T.

The same dimensional expression was obviously found by the authors quoted in the Introduction. However, Fabian and Rees[1] used a general limit on the bolometric luminosity, due to the efficiency of gravitational engines, and ended up with the mass of the proton instead of the electron. Cavallo and Rees[2] had the electron mass in their formula, because they took into account the photon-photon interaction, as we also did; but their limit applies only to the compactness parameter at the reaction threshold, $\nu \simeq mc^2/h$, where the extragalactic observational material is exceedingly scarce.

On the contrary, our considerations involve the lighter particle - i.e., they result in a stronger constraint - and are applicable to the medium-to-hard X-ray range, where interesting observations already exist[6].

For instance, a possible model for the AGN continuum radiation invokes repeated scatterings of the primary synchrotron photons into successive Compton orders[7-9]; when this idea is developed within our assumptions, the X-rays become Compton-scattered optical photons, so that $R_o \simeq R_x$ and $p = 1+2\alpha \simeq 2$. Even with the most favorable choice $\gamma_1 \simeq 1$, Eq. 9 would give a steady-state limit which is known to have been violated in certain outbursts[10-13].

If only a fraction f of the pairs produced can be recycled as emitters, the limit of Eq. 9 is raised by 1/f. All this indicates that efficient inverse-Compton scatterings require either a structure of the active region more complex than a simple sphere (uniform, isotropic, steady), or else, relativistic bulk motions.

An alternative, attractive possibility is that no steady state is available to the system. A diverging production of pairs may eventually quench the reacceleration mechanism, if the power output of the latter is limited; the typical particle energy is then expected to drop to a level, where pairs are no longer produced, and the system is perhaps restored to the initial conditions.

On the observational side, we stress the importance of monitoring the AGN's in hard X-rays; a rise in the observed value of ν_2 would lower the right-hand side of Eq. 9, while, at a constant R_x, the left-hand side is expected to increase because of the general flatness of AGN spectra. Obviously, a conflict between observations and Eq. 9 in its more conservative form ($p \simeq 3$) would have far-reaching, almost model-independent implications.

REFERENCES

1. A. Fabian and M.J. Rees, X-ray Astronomy, XXI COSPAR Plenary, W.A. Baity and L.E. Peterson eds. (1979), p. 381.
2. G. Cavallo and M.J. Rees, M.N.R.A.S. 183, 359 (1978).
3. A. Cavaliere, International School and Workshop on Plasma Astrophysics, ESA SP-161 (1982), p. 97.
4. J.R.P. Angel and H.S. Stockman, Ann. Rev. Astron. Astrophys. 18, 321 (1980).
5. G.R. Burbidge, Phys. Scripta 17, 281 (1978).
6. K.A. Pounds, Variability in Stars and Galaxies, V European Regional Meeting (1980), p.C.5.1.
7. R.F. Mushotzky, P.J. Serlemitsos, R.H. Becker, E.A. Boldt and S.S. Holt, Astrophys. J. 220, 790 (1978).
8. F. Pacini and M. Salvati, Astrophys. J. 225, L99 (1978).
9. A. Cavaliere and P. Morrison, Astrophys. J. 238, L63 (1980).
10. H. Tananbaum et al., Astrophys. J. 234, L9 (1979).
11. H. Tananbaum, X-ray Astronomy, Giacconi and Setti eds. (1980), p.291.
12. A.F. Tennant, R.F. Mushotzky, E.A. Boldt and J.H. Swank, Astrophys. J. 251, 15 (1981).
13. T. Matilsky, C. Shrader and H. Tananbaum, Astrophys. J. 258, L1 (1982).

ELECTRON-POSITRON PROCESSES AND SPECTRAL EVOLUTION IN BLACK HOLE ACCRETION DISK DYNAMO MODELS FOR AGN SOURCES OF THE COSMIC X-RAY AND GAMMA RAY BACKGROUNDS

Darryl Leiter[*]
Laboratory for High Energy Astrophysics
NASA Goddard Space Flight Center, Greenbelt, Maryland 20771

INTRODUCTION TO THE AGN SPECTRAL EVOLUTION MODEL

This work discusses a black hole accretion disk dynamo model for Active Galactic Nuclei (AGN) sources of the cosmic X-ray and gamma ray backgrounds which involves both thermal and nonthermal accretion disk processes around $\gtrsim 10^8 M_\odot$ Kerr black holes (Leiter and Boldt (1) referred to henceforth as Paper I). Before black hole spin-up to the Kerr metric state, the large value of the compactness parameter L(luminosity)/r(size of emitting region) $> 10^{30}$ erg/cm-sec associated with the $L/L_{Edd} \lesssim 1$ luminosity ratio in Precursor Active Galaxies (PAG) suppresses all nonthermal emission mechanisms. In this PAG state the resulting emission is predominantly thermal and is due to Comptonization of soft photons by an electron-positron plasma, generated within the hot accretion disk region by $\gamma+\gamma \longleftrightarrow e^\pm$ processes in the transrelativistic regime. While the underlying plasma in the PAG accretion disk hot inner region may be optically thin initially, the overall effect of the copious $\gamma+\gamma \longleftrightarrow e^\pm$ generated electron-positron plasma is to push the overall optical depth to $\tau \gtrsim 1$. This has two main effects: a) it causes the resulting Comptonized spectrum of X-radiation from PAG to be associated with a flat spectral index comparable to that of the residual Cosmic X-ray Background (CXB), and b) the copious $\gamma+\gamma \longleftrightarrow e^\pm$ within the hot accretion disk region play the role of a phase transition thermostat, and act to maintain the temperature of the hot inner region at $\gtrsim 10^9$ °K.

After the $\gtrsim 10^8$ year time period required for the $L/L_{Edd} \lesssim 1$ Eddington limited black hole spin-up into a Kerr metric, the subsequent growth of extended jet structure (or the transition to a supply limited accretion regime) reduces the compactness parameter to $L/r < 10^{30}$ erg/cm-sec. Then nonthermal accretion disk dynamo emission mechanisms

(*)Now at Sciences Division, FSTC, Charlottesville, VA.

0094-243X/83/1010337-06 $3.00 Copyright 1983 American Institute of Physics

are switched on and act to generate a broad band of nonthermal emission extending into the gamma ray region at the expense of cooled thermal disk emission. This results in a pronounced spectral and luminosity evolution of the radiation emitted by the black hole accretion disk dynamo system, causing it to become typical of the steep power laws seen for AGN in the present epoch. This model leads to a unified picture of the X-ray and gamma ray backgrounds, in which a superposition of PAG sources can account for that major portion of the CXB remaining after the contribution of usual AGN are considered, while a superposition of AGN sources $z<1$ can account for the gamma ray background.

THE $\gamma+\gamma \leftrightarrow e^{\pm}$ THERMAL SPECTRUM THERMOSTAT IN PRECURSOR ACTIVE GALAXIES (PAG)

A basic assumption of Paper I is that the cosmic X-ray background radiation is associated with supermassive black holes, $M \gtrsim 10^8 M_\odot$, which acquire accretion disks for the first time during the epoch of galaxy formation $z \sim 4$. The black holes are generated by primordial processes preceding galaxy formation. However, after the acquisition of accretion disks during galaxy formation, they become the seeds which distinguish PAG from galaxies which will evolve normally. In PAG the luminosity ratio $L/L_{Edd} \lesssim 1$ causes the compactness parameter of the accretion disk dynamics to have the value $L/r > 10^{30}$ erg/cm-sec. Hence the hot inner region of the PAG accretion disk, to which this value of the compactness parameter applies (i.e., $r \lesssim 100 r_g$ where $r_g = GM/c^2$) produces only a thermal emission spectrum. As shown by Cavaliere and Morrison (2), the reason for this is due to the fact that in such a situation the energy losses due to inelastic particle production processes will exceed the energy input from nonthermal accretion disk dynamo mechanisms (see Blandford (3), Lovelace et. al. (4), Thorne et. al. (5)).

Under these circumstances the radiated thermal spectrum will be due predominantly to Comptonization of soft photons by a transrelativistic electron-positron plasma in the hot inner region of the accretion disk, where copious $\gamma+\gamma \leftrightarrow e^{\pm}$ processes act as a phase transition thermostat, which maintains the temperature of the hot inner region at $\gtrsim 10^9$ °K.

Lightman (6), Araki and Lightman (7) and Takahara

and Tsuruta (8), have studied cases involving hot inner disk regions where bremsstrahlung and cyclotron processes are the source of soft photons to be Comptonized. Leiter and Boldt (Paper I) and Liang (9) have studied the case where the soft photons are generated by the cool external regions of the accretion disk.

In the most general case we expect that the soft photons will be supplied by all three sources (bremsstrahlung, cyclotron and cool external disk) in the PAG accretion disk system. Hence in the following we will show that all three cases have parameter spaces compatible with the observed residual (CXB) spectrum given by the very flat spectrum $I_E \sim E^{-\alpha}, \alpha \lesssim 0.2$ at energies well below the source temperature $kT \lesssim 150$ KeV.

Let us first consider the case of bremsstrahlung and cyclotron soft photon sources within the hot inner regions of the PAG accretion disk. In this case the main parameters controlling the spectrum are: the size of the hot inner region $r(cm)$, the magnetic field present in the hot inner region $B(gauss)$, and the overall optical depth τ in the electron-positron dominated plasma produced by the $\gamma + \gamma \leftrightarrow e^{\pm}$ process. The formula for this latter optical depth is given by

$$\tau = \tau_{es}\left((N_+ + N_-)/N\right) \quad (1)$$

where N, N_+ and N_- are the proton, positron and electron densities respectively, $\tau_{es} = \sigma r N$ is the electron scattering optical depth. Initially the underlying plasma is taken to be optically thin, consistent with the hot inner region of the accretion disk, such that $\tau_{es} < 1$. As the luminosity parameter (L/L_{Edd}) approaches $\lesssim 1$, it is found (6,7) that the Eddington limited range of energy productions lies within the broad range of energies covered by the transrelativistic regime. There it is found that copious $\gamma + \gamma \leftrightarrow e^{\pm}$ processes dominate the plasma, causing $\tau \gtrsim 1$ to occur while keeping the temperature within the range $0.1 \lesssim kT/mc^2 \lesssim 1.0$. Additional magnetic fields and/or external sources of soft photons tend to favor the lower end of this temperature range. Since the Comptonized spectrum is independent of whether normal plasmas or electron-positron plasmas are doing the Comptonization (the only difference being the possible presence in the latter case of a weak annihilation line contribution), we can use the work of McKinley and Ramaty (10), Lorentz (11), and Pozdnyakov (12) to deduce the fact that the Comptonized spectrum from an $e^{\pm}$ plasma with $\tau \gtrsim 1$ and

$kT/mc^2 \gtrsim 0.1$ will have an effective spectral index $\alpha \leq 0.2$ for $E < kT$.

Secondly we consider cool external disk sources of soft photons. The case of the soft photon source coming from the cool outer regions of the accretion disk, which surround the hot inner region where the Comptonization takes place, was discussed in Paper I. From that work it can be shown that for $L/L_{Edd} \leq 1$ and Kompaneets parameter $y = (4kT/mc^2) \text{Max}(\tau, \tau^2) \geq 3$ in a hot inner region of size $r \leq 10^{15}$ cm within an accretion disk of a PAG, the Comptonized spectrum will again be associated with a spectral index for $E < kT$ of $\alpha \leq 0.2$. Similarly, the effects of copious $\gamma + \gamma \leftrightarrow e^{\pm}$ production act as a thermostat to maintain the temperature $kT/mc^2 \gtrsim 0.1$.

Hence, if soft photons are supplied by bremsstrahlung, cyclotron and external sources of soft photons in a PAG accretion disk operating at $L/L_{Edd} \leq 1$, then the resulting Comptonized spectrum associated with the copious $\gamma + \gamma \leftrightarrow e^{\pm}$ dominated plasma in the hot inner region of the PAG accretion disk will have a thermostatically controlled temperature $kT/mc^2 \gtrsim 0.1$ and a spectral index $\alpha \leq 0.2$ for $E < kT$, consistent with that of the residual CXB. In this way a superposition of these PAG objects can account for the residual CXB observed (see Paper I).

THE (LUMINOSITY/SIZE) NONTHERMAL SPECTRAL SWITCH TO THE AGN STAGE

In a time period of $\geq 10^8$ years, the $L/L_{Edd} \leq 1$ PAG black hole accretion disk system will spin up into an $a/M = 0.998$ Kerr metric limit. This initiates the Penrose Compton Scattering of accretion disk X-rays into MeV gamma rays above and below the X-ray emitting accretion disk (Leiter, 13). The resultant $\gamma + X \rightarrow e^{\pm}$ above and below the disk act to trigger and surge accretion disk dynamo mechanisms, such as that of Blandford (3), Lovelace (4), Thorne (5), Rees (14) and Phinney (15). This sets the stage for the development of a broad band of nonthermal emission from the accretion disk dynamo system as the compactness parameter L(luminosity)/r (size) is reduced by the growth of extended jet structure, or by the transition of the system into a supply limited accretion regime. The former case would be associated with elliptical PAG evolution into quasars, while the latter would be associated with the spiral PAG evolution into Seyfert galaxies (see Paper I).

In either case the compactness parameter reduction to $(L/r) < 10^{30}$ erg/cm-sec increases the efficiency of the nonthermal emission of the accretion disk dynamo, switching the spectral output to a nonthermal AGN type at the expense of cooled thermal disk emission. The spectral switching time for black hole accretion disk dynamo PAG ranges from $\sim 10^8$ years to $\sim 10^9$ years, depending on the morphology and the dynamics of the PAG involved (see Paper I).

Since the nonthermal emission extends into the gamma ray band, a superposition of AGN in this nonthermal phase can account for the extragalactic gamma ray background. A unique prediction of this model is a ≤ 1 day variability in the $\leq$ 3MeV gamma ray background, which may be observable with the GRO Compton telescope. This would represent a possible test of the theory in the gamma ray region. Another test would be to search for optical signatures of the precursor states of AGN (associated with the optical emission lines generated by the flat $\alpha \leq 0.2$ X-ray spectrum they emit) before spectral switching occurs (Boldt and Leiter, 16). Ultimately the best test for the validity of the spectral evolution concept would require imaging the precursor AGN directly and would involve the AXAF X-ray telescope projected for the 1990's.

ACKNOWLEDGEMENT

The author thanks Elihu Boldt of the Laboratory for High Energy Astrophysics, NASA Goddard Space Flight Center, for his many useful suggestions in the development of the ideas in this manuscript.

REFERENCES

1. Leiter, D., Boldt, E., AP.J., 260, 1, (1982); see also Boldt, E., Leiter, D., NATURE, 290, 438, (1981).
2. Cavaliere, A., Morrison, P, AP. J. LETT., 238, L63, (1980).
3. Blandford, R., Active Galactic Nuclei, (eds. Hazard and Mitton), (1979), 241.
4. Lovelace, R., Proc. La Jolla Inst., #53, (1979), 399.
5. Thorne, K.S., MacDonald, D., MNRAS, 198, 345, (1982).
6. Lightman, A., AP. J., 253, 842, (1982); see also Lightman's contribution in this volume.

7. Araki, S., Lightman, A., Preprint, "Center for Astrophysics," #1712, (1983).
8. Takahara, F., Tsuruta, S., PROG. THEO PHYS., $\underline{67}$, 485, (1982).
9. Liang, E.P.T., AP. J., $\underline{234}$, 1105, (1979).
10. McKinley, J., Ramaty, R., AP. J., in preparation, (1983).
11. Lorentz, M., Preprint, "Max Plank Institute," ISSN 0340-8922, (1981).
12. Pozdnyakov, L.A., et. al., SOVIET ASTRONOMY, $\underline{21(6)}$, 708, (1977).
13. Leiter, D., AST. & AP., $\underline{89}$, 370, (1980).
14. Rees, M.J., Begelman, M.C., Blandford, R.D., Phinney, E.S., NATURE, $\underline{295}$, 17.
15. Phinney, E.S., <u>Proc. Torino Workshop on Astrophysical Jets</u>, (1982), to be published by Reidel.

16. Boldt, E., Leiter, D., AP.J., in preparation (1983).

e^+-e^- ANNIHILATION AND THE COSMIC X-RAY BACKGROUND

Demosthenes Kazanas and Richard Arrick Shafer
NASA/Goddard Space Flight Center
Greenbelt, MD. 20771
and
Department of Physics
University of Maryland

ABSTRACT

The possibility that the processes responsible for the Cosmic X-ray Background (CXB) would also produce an e^--e^+ annihilation feature is examined. Under the assumption that these processes are thermal, the absence of a strong e^--e^+ annihilation feature places constraints on the compactness (L/R ratio) of these sources. Observations favor sources of small compactness ratio.

INTRODUCTION

The fact that the X-ray sky is dominated by an isotropic component (the so called Cosmic X-ray Background, hereafter CXB) had been established by the earliest X-ray astronomy observations[1]. The subsequent satellite X-ray observations, especially by the A-2 and A-4 experiments on HEAO 1[2,3], allowed the detailed spectral determination of CXB. The observed spectrum in the region 3-150 keV, along with the higher energy data is shown in Fig 1. The HEAO 1 experimenters have found that thermal bremsstrahlung from an optically thin plasma of temperature 40 ± 5 keV provides a remarkably good fit to the data from 3 to 100 keV. Interestingly enough, no studied population of sources is known to have a thermal spectrum with the required properties.

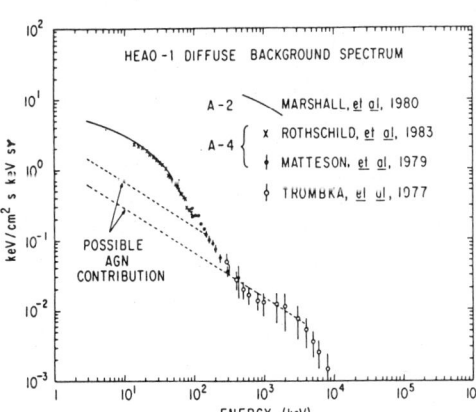

Fig. 1 - The unresolved X and gamma-ray background. (From refs [2,3,12,13])

One can of course contrive to combine sources with a variety of spectra emitting over a range of red shifts to produce the observed total background spectrum, even if the individual spectra are different than that of CXB[4]. The shape of of the spectrum however clearly suggests a thermal distribution of rather specific temperature and it would be more natural if the CXB could be explained as such.

A thin thermal bremsstrahlung from a heated intergalactic medium will provide such a spectrum [5,6]. However, the energy required to heat a diffuse uniform medium to such a temperature is quite large. Also, the low densities and consequent long electron-ion coupling times provide difficulties in attaining and maintaining a Maxwellian distribution for the medium[7]. Clumping of the medium would reduce the input energy requirements and the coupling time scales. Taken to extremes, the clumps might be reduced to a size comparable to galaxies, or smaller, becoming "compact" sources of X-rays.

For either the heated intergalactic medium or the "compact source" models for the CXB, the bulk of the emission originates at redshifts $\sim$ 2-3 (or even larger) so that the corresponding source temperature would be $kT \gtrsim$ 100-200 keV. For these temperatures a sufficiently compact source will produce electron-positron pairs from the tails of the photon and particle thermal distributions. Under certain conditions the positron abundance would be sufficient to produce an observable e^+-e^- annihilation feature in the CXB.

THE POSITRON ABUNDANCE

In a thermal plasma of temperature $kT \gtrsim$ 100 keV it is possible to produce positrons at significant abundances by ee, eγ or $\gamma\gamma$ collisions since a non-negligible fraction of the particles and photons at the tails of the distributions fulfills the pair production threshold condition. Their steady state abundance is determined by the balance between pair production and annihilation reactions (for a detailed treatment see Ref. 8).

In the cases of interest for the CXB the dominant positron production is due to $\gamma\gamma$ reactions so the ee, eγ reactions will not be considered further. (Their cross sections are correspondingly smaller by α^2 and α). The approximate expressions for the production and annihilation rates are:

$$R_{\gamma\gamma} \simeq \frac{3}{16\pi} \sigma_T c K_\gamma^2 f(x) , \qquad (1)$$

$$R_{+-} = \frac{3}{8} \sigma_T c\, n_+ n_- ,$$

where n_+, n_-, are the positron and electron number densities, σ_T the Thomson cross section and $x = 2\, m_e c^2/kT$. $f(x)$ is a function of the temperature only, resulting from the averaging of the photon-photon pair production reaction rate over the photon distribution functions. It is given by

$$f(x) = c_1 \frac{e^{-x}}{x(x+c_2)},$$

where $c_1 = 1.143$ and $c_2 = 3.63$ are constants. K_γ is the normalization of the bremsstrahlung photon distribution function determined by the condition

$$\int_0^\infty \frac{K_\gamma}{E} e^{-E/kT} E \, dE = \frac{L(1+\tau)}{\pi R^2 c} \text{ or}$$

$$K_\gamma = \frac{L(1+\tau)}{kT \pi R^2 c} ,$$

(2)

where L, R, τ, T are respectively the luminosity, radius, optical depth and temperature of the sources. The steady state condition $R_{\gamma\gamma} = R_{+-}$ gives

$$\frac{1}{2\pi} \left(\frac{K_\gamma}{n_+ + n_-}\right)^2 f(x) = \frac{n_+ n_-}{(n_+ + n_-)^2} .$$

(3)

Since the compact sources presently considered are presumably gravitationally bound, one can scale L and R by their gravitational units i.e. the Eddington luminosity ($L_E = 10^{38}$ M erg s^{-1}) and the Schwarzschild radius ($R_s = 3 \times 10^5$ M cm) where M is the mass of the sources in solar masses. For this we introduce two parameters F and f, both with expected values between 0 and 1, defined as

$$F \equiv L/L_E \text{ and } f \equiv R_s/R .$$

Their product Ff ∝ L/R is a mass independent measure of their compactness. A high value of Ff corresponds to a high density of photons and hence e^+e^- pairs. The additional factor of R in K_γ can be expressed in terms of the optical depth τ, since $(n_+ + n_-) \sigma_T R \simeq$ ≃ sτ where s is the aspect ratio of the source, i.e. the ratio of its largest to its shortest dimension. We can therefore reexpress eq (3) in terms of the positron abundance $\lambda \equiv n_+/n_-$ by:

$$2 \times 10^3 \text{ Ff } e^{-x/2} \left(\frac{x}{x+c_2}\right)^{1/2} \frac{\tau+1}{s\tau} = \frac{\lambda^{1/2}}{1+\lambda}.$$

(4)

COMPARISON TO OBSERVATION

Equation (4) can be directly related to observations of the CXB. Let A be the ratio of the annihilation to the bremsstrahlung spectral luminosity at the peak energy of the annihilation feature, $E \simeq m_e + kT$. Since no obvious annihilation feature is observed in the CXB A $\lesssim$ 1. Using the results of Ref. 9, one can relate λ to A and x by

$$\frac{\lambda^{1/2}}{1+\lambda} \lesssim 0.134 \, A^{1/2} \, x^{-1/4} \, e^{-x/4} .$$

(5)

Elimination of λ between (4) and (5) gives

$$\text{Ff} \lesssim 1.7 \times 10^{-4} \, e^{x/4} \cdot \frac{(x+c_2)^{1/2}}{x^{3/4}} A^{1/2} \frac{\tau s}{\tau+1} .$$

(6)

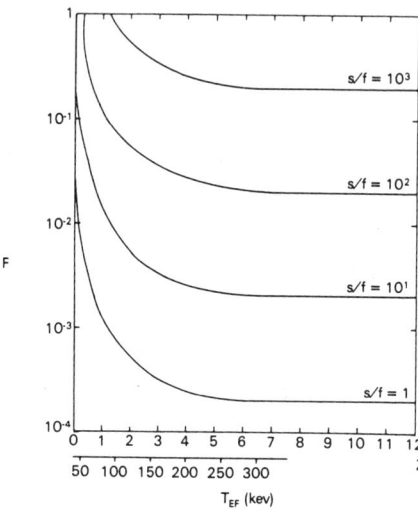

Eq (6) is independent of the mass of the sources and limits (as a function of the redshift Z at which the CXB was produced) their L/R ratio to be compatible with the absence of a prominent annihilation feature in the CXB. The optical depth τ of the sources is unknown, however it is constrained to be $\tau \lesssim 3$ otherwise the corresponding self-Comptonization Wien peak should be apparent in the spectrum[10]. Another unknown parameter of the sources is their aspect ratio s. For spherical sources $s \simeq 1$, although it could be s>>1 for sources of thin disk geometry. It appears, however, that $s \simeq 1$ for sources emitting close to their Eddington luminosity[11]. The constraints imposed by Eq. (6) are shown in Fig 2. For example, the absence

Fig. 2. Limits on source luminosity F (in units of L_E) vs. redshift of CXB production, z, for different source effective sizes, s/f (in units of R_s). The temperature is the source temperature in the emitter frame.

of a prominent annihilation feature at 100-150 keV, constrains compact sources ($f \simeq 0.1$) operating at $z \simeq 4$ to emit at a small fraction of their Eddington luminosity. Due to the absence of positive detection of such a feature, all the derived constraints are in the form of inequalities which however may prove important in understanding the nature of the sources of the CXB. Fig. 3 shows the superposition of the energy spectrum of thermal bremsstrahlung with T = 200 keV and an annihilation feature of the same temperature and A = 1. Such a feature appears detectable, though unfortunately in this example it peaks in the observer frame energy range $E \gtrsim 100$ keV, which is probably dominated not by the thermal component but by the contribution of active galaxies. This is currently very poorly known and we may always have to be content with the existing upper limits. A positive detection of an annihilation feature in the CXB spectrum, rather than the above qualitative upper bound, would provide an important contribution to the understanding of the CXB's origin. Such a detection will determine precisely the emitter-frame temperature (and hence the redshift) of the sources, it will signify that they are compact and it will provide a consistency check for the temperature of the thermal component. If this feature is not detectable in the CXB spectrum as a whole, it may someday be studied in individual spectra of compact objects, if such objects make up the background. However such studies will probably require high energy imaging capabilities beyond those planned for the next generation of experiments.

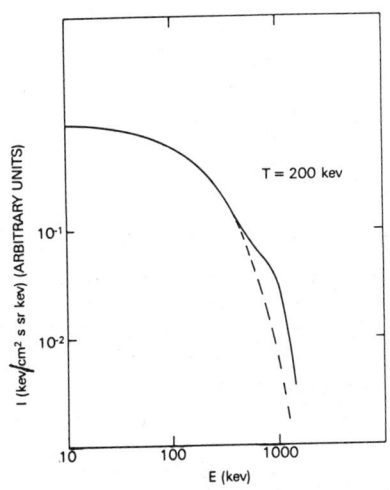

Fig. 3. Superposition of bremsstrahlung and annihilation radiation energy spectrum for A = 1 and T = 200 keV.

REFERENCES

1. R. Giacconi, J. Gursky, F. Paolini, B. Rossi, Phys. Rev. Lett., 9, 439 (1962).
2. F. E. Marshall, E. A. Boldt, S. Holt, R. Miller, R. F. Mushotzky, L. A. Rose, R. E. Rothschild and P. Serlemitsos, Ap. J., 235, 377 (1980).
3. R. E. Rothschild, R. F. Mushotzky, W. A. Baity, D. E. Gruber and J. L. Matteson, Ap.J., (1983) (in press).
4. G. DeZotti, E. A. Boldt, A. Cavaliere, L. Danese, A. Francischini, F. E. Marshall, J. H. Swank and A. J. Szymkoviak, Ap. J., 253, 47 (1982).
5. R. Cowsik and E. J. Kobetich, Ap.J., 177, 585 (1972).
6. G. B. Field and S. Perrenod, Ap.J., 215, 717 (1977).
7. A. C. Fabian and A. K. Kembhavi, in Heeschen and Wade (eds.) IAU Symposium 97 on Extragalactic Radio Sources, (Reidel; Dordrecht, 1982) p. 453.
8. R. Svensson, Ap.J., 258, 335 (1982).
9. R. Ramaty and P. Meszaros, Ap.J., 250, 384 (1981).
10. A. D. Lightman and D. L. Band, Ap. J., 251, 715 (1981).
11. J. Eilek, Ap.J., 236, 664 (1980).
12. J. L. Matteson, D. E. Gruber, P. L. Nolan, L. E. Peterson and R. L. Kinzer, BAAS, 11, 653 (1979).
13. J. I. Trombka, C. S. Dyer, L. G. Evans, M. J. Bielfeld, S. M. Seltzer and A. E. Metzger, Ap.J., 212, 925 (1977).

HIGH ENERGY SPECTRUM OF SPHERICALLY ACCRETING BLACK HOLES

P. Mészáros*
Harvard-Smithsonian Center for Astrophysics,
Cambridge, Massachusetts 02138

J. P. Ostriker
Princeton University Observatory,
Princeton, New Jersey 08540

ABSTRACT

Spherically accreting black holes may sustain strong collisionless shocks, downstream of which the fluid approximation is not valid. The proton-electron Coulomb exchange provides for the downstream matter diffusion into the hole. Energy conversion efficiencies upward of 10-30% are obtained, with most of the luminosity in hard X-rays and γ-rays. We discuss the whole spectrum and its application for radio-quiet QSO's and galactic X- and γ-ray sources.

INTRODUCTION

The gas accreted by black holes may often not be able to form a disk. In this case quasi-spherical infall occurs, and the efficiency of radiation is $<10^{-11}$, far below disk efficiencies, if there is no dissipation.[1,2] The presence of equipartition magnetic fields in the accretion flows would ensure $\gtrsim 10\%$ efficiency,[3] but it is not clear that fields grow to that strength. We explore here the complementary possibility, that magnetic fields disappear so efficiently that they do not play any role either in the spectrum or the dynamics.

We consider a situation with supersonic freefall at large distances from the horizon. Any small temporary perturbation or instability will cause a shock to arise in the flow. This converts the upstream <u>radial</u> energy of the protons into <u>random</u> energy downstream. If magnetic fields are very small, the protons and electrons below the shock will not behave as a fluid but as particles (as in the problem of a star cluster with a central black hole). If the shock distance is a large multiple of the horizon radius, only a small fraction of the heated protons will be captured by the hole right away, and the rest will orbit around the hole with different eccentricities, but with a maximum elongation $\lesssim$ initial shock distance. This heating and randomization will tend to sphericize the shock, if it was not already so initially. A self-sustaining situation can be achieved, since the random downstream motions provide an effective quasi-collisionless pressure "p" $= \rho \overline{v}^2_{rand}$, as in stellar dynamics. This back-pressure, satisfying the shock jump conditions, holds the shock at a particular radius r_s. An initial shock at a different radius would tend towards this equilibrium radius. Thus,

---

*Visiting Scientist, Smithsonian Astrophysical Observatory, supported by NASA grant NAGW-246. On leave from Max-Planck-Institut fuer Physik und Astrophysik MPA.

while a black hole presents no solid surface to stop the gas, a shock can be sustained nonetheless, because the particles downstream have mean free paths much larger than the horizon, and most of them miss it. They can only drift inwards on the Coulomb diffusion time, which is much longer than the upstream free-fall time, causing a pile-up below the shock.

INFALL, SHOCK, AND DIFFUSION

Below the accretion radius $r_a \sim 2GMc_s^{-1}$, where c_s is sound speed at $r \gg r_a$, the matter is in approximate spherical infall. The upstream random interstellar magnetic fields may either be left behind, due to buoyancy of field loops, or if entrained, we shall assume them to be signficantly below equipartition with respect to the infall kinetic energy, so that the motion is supersonic. The upstream gas ($r_s < r < r_a$) may or may not, depending on the field strength and temperature, satisfy the fluid approximation. The essential thing is that it will be in supersonic (radial) free-fall. Since at postshock temperatures the Coulomb mean free path $\lambda_{pe} \gtrsim r$, the shock must be a collisionless shock. Electrostatic instabilities may cause such a shock at low Mach numbers, while other instabilities have been considered at higher Mach numbers.[4] The shock transition width is $l_s \sim v_{ff}\Omega_{op}^{-1}$, where Ω_{op} is the proton plasma frequency, so $l_s \lll r$. Turbulent electric and magnetic fields will exist at the shock, varying over lengthscales l_s, of magnitude[5] (Gaussian units)

$$E_s \sim (c/v_e)B_s \sim \delta\phi\, l_s^{-1} \sim 3kT_e(el_s)^{-1} \sim 3\ 10^4\ m_8^{-1/2}\ \dot{m}^{1/2}\ \tilde{r}^{-5/4}. \quad (1)$$

Here $\tilde{r} = r/(2GMc^{-2})$, $m_8 = (M_{BH}/10^8\ M_\odot)$, $\dot{m} = (\dot{M}c^2/2L_{ED})$. The conductivity is $\sigma = jE^{-1} \lesssim n_e ecE^{-1} \sim cv_p^{-1}\Omega_{op}$, so the dissipation time for the fields caused by the shock is $t_B \sim 4\pi\sigma l_s^2/c^2 \sim 10^{-9}\ m_8^{1/2}\ \dot{m}^{-1/2}\ \tilde{r}^{-1/4} \ll t_{dyn}$. Similarly, if the upstream gas carried any large scale magnetic field, these would be well below equipartition by hypothesis, whereas the turbulent E_s, B_s are by definition at equipartition. The turbulent currents will then twist and impose a structure of scale l_s to the large scale field, which allows reconnection and dissipation on the same timescale $t_B \ll t_{dyn}$. Therefore, the downstream material will be essentially field-free, and the nonfluid approximation downstream is justified.

Below the shock, protons and electrons come up to the adiabatic postshock temperature, with

$$T_p \sim 2\ 10^{12}\ \tilde{r}^{-1}\ (K)\ . \quad (2)$$

The electrons rapidly cool from $T_e \sim T_p$ to $T_e \ll T_p$ due to synchrotron and Compton losses in the shock. Behind the shock, the electrons seek temperature equilibrium between Coulomb heat gains from protons and bremsstrahlung losses, stabilizing near

$$T_e \sim \theta_e(m_ec^2)/k \sim 6\ 10^9\ \theta_e\ (K)\ , \quad (3)$$

with $\theta_e \lesssim 1$. The onset of enhanced losses due to relativistic effects, Comptonization, and pair formation ensures that T_e remains below (3). Electron-electron heat conduction ($t_{ee} \ll t_{dyn} \sim t_{pe}$) ensures practically uniform T_e throughout $\tilde{r} < \tilde{r}_s$ = shock radius, while T_p obeys (2). The relaxation zone Δr_t over which T_e goes from the shock value to (3) is of width $(\Delta r_t/r_s) \sim T_e/T_p \ll 1$ [cf. equations (2) and (3)].

Below $r_s - \Delta r_t \sim r_s$ the matter is quasi-collisionless and diffuses inward on a timescale $t_{dyn} \sim \min(t_{pp}, t_{pe})$. For $\tilde{r}_s < 10^2$, as we find, and T_p, T_e given by (2),(3) the shortest timescale is the proton-electron Coulomb time[6]

$$t_{pe} \sim \frac{m_p c^2 \theta_e}{n_e \sigma_T m_e c^3 \ln \Lambda} = 1.5\ 10^{-8}\ \theta_e\ \rho^{-1}\ s \quad . \quad (4)$$

We apply jump conditions across the relaxation region Δr_t, assuming that downstream the random velocities $\overline{V_p^2}$ are characterized by (2), and the diffusion (bulk) velocity $u \ll V_p$ is $u \sim r\,t_{pe}^{-1}$. The jump conditions yield the shock radius

$$\tilde{r}_s \simeq 50\ \dot{m}^{1/2}\ \theta_2^{-3/2} \quad . \quad (5)$$

After undergoing a jump in density across Δr_t of $\rho_2/\rho_1 = V_{ff}/u_2 \simeq 1.7\ 10^1\ \dot{m}^{-1/2}\ \theta_2^{1/2}$, the downstream gas density is $n_p(\tilde{r}) = 8\ 10^{11}\ m_8^{-1} \dot{m}^{1/2}\ \theta_2^{1/2}\ \tilde{r}^{-3/2}$ and the downstream Thomson optical depth is $\tau_T \sim n_p \sigma_T r_s \sim 2.5\ \dot{m}^{1/4}\ \theta^{5/4}$, of order unity.

OBSERVABLE CONSEQUENCES

The efficiency of conversion of gravitational energy into radiation, ε, is close to unity. This is inherent in the fact that the protons move inward by giving up their energy to the electrons, which immediately radiate it away. Since our classical diffusion treatment probably breaks down at some $\tilde{r}_L \gtrsim 3$, due to G.R. and other effects (cf. below), the actual efficiency should be $\varepsilon \sim 0.1-0.3$.

The spectrum will extend from the IR-optical range to the far γ-ray region. The shock transition gives rise to synchrotron photons, of frequency characterized by B_s (eq. 1) and the adiabatic electron Lorentz factor $\gamma \sim (kT_e/m_ec^2)_s \sim kT_p/m_ec^2$, producing photons at

$$\nu_m \sim (eB/2\pi m_e c)\gamma^2 \gtrsim 10^{12}\ m_8^{-1/2}\ \dot{m}^{-9/8}\ (Hz) \quad . \quad (6)$$

Since half of these escape downwards and are comptonized ($\tau_T \gtrsim 1$) on the hot ($\theta_e \sim 1$) subshock electrons, a power law $\nu^{-\alpha}$, with $\alpha \sim 0.7-1.5$ is expected bluewards of ν_m. A fraction $\sim (r_L/r_s)$ of the total luminosity comes out in this component. An additional thermal UV component ($T_{bb} \sim$ few 10^4) may appear superposed on this, due to infalling clouds, which further out gave rise to lines and became opaque closer in.

X-rays with $h\nu \lesssim 130$ keV will be produced by the upstream freely falling gas, which has $\rho \sim r^{-3/2}$ and $T_e \sim T_1(r_s/r)$. This gas cools by bremsstrahlung, and is heated by Compton collisions with hard photons coming from $r < r_s$, $\langle h\nu_\gamma \rangle \sim 500$ keV, so $T_1 \simeq \langle h\nu_\gamma \rangle/k \sim 130$ keV. The spectrum produced has a low frequency asymptote (several decades below T_1, i.e., below 1 keV) going as $\nu^{0.5}$. Between this asymptote and $h\nu \sim kT_1 \sim 130$ keV there is a very gradual bendover towards an exponential fall-off which sets in above 150-200 keV. This bendover is so gradual that over the range 1-100 keV the spectrum that would be inferred is, to a very good approximation, a power law $F_\nu \sim \nu^{-\alpha}$. Depending on the window sampled and the exact T_1, α is 0.6-0.7, as observed for most QSO's and Seyferts. Since the fraction of hard photons scattered is τ_T(upstream) $\sim 2\dot{m}\tilde{r}_s^{-1/2}$, this is also the fraction of the total energy in this component, $(2/7)\dot{m}^{1/2}$.

Another hard X-ray component will be produced by the downstream hot gas, at $T_e \sim \theta_e(m_e c^2/k) \lesssim 511$ keV $\sim$ constant, via ee and ep bremsstrahlung. This will be comptonized, since $\theta \sim 1$, $\tau_T \gtrsim 1$, consisting mostly of photons at $\langle h\nu \rangle \sim (1-3)\theta_e m_e c^2 \sim 0.5$-$1.5$ MeV. A broad excess above 0.5 MeV is also expected due to annihilation of pairs (e.g., Reference 7), which may represent a fraction of the total electron density. Such an excess also may appear due to photon-photon absorption of the harder radiation discussed below.

A high energy γ-ray component will be produced due to inelastic proton-proton collisions, which become important below about $\tilde{r} < 3$, cf. equation (2). At these temperatures $p + p \rightarrow \pi^0 \rightarrow 2\gamma$, or $\rightarrow \pi^+ \rightarrow +e^+ + \ldots$. The rates[8,9] give $t_\pi \lesssim t_{pe}$ for $\tilde{r} < 3$, and the spectrum has a broad maximum about 20 MeV, with 10% of the photons above 100 MeV. The γ-ray luminosity would be comparable to that in the hard X-ray component, but loss cone effects may reduce this. Furthermore, this component will be visible only when $\tau_{\gamma\gamma} \sim 10^2 \dot{m}\tilde{r}^{-1}$, i.e., for $\dot{m} < 10^{-2}$, otherwise the photons are degraded and contribute to a broad annihilation hump.

CONCLUSIONS

Radio-quiet QSO's represent about 90% of all quasars, and their observed characteristics can be well described by our model. The model goes further, in that it predicts a hard X-ray and γ-ray spectrum, as yet seen in only a few objects (e.g., 3C273). The optical radiation is expected to vary on the freefall timescale at $\tilde{r}_s$, i.e., $t_o \sim t_{ff}(r_s) \sim 4 m_8 \dot{m}^{3/4}$ days ($\Delta L/L \sim 1$). The X-rays should vary on the slower Coulomb timescale, $t_x \sim t_{pe}(<r_s) \sim 4\,10^1 m_8 \dot{m}^{1/4}$ days (for $\Delta L/L \sim 1$). Smaller amplitude variations may occur on shorter times, $t_x \sim t_{ff}(>r_s)$ for 1-10^2 keV, and jitters with $\Delta L/L \ll 1$ on the light travel time $t_L \sim 1.4\,10^1 m_8 \dot{m}^{1/2}$ hours. These variations would be superposed on the longer timescale variations[10,11] arising from $r \gg r_s$. The same model, scaled down to $m \sim 1$-10 may apply to galactic X-ray sources such as Cyg X-1. The X-ray spectrum in these should be similar, but the shock radiation component [cf. equation (6)] would peak in the UV.

The spherically symmetric flow with a shock thus provides a model for QSO's and other active galactic nuclei, as well as galactic sources, where radio emission, high polarization, and jets are absent. Possibly the latter may be associated with those sources where there is enough angular momentum to produce a disk, or where $\dot{m} > 1$ and radiation instabilities occur, both of which may be favorable for jet formation, radio emission, and large variable polarization.

We conclude that spherical accretion can be very efficient, with $\varepsilon > 0.1$, and it provides a self-consistent model for the bulk of QSO's and X-ray active galactic nuclei, capable of extensive predictions for testing and comparison.

Acknowledgments:

This research has been partially supported by NASA grant NAGW-246 (PM) and NSF grant AST 80-22785 (JPO). We are grateful to R. Kulsrud, A. P. Lightman, D. Q. Lamb, R. Lovelace, and C. Max for comments.

REFERENCES

1. V. F. Shvartsman, Astron. Zh. [Sov. Astron.-AJ] 15, 377 (1971).
2. S. L. Shapiro, Astrophys. J. 180, 531 (1973).
3. P. Mészáros, Astron. Astrophys. 44, 59 (1975).
4. C. Max et al., Bull. Am. Astron. Soc. 14, 937 (1983).
5. Ya. B. Zeldovich and Yu. P. Raizer, Physics of Shock Waves and High Temperature Hydrodynamic Phenomena (Academic Press, N. Y., 1967).
6. R. J. Gould, Phys. Fluids 24, 102 (1981).
7. R. Ramaty and P. Mészáros, Astrophys. J. 250, 384 (1981).
8. G. A. Dahlbacka et al., Nature (Lond.) 250, 36 (1974).
9. P. I. Kolykhalov and R. A. Sunyaev, Astron. Zh. [Sov. Astron.-AJ] 23, 189 (1979).
10. J. P. Ostriker et al., Astrophys. J. Lett. 208, L61 (1976).
11. L. L. Cowie et al., Astrophys. J. 226, 1041 (1978).

THE HIGH ENERGY SPECTRUM OF HOT ACCRETION DISKS*

J. A. Eilek
New Mexico Institute of Mining and Technology, Socorro, NM 87801

M. Kafatos
George Mason University, Fairfax, VA 22030
NASA/Goddard Space Flight Center, Greenbelt, MD 20771

ABSTRACT

One possible thermal state for matter accretion onto a black hole is the two temperature ($T_i \gg T_e$) model of Shapiro, Lightman and Eardley (1976). This state may arise if strong ion heating and strong radiative electron cooling combine, and if the ion-electron coupling is due to Coulomb-type processes. In such a case, extremely high ion temperatures are predicted for optically thin accretion onto a Kerr black hole.

We investigate the high energy particle and photon spectrum such a flow would produce. Proton-proton reactions lead to primary electron, positron and γ ray production; the primary electrons (with E $\sim$ 100 MeV) interact with the accretion flow to produce secondary X and γ rays. We model the flow with the standard spatially and optically thin Kerr accretion disk with α-viscosity and sub-Eddington accretion flow. We find that ion temperatures 10^{12}K $\lesssim T_i \lesssim 10^{13}$K (in which range the ions are just subrelativistic) are reached in these models with mass accretion rates $\dot{M}/M < 10^{-9} \alpha$ yr^{-1}. Spectrum calculations for ion temperatures in this range result in a fraction, on the order of ten percent, of the total disk luminosity emerging in hard X rays and γ rays.

We also find that the high-$\dot{M}/M$ models are optically thick to $\gamma\gamma$ pair production. While full transfer/cascade calculations have not yet been done, we expect this scattering to degrade the > MeV photon and particle luminosity to $\lesssim$ 1 MeV.

*This paper appeared in the Astrophysical Journal, August 15, 1983.

HOT ACCRETION DISKS AND γ-RAY COSMIC SOURCES

M. Kafatos
George Mason University, Fairfax, VA 22030

J.A. Eilek
New Mexico Institute of Mines and Technology, Soccoro, NM 87801

ABSTRACT

Hot accretion disks around rotating black holes produce copious amounts of X-rays, γ-rays and e^+e^- pairs. These Comptonized models with an energy flux spectral index of the X-ray spectrum ~ 1 are optically thick to γγ pair production. The resultant high energy spectrum steepens above a few MeV. The few γ-ray spectra of active galactic nuclei that are available show this steepening as well as the γ-ray diffuse background. The same processes that take place in a hot accretion disk around a supermassive black hole at the center of an active nucleus also operate in a hot accretion disk around a stellar mass black hole. We here examine the implications for the two galactic sources Cygnus X-1 and particularly the positron source at the galactic nucleus.

INTRODUCTION

Three active galaxies have been detected up to this date at γ-ray energies above 1 MeV, Centaurus A, NGC 4151 and 3C 273. Only the last one has been detected up to energies of a few hundred MeV [1], with the possibility of 300 GeV emission from Cen A [2]. A steepening of the γ-ray continua for all three objects must occur at a few MeV as pointed out by Bignami et al. [1]. The spectrum of 3C 273, for example, can be fit with the empirical law $0.016\ E^{-1.4}\{1+(E/2\times10^3)^{1.3}\}^{-1}$ photons $cm^{-2}\ s^{-1}\ keV^{-1}$ from the X-ray to the γ-ray range, with upper energies of hundreds of MeV. The γ-rays above some hundreds of keV make up at least 10 - 50% of the bolometric luminosity for these active galactic nuclei (AGN) [3]. Three AGN is not, obviously, a large sample to make strong statements. The observations up to now leave open the possibility that many AGN will be shown to be strong γ-ray continuum emitters.

In contrast to the situation above 1 MeV, many AGN have been observed in the range a few keV to hundreds of keV [4]. The mean energy flux spectral index is in the range 0.6 - 0.7 [4]. Some steepening is indicated above about 100 keV.

The high energy background may be attributed to discrete sources which are usually thought to be AGN [1,4]. The contribution of AGN to the diffuse high energy background is uncertain at X-rays, although it is generally estimated to be in the range 20 - 30% [4]. At photon energies in the range 1 - 150 MeV, AGN (specifically Seyferts) could account for all the background [1].

Besides γ-ray emission from AGN, which may contain supermassive black holes, high energy radiation has been detected from at least one black hole candidate in our galaxy, Cygnus X-1 [5]. A significant excess above a single temperature Comptonized spectrum is seen at high energies (E > 300 keV) [5]. High energy continuum radiation has also been observed from the positron annihilation source at the galactic center [6]. The majority of the radiation from the galactic center source may be in the range 100 keV - 10 MeV, of the order of a few x 10^{38} erg/s.

In contrast to the γ-ray continuum emission, few of the above sources show any emission of γ-ray lines. The only AGN that may have been observed to emit lines is Cen A [7]. No positron annihilation line has been observed from any active nucleus. The galactic nucleus source remains unique in its strength of the annihilation line; as much as 50% of the total power radiated above 70 keV may be required to produce the observed annihilation radiation [8]. Cygnus X-1 may also be emitting a broad 511 keV annihilation line [9].

HOT ACCRETION DISK SPECTRA

Eilek and Kafatos [10,11] have developed models of hot accretion disks around black holes. In these models the ion temperature is much higher than the electron temperature [12]. The models are applicable to both Schwarzchild and Kerr metrics. Gamma-ray emission is assured from the pion production as well as the high energy e^+e^- production, both products of the high ion temperatures. The X-rays are produced by Comptonization of soft photons and the hot, thermal electrons with $T_e \sim 10^9$ K. These Comptonized X-rays have been suggested as the best explanation for the hard X-rays emitted by Cygnus X-1. The behavior of the γ-rays is dependent on the value of the Comptonization parameter y, where y is defined as the product of the mean fractional energy change per scattering times the mean number of scatterings. Large amplification of the incoming soft photon flux occurs when $y \sim 1$ and the resulting energy losses of the electrons act as a thermostat to keep y near unity [12]. The y parameter is related to the energy flux spectral index (where the energy flux is in keV cm^{-2} s^{-1} keV^{-1}). The spectral index is approximately inversely proportional to y and equal to ~ 0.72 for $y = 1$ [3]. It follows that for $y \gtrsim 1$, the spectral index is close to 0.7. Models with $y \gtrsim 1$ are optically thick to $\gamma\gamma$ scattering producing pairs [10]. Gamma-rays of energy greater than a few MeV are degraded to lower energy radiation. The photon spectra of thick sources with a large number of pairs have not yet

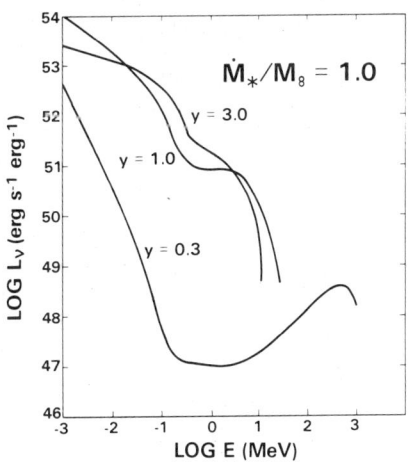

Fig. 1. Theoretical hot accretion disk spectra for different y's and for canonical Kerr metrics.

been computed, although the first steps in such a complicated calculation have been taken [13]. In Figure 1 we show spectra of two-temperature disks for the canonical Kerr metric and for three values of y. $\dot{M}_*$ is the mass accretion rate in $M_\odot$/yr and M_8 the central black hole mass in $10^8 M_\odot$. For different black hole masses the curves are identical as long as the accretion rate changes proportionally to the change in the hole's mass. The spectra for y = 1, 3 are approximate above a few hundred keV. The general behavior of the spectra is as follows: the higher the value of y, the harder the X-ray spectra but also the lower the γ-ray energy for which the scattering $\gamma + \gamma \rightarrow e^+ + e^-$ becomes important. These general results are used in what follows. We emphasize that γ-ray production is efficient as long as the accretion rate is close to the critical accretion rate, producing a total luminosity close to the Eddington value [10].

ACTIVE GALACTIC NUCLEI

The mean AGN spectrum can be found in Rothschild et al. [4]. Good fits to the mean spectrum can be obtained for values of y in the approximate range 1 - 3 [3]. It follows that the spectrum of a typical active nucleus should steepen above a few MeV. This is an important prediction for future γ-ray observations of Seyferts, particularly with the Gamma Ray Observatory. Moerover, it is easy to understand why AGN seem to have a mean energy flux spectral index in the range 0.6 - 0.7 [4]: This is equivalent to $y \gtrsim 1$, the stable condition. If our model is correct, it would mean that AGN are accreting close to the Eddington limit.

The specific high energy spectra of the three AGN which show MeV emission have been investigated in the light of our model. We expect a high energy tail even in thick spectra [14], although this would probably not explain the 500 MeV emission from 3C 273. Values of y near unity can explain the X-ray spectra of Cen A and 3C 273 with a steepening of the spectra above a few MeV as the observations imply. Somewhat higher values may be required for NGC 4151, say $y \sim 3$, again in agreement with the lack of emission of γ-rays more energetic than a few MeV. As is usually the case, the viscosity parameter has a reasonable value, say $\alpha \sim 0.1$. Fitting the observed X-ray fluxes with theoretical spectra of disks having $y \gtrsim 1$, $\alpha \sim 0.1$ and $\dot{M}_* / M_8 \sim 1$ yields $M_8 \sim 0.01$ for NGC 4151 (distance=20 Mpc) and $M_8 \sim 5$ for 3C 273 (distance=900 Mpc). These values are only suggestive of the order of magnitudes involved.

Kafatos and Eilek [15] have explored the possibility that the high energy (X-ray, γ-ray) diffuse background results from discrete sources accreting close to the Eddington limit. A unique mechanism and type of sources could thus help to explain the high energy background. The sources would be numerous but faint and the average mass of the central object $M \sim 10^4 M_\odot$. The lack of observable positron annihilation line either in the background or in the observed spectra of AGN implies that annihilation is taking place in media of ambient temperatures $\gtrsim 10^7$ K [10].

Relating the X-ray luminosity to the optical luminosity also requires accretion rates close to the Eddington value [16] for quasars.

THE POSITRON ANNIHILATION SOURCE AT THE GALACTIC NUCLEUS

Few galactic sources have been detected above 100 keV except for the transient and bursting sources. Our model would not apply to these sources [10]. Single temperature Comptonization models are the most favored to explain the X-ray emission from Cygnus X-1, a black hole candidate in our galaxy. The relatively flat X-ray slope and the excess emission above 100 keV, with a cutoff at a few MeV, agree with the predictions of our optically thick ($y \gtrsim 1$) models [10].

The observations of the positron annihilation line source at the galactic nucleus have been summarized by Matteson [6]. The source seems to be variable and shows strong continuum X-rays extending to the γ-ray region. Soft X-ray emission is, though, very weak. The 1977 observations are shown in Figure 2. We can fit the observed continuum below a few hundred keV very well with our models having y near unity (the excess points below 500 keV may be due to the three photon positronium continuum [17]). Even near a few MeV where our calculations are approximate due to the optical thickness effects the general agreement is obvious.

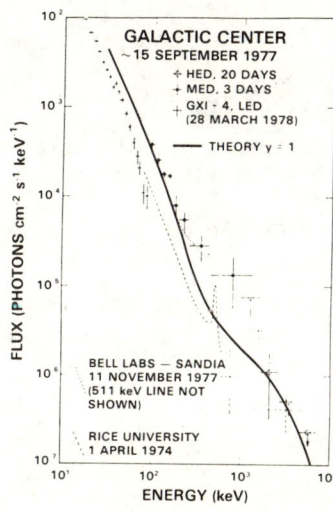

Fig. 2. Observations of the galactic center continuum to our theoretical, near Eddington models with y near unity.

We have compared the observed ratio of the line power to the continuum power. We write the pair annihilation rate in terms of N_{43} (in units of 10^{43} pairs/s) and the γ-ray luminosity in terms of L_{38} (in units of 10^{38} erg/s). The observed ratio N_{43}/L_{38} is then compared to expected values from our models [10]. The observed ratio is near unity. In our models we estimate that the number of secondary pairs is about ten times the number of primary relativistic pairs in accordance with optically thick source estimates [18]. The ratio N_{43}/L_{38} approaches unity for luminosities near the Eddington value but even for values much less than the Eddington value the ratio is never less than about 0.4. The total luminosity of the positron source is hard to estimate due to uncertainties in the soft X-ray and infrared luminosities. If our picture is correct, we predict a bolometric luminosity in the range 10^{38}-10^{39} erg/s, primarily in the form of UV and soft X-rays. The required black hole mass can then be estimated from the value of the diameter of the positron source which is $\leq 3\times10^8$ cm [8]. The majority of pairs and γ-rays are emitted-according to our models-at distances 10-20 gravitational radii from the black hole. This would mean that the upper limit of the mass of the black hole is 50-100 $M_\odot$. Such a small black hole is, of course, stellar and negligible compared to the dynamical mass of 10^6 $M_\odot$ for the central region of our galaxy [19]. We find that the required mass accretion rate is $\sim 5 \times 10^{-7}$ $M_\odot$/yr. Such rates could easily result if the black hole is a member of a binary star system and accretes matter from an O or B-type star or an M-type giant. In this scheme time-dependence would result from the behavior of the orbit and would have nothing to do with the physics of accreting matter near the black hole.

REFERENCES

1. G.F. Bignami, C.E. Fichtel, R.C. Hartman, and D.J. Thompson, Ap.J. **232**, 649 (1979).
2. J.E. Grindlay, H.F. Helmken, R. Hanbury Brown, J. Davis, and L.R. Allen, Ap.J. (Letters) **197**, L9 (1975).
3. M. Kafatos, in Highlights of Astronomy, Vol.6, edit. R.M. West (D. Reidel, Dordrecht, 1983).
4. R.E. Rothschild, R.F. Mushotzky, W.A. Baity, D.E. Gruber, and J.L. Matteson, Ap.J. in press (1983).
5. P.L. Nolan, et al., Nature **293**, 275 (1981).
6. J.L. Matteson, in the Galactic Center, ed. G.R. Riegler and R.D. Blandford (American Institute of Physics, N.Y. 1982) p. 109.
7. R.D. Hall, C.A. Meegan, G.D. Walraven, F.T. Djuth, and R.C. Haymes Ap.J. **210**, 631 (1976).
8. R.E. Lingenfelter, and R. Ramaty, in the Galactic Center, ed. G.R. Riegler and R.D. Blandford (AIP, N.Y. 1982) p. 148.
9. P.L. Nolan, R.E. Lingenfelter, and J.L. Matteson, Ap.J., submitted.
10. J.A. Eilek, and M. Kafatos, Ap.J. Aug. 15 issue (1983).
11. J.A. Eilek, and M. Kafatos, present volume.
12. S.L. Shapiro, A.P. Lightman, and D.M. Eardley, Ap.J. **204**, 187 (1976).
13. L.J. Caroff, and J.A. Eilek, present volume.
14. G. Cavallo, and M.J. Rees, M.N.R.A.S. **183**, 359 (1978).
15. M. Kafatos, and J.A. Eilek, in Early Evolution of the Universe and its Present Structure, IAU Symposium 104, ed. G.O. Abell and G.L. Chincarini (D. Reidel, Dordrecht, 1983).
16. W.H. Tucker, Ap.J. submitted.
17. M. Leventhal, C.J. MacCallum, and P.D. Stang, Ap.J. (Letters) **225**, L11 (1978).
18. P.D. Noerdlinger, present volume.
19. J.H. Lacey, C.H. Townes, T.R. Geballe, and D.J. Hollenbach, Ap.J. **241**, 132 (1980).

Chapter VI   Physical Processes in Plasmas

FUNDAMENTAL PROCESSES IN PAIR PLASMAS

Alan P. Lightman
Harvard-Smithsonian Center for Astrophysics,
Cambridge, Massachusetts 02138

ABSTRACT

We first review the various processes that produce and destroy electron-positron pairs, and then compare the timescales of these processes to thermalization, accretion, and cooling timescales. We next consider the various radiation spectra produced by relativistic, thermal plasmas. Finally, we review recent results for the equilibria available to finite, thermal relativistic plasmas with and without embedded magnetic fields. Such plasmas, in steady state, have maximum temperatures, luminosities, and field strengths--useful diagnostics for interpreting quasars and active galaxies.

FUNDAMENTAL REACTIONS

To lowest order in the fine structure constant α and in the number of interacting particles, pairs are produced in the reactions

$$\gamma\gamma \rightarrow e^+e^- \qquad o(n_\gamma^2) \quad , \qquad (1a)$$

$$\gamma p \rightarrow e^+e^-p \qquad o(n_\gamma N\alpha) \quad , \qquad (1b)$$

$$\gamma e \rightarrow ee^+e^- \qquad o(n_\gamma n_\pm \alpha) \quad , \qquad (1c)$$

$$ee \rightarrow eee^+e^- \qquad o(n_\pm^2 \alpha^2) \quad . \qquad (1d)$$

Here γ, p, e^+, e^- denote photon, proton, positron, and electron, respectively, and e may be either electron or positron; n_γ, N, n_+, and n_- are the corresponding number densities. At the right of each equation above is the dependence of the reaction rate on particle densities and the fine structure constant, omitting the detailed energy dependences.

To lowest order, pairs are annihilated by the single reaction

$$e^+e^- \rightarrow \gamma\gamma \qquad o(n_+n_-) \quad . \qquad (2)$$

Photons are produced in the reactions

$$ee \rightarrow ee\gamma \qquad o(n_\pm^2 \alpha) \quad , \qquad (3a)$$

$$ep \rightarrow ep\gamma \qquad o(n_\pm N\alpha) \quad , \qquad (3b)$$

$$e^+e^- \rightarrow \gamma\gamma \qquad o(n_+n_-) \quad , \qquad (3c)$$

$$e\gamma \rightarrow e\gamma\gamma \qquad o(n_\pm n_\gamma \alpha) \quad , \qquad (3d)$$

$$e \underset{B}{\rightarrow} e\gamma \qquad o(n_-) \quad . \qquad (3e)$$

The first two of these reactions are bremsstrahlung, equation (3c) is annihilation radiation, equation (3d) is double Compton radiation, and equation (3e) is synchrotron radiation.

For all applications, relativistic plasmas will be optically thin, so we can ignore photon destruction. However, the alteration of photon frequency by inverse Compton scattering can be important

$$e\gamma \to e\gamma \quad . \qquad (4)$$

TIMESCALES

A. Thermalization versus Bremsstrahlung

Ultrarelativistic electrons thermalize by electron-electron collisions on a timescale t_{ee} given by[1]

$$t_{ee} \approx 8\, T_*^2\, (n_+ + n_-)^{-1}\, (c\sigma_T \ln \Lambda)^{-1} \quad , \qquad (5)$$

where $\ln \Lambda \approx 30$ is the Coulomb logarithm, σ_T is the Thomson cross section, and

$$T_* \equiv kT/mc^2 \quad . \qquad (6)$$

Such plasmas radiate by electron-electron bremsstrahlung on a timescale t_{brem} given by[1,2]

$$t_{brem} \approx (n_+ + n_-)^{-1} (\alpha c \sigma_T)^{-1} (\ln 4T_*)^{-1} \quad . \qquad (7)$$

Thus, bremsstrahlung alone will prevent the establishment of thermal pair distributions for $T_* \gtrsim 12$.

B. Pair Production versus Accretion Timescales

With reaction (1c) as an illustration, the rate of electron production is[3]

$$\dot{n}_- \approx c\, \sigma_T\, \alpha\, n_\gamma\, (n_+ + n_-) \quad , \qquad (8)$$

where we have omitted a slowly varying logarithmic function of photon frequency and assumed essentially all photons are above the pair production threshold. The inward drift timescale in an accretion flow, evaluated at 10 Schwarzschild radii ($r = 20\, GM/c^2$) from a central mass M, is

$$t_r = 3 \times 10^5 \text{ s } M_8/\beta \quad , \qquad (9)$$

where $M_8 \equiv M/10^8 M_\odot$ and β is the ratio of Keplerian orbital time to inward drift time, always less than unity. Now, defining a pair production timescale t_p by $t_p \equiv n_-/\dot{n}_-$, relating the photon density to luminosity L and mean photon energy $\langle h\nu \rangle$ by $n_\gamma = L/(4\pi r^2 c \langle h\nu \rangle)$, and using the Eddington luminosity $L_{EDD} \equiv 4\pi GMcm_p/\sigma_T$ as normalization, equations (8) and (9) may be combined to yield[4]

$$t_p/t_r \simeq 0.05\beta \langle h\nu/mc^2 \rangle (L/L_{EDD})^{-1} n_-/(n_+ + n_-) \quad . \quad (10)$$

Thus electron-photon interactions should have time to produce pairs in accretion flows, unless the luminosity is several orders of magnitude or more below the Eddington limit and the flow is spherical, $\beta \approx 1$. Zdziarski[5] has also considered the above ratio, but evaluated explicitly for a bremsstrahlung-dominated luminosity, at a radius $r = GM/c^2$, and with $\beta = 1$.

C.   Cooling versus Annihilation in Nonsteady Plasmas

In nonsteady plasmas, narrow annihilation lines will be seen only if high energy pairs with Lorentz factors $\gamma \gg 1$ have time to cool to $\gamma \sim 1$ before they annihilate. The annihilation time t_{ann} for $\gamma \gg 1$ is[1]

$$t_{ann} \approx 0.5 \gamma^2 (\sigma_T c n_- \ln \gamma)^{-1} \quad . \quad (11)$$

An illustrative cooling mechanism, always important for fairly modest magnetic fields B, is synchrotron cooling, on a timescale

$$t_{syn} = 6\pi mc(\gamma B^2 \sigma_T)^{-1} \quad . \quad (12)$$

If we now relate particle densities to gas pressure P_G by $P_G = 1/3 \gamma mc^2 (n_- + n_+)$, and magnetic fields to magnetic pressure P_B by $P_B = B^2/8\pi$, then equations (11) and (12) yield the result[4] (valid in the relativistic domain)

$$t_{syn}/t_{ann} \approx 5 \left(\frac{\ln\gamma}{\gamma^4}\right)\left(\frac{n_-}{n_- + n_+}\right)\left(\frac{P_G}{P_B}\right) \quad . \quad (13)$$

Thus, adequate synchrotron cooling for line production, i.e., $t_{syn} < t_{ann}$ at $\gamma \sim 1$, requires magnetic fields of equipartition strength or higher.

SPECTRA IN THERMAL PAIR PLASMAS

Annihilation spectra in thermal plasmas have recently been calculated by a number of investigators[6,7,8]. In the nonrelativistic limit the radiation intensity I_ν (erg s^{-1} cm^{-2} Hz^{-1} ster^{-1}) has the form

$$I_\nu \propto \nu \exp\left[\frac{-mc^2}{kT}\left(\frac{h\nu}{mc^2} - 1\right)^2\right] \quad . \quad (14a)$$

In the relativistic limit, the form is

$$I_\nu \propto \nu^2 e^{-h\nu/kT} \quad . \tag{14b}$$

A recent analytic fit to bremsstrahlung spectra in the transrelativistic regime has been tabulated by Stepney and Guilbert[9]. In the relativistic regime, the intensity is approximately[10]

$$I_\nu \propto \ln\left[\frac{(kT)^2}{mc^2 h\nu}\right] e^{-h\nu/kT} \quad , \tag{15}$$

very similar in shape to the nonrelativistic expression. Bremsstrahlung always dominates annihilation radiation at low frequencies, and also at $h\nu \sim kT$ for $T_* \gtrsim 3$.[11]

A third important spectrum is that of Comptonized synchrotron radiation. Pozdnyakov, Sobol, and Sunyaev[12] first showed that repeated scattering by relativistic thermal electrons of a monochromatic source of photons of frequency $h\nu_0 \ll kT$ and intensity I_0 leads to a spectrum (for $\nu > \nu_0$)

$$I_\nu = I_0 \left(\frac{\nu}{\nu_0}\right)^{-s} e^{-h\nu/kT} \quad . \tag{16}$$

Here the spectral index s is related to the temperature and Thomson scattering depth

$$\tau_{th} \equiv (n_+ + n_-) \sigma_T R \tag{17}$$

of a medium of size R by

$$s = \frac{\ln 1/\tau_{th}}{\ln 16\, T_*^2} \quad . \tag{18}$$

When sufficiently strong magnetic fields are present, synchrotron photons, self-absorbed up to a frequency ν_{ts}, constitute a source of soft photons I_0. With $\nu_0 = \nu_{ts}$, the appropriate relations to be substituted into equation (16) are[13,14]

$$\nu_{ts} = \frac{2T_*^2}{9} \nu_c \left\{ \ln\left[\frac{\pi \tau_{th} mc^2}{4(2)^{1/2} \alpha T_*^3 h\nu_{ts}}\right] \right\}^3 \quad , \tag{19}$$

where $\nu_c \equiv eB/(2\pi mc)$, and

$$I_0 = 2\nu_{ts}^2 kT/c^2 \quad . \tag{20}$$

For values of s between 0.5 and 1.0, appropriate to active galactic nuclei, Comptonized synchrotron will dominate bremsstrahlung both in emergent radiation (at $h\nu \sim kT$) and in the production of pairs unless B is several orders of magnitude or more below its equipartition value.[14]

CONDITIONS FOR STEADY STATE THERMAL PAIR PLASMAS

Although the relativistic plasmas in gamma ray burst sources are clearly not steady and the relativistic plasmas in active galactic nuclei and quasars may not be thermal, the steady state, thermal relativistic plasma is clearly an important case--both on theoretical grounds and as a first step toward providing constraints for actual relativistic plasmas in nature.

A. Without Magnetic Fields

Lightman[15] and Svensson[16] have considered the problem of a spherical, homogeneous plasma of specified temperature T, radius R, and proton number density N, as shown in Figure 1 below. The electrons

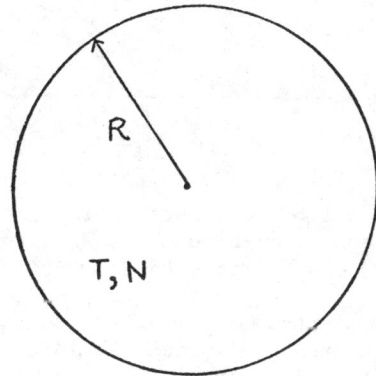

Fig. 1. Fixed parameters of a relativistic thermal plasma without magnetic fields.

and positrons are assumed to have Maxwell-Boltzmann distributions, and the plasma is optically thin. The problem then is to determine the equilibrium pair densities and radiation field self-consistently, balancing pair production against pair annihilation. The dimensionless solutions are determined by the two dimensionless parameters T_* and

$$\tau_N \equiv N \sigma_T R \quad . \tag{21}$$

Aspects of the solutions are illustrated in Figure 2. The curve labeled $\tau_N = 0$, BKZS is the result (corrected for an error) of Bisnovatyi-Kogan, Zeldovich, and Sunyaev[17], which neglects all photon-pair-production processes. Note that (1) there is a maximum value of T_*, $T_{*max}(\tau_N)$, at each τ_N, above which no equilibrium solutions exist (pairs are produced more rapidly than they can be destroyed); (2) as $\tau_N \to 0$, $T_{*max} \to 25$; (3) for each $T_* < T_{*max}(\tau_N)$, there are two equilibrium solutions, an upper branch (dominated by photon processes)

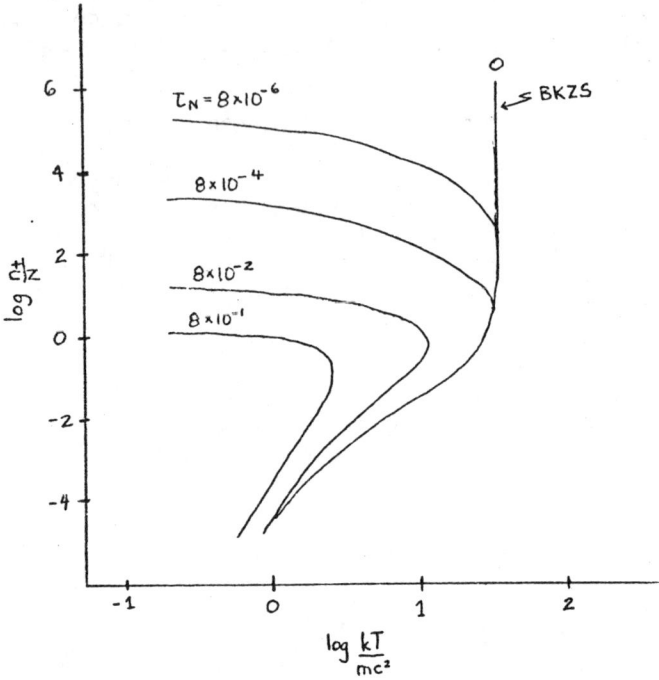

Fig. 2. Equilibrium solutions for the dimensionless positron density as a function of T_* and τ_N, adapted from Reference 16.

and a lower branch (dominated by particle processes). Corresponding to the maximum temperatures are maximum luminosities $L_{max}(T_*,R)$. There is a large range of luminosities, connecting optically thin and thick plasmas, over which the temperature remains in the transrelativistic domain, $0.1 \lesssim T_* \lesssim 1$.

B. With Magnetic Fields

Araki and Lightman[14] and Kusunose and Takahara[18] have extended the above treatments to include magnetic fields, with Comptonized synchrotron photons playing a significant role in producing pairs. The dimensionless solutions may be determined in terms of three dimensionless parameters: T_*, s (of equation (18)), and a magnetic parameter

$$\beta \equiv \sigma_T R \left(\frac{mc}{h}\right)^3 \left(\frac{h\nu_{ts}}{kT}\right)^{2+s} . \qquad (22)$$

For each set of values of (T_*,s) there is a maximum value of β, $\beta_{max}(T_*,s)$, above which no equilibria exist. For $\beta > \beta_{max}$, pairs are produced (by the abundance of Comptonized synchrotron photons) more rapidly than they can be destroyed. By choosing a value of R,

$\beta_{max}(s, T_*)$ may be translated into a $B_{max}(s, T_*, R)$. This is shown in Figure 3 for s = 0.7, a value for the spectral index observed[19] for many active galactic nuclei, and for $R = 10^{13}$ cm, a characteristic size for the inner emission region of AGN and QSOs suggested by time variabilities.[4]

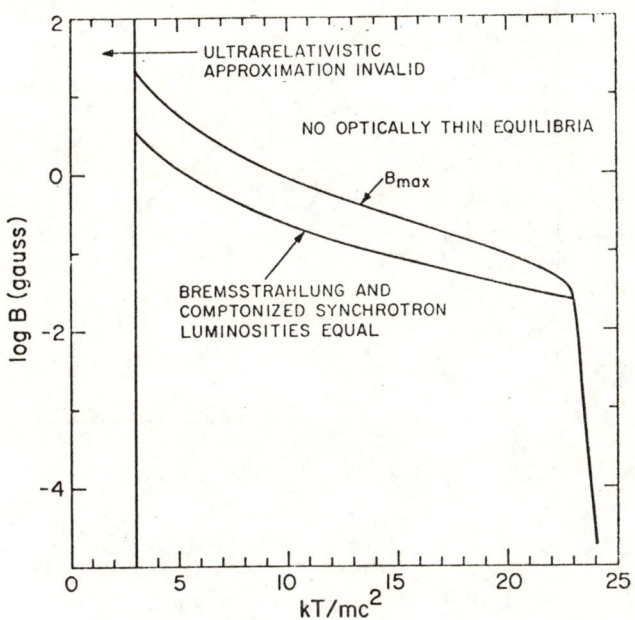

Fig. 3. Parameter space in the magnetic field-temperature plane, with s = 0.7 and $R = 10^{13}$ cm. From Reference 14.

Corresponding to the maximum values of B in Figure 3 are maximum luminosities $L_{max}(T_*, s, R)$, shown in Figure 4. An approximate analytic expression for L_{max}, in the region $3 \lesssim T_* \lesssim 20$, is[14]

$$L_{max} \approx 0.06 \left(\frac{mc^3}{\alpha\sigma_T}\right) R \; T_*^{-1-4s}$$

$$= 3 \times 10^{42} \text{ erg s}^{-1} \left(\frac{R}{10^{13} \text{ cm}}\right) T_*^{-1-4s} \quad . \quad (23)$$

Equation (23), in fact, is a general expression for the maximum luminosity of a steady-state plasma limited by pair production, with the specific radiation process entering only in the exponent of T_* and slightly in the overall numerical coefficient.

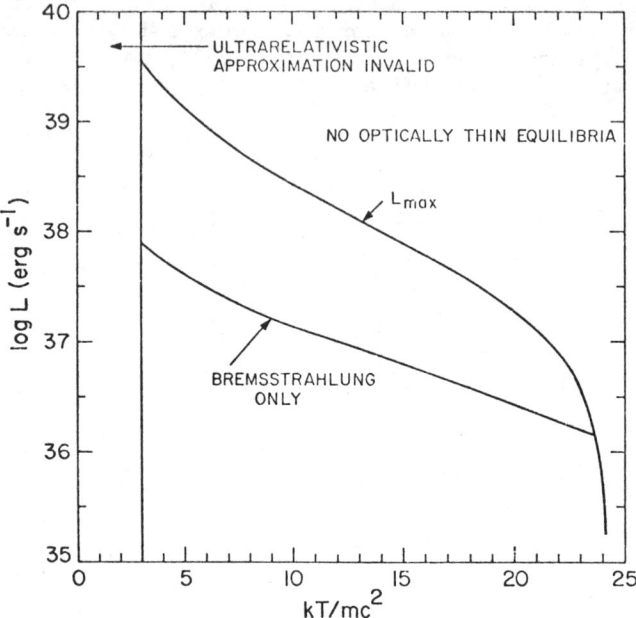

Fig. 4. Parameter space in the luminosity-temperature plane, with s = 0.7 and R = 10^{13} cm. From Reference 14.

Since AGN and QSOs have observed luminosities in the range 10^{42}-10^{47} erg s^{-1}, we conclude that such objects cannot support thermal plasmas in equilibrium at relativistic ($T_* \gtrsim 3$) temperatures. If emission from these objects is observed at energies exceeding a few MeV, then the emission must be either nonthermal or nonsteady.

Future theoretical work in this subject should concern time-dependent, nonthermal plasmas--possibly as produced by chaotic, high energy shock waves.

This work was partly supported by NASA grant NAGW-246.

REFERENCES

1. S. Stepney, Mon. Not. Roy. Astr. Soc. 202, 467 (1983).
2. J. Stickforth, Z. Physik 164, 1 (1961).
3. J. M. Jauch and F. Rohrlich, The Theory of Photons and Electrons (Springer-Verlag, N. Y., 1976).
4. A. P. Lightman, Sp. Sci. Rev 33, 335 (1982).
5. A. Zdziarski, Astron. Astrophys. 110, L7 (1982).
6. A. Zdziarski, Acta Astr. 30, 371 (1980).
7. R. Ramaty and P. Mészáros, Astrophys. J. 250, 384 (1981).
8. R. Svensson, Astrophys. J. 258, 321 (1982).

9. S. Stepney and P. W. Guilbert, Mon. Not. Roy. Astr. Soc., in press (1983).
10. M. Alexanian, Phys. Rev. 165, 253 (1968).
11. A. P. Lightman and D. Band, Astrophys. J. 251, 713 (1981).
12. L. A. Pozdnyakov, I. M. Sobol, and R. A. Sunyaev, Soviet Astr. 21, 708 (1977).
13. F. Takahara and S. Tsuruta, Progr. Theor. Phys. 67, 485 (1982).
14. S. Araki and A. P. Lightman, Astrophys. J., in press (1983).
15. A. P. Lightman, Astrophys. J. 253, 842 (1982).
16. R. Svensson, Astrophys. J. 258, 335 (1982).
17. G. S. Bisnovatyi-Kogan, Ya. B. Zeldovich, and R. A. Sunyaev, Soviet Astron.-AJ 15, 17 (1971).
18. M. Kusunose and F. Takahara, preprint (1983).
19. R. F. Mushotzky, Astrophys. J. 256, 92 (1982).

TEMPORAL EVOLUTION OF ELECTRON-POSITRON PLASMAS

W. Brinkmann
Max-Planck-Institut für Physik und Astrophysik
Institut für extraterrestrische Physik
8046 Garching, Federal Republic of Germany

INTRODUCTION

Recently it has become increasingly clear that e^-e^+ pairs can play an important role in various high-energy astrophysical phenomena and there has been growing interest in understanding these plasmas.

In spherical accretion flows[1] and in accretion discs[2] the pairs created close to the Schwarzschild radius have a strong influence on the overall dynamics of the flow, and the spectrum of the emitted high energy radiation shows the typical signature of the pair component. In gamma-ray bursters, where the physics of the primary energy release is not fully understood[3,4] the observed radiation reflects the temporal evolution of the trans-relativistic pair plasma[5].

Various authors have investigated the properties of e^-e^+ plasmas, but although most of the relevant elementary cross-sections have been known for some time, their enormous complexity makes a rigorous analytical treatment impossible.

Zdziarski[6] and Ramaty and Mészáros[7] performed Monte Carlo calculations to obtain the annihilation radiation spectra from relativistic plasmas. Gould[10,11] and Lightman[12] determined the equilibrium properties and relaxation rates of e^-e^+ pair plasmas qualitatively and Svensson[8,9] was able to give analytical expressions for the differential cross sections for annihilation rate and emissivity in isotropic plasmas, leaving only the integrations over the distribution functions of the particles.

Nearly all these papers dealt with plasmas having (thermal) Maxwellian distribution functions. It is known from classical plasma physics that this assumption allows to obtain fairly good estimates of the relevant plasma parameters. Nevertheless, for a detailed description of a non-equilibrium plasma one has to solve the transport equation for the system. In particular, in situations where the dynamics of the whole system depends on the actual form of the particle distribution functions (e.g. in accretion flows), a better knowledge of the temporal evolution of pair plasmas is required.

THE TRANSPORT EQUATION

A numerical solution of the relativistic transport equation for a plasma consisting of electrons, positrons, heavy particles (protons), and photons is today still outside our computational capabilities. Therefore, simplifications of the general equations cannot be avoided. For the investigation of time scales and radiation properties of e^-e^+-plasmas reasonable simplifications are the assumption of spatial homogeneity and isotropy (in energy phase space) of the distribution functions. The equations to be solved are of the general form

$$\frac{\partial f_i}{\partial t} = \sum_{i,k} \frac{\delta f_{i,k}}{\delta t}\bigg|_{sc} + S_i(E) + L_i(E) \qquad \text{where } i = \begin{cases} p \\ e^+ \\ e^- \\ h\nu \end{cases} \qquad (1)$$

that is, the temporal change in the energy-dependent distribution function $f_i(E)$ of the i-th species is determined by mutual scatterings of the particles plus source and loss terms which depend as well on the other particle species of the system.

The solution of this coupled set of integro-differential equations is still a substantial effort. Thus, in a first approach the plasma is assumed to be non-relativistic and optically thin, i.e., electron-photon scatterings are neglected. In this case the Boltzmann-collision integral can be written as a (differential) Fokker-Planck operator. At first glance the investigation of the temporal evolution of a non-relativistic e^-e^+-plasma seems rather unexciting Binary collisions of the particles will establish thermal equilibrium on a time scale of

$$T_c = 0.267 \; T^{3/2} \; / \; n \ln(\lambda) \; \text{sec} \qquad (2)$$

where $\ln(\lambda)$, the Coulomb logarithm[13], typically has a value of 20. On the other hand the (non-relativistic) time scale for pair annihilation is much longer than T_c

$$T_a = (\pi c r_0^2 n)^{-1} \; \text{sec} \qquad (3)$$

$$\approx 1.4 \cdot 10^{14} \; / \; n \; \text{sec.}$$

Therefore, the pair plasma will evolve rapidly towards a thermal equilibrium configuration and then, on a much longer time scale, annihilate. Nevertheless, the above estimates of the thermalization time scale are only valid for the bulk of the particles, whereas numerical calculations show that the evolution of a plasma towards a Maxwellian typically takes about ten times this Spitzer-time and even much longer to fill up the tail of the distribution function[10,14]. Further, all radiation processes are quite sensitive to the exact form of the distribution functions and it is thus not at all clear how well the annihilation spectrum is represented by its thermal form.

THE NUMERICAL CALCULATIONS

The non-relativistic, isotropic Fokker-Planck equation for each particle species (electrons, positron, or protons) reads

$$\frac{\partial f}{\partial t} = A \frac{\partial^2 f}{\partial v^2} + B \frac{\partial f}{\partial v} + Cf + D \qquad (4)$$

where A, B, and C are the time, velocity and distribution function dependent Rosenbluth coefficients[15] and $D(f_j, f_k)$ is the loss term due to annihilation of electrons and positrons.

Equation (4) was solved by a fourth order (in space and time) fully explicit scheme to ensure particle conservation (losses $\leq 10^{-5}$ per time step). The most time-consuming parts were the calculations of the $D(f_j, f_k)$ and the determination of the emitted photon spectrum.

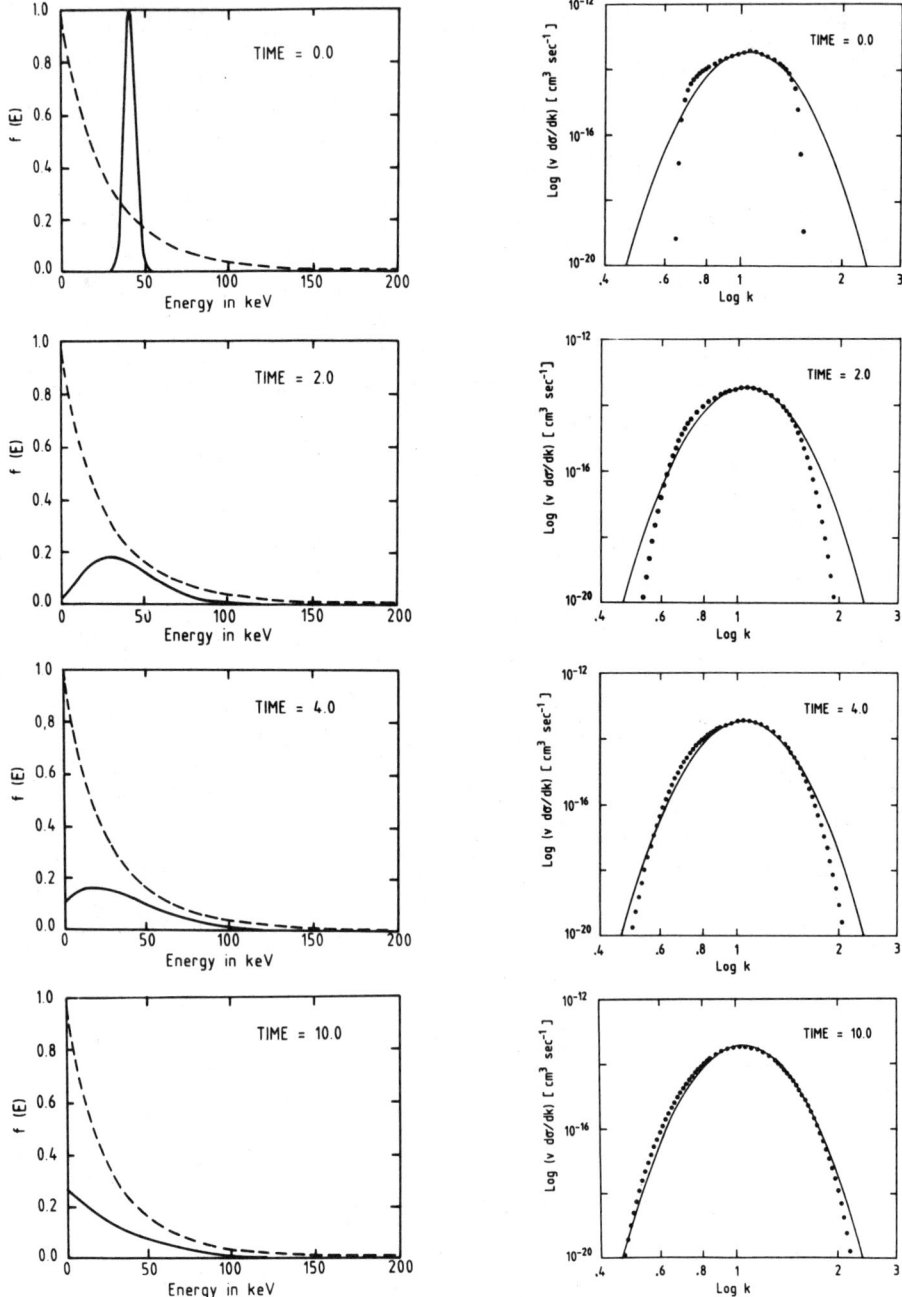

Fig.1: Temporal evolution of a pure e^-e^+-plasma and the emitted annihilation spectra (see text)

The loss term $D(f_i,f_k)$ was obtained by integrating the angle averaged reaction rate (formula 18 of Svensson[8]) over the actual electron and positron distribution functions. The emergent photon spectrum was calculated by using Svensson's angle-averaged emissivity (formula 55), again integrated over both distribution functions.

Figure 1 shows as an example the evolution of a pure e^-e^+-plasma $(n^+ = n^-)$ released at t=0 as a Gaussian, centered at $E_0 = 40$ keV. Astrophysically such low temperature pair plasmas might be created in accretion flows by photon-photon collisions close to the Schwarzschild radius. The dashed line represents a Maxwellian with the same total thermal energy (temperature). Both functions are normalized to 1.0 at t=0 (t in units of the Spitzer time). The right hand side shows, as dots, the annihilation emissivity per particle as function of the dimensionless energy $k = h\nu/m_e c^2$, whereas the full lines represent the emissivity of the corresponding Maxwellian distribution function.

It can clearly be seen how the distribution functions evolve to their thermodynamic equilibrium form, but even after ten Spitzer times thermalization has not occurred. In particular the tails remain unpopulated. The particle losses due to annihilation are minute and do not play a role on these time scales. The emitted spectrum at first differs strongly from a thermal behaviour, but fairly soon becomes observationally undistinguishable from it.

CONCLUSION

Pair annihilation processes do not influence the temporal evolution of non-relativistic e^-e^+-plasmas to a large degree. This is mainly due to the short thermalization time scale and the fact that the non-relativistic annihilation cross section is nearly energy- independent. Nevertheless, the emitted spectrum might differ at first drastically from a thermal equilibrium form and observations of the annihilation radiation with high temporal and spectral resolution will give valuable information about the state of the system.

If the plasma contains protons at a different temperature, the pair component evolves as quasi Maxwellian on a much longer time scale $(T_c \cdot m_i/m_e)$ towards total thermal equilibrium.

In relativistic plasmas where the energy dependence of all cross sections is much more pronounced things will be different, but the strongest influence on the evolution of the plasma is expected from Compton scattering of the annihilation photons on the pairs.

REFERENCES

1. R.Z.Yahel and W.Brinkmann, Ap.J. Lett. 244, L7 (1981)
2. E.P.T.Liang, Ap.J. 234, 1105 (1979)
3. R.Ramaty and R.E.Lingenfelter, Ann.Rev.Nucl.Part.Sci. 32, 235 (1982)
4. S.Woosley, in: Accreting Neutron Stars (W. Brinkmann, J.Trümper, eds.),p. 189, Max-Planck report 177, München 1982
5. G.Cavallo and M.J.Rees, MNRAS 183, 359 (1978)
6. A.A.Zdziarski, Acta Astronomica 30, 371 (1980)
7. R.Ramaty and P.Mészáros, Ap.J. 250, 384 (1981)
8. R.Svensson, Ap.J. 258, 321 (1982)
9. R.Svensson, Ap.J. 258, 335 (1982)
10. R.J.Gould, Phys.Fluids 24, 102 (1981)

11. R.J.Gould, Ap.J. 254, 755 (1982)
12. A.P.Lightman, Ap.J. 253, 842 (1982)
13. L.Spitzer, Physics of Fully Ionized Gases (Interscience Publ., New York), 1956
14. D.C.Montgomery and D.A.Tidman, Plasma Kinetic Theory (McGraw Hill, New York), 1964
15. M.N.Rosenbluth, W.M.MacDonald and D.L.Judd, Phys.Rev. 107, 1 (1957)

PAIR PRODUCTION IN THERMAL PLASMAS: A COMPUTER MODEL

Susan Stepney
Institute of Astronomy, Madingley Road, Cambridge, England.

ABSTRACT

A computer code has been developed to follow the processes of electron-positron pair production, annihilation, bremsstrahlung and Comptonization in a slab of mildly relativistic thermal plasma. The resulting equilibrium solutions are compared with the semi-analytic calculations of Svensson[1].

INTRODUCTION

The various processes occuring in relativistic plasmas have been discussed elsewhere[2]. Since the electron thermalization timescale[3] is less than the bremsstrahlung cooling timescale for $kT \lesssim 10\ m_e c^2$, it is self-consistent to consider thermal plasmas at mildly relativistic temperatures. In this paper I will discuss a computer code that has been developed to model slabs of thermal plasma at temperatures of $kT \lesssim m_e c^2$ and various optical depths.

The starting point was a one-D radiative transfer code, written by Guilbert[4], to study the Compton cooling of hot gas by an external source of soft photons. It correctly treats Comptonization at mildly relativistic temperatures. I have extended the program to include internal sources and sinks of photons, and pair production.

The processes considered are:
(1) Comptonization
(2) Thermal Bremsstrahlung:
(i) electron-proton: This is very straight-forward to calculate, being merely a single integral over the cross section.
(ii) electron-electron ($e^{\pm}e^{\pm}$): This emission is comparable to (i) at temperatures $\gtrsim 50$ keV, but the calculation of the thermal spectrum requires the evaluation of a 5-D integral. I use the fit recently calculated by Guilbert[5].
(iii) electron-positron: The spectrum is still unknown at mildly relativistic temperatures. I have estimated its contribution by putting it equal to the e-p rate at low temperatures (since we are interested in photons with energies $\gtrsim 0.5$ MeV, one of the radiating electrons will be relativistic, as in the case of the high energy tail of e-p bremsstrahlung), twice the e-e rate at high temperatures (the ultra-relativistic limit), and joining smoothly in between. Since the equilibrium pair densities are low, the error made by using this approximation is small.
(3) Pair production: The dominant production mechanism, and that which takes most of the computing time, is that of photon-photon collisions. Photon-electron collisions account for 10-20% of the pair production. The other processes (photon-proton and particle-particle) are included for completeness, since they take a negligable amount of computing time.

(4) Annihilation: Svensson's[6] one parameter expression for the spectrum is used.

THE MODEL

Equilibrium models are characterized by two parameters: the temperature and an optical depth. Since the pair density is the quantity to be determined, the electron scattering depth is not a suitable parameter to chose. Instead the "proton optical depth" $\tau_p = N_p \sigma_T R$ is used. τ_p is related to the Thomson optical depth τ_{es} by $\tau_{es} = (1+2z)\tau_p$, where z is the number of electron-positron pairs per proton.

I ran models for various combinations of T and τ_p. Initially the slab has no photons or pairs present. At each timestep the photon and pair densities are incremented as required, and the photons are Comptonized and transported spatially. The pairs are assumed to stay where they are produced, to be instantly thermalized and to have an isotropic distribution. The model is then allowed to evolve either to an equilibrium solution, or until the pair density diverges.

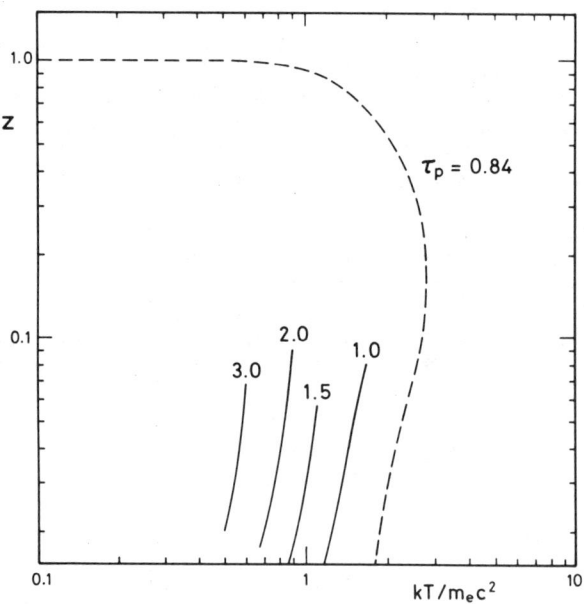

Figure 1. Equilibrium pair densitiy as a function of temperature. The curves are labelled by $\tau_p = N_p \sigma_T R$. The dotted curve is Svensson's result for $\tau_p = 0.84$.

It is well known that there are some combinations of temperature and optical depth for which no equilibrium solution exists. This is due to the fact that the annihilation cross section decreases at high energies, but that the pair production rate increases. At high enough temperatures the pairs are created faster than they can be destroyed. Physically, any extra energy goes into increasing the total number of particles, rather than their mean energy. The maximum temperature, in the limit of zero optical depth, is $kT \simeq 24\ m_ec^2$.

RESULTS

The results of these runs are shown in the figures, along with Svensson's[1] semi-analytical results for comparison. Svensson calculated the equilibrium solutions by assuming isotropic, homogeneous particle and photon distributions, and neglecting Comptonization. For each value of temperature and optical depth there are either two or no solutions for the equilibrium pair density, z. In the region where there are two solutions the higher one represents an unstable equilibrium - the plasma has negative specific heat. This solution will not be found by the computer program.

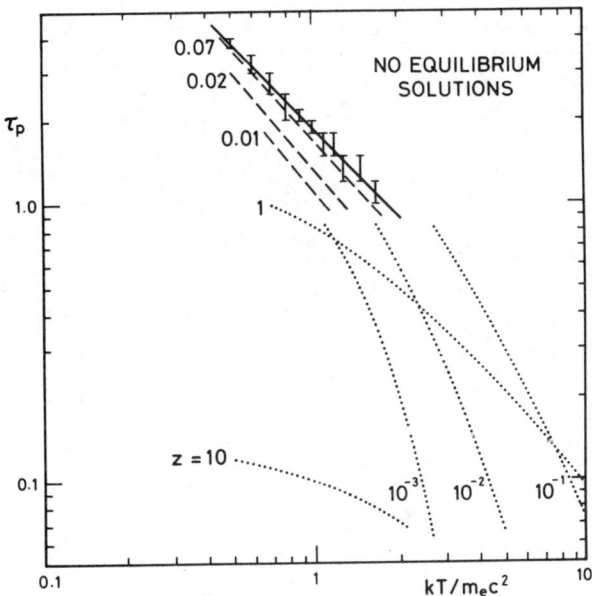

Figure 2. Proton optical depth as a function of temperature. The curves are labelled by z. The bottom of an error bar represents a model which converged, the top, one which diverged. The dotted curves are Svensson's results, adapted from his figure 7.

In figure 1 it can be seen that, for a fixed proton optical depth, the pair density increases rapidly with temperature. The slopes of the lines are consistent with Svensson's low pair density solutions. The solid line in figure 2 shows the boundary between models which had equilibrium solutions and those which diverged. The slope of this line is also consistent with the envelope of the semi-analytic constant z models. It is, however, somewhat lower. This is the result of including Comptonization and inhomogeneous distributions. Since the pair and photon densities are higher at the centre of the slab, this mimics a higher optical depth (or temperature) and so the runaway occurs sooner. In fact, the two methods are complementary. Svensson's results can be found relatively quickly, and also the method finds the high pair density branch, but it is only valid for small optical depths. The model discussed here requires much more computing time, but is valid for optical depths greater than unity.

The use of a thermal electron distribution introduces the minimum number of free parameters, and so leads to the simplest models. However, as Brinkmann[7] pointed out, the distribution might be truncated to some extent. If this is the case, it will reduce the cooling, (essentially by mimicking a lower temperature) and so move the boundary line up.

REFERENCES

1. R. Svensson, Ap.J. 258, 355 (1982).
2. A.P. Lightman, (these proceedings).
3. S. Stepney, MNRAS, 202, 467 (1983).
4. P.W. Guilbert, MNRAS, 197, 451 (1981).
5. S. Stepney & P.W. Guilbert, MNRAS, (in press).
6. R. Svensson, preprint, (1982).
7. W. Brinkmann, (these proceedings).

MONTE CARLO CALCULATIONS OF PAIR ANNIHILATION AND ITS INVERSE

Peter D. Noerdlinger
Los Alamos National Laboratory
Computational Physics Group
Applied Theoretical Physics Division
Los Alamos, New Mexico 87545

ABSTRACT

The reaction rates and product energy spectra were evaluated by Monte Carlo methods for the processes of electron positron pair annihilation and the inverse (pair production by the collision of gamma rays) in a relativistic, low density plasma. The results were applied to study the collision of isotropic power law spectra of pairs or gamma rays, respectively, as a model for processes of interest in winds from active galactic or QSO nuclei. As emphasized by Cavallo and Rees,[1] the annihilation spectra favored photons with a strong peak around the electron rest mass; extended tails persisted, however, with slope about 80% that of the incident particle spectra. In the case of pair production by gamma ray collision, the resulting pair spectra had much broader peaks near 5 to 10 electron rest masses, rather insensitive to the form of the photon spectra. If the two processes feed each other repeatedly, as expected in a wind flow, one expects substantial degradation of the emergent gamma ray spectrum toward the 0.5- 1.0 MeV region.

INTRODUCTION

As part of the joint work with J. Eilek and L. Caroff on Compton Thick Sources (e.g. outflows from active galactic or QSO nuclei), reported elsewhere in this conference, I evaluated the reaction rates in the fluid frame and the energy spectra of the reaction products for two typical processes in energetic, relativistic wind regions: electron-positron pair annihilation and pair production by the mutual collision of gamma rays, a process first suggested to be astrophysically important by Jelley.[2] The collision cross section and kinematics are identical with those of Ramaty and his co-workers.[3,4] Since the wind regions we have in mind are considerably less dense than the burst sources of primary interest to these workers, however, we could not assume a thermal plasma. Rather, the reaction rates and the distribution of the products in energy were calculated by Monte Carlo integration with finite energy bins. In the detailed simulations of winds the photon and particle spectra evolve according to the processes reported here and others, such as Compton and bremsstrahlung. In order to present a preliminary study of only the two pair processes, I have assumed power law spectra, suggested by cosmic ray and radio spectra.

METHOD

The Monte Carlo procedure was applied to the angles θ_{12}, θ, and $\emptyset$ of ref.(4), but was not used for determining the incoming energies. Rather, bins distributed uniformly in the logarithm of the kinetic energy (total energy for photons) were used, and the integrations were done with the particle energy fixed at a fiducial point within each bin. Since the particle spectra are not known when the integrals are done, it is necessary to pick the fiducial point somewhat arbitrarily. For example, picking the bin center amounts to assuming a flat spectrum. The choice implemented was to use the mean energy for the bin calculated as if the spectrum went like $E^{-\frac{1}{2}}dE$. The bin sizes were chosen small enough that the results were not sensitive to this choice. The energies of the reaction products were binned. This procedure, of course, destroys energy conservation if the products are then assumed to lie at the fiducial bin energies. The error is not significant in a single pass calculation such as that reported here. For the simulations of actual flows, the reaction products are re-used over and over, so that catastrophic energy errors could cumulate. For that case, a procedure for rebinning was devised that conserves particle number and energy. This procedure will be reported elsewhere.

For pair production, I used 18 bins with fiducial bin "centers" at 0.046 to 6000 $m_e c^2$. For annihilation I used 23 bins "centered" at 0.024 to 3000 $m_e c^2$.

Finally, a few caveats must be observed in interpreting the results presented here. The largest error probably arises from my using "in flight" cross sections for annihilation, although the assumed pair spectra reach such low energies that positronium formation could be significant. Ramaty and his co-workers avoided this problem by assuming a hot, <u>thermal</u> plasma. Secondly, the normalization is sensitive to the low energy cutoff (for spectra steeper than E^{-1}), or the high energy cutoff (for spectra flatter than E^{-1}), or to both cutoffs (1/E spectra). For pair production, this problem is ameliorated by the threshold. For annihilation, the height of the bump around 0.511 MeV is sensitive to the low energy cutoff.

RESULTS

The results are presented in Figs. (1) and (2), which show the spectra of the reaction products in the cases of annihilation and pair production, respectively. In each case, any multiplier or divisor shown has been applied to the data before plotting (i.e., reaction rates are of the order of 10^{-18} cm^3/sec in a unit increment of the natural logarithm of the product energy.)

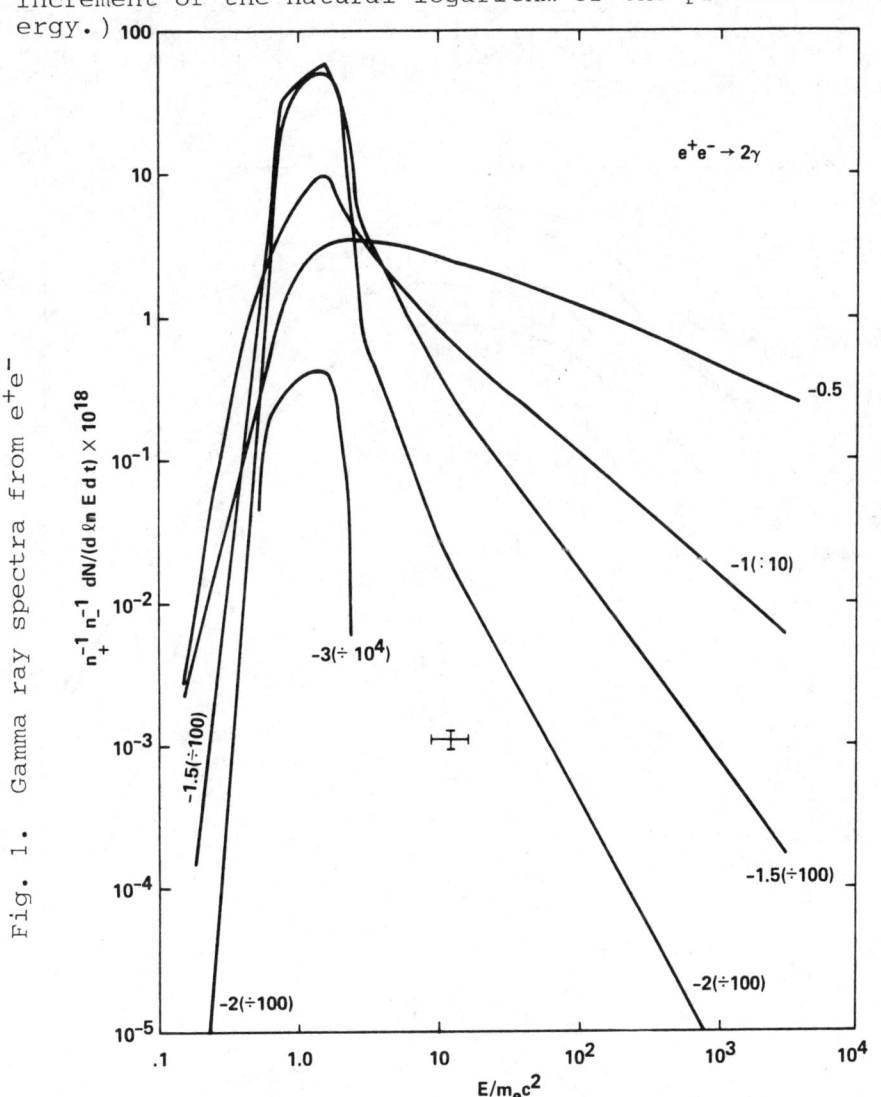

Fig. 1. Gamma ray spectra from e^+e^-

The production curves are labeled with the power law of the input energy spectrum, and, if necessary, a scale factor. For example, in Fig. 1, the curve labeled "-1.5($\div$ 100)" refers to an input spectrum $\sim E^{-1.5}$ dE, and the values have been divided by 100 before plotting. The error bars indicate on the ordinate the Monte Carlo noise, and on the abscissa the width of the energy bins. Since the widths of the annihilation peaks at 0.5-1.0 MeV were hardly greater than the bin width, additional runs were done at 60 times better energy resolution in the energy range 0.024 to 50 $m_e c^2$. These runs confirmed the width and shape of the peaks.

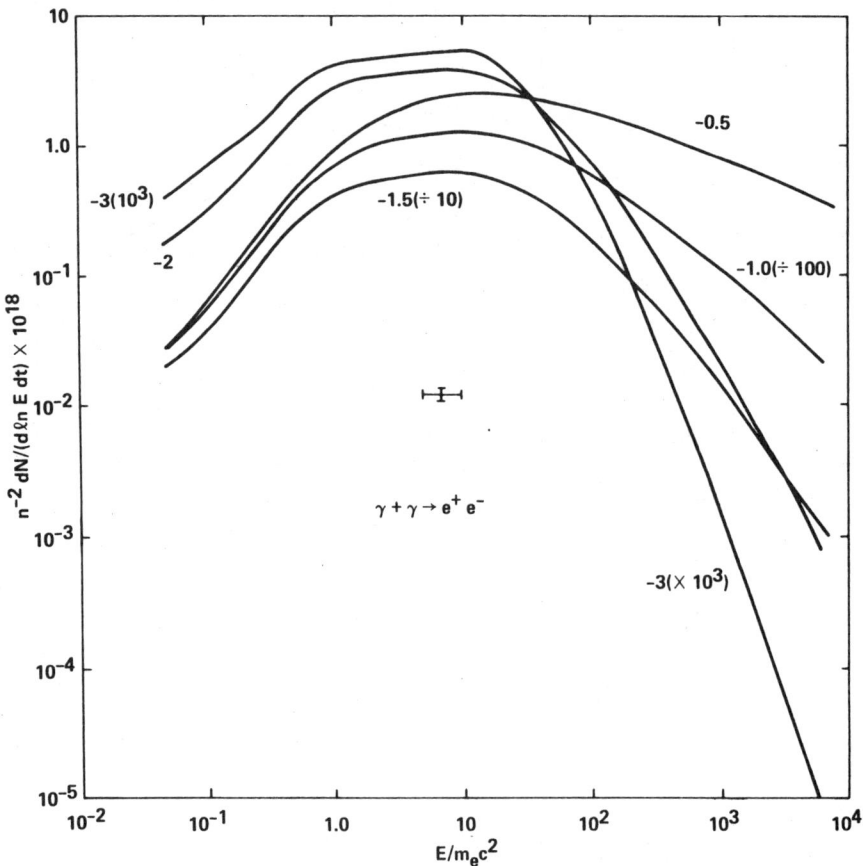

Fig. 2. Pair spectra from γ - γ interactions.

The extreme tails in Fig. 1 have an average slope about 80% that of the incident spectra. Collisions between a particle near the low energy cutoff and its anti-particle in the tail of the incident spectrum must be the main source of products in the high energy tails; if collisions among particles of comparable energy were responsible the slopes should be roughly double those of the incident spectra.

Fig. 2 shows the pair spectra from mutual collision of gamma rays. The threshold condition for gamma rays of frequency f_1 and f_2 colliding at angle θ to make a pair is $E_{cm} \gtrsim 2m_e c^2$, where

$$E_{cm} = h \left[2f_1 f_2 (1-\cos \theta) \right]^{1/2}.$$

This condition, along with the variation of the cross section with energy above threshold, conspire to produce the broad, featureless peaks in Fig. 2, rather insensitive to the specific power law distribution assumed. Although the peaks are very broad, their presence at a few times $m_e c^2$ can only speed the degradation of broad pair spectra to the 0.5-1.0 MeV range, when this process is combined with that of annihilation.

ACKNOWLEDGEMENT

This work was performed at the NASA Ames Research Center in 1979-80 under the support of a National Research Council Senior Resident Research Associateship. I am also indebted to L. Caroff and J. Eilek for valuable conversations.

REFERENCES

1. G. Cavallo and M.J. Rees, Mon. Not. Royal Astron. Soc. <u>183</u>, 359 (1978).
2. J.V. Jelley, Nature <u>211</u>, 472 (1966).
3. R. Ramaty and P. Mészáros, Astrophys. J. <u>250</u>, 384 (1981).
4. R. Ramaty, J.M. McKinley, and F.C. Jones, Astrophys. J. <u>256</u>, 238 (1982).

THE PARTICLE AND PHOTON SPECTRUM
OF AN OPTICALLY THICK RELATIVISTIC WIND

L. J. Caroff
NASA Ames Research Center, Moffett Field, CA 94035

J. A. Eilek
New Mexico Institute of Mining and Technology, Socorro, NM 87801

P. D. Noerdlinger
Los Alamos National Laboratories, Los Alamos, NM 87545

ABSTRACT

Compact astrophysical sources of energetic particles and photons are somewhat optically thick to particle-particle, particle-photon and photon-photon interactions. These interactions include pair production and annihilation, Compton and inverse Compton scattering, and electron-electron collisions (including bremsstrahlung). We present preliminary results from a calculation in which we evaluate the modulation of an initial source spectrum by these processes in an expanding relativistic electron-photon wind.

INTRODUCTION

Compact astrophysical sources of high energy particles and γ ray photons are often small and highly luminous. The particle density in these sources is high enough to make scattering and absorption processes important. Galactic sources -- stellar and accretion disk X ray sources, γ ray bursters, possibly the galactic center -- may be small and luminous enough that the mean free path of energetic photons or particles is several times smaller than the source size. Extragalactic sources -- active galactic nuclei and quasars -- produce relativistic particles and, often, X and γ rays, in a region small enough that these quanta should scatter or be absorbed before escaping. Theoretical models of radiation and acceleration processes near compact objects also predict generous production of energetic quanta in a small volume.

Thus, observations suggest that these sources are too small and luminous for scattering to be unimportant. However, the sources are not extremely optically thick to photon creation and absorption processes; if they were, we would find only cool, black body spectra at temperatures $T \sim (L/R^2 \sigma_{SB})^{1/4} \sim 10^4 - 10^7$ K. Conversely, despite the indications that the mean free path to photon destruction by γγ pair production is short, γ rays are observed from some sources. Despite the short path length for inverse Compton scattering of electrons against soft photons in compact radio sources, energetic electrons do survive, as implied by the radio emission. These facts suggest that quanta undergo a few interactions between production and escape, in a region where the optical depth is neither negligible nor very large. The facts also, perhaps, suggest that energetic photons and particles are produced in the sources, so that particle-photon cascades are as important as scattering processes and soft photon emission in producing the observed spectra.

We are therefore investigating particle-photon transfer and describe the model and present only preliminary results here; the full calculations will be presented elsewhere.[1]

THE MODEL

We consider an expanding plasma composed of electrons, positrons and photons. We assume that the high internal energy of the plasma, along with the particle/photon scattering, allows the entire plasma to expand in a relativistic wind. The particles and photons are well coupled, so that outward energy transport is by advection. This is similar to the system considered by Burns and Lovelace.[2] Cavallo and Rees[3] considered photons escaping from a plasma which is itself expanding subrelativistically. A different approach -- that of energetic photons diffusing through a proton-electron plasma which is held static by gravity -- has been taken recently by Lightman[4] and by Svensson.[5]

We consider a steady central energy source which produces some initial particle and photon spectrum. Two-quantum interactions will modulate this "source" spectrum while the fluid expands. Eventually the expansion will make the plasma thin to all interactions; at this point the spectrum will "freeze out". The photon spectrum observed from such an object will be the photon spectrum at freezeout, plus any subsequent radiation from the escaping electrons.

In the limit where the particles and photons are coupled and expand together, the spatial evolution of the particle distribution in a small region of the plasma depends only on local processes: scattering, creation and destruction of quanta. In this limit the problem is not a transfer problem, but rather involves the evolution of the initial distribution function towards thermalization in an expanding medium. If the expansion rate is faster than the thermalization rate, the distribution function will evolve from its initial state but may not fully thermalize. The local nature of the problem also means that the geometry of the source -- the expansion speed, $v(r)$, and whether the expansion is spherical, as in a wind, or linear, as perhaps in a jet -- is reflected simply in the local expansion timescale.

The most important two-quantum processes, with their total cross sections in the relativistic limit, are

a) Inverse Compton scattering: $\sigma_{IC} \sim \sigma_T$;

   Compton scattering: $\sigma_C \sim \frac{3}{8} \sigma_T \frac{1}{x} (\ln 2x + 1/2)$

b) Pair production:

   $\gamma\gamma \to e^+e^-$; $\sigma_{\gamma\gamma} \sim \frac{3}{8} \sigma_T \frac{1}{x^2} (2\ln 2x - 1)$

   $\gamma e \to e\, e^+e^-$; $\sigma_{\gamma e} \sim \frac{3}{8\pi} \alpha\, \sigma_T \left(\frac{28}{9} \ln 2x - \frac{218}{27}\right)$

c) Pair annihilation: $\sigma_{ee} \sim \frac{3}{8} \sigma_T \frac{1}{\gamma} (\ln 2\gamma - 1)$

d) Bremsstrahlung: $\sigma_{ff}^{em} \sim \frac{3}{2\pi} \alpha \sigma_T (\ln 2\gamma - 1)$

Here, σ_T is the Thomson cross section; α is the fine structure constant; $x = h\nu/m_e c$ and $\gamma = E_e/m_e c^2$.

The soft photon source will have a strong effect in modulating the spectrum through inverse Compton scattering and by degrading the hard photons in pair production. We omit free-free absorption in the calculation; the limit when $\tau_{ff}^{abs} \gg 1$ is the extreme, fully thermalized Planckian limit, and is not of interest for these calculations. The free-free optical depth can, in fact be still quite small in conditions where the optical depths to the other processes listed above are interestingly large (say, on the order 1 - 10). Also, in this range of optical depth, the limit of thermal equilibrium in pair creation and annihilation (cf. Landau and Lifshitz[6]) is never reached.

The spatial evolution of the density of quanta of energy E in a spherical flow is given by

$$\frac{1}{r^2} \frac{d(r^2 n(E,r) v(r))}{dr} = \sum_{\alpha,\beta} G_{\alpha\beta}(r) - n(E) \sum_{\alpha} L_{\alpha}(r) \quad (1)$$

where the gain function due to species α and β (with relative velocity $v_{\alpha\beta}$) is

$$G_{\alpha\beta} = \iint n_{\alpha}(E_{\alpha}) n_{\beta}(E_{\beta}) \sigma_{\alpha\beta}(E_{\alpha}, E_{\beta}; E) v_{\alpha\beta} dE_{\alpha} dE_{\beta} \quad (2)$$

and $\sigma_{\alpha\beta}(E_{\alpha}, E_{\beta}; E)$ is the differential cross section for a particle of species α with energy E_{α} and species β with energy E_{β} to make the particle in question with energy E. Similarly, the loss function is

$$L_{\alpha} = \int n_{\alpha}(E_{\alpha}) \sigma_{\alpha}(E_{\alpha}, E) v_{\alpha} dE_{\alpha} \quad (3)$$

with $\sigma_{\alpha}(E_{\alpha}, E)$ being the total cross section for destruction of the particle at energy E by interaction with a particle of species α at energy E_{α}.

We discretize this nonlinear integrodifferential system by dividing the energy ranges considered for the photons and electrons into a number of discrete bins. Within each bin we estimate the integrals involved in equations (2) and (3), using the differential form of the cross section and using smooth intrabin distribution functions; this calculation gives discrete values for σ_{ij}^k, namely,

$$\sigma_{ij}^k = \left\langle \frac{1}{n(E_i) n(E_j)} \right\rangle \iint n(E_i) n(E_j) \sigma(E_i, E_j; E_k) dE_i dE_j \quad (4)$$

The differential equation now becomes, for the ith and jth energy bins,

$$\frac{1}{r^2}\frac{d(r^2 n_i(r)v(r))}{dr} = \sum_{i,j} \sigma^i_{jk} n_j n_k v_{jk} - \sum_j \sigma_{ji} n_j n_i v_{ij} \qquad (5)$$

We can write this in dimensionless variables, using $x = r/r_o$, $c^i_{jk} = \sigma^i_{jk} v_{jk}/\sigma_T c$, $s_{ji} = \sigma_{ji} v_{ji}/\sigma_T c$ and $f_i = r^2 n_i v(r)/F_o$, where σ_T is the Thompson cross section, and r_o and F_o are fiducial values of the radius and number flux. We then get

$$\frac{df_i}{dr} = \frac{\eta}{x^2 \gamma^2 \beta^2} \sum_{j,k} c^i_{jk} f_j f_k - \sum_j s_{ji} f_j f_i . \qquad (6)$$

The wind flow is described by the functions $\beta(r) = v(r)/c$; $\gamma(r) = (1 - \beta(r)^2)^{-1/2}$. The source size and particle flux have been combined in the optical depth parameter,

$$\eta = F_o \sigma_T / 4\pi r_o c . \qquad (7)$$

The set of equations (6) can be solved numerically for $\{f_i(x)\}$, given any initial spectrum, $\{f_i^o\}$, and any value of η. The wind dynamics can be specified externally through the $(\beta(r), \gamma(r))$ field.

THE EVOLUTION OF THE WIND

Thus far we have completed only preliminary calculations. We therefore will comment here on the physics expected in these models; the full calculations will be reported elsewhere.[1] Our discussion should complement those of Cavallo and Rees,[3] and Noerdlinger.[7]

There are two important features of the model: the initial optical depth for interaction of a species j particle with a species i particle, $\tau_i = \eta f_i s_{ji}$; and the source spectrum, $\{f_i^o\}$. Knowledge of the cross sections for the different interactions allows us to predict several important limits, as follows.

a) <u>The totally thin case</u> ($\tau_i \ll 1$ for all i). The source spectrum should be directly observable, with no modulation by the wind.

b) <u>A hard source function, with little soft photon emission in the wind</u> (most source quanta above ~ 1 MeV; $\tau^{em}_{ff} < 1$). The dominance of $e^+ e^- \to \gamma\gamma$ and its inverse process (or $\gamma e \to e\, e^+ e^-$ at high photon energy) will cause these two processes to modulate the initial source function towards a balance in which $f_e/f_\gamma \sim (\sigma_{\gamma\gamma}/\sigma_{ee})^{1/2} \sim 1/3 - 1$. With only a few scatterings before freezeout, the electrons and photons will reach this balance rapidly and emerge at nearly their initial energy; with many such scatterings, adiabatic losses will reduce the energy per particle before escape.

c) <u>A steep source function with little soft photon emission in the wind</u> (an initially large number of low energy/subrelativistic particles; $\tau^{em}_{ff} \lesssim 1$ still). In the presence of cold quanta in moderate quantities, the still-dominant pair production and annihilation reactions (as in case b) will degrade the initial energy per particle to ~ 1 MeV, at which point the photon-based pair production shuts off.[7] Thus, with a few scatterings before freezeout, the emergent spectrum will have a peak or turnover around 1 MeV. Again, if there are a larger number of scatterings before escape, adiabatic losses will lower the energy per particle below ~ 1 MeV.

d) <u>A strong soft photon source in the wind</u> (e.g., $\tau_{ff}^{em} > 1$). A large number of soft photons, as from bremsstrahlung, will cause inverse Compton (cold photon - hot electron) and pair production (cold photon - hot photon) reactions to dominate. As long as the hot photons have energy $\gtrsim$ 1 MeV, these two reactions will lead to a balance in which $f_e/f_\gamma \sim \sigma_{\gamma\gamma}/\sigma_{TC} \sim 0.1 - 0.3$. However, as the number of scatterings before escape increases, the adiabatic decay and also the pair production/annihilation degrading (as in case c) reduces the hard photon energy below 1 MeV, thus shutting off the electron-regenerating pair production reactions.

e) Free-free absorption important ($\tau_{ff}^{abs} \gg 1$). In this extreme limit the source should be fully thermalized. This limit is not seen in practice.

These optical depth effects are clearly dominant in determining the escaping photon and particle spectrum. The "interesting" range -- cases (b), (c) and (d) above -- represent effective optical depths ~ 1 - 10, in general. However, the rate of expansion, and hence the source geometry, is also important. A spherical flow expands rapidly, so that freezeout occurs quickly and the effective optical depth is therefore reduced. Fairly high initial optical depths in a spherical flow may still produce spectra at freezeout which are not fully thermalized. On the other hand, a constrained flow (such as a jet) expands more slowly, and lower initial optical depths will be required in order to produce observationally interesting freezeout spectra.

ACKNOWLEDGEMENTS

We wish to thank Mike Dove for extensive computing assistance. J.A.E. was partially supported by NASA-Ames University Consortium NCA2-OR511-101. P.D.N. is grateful for an NRC Senior Postdoctoral Research Associateship.

REFERENCES

1. Caroff, L. J., Eilek, J. A. and Noerdlinger, P. D., 1983, in preparation.
2. Burns, M. L. and Lovelace, R. V. E., 1982, Ap.J., 262, 87.
3. Cavallo, G. and Rees, M. J., 1979, M.N.R.A.S., 183, 359.
4. Lightman, A. P., 1982, Ap.J., 253, 842.
5. Svensson, R., 1982, Ap.J., 258, 335.
6. Landau, L. D. and Lifshitz, E. H., 1980, Statistical Physics (New York:Pergamon).
7. Noerdlinger, P. D., this proceedings.

PAIR PRODUCTION AND ANNIHILATION IN STRONG MAGNETIC FIELDS

J. K. Daugherty
University of North Carolina-Asheville, Asheville, NC 28804

A. K. Harding
NASA/Goddard Space Flight Center, Greenbelt, MD 20771

ABSTRACT

Electromagnetic phenomena occurring in the presence of strong magnetic fields are currently of great interest in high-energy astrophysics. In particular, the process of pair production by single photons in the presence of fields of order 10^{12} Gauss is of importance in cascade models of pulsar gamma ray emission, and may also become significant in theories of other radiation phenomena whose sources may be neutron stars (e.g., gamma ray bursts). In addition to pair production, the inverse process of pair annihilation is greatly affected by the presence of superstrong magnetic fields. The most significant departures from annihilation processes in free space are a reduction in the total rate for annihilation into two photons, a broadening of the familiar 511-keV line for annihilation at rest, and the possibility for annihilation into a single photon (which dominates the two-photon annihilation for $B > 10^{13}$ Gauss). The physics of these pair conversion processes, which is reviewed briefly, can become quite complex in the teragauss regime, and can involve calculations which are technically difficult to incorporate into models of emission mechanisms in neutron star magnetospheres. However, recent theoretical work, especially in the case of pair annihilation, also suggests potential techniques for more direct measurements of field strengths near the stellar surface.

INTRODUCTION

The observational discovery of pulsars in the late sixties was rapidly followed by their identification with rotating magnetic neutron stars. Early models of these objects, in which the magnetic fields were postulated as the means by which the stellar rotational energy could be converted to electromagnetic radiation, in fact led to enormous estimates of the surface field strengths, on the order of 10^{12} Gauss or even more. Macroscopic fields of such intensity are roughly six orders of magnitude greater than the strongest attainable laboratory fields (which are currently the megagauss fields generated by implosive flux-compression techniques[1]). In fact, the field strengths thought to exist in the interiors and magnetospheres of neutron stars may well be the highest values occurring in nature.

In light of the fact that teragauss magnetic fields are so exotic by terrestrial standards, it is not surprising that the theory of electromagnetic phenomena occurring in such an environment is far from complete. Although many of the fundamentals in this area

0094-243X/83/1010387-13 $3.00 Copyright 1983 American Institute of Physics

were investigated well before the discovery of pulsars[1], even these early investigations have only gradually become well known to astrophysicists. Hence a considerable amount of effort has been spent, especially in the last decade, on efforts both to increase our understanding of quantum electrodynamics in strong fields, and to incorporate this knowledge into specific models of neutron star magnetospheres and their emission mechanisms.

In this discussion we will review some recent developments in these areas, especially those relating directly to electron-positron pair conversion processes. The treatment is brief and incomplete, but is intended to convey roughly the level of our current knowledge of these processes, and to indicate some of the difficulties which this sort of physics presents to those interested in building models of pulsar cascade showers and other neutron star emission mechanisms.

QUANTUM ELECTRODYNAMICS IN STRONG MAGNETIC FIELDS

Although on a scale of the dimensions of a neutron star the magnetic field is certainly nonuniform, the length scale of interest in quantum electrodynamics is so much smaller (on the order of the Compton wavelength) that in calculations of electron-photon interactions the field may be considered perfectly uniform and infinite in extent. It may be noted that this is already a good approximation for typical accelerator magnets, and should in fact be much better for neutron star dimensions, whether these fields might be simply dipolar or of some more complex multipolar form. Moreover, the field is believed to be so intense that its treatment as a classical or prescribed "external" field (which is not itself influenced by the particle interactions) is also an excellent approximation. Hence the fundamental tool needed for calculating electromagnetic processes in neutron star magnetospheres is the quantum mechanical solution for electron/positron motion in a constant, uniform magnetic field. Fortunately this is one of the cases for which exact solutions of the relativistic Dirac equation are available[2]. We will not discuss the details of these wavefunctions here, but will concentrate on the energetics and kinematics associated with them.

The energy levels of a Dirac electron moving in a uniform magnetic field $\underline{B}$ may be written in the form

$$E_n = [c^2 p^2 + m^2 c^4 (1 + 2nB/B_{cr})]^{1/2} \qquad (1)$$

where p denotes the component of momentum parallel to the field axis, $B_{cr} = m^2 c^3 / e\hbar = 4.414 \times 10^{13}$ Gauss, and n = 0, 1, 2, ... The form of this equation shows that the parallel momentum is not affected by the presence of the field, while the transverse motion is quantized. The critical field strength B_{cr} is seen to be a combination of fundamental constants whose value is such that a transition between adjacent orbitals produces an energy change comparable to the rest mass of the electron.

It is worth noting that the fully relativistic Dirac solutions to the equations of motion, which correspond to the energy level formula given above, should be preferred to the corresponding nonrelativistic Schrödinger wavefunctions in most applications to neutron star astrophysics. This is true in particular when either the parallel momentum p or the product nB/B_{cr} becomes significant compared to the rest energy mc^2. For fields far below B_{cr}, the transverse energy becomes large only for enormous values of n (in which case the transitions to lower levels produce what is usually termed synchrotron radiation). However, at field strengths comparable to B_{cr}, even moderate values of n (normally associated with nonrelativistic "cyclotron" radiation) can produce effectively relativistic behavior.

Some obvious characteristics of the relativistic motion may be inferred from Figure 1, which shows the first few energy levels in a field of 5 teragauss. (Here the parallel momentum p has been set to zero.) As the figure indicates, each energy level above the ground state E_0 actually corresponds to two distinct spin states, which may be thought of as "up" or "down" along the field direction. It may be found from the relativistic Dirac wavefunctions that transitions involving spin flips are less probable than the corresponding "no-flip" transitions. However, the full set of spin states must be considered in determining the relative populations of excited states in neutron star magnetospheres. In addition, the decreasing spacings between levels for increasing quantum numbers n implies that multi-level populations will produce not only multiple harmonics of emission lines, but also line-splitting effects at each harmonic.

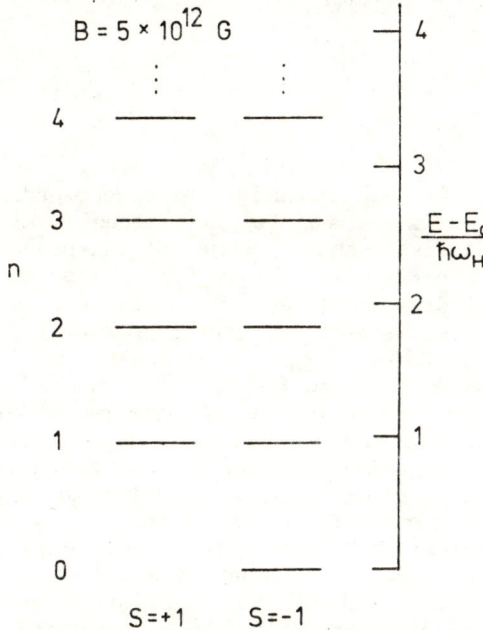

Fig. 1. Energy-level diagram for a Dirac electron in a uniform magnetic field of 5 teragauss. Here n is the orbital quantum number as given in eq. (1), and the parallel momentum p = 0. Scale at right is the energy spacing in units of cyclotron energy $\hbar\omega_H = e\hbar B/mc$.

In terms of perturbation theory calculations, pair production and annihilation effects in strong magnetic fields are described not only by the familiar second-order Feynman diagrams for these processes[3], but also by first-order diagrams involving only a single photon vertex. However, the meaning of "first-order" here must be clarified. In contrast to quantum electrodynamics in free space (where in fact all first-order processes are kinematically forbidden), the Dirac electron wavefunction itself fully describes the interaction with the constant, uniform magnetic field to all orders, while transitions induced by radiation-field photons are described by perturbation theory. The distinction is shown in Figure 2, which compares the second-order process of Bremsstrahlung in field-free space with the "first-order" process of magnetic Bremsstrahlung (synchrotron radiation) in a uniform magnetic field. (Note that there is a second contribution to case (a), in which the vertices are interchanged.) In (b) the electron interaction with the macroscopic B-field is depicted as an infinite number of zero-frequency photons, as opposed to the single "photon" associated with the microscopic Coulomb field of a nucleus in case (a).

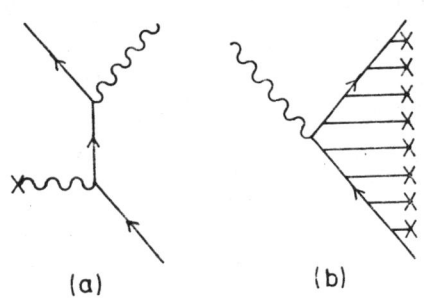

Fig. 2. Comparison of second-order diagram (a) for Bremsstrahlung in free space with "first-order" diagram (b) for magnetic Bremsstrahlung (synchrotron radiation) in external field.

Of the four possible ways to draw Feynman diagrams for first-order transitions, two represent just the familiar processes of synchrotron radiation and absorption. Obviously these effects are observable for fields far below the critical value B_{cr}, although it should be noted that the behavior of these effects in the strong-field regime is quite different from that associated with interstellar or accelerator fields[4-8]. However, the remaining two first-order processes, which are seen to be pair production and annihilation, are essentially quantum-mechanical effects and as such become significant (i.e., observable) only for fields approaching B_{cr}.

The kinematics of all the first-order transitions are determined by two equations, one for conservation of energy and one for conservation of parallel momentum:

$$\hbar\omega = E_j + E_k \qquad (2a)$$

$$\hbar\omega \cos\theta/c = p + q \qquad (2b)$$

where (E_j,p) and (E_k,q) are the total energy and parallel momenta of the positron and electron respectively, and θ is the angle between the photon wave vector $\underline{k}$ and the field direction. There is no equation for the conservation of transverse momentum because the field itself participates in the transverse momentum transfer, and since the field lines are assumed to be infinitely "rigid" it is not possible to determine the transverse momentum exchange suffered by the particles. It is perhaps worth noting, however, that the rigidity of the field lines is also required to guarantee the conservation of total energy among the particles (equation 2a). In the case of neutron star magnetic fields, the assumption of infinitely rigid field lines is an extremely good approximation, since the field is both superstrong and and "anchored" (mechanically supported) by the mass of the star itself.

PAIR PRODUCTION

The process of magnetic pair production (the conversion of a single photon into an electron-positron pair in the presence of an external magnetic field) has been investigated by a number of authors[9-13], and the essential results have been known (although in some cases available only as unpublished doctoral theses) since the early fifties. The quantity of physical interest in dealing with this process is the attenuation coefficient (inverse of the mean free path) for a photon of specified energy ω and polarization vector $\underline{\varepsilon}$, propagating at some angle Θ to the uniform magnetic field $\underline{B}$. (Henceforth natural units, in which $\hbar = c = 1$, will be assumed.) The actual calculation of the attenuation coefficient, based on the first-order Feynman diagram discussed above, may be performed in the Lorentz frame for which the photon motion is perpendicular to the field direction, with the understanding that the results (for unpolarized photons at least) may be generalized to arbitrary directions of propagation by making Lorentz transformations parallel to $\underline{B}$. We will follow this approach, but we must remember that the final step of transforming the results must eventually be performed in applications where $\underline{k} \cdot \underline{B} = 0$.

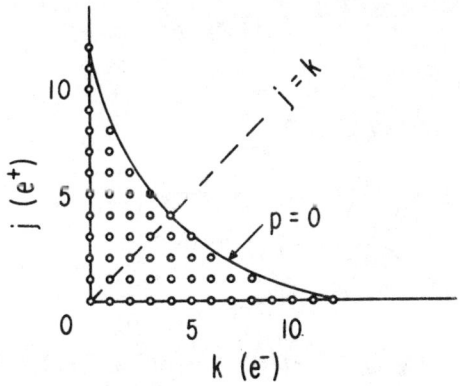

Fig. 3. Kinematically allowed states for the created e^+/e^- pair with quantum numbers (j,k).

Before we give the results of the dynamical calculations, it will be useful to consider the kinematic aspects of the pair production process. The final states of this process (namely the pair) may be labeled by two quantum numbers (j,k) describing the energy levels of the electron and positron respectively. As discussed in the preceding section, the kinematical requirements may be expressed as equations (2a,b). Now for a given photon energy and field strength B, these equations may be solved for the parallel momentum p:

$$p = p(j,k) = \pm m \left[\omega'^2 - 1 - (j+k)B' + (j-k)^2 \frac{B'^2}{4\omega'^2}\right]^{1/2} \quad (3)$$

where $\omega' = \omega/2m$ and $B' = B/B_{cr}$. The energy-momentum restrictions are equivalent to the requirement that p^2 be a nonnegative number; for given ω and B, any pair states (j,k) which meet this requirement will be allowed. As shown in Figure 3, the region of allowed states in the (j,k) plane is thus enclosed within the line corresponding to the equation $p(j,k;\omega,B) = 0$. The sum of the individual transition rates for these final states will yield the attenuation coefficient for the pair conversion process. In addition, the relative probabilities for the various transitions over the (j,k) plane determine the energy distribution of the emergent pair, which is also of importance in applications (e.g., pulsar cascade models).

The explicit calculation of the attenuation coefficients in terms of the Dirac wavefunctions has been fully discussed in the references mentioned above and will not be repeated here. The results for the cases of photons whose electric vectors are polarized parallel and perpendicular to the field direction respectively are given by the expressions

$$R_{\parallel}(\omega',B') = \frac{\alpha_o}{2\xi} \sum_j \sum_k \frac{1}{|p_{jk}|} \{(E_j E_k + m^2 - p^2)(|M(j,k)|^2 + |M(j-1,k-1)|^2)$$

$$+ 2\sqrt{jk}\, B'm^2\, [M^{\dagger}(j,k)M(j-1,k-1) + M^{\dagger}(j-1,k-1)M(j,k)]\} \quad (4a)$$

$$R_{\perp}(\omega',B') = \frac{\alpha_o}{2\xi} \sum_j \sum_k \frac{1}{|p_{jk}|} \{(E_j E_k + m^2 + p^2)(|M(j-1,k)|^2 + |M(j,k-1)|^2)$$

$$- 2\sqrt{jk}\, B'm^2\, [M^{\dagger}(j-1,k)M(j,k-1) + M^{\dagger}(j,k-1)M(j-1,k)]\} \quad (4b)$$

where $\xi = \frac{2\omega'^2}{B'}$, $M(j,k) = (-1)^{G-S} \sqrt{\frac{S!}{G!}} e^{-\xi/2} \xi^{\frac{G-S}{2}} L_S^{G-S}(\xi)$,

and $G = \max(j,k)$, $S = \min(j,k)$..

At first glance these equations are not very illuminating, but from them we can immediately infer one important aspect of the behavior of the attenuation coefficients as functions of photon energy and field strength. The appearance of the momentum term p in the denominator of each summand implies that if any p vanishes, the entire expression becomes singular. Now from equation (3) it may be seen that for each integer pair (j,k), only certain combinations of ω and B can make p vanish. This in turn implies that for fixed B the attenuation coefficients R(ω,B) are singular at a discrete sequence of energies ω_{jk}, with each singularity resulting from

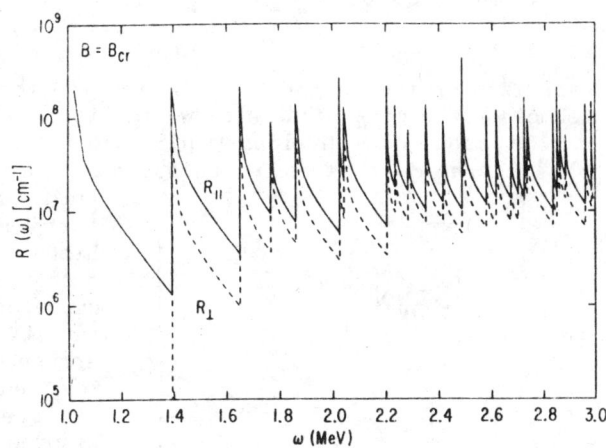

Fig. 4. Exact attenuation coefficients for photons propagating at right angles to the field direction and with polarizations parallel and perpendicular to the field, plotted vs. energy for fixed B (here chosen to be B_{cr}).

particular (j,k) values for which p = 0. In terms of Figure 3, the attenuation coefficients are singular whenever ω and B are such that the line p(j,k; ω,B) = 0 intersects an integer pair (j,k). Hence a plot of R(ω,B) vs. ω shows the sort of sawtooth behavior which is depicted in Figure 4. The energies at which the singularities occur are readily found from equation (3), and by plotting these energies it is found that the average spacing between peaks rapidly becomes smaller as the field strength is decreased, and that for fixed B the peaks in successive fixed-length energy intervals dω become more numerous.

This complex behavior was noted even by the earliest authors who investigated this process[9,10], but for the maximum field strengths then considered to be of conceivable practical interest, the density of singularities in measurable energy intervals would be so great that only smoothed-out, average values taken over each interval were considered to be of physical interest. Hence asymptotic expressions for the attenuation coefficients, valid in the limiting regimes $\omega \gg 2m$ and $B \ll B_{cr}$, were derived from expressions (4). The crucial steps involved in this derivation are the replacement of the (j,k) summations by integrations over suitably chosen continuous variables, and the determination of appropriate asymptotic forms for the generalized Laguerre polynomials[9,10]. (For a recent discussion of this derivation, see reference[13].) The final results are expressible in the relatively simple form

$$\bar{R} \sim 0.23 \frac{\alpha_o}{\lambda_c} B' \exp\left(-\frac{4}{3\chi}\right) \quad , \chi = \omega'B' \ll 1 \tag{5}$$

A comparison of the asymptotic results with the exact forms (4) for two sample field strengths in shown in Figure 5.

In most astrophysical applications to date, especially in models of cascade showers in pulsar magnetospheres[14,15], the last form of the attenuation coefficient has been used. Since the typical photon energies produced by curvature or synchrotron radiation in a cascade sequence are well above the MeV range, and the ambient field strengths are usually assumed to be no more than a few teragauss, the complex behavior for near-threshold energies and fields $B < B_{cr}$ has essentially been ignored. Moreover, the energy distribution of the pair has usually been assumed to be given simply by $E_\pm = \omega/2$. (It should also be noted that only unpolarized attenuation coefficients are used in these models.)

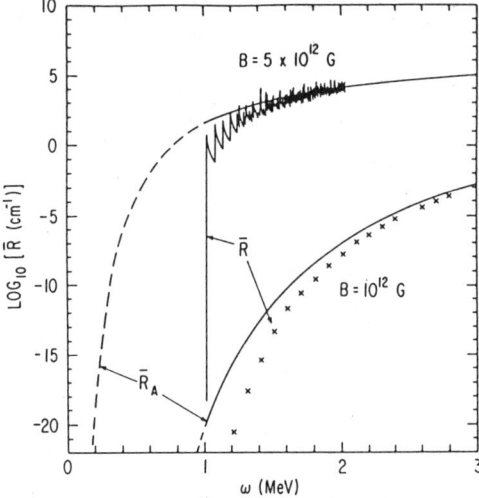

Fig. 5. Comparisons of exact (unpolarized) photon attenuation coefficients vs. asymptotic forms for 1- and 5-teragauss fields.

However, it turns out that this asymptotic form is often not a good approximation at all, since (as mentioned above) it is really necessary to Lorentz-transform the attenuation coefficients before they may be applied in frames for which $\underline{k}\cdot\underline{B}$ does not vanish. Now it turns out that the transformation law of the attenuation coefficients may be expressed as

$$\bar{R}(\omega, \underline{B}) = \sin\theta \, \bar{R}_o(\omega \sin\theta, \underline{B}) \tag{6}$$

This law expresses the fact that energetic photons, propagating at an angle $\theta \simeq 1/\omega$ to the field direction, may be Lorentz-transformed down to energies near (or even below) threshold in the "transverse" frame for which $\underline{k}\cdot\underline{B} = 0$. But it is just such angles of propagation which are typical for the curvature-radiation photons which should inititate the pulsar cascade process. Hence a proper treatment of pair production by such photons must take into account the near-threshold effects, and the resulting modifications to the estimated multiplicities (ratios of secondary pairs to the number of radiating primary electrons drawn from the stellar surface) may have a significant effect on the entire cascade development. Moreover, it turns out that the energy distributions of the pairs in the near-threshold or high-field regimes, as determined by the individual contributions to expressions (4), can broaden considerably and even show double-peaked behavior as shown in Figure 6. (The double peaks imply that one member of the pair tends to get most of the photon energy, while the other emerges much closer to its ground state.) Hence the usual assumption that $E_\pm = \omega/2$ is also likely to be a poor approximation. This is unfortunately only one example where more careful treatments of the underlying processes in neutron star magnetospheres may force significant revisions to current models of their emission mechanisms.

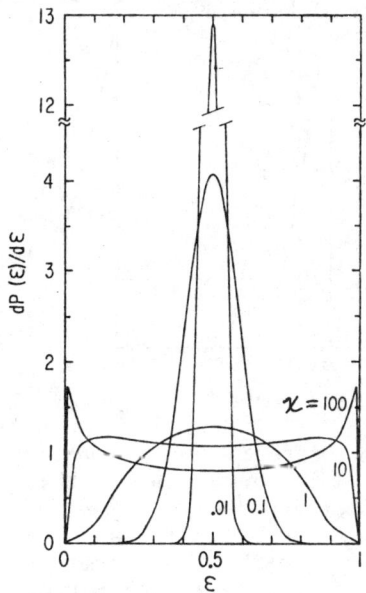

Fig. 6. Energy distributions for one member of the created pair, computed from an integration over the (j,k) probability distributions. The distributions shown are all normalized to unity, and are plotted vs. pair energy divided by photon energy.

PAIR ANNIHILATION

The inverse process of first-order pair production, namely pair annihilation into a single photon[16,17], is also of interest in neutron star astrophysics, but in this case the process becomes significant only for fields very close to B_{cr}. In weaker fields the first-order process is dominated by the analog of annihilation in free space, in which two or more photons make up the final state. The two-photon annihilation process, which will be discussed later below, is itself strongly modified by ambient fields of a few teragauss or more.

The kinematics for one-photon annihilation are similar to equations (2) for pair production, where the roles of initial and final states are now reversed. These equations imply that for annihilation by pairs at rest, the emergent radiation must form a flat fan beam at right angles to the field direction. For thermally or otherwise broadened e^+/e^- parallel momentum distributions, the fan beam would fill out to a form such as that shown in Figure 7. Corresponding to the angular distribution, the spectral distribution of the photon, assuming emission at specific angles, have characteristic asymmetrical shapes as shown in Figure 8. Both the overall shift toward higher energies and the increased broadening are due essentially to the Doppler effect caused by annihilation of pairs with nonzero net parallel momentum.

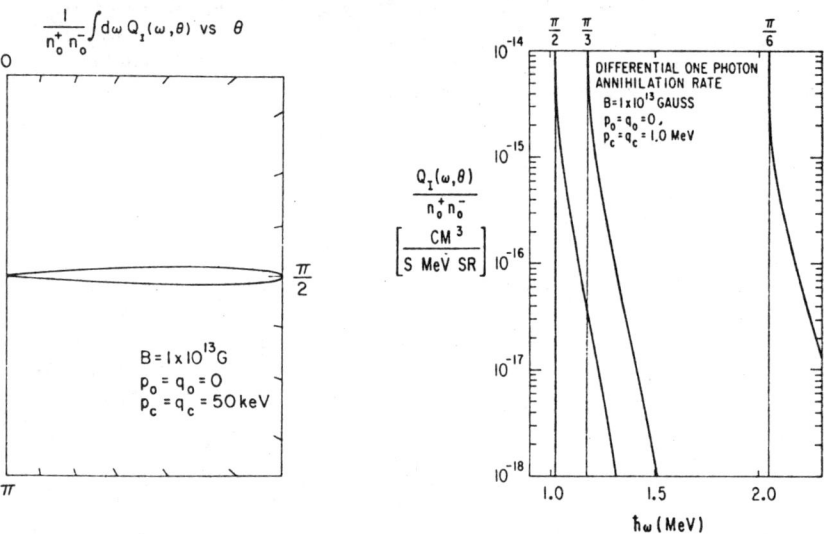

Fig. 7. Angular distribution of the one-photon annihilation radiation resulting from Gaussian electron-positron distributions with momentum widths of 50 keV, in a field of 10 teragauss (see reference 17).

Fig. 8. Differential one-photon annihilation spectrum emitted from a ground-state gas of electrons and positrons with Gaussian parallel momentum distributions, here with widths of 1 MeV (reference 17). Separate spectra are shown for three angles of emission.

The kinematics for two-photon annihilation are more complex but are still characterized by a loss of transverse momentum conservation. As in the case of the first-order processes, the electron wavefunctions (and here the propagator as well) correspond to the Dirac constant-field wavefunctions. Several dynamical and kinematical aspects of two-photon annihilation turn out to be of special interest for neutron star astrophysics. The first is a reduction in the total annihilation rate as compared with the free-space value, accompanied by a flattening of the isotropic angular distribution of the emitted photons (tending toward a fan beam perpendicular to the field direction)[16]. In addition, the role of the field as a transverse momentum absorber leads to a sharply field-dependent broadening of the annihilation spectrum[17]. In particular, for annihilation at rest, the spectrum is no longer constrained by momentum conservation to the familiar 511-keV line, but is broadened to an extent which is quite sensitive both to field strength and angle of emission (see for example Figures 9,10). The line broadening is especially interesting in that it might become observable to detectors with high energy resolution for fields of only a few teragauss. Hence as a potential means for direct measurement of field strengths, the line-broadened two-photon annihilation spectrum might have wider usefulness. In this context it should be also be noted that the angular dependence of the broadening effects should, in beams emerging from local "hot spots" of rotating neutron stars, translate to a temporal dependence of the pulsed emission.

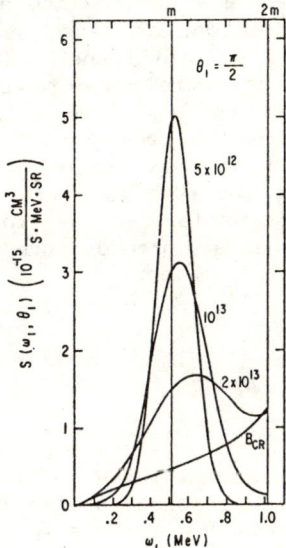 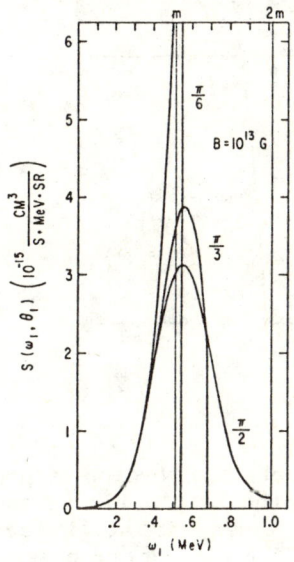

Fig. 9. Two-photon annihilation spectrum for pairs at rest, observed perpendicular to the field direction, for several field strengths above one teragauss. The sharply field-dependent line broadening is evident and is here due entirely to the loss of transverse momentum conservation in the intense field (i.e., no thermal broadening is included).

Fig. 10. Two-photon spectrum for annihilation at rest in a field of 10 teragauss, as viewed from several polar angles.

Finally, it should be noted that the range of field strengths over which the line broadening and overall rate decrease of the two-photon annihilation process becomes observable is really quite small, on the order of only one teragauss. The same statement holds for the "crossover" field strength (about 10 teragauss) at which the total one-photon annihilation rate begins to dominate the two-photon rate, as shown in Figure 11. Given this extreme sensitivity to the ambient field strength, it may eventually become possible for detectors with both high spectral and temporal resolution to provide accurate local values or even maps of the surface fields through the measurement of annihilation radiation.

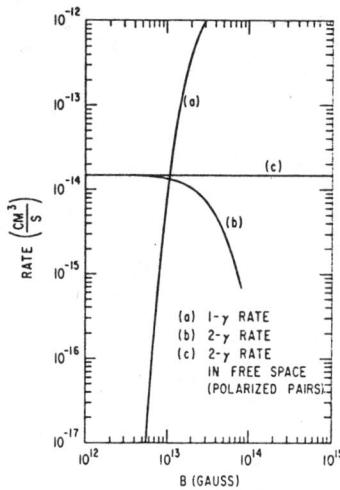

Fig. 11. Comparison of the total rates for annihilation into one and two photons at rest, as a function of field strength. The dashed line corresponds to the limit for two-photon annihilation at rest in free space, taking into account the fact that the electrons are in this case polarized as they would be in the ground state if the field is present.

CONCLUSION

The detailed physics of processes in teragauss magnetic fields, including the simple first- and second-order processes of the type discussed above, is still far from complete. The calculations of even the first-order processes involve a great deal of technical difficulties which are especially troublesome in the context of neutron star astrophysics, where often time-consuming calculations are already required to estimate emission spectra from the models. Indeed, it appears in many cases impractical to incorporate "exact" expressions such as equation (4) into models of pulsar cascades or gamma-ray burst sources, at least with computing resources that are less than what currently fall into the category of supercomputers. On the other hand, the analytical difficulties and pitfalls associated with attempts to improve or extend the asymptotic results for these processes are equally formidable, and the reliability of the results is open to question.

These same problems are only aggravated if we begin to consider various other fundamental processes, most of them intrinsically more complex than the simple pair conversion effects discussed here, which may also play significant roles in neutron star astrophysics. Among these may be mentioned such effects as the field-induced index of refraction[18,19], photon splitting[20], trident production[1], and such complications to the pair annihilation process as "magnetic" positronium formation and annihilation of positrons with the atomic electrons in the surface layers of the neutron star. At present, it is safe to say that the theoretical problems in this area will only slowly be put to rest.

At the same time, it is encouraging to note that even fairly elementary models of such processes as pulsar cascade showers[2,1] have already yielded reasonable agreement with the currently available observational data, in spite of the fact that the fundamental conversion processes have been treated in rather crude ways to date. Hence it may be that refinements to the theoretical treatments of these processes will only improve this agreement and not produce drastic or qualitative changes in the existing models. At present, however, it appears necessary not only to make the usual plea for more observational data, but also to emphasize the need for further theoretical investigations of the elementary processes of high-field physics and their careful treatment in models of pulsars, pulsating X-ray sources, and gamma-ray bursters.

REFERENCES

1. T. Erber, Rev. Mod. Phys. 38, 626 (1966).
2. See for example M. H. Johnson and B. A. Lippmann, Phys. Rev. 76, 828 (1949).
3. J. D. Bjorken and S. D. Drell, Relativistic Quantum Mechanics, (McGraw-Hill, N. Y.).
4. A. A. Sokolov and I. M. Ternov, Synchrotron Radiation, (Berlin: Akademie-Verlag), 1968.
5. H. G. Latal and T. Erber, Ann. Phy. (N.Y.), 108, 408 (1977).
6. D. White, Phys. Rev. D 9, 868 (1974).
7. D. White, Phys. Rev. D 13, 1791 (1976).
8. J. K. Daugherty and J. Ventura, Phys. Rev. D 18, 1053 (1978).
9. J. S. Toll, Ph.D. thesis, Princeton University (1952).
10. N. P. Klepikov, Zh. Eksp. Teor. Fiz. 26, 19 (1954).
11. M. E. Rassbach, Ph.D. thesis, California Institute of Technology (1971).
12. Wu-yang Tsai and T. Erber, Phys. Rev. D 10, 492 (1974).
13. J. K. Daugherty and A. K. Harding, Astrophys. J., in press.
14. P. A. Sturrock, Astrophys. J. 164, 529 (1971).
15. M. A. Ruderman and P. G. Sutherland, Astrophys. J. 196, 51 (1975).
16. G. Wunner, Phys. Rev. Lett. 42, 79 (1979).
17. J. K. Daugherty and R. W. Bussard, Astrophys. J. 238, 296 (1980).
18. Wu-yang Tsai and T. Erber, Acta Phys. Austriaca 45, 245 (1976).
19. A. E. Shabad, Ann. Phys. (N. Y.) 90, 166 (1975).
20. S. L. Adler, Ann. Phys. (N. Y.) 67, 599 (1971).
21. J. K. Daugherty and A. K. Harding, Astrophys. J. 252, 337 (1982).

EQUILIBRIUM PAIR DENSITY IN A RELATIVISTIC PLASMA WITH MAGNETIC FIELDS

F. Takahara
Nobeyama Radio Observatory, Tokyo Astronomical Observatory
University of Tokyo, Nobeyama, Minamimaki-mura
Minamisaku-gun, Nagano 384-13, Japan

M. Kusunose
Department of Astronomy, University of Tokyo,
Tokyo 113 Japan

ABSTRACT

Equilibrium e^+-e^- pair density in an optically thin relativistic plasma with magnetic fields is numerically calculated. We examine how the properties of a plasma are affected by hard photons produced by the multiple inverse Compton scatterings of synchrotron photons. The effect of magnetic fields, i.e., copious production of synchrotron photons, turns out to be significant even if the energy density of magnetic fields is below the equipartition value by several orders of magnitude. Both maximum temperature attainable and allowable maximum thickness of a plasma are shown to decrease rapidly with the increase of magnetic fields.

INTRODUCTION

Recently much attention has been paid to relativistic plasmas in quasars and active galactic nuclei. Major emission mechanisms in such a plasma are bremsstrahlung, cyclotron-synchrotron radiation and the inverse Compton scattering. In a relativistic plasma many electron-positron pairs may be created, which in turn affects the physical state of a plasma. However, the effect of magnetic fields has not been taken into account in earlier papers[1,2,3] examining the pair equilibrium state. In the presence of magnetic fields copious soft photons are produced by synchrotron radiation and those photons become so hard by the multiple inverse Compton scatterings that they should contribute to pair creation. Here we present the numerical results of the equilibrium pair density in a relativistic plasma with magnetic fields. It is to be noted that a similar work has been done independently by Araki and Lightman[5] and they obtained consistent results with ours.

FORMULATION

We consider a static and uniform plasma with finite radius R and proton number density N. The thickness of a plasma is defined as $\tau_N = N\sigma_T R$, where σ_T denotes the Thomson cross section. Magnetic fields are assumed to be randomly oriented and their average magnitude is denoted by B. Electrons and positrons are assumed to take the Maxwellian distribution with the same temperature

$T_e \equiv T_* m_e c^2 / k_B$. A plasma keeps the charge neutrality $n_- = n_+ + N$, where n_- and n_+ are number densities of electrons and positrons, respectively.

There are three pair creation processes, i.e., particle-particle (p - p), particle-photon (p - γ) and photon-photon (γ - γ). These creation rates are given by Lightman[2] and Svensson[3] as

$$\frac{dn_+^{p-p}}{dt} = \frac{7}{18\pi^2} \alpha^2 \sigma_T c N^2 \{1+4(2y-1)\} \cdot (2y-1)(\ln T_*)^3, \qquad (1)$$

$$\frac{dn_+^{p-\gamma}}{dt} = \frac{1}{3} \alpha \sigma_T c N (2y-1) \int_{T_*^{-2}}^{\infty} n_x \ln(4T_*^2 x) \, dx \qquad (2)$$

and

$$\frac{dn_+^{\gamma-\gamma}}{dt} = \frac{3}{16} \sigma_T c T_*^{-2} \int_0^{\infty} x_1^{-1} n_{x_1} dx_1 \int_{(x_1 T_*^2)^{-1}}^{\infty} x_2^{-1} n_{x_2} \ln(4T_*^2 x_1 x_2) dx_2, \qquad (3)$$

where α denotes the fine structure constant, y and x are defined as $y \equiv n_- / N$ and $x \equiv h\nu / k_B T_e$, respectively and n_x denotes the number density of photons per unit x. Finally pair annihilation rate is given by

$$\frac{dn_+^{ann}}{dt} = -\frac{3}{16} \sigma_T c N^2 y(y-1) T_*^{-2} \ln T_*. \qquad (4)$$

To solve the pair equilibrium equation

$$\frac{dn_+^{p-p}}{dt} + \frac{dn_+^{p-\gamma}}{dt} + \frac{dn_+^{\gamma-\gamma}}{dt} + \frac{dn_+^{ann}}{dt} = 0, \qquad (5)$$

we must calculate n_x simultaneously. The photon spectrum of synchrotron radiation and bremsstrahlung is calculated by

$$n_x = 8\pi \left(\frac{m_e c}{h}\right)^3 \frac{T_*^3 x^2}{e^x - 1} (1 - e^{-\tau_x}), \qquad (6)$$

where $\tau_x = (\kappa_x^{syn} + \kappa_x^{bre}) R$, κ_x^{syn} and κ_x^{bre} being the absorption coefficients of synchrotron radiation and bremsstrahlung of Maxwellian plasma, respectively. For details see the paper by Kusunose and Takahara.[4] As for the inverse Compton scattering, we follow the simplified method by Lightman and Band[6]. A photon is scattered with a probability $1 - e^{-\tau_{th}}$ and its energy is boosted by a factor of $16 T_*^2$, where τ_{th} is defined as $\tau_{th} \equiv (n_+ + n_-) \sigma_T R$. For the Wien spectral region, an appropriate separate treatment is done.

NUMERICAL RESULTS AND CONCLUSION

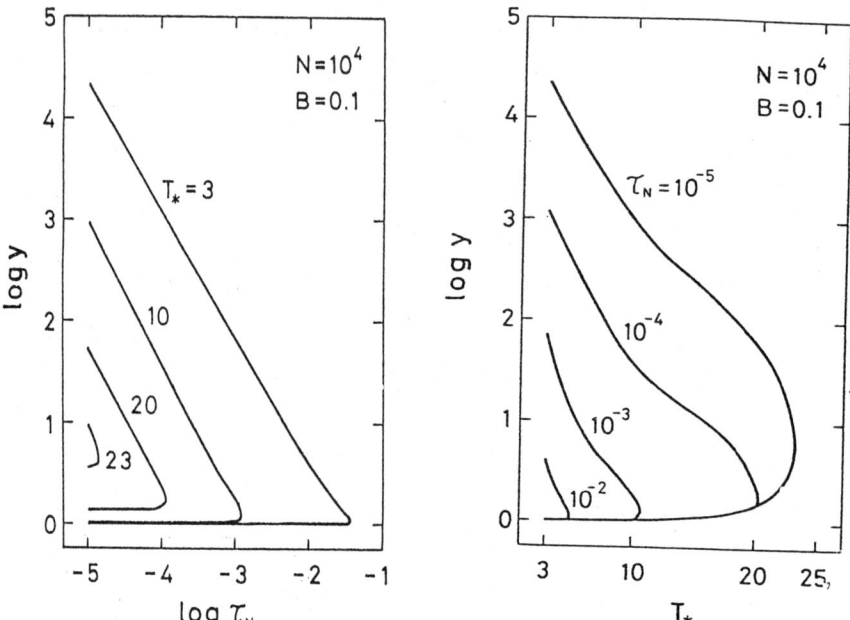

Fig.1. The behavior of y versus τ_N with fixed T_* in the case B=0.1G and N=10^4cm^{-3}.

Fig.2. The behavior of y versus T_* with fixed τ_N in the case B=0.1G and N=10^4cm^{-3}.

In Figs. 1 and 2 are shown the equilibrium pair density for various values of T_* and τ_N for the case N=10^4cm^{-3} and B=0.1G. As is seen qualitative features of solution are similar to those of the case without magnetic fields, where only bremsstrahlung photons are taken into account. Two branches of solution exist and the upper branch solution is dominated by photon processes while the lower branch solution is dominated by particle process. There exist the maximum temperature attainable T_*^{max} and the allowable maximum thickness τ_N^{max}.

In Figs.3 and 4, solutions for various values of magnetic fields are plotted in the cases of T_*=10 and τ_N=10^{-3}, respectively. The effect of magnetic fields is crearly seen; both T_*^{max} and τ_N^{max} decrease rapidly with the increase of B. The pair density of upper branch solution decreases, while one of lower branch solution slightly increases with B. Thus we conclude that the effect of magnetic fields turns out to be significant even when the energy density of magnetic fields is below the equipartition value by several orders of magnitude.

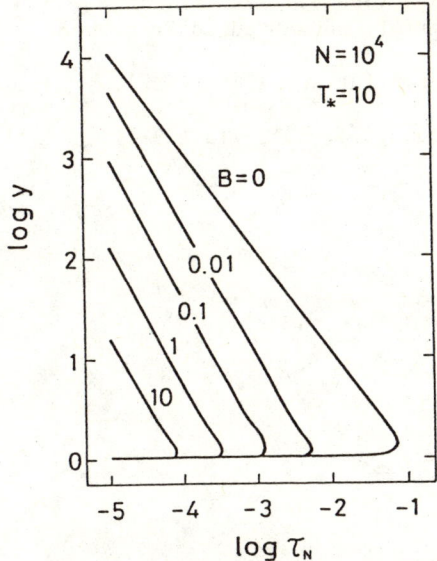 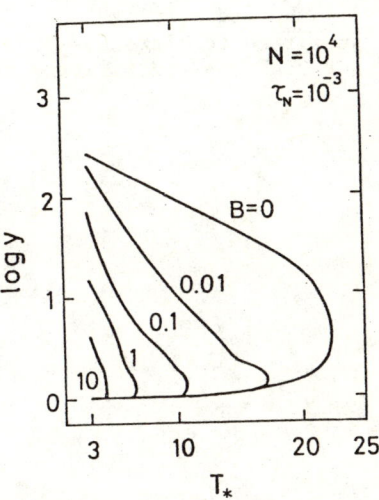

Fig.3. The behavior of y versus τ_N for various values of B in the case $T_*=10$ and $N=10^4 cm^{-3}$.

Fig.4. The behavior of y versus T_* for various values of B in the case $\tau_N=10^{-3}$ and $N=10^4 cm^{-3}$.

DISCUSSION

When we compare our results with observation of compact radio sources, we note two important points. One is that some sources show brightness temperature of about 100 $m_e c^2$, which is much higher than T_*^{max} in a pair equilibrium state. The other is that optical and X-ray spectrum seems to be a fairly pretty power law, while calculated spectrum of Maxwellian synchrotron sources is a superposition of several broad components and never becomes a power law. So plasmas in compact radio sources may contain a significant amount of nonthermal particles and / or may be in a nonequilibrium state to pair processes.

ACKNOWLEDGEMENTS

One of the authors, F.T., thanks Yamada Science Foundation for financial aid to participate in the workshop.

REFERENCES

1. G.S.Bisnovatyi-Kogan, Ya.B.Zeldovich and R.A.Sunyaev, Soviet Astron. **15**, 17 (1971).
2. A.P.Lightman, Astrophys.J. **253**, 842 (1982).

3. R.Svensson, Astrophys.J. 258, 335 (1982).
4. M.Kusunose and F.Takahara, preprint, submitted to Prog.Theor. Phys. (1983)
5. S.Araki and A.P.Lightman, CFA preprint No. 1712 (1982), submitted to Astrophys.J.
6. A.P.Lightman and D.L.Band, Astrophys.J. 251, 713 (1981).

ANNIHILATION LINES FROM CONFINED PLASMAS

P. W. Guilbert

Joint Institute for Laboratory Astrophysics
University of Colorado and National Bureau of Standards
Boulder, Colorado 80309

ABSTRACT

We investigate confined electron positron pair plasmas produced by the two photon process and by the magnetic single photon process. Two photon pair production is shown to be incapable of producing an annihilation line, however, sufficiently large magnetic fields cause single photon pair production which can give large luminosities in the annihilation line.

INTRODUCTION

Gamma-ray bursts are now generally thought to originate at the surface of a neutron star. The compact nature of the emission region implies very high energy densities and radiation efficiencies unless the sources are very close. This has led to the investigation of confined plasmas as a way of circumventing the problem of obtaining high radiation efficiency together with low optical depth, so that the electron positron lines can be observed. In this paper we investigate pair production and annihilation in confined plasmas and show that two photon pair production cannot be responsible for producing the bulk of the pairs. Magnetic single photon pair production is shown to be capable of giving observable lines and naturally leads to confinement.

TWO PHOTON PAIR PRODUCTION

We shall assume that some mechanism produces photons with an initial spectrum:

$$N(\varepsilon) = \frac{L}{(\varepsilon_M m_e c^2)^2} (2-\alpha)\left(\frac{\varepsilon}{\varepsilon_M}\right)^{-\alpha} \quad \begin{array}{l} \varepsilon_0 \leq \varepsilon \leq \varepsilon_M \\ 2 < \alpha < 1 \\ \varepsilon_0^{2-\alpha} \ll 1 \ll \varepsilon_M^{2-\alpha} \end{array} \quad (1)$$

where L is the luminosity and $\varepsilon \equiv h\nu/m_e c^2$ and that the geometry is slab like with dimensions $R^2 h$. An accurate calculation of the two photon pair production optical depth requires a numerical integration over the cross section[1], however we shall write:

$$\tau_{\gamma\gamma}(\varepsilon) \equiv a\sigma_T h \int_{\sqrt{2}/\varepsilon}^{\varepsilon_M} \frac{N(\varepsilon_1)}{R^2 c} \left(\frac{\sqrt{2}}{\varepsilon\varepsilon_1}\right) d(\varepsilon_1 m_e c^2) \qquad (2)$$

where a is actually a function of ε_M and the geometry. The maximum cross section is less than $\sigma_T(\sim\sigma_T/4)$ and so a < 1. The optical depth is thus:

$$\tau_{\gamma\gamma}(\varepsilon) = \frac{\ell h_{\gamma\gamma}}{R} \frac{2-\alpha}{\alpha} \frac{a}{\varepsilon_m^{2-\alpha}} \left(\frac{\varepsilon}{\sqrt{2}}\right)^{\alpha-1} \qquad (3)$$

where $\ell = L\sigma_T/Rm_e c^3$ is a dimensionless luminosity such that if every $m_e c^2$ of photons were converted into an electron the Thomson length $N_e \sigma_T R$ would be one. For the pair region to be confined ($h_{\gamma\gamma} < R$) we require:

$$\ell > \frac{\alpha}{2-\alpha} \frac{\varepsilon_m^{2-\alpha}}{a} \qquad (4)$$

Now:

$$\ell = 2\pi \frac{m_p}{m_e} \frac{R_s}{R} \frac{L}{L_{Edd}} \simeq 10^4 \frac{R_s}{R} \frac{L}{L_{Edd}} \qquad (5)$$

where R_s is the Schwarzschild radius of the source and L_{Edd} is its Eddington luminosity. For a neutron star, therefore, (4) is not too difficult to satisfy provided ε_m is not too large. In equilibrium the Thomson depth of the pairs is given by:

$$\tau_{pair}^2 = 4x \frac{\ell h_{\gamma\gamma}}{R} \frac{1}{b\varepsilon_m} \frac{2-\alpha}{\alpha-1} \left(\frac{\varepsilon_m}{\sqrt{2}}\right)^{\alpha-1} \simeq \frac{4x}{ab} \frac{\alpha}{\alpha-1} \qquad (6)$$

where the annihilation rate is $b\sigma_T c$, b < 1, and x is the average number of pairs produced per gamma-ray photon, x ≥ 1. Therefore even for the minimum (energy to pairs) conversion efficiency, x=1, the Thomson depth of the slab is greater than one. We note that since $\tau_{pair} > 1$ we do not need a magnetic field to confine the plasma since the production and annihilation timescales equal $h_{\gamma\gamma}/(c\tau_{pair})$. We do however need a means of cooling the pairs before they annihilate so that the line is as narrow as possible (and also so that b is maximized[2]). Compton scattering of soft photons is no use since this increases x and so magnetic bremsstrahlung is the only radiation mechanism fast enough [$t_{cool} \ll h_{\gamma\gamma}/(c\tau_{pair})$]. If the line could be made arbitrarily narrow then with sufficient resolution it could always be observed; however the minimum temperature of the pairs is the black body temperature:

$$\theta_{BB} \equiv \frac{\kappa T_{pair\ BB}}{m_e c^2} = \left(\frac{45}{2\pi^3 \alpha_f^3} \frac{\ell r_0}{R}\right)^{1/4} \quad (7)$$

where α_f is the fine structure constant and r_0 is the classical electron radius. The minimum temperature is thus:

$$\theta \gtrsim \left(\frac{45}{2\pi^3 \alpha_f^3} \frac{\alpha}{2-\alpha} \frac{\varepsilon_m^{2-\alpha}}{a} \frac{r_0}{h_{\gamma\gamma}}\right)^{1/4} \simeq 10^{-2} \left(\frac{\alpha}{2-\alpha} \frac{\varepsilon_m^{2-\alpha}}{a}\right)^{1/4} h_{\gamma\gamma}^{-1/4} \quad (8)$$

where $h_{\gamma\gamma}$ is in meters. This implies minimum line widths on the order of 100 keV[2]. We note that for the Ramaty, Lingenfelter and Bussard[3] model of the March 5, 1979 burst, if associated with N49, two photon pair production can be ruled out on the grounds of temperature alone. Numerical calculations using the radiative transfer program developed by Stepney and Guilbert[4] confirm the implications of equation (6) and it would seem to be impossible to produce an observable line, except perhaps for very hard primary spectra $\alpha \lesssim 1$. As we shall see below, however, spectra for which $\alpha \lesssim 1$ would be completely dominated by single photon pair production at large magnetic flux densities.

MAGNETIC PAIR PRODUCTION

The optical depth to magnetic pair production is given by[5]:

$$\tau_\gamma(\varepsilon) = \frac{h_\gamma}{2} \frac{\alpha_f^2}{r_0} B_+ T(\chi) \quad \varepsilon \geq 2 \quad (9)$$

where $\chi = \varepsilon B_+/2$ and:

$$T(\chi) = \begin{cases} 0.46 \exp(-4/3\chi) & \chi \ll 1 \\ 0.6 \chi^{-1/3} & \chi \gg 1 \end{cases}$$

B_+ is the magnetic flux density perpendicular to the direction of the photon measured in units of the critical value $m_e^2 c^3/\hbar e$. For $\chi = 0.1$ the mean free path is $\sim 10^{-4} B_+^{-1}$ m and for $\chi = 0.05$ it is $10^2 B_+^{-1}$ m. The energy at which $\tau_\gamma = 1$ is thus very insensitive to h_γ and we shall assume that it occurs at $\chi = 0.1$. If the region where the primary energy supply is released has a width greater than $10^{-4} B_+$ then this determines h_γ. The Thomson depth of the pairs is then:

$$\tau_{pair}^2 \simeq 4x \frac{\ell h_\gamma}{R} \frac{1}{b\varepsilon_m} \frac{2-\alpha}{\alpha-1} \left(\frac{\varepsilon_m}{\varepsilon_\ell}\right)^{\alpha-1} \quad (10)$$

where

$$\varepsilon_\ell \sim \text{Max}(2, \frac{0.2}{B})\ .$$

The pairs will typically have an energy $\gamma \sim \varepsilon/2$ when produced and so since $B\varepsilon/2 > 0.1$ most of the energy is radiated by the pairs as magnetic bremsstrahlung in the quantum regime where the spectrum peaks at $\varepsilon \sim \gamma$. We therefore expect the efficient conversion of energy into pairs and so x will take approximately its maximum value giving:

$$\tau_{pair}^2 \sim \frac{2\ell h_\gamma}{bR}\ . \quad (11)$$

The region of magnetic pair production (henceforth the single photon region) will be optically thin to Thomson scattering provided that $\ell < bR/2h_\gamma$ and so can be made optically thin for arbitrary ℓ by making h_γ/R small enough (providing the magnetic pressure is large enough, which requires $B^2 > 9\ell/\alpha_f^2\ r_0/R$). Since $\gamma-1 \gg 1$ for most of the uncooled pairs they must be cooled parallel as well as perpendicular to B otherwise the transverse Doppler shift will broaden the line and so we require:

$$n(\varepsilon<1)\sigma_T h > \tau_{pair} \quad (12)$$

since at least one scattering with a soft photon is required to cool them before they annihilate. This is not guaranteed by the primary spectrum since

$$n(\varepsilon<1)\sigma_T h = \frac{\ell h_\gamma}{R}\frac{2-\alpha}{\alpha-1}(\frac{\varepsilon_0}{\varepsilon_m})^{1-\alpha}\frac{1}{\varepsilon_m}\ ;\quad (13)$$

however, if (13) is less than τ_{pair} this does not mean the pairs cannot cool. The soft photon density will be much higher than that due to the primary spectrum due to soft photon production by curvature radiation and cyclotron from pairs cooled perpendicular to B. $\varepsilon_\ell > 2$ and so if $\ell_\gamma(\varepsilon>1) > \alpha/(2-\alpha)\ \varepsilon_\ell^{2-\alpha}/a$; where $\ell_\gamma(\varepsilon>1)$ is the luminosity from the primary spectrum plus the cooling pairs; there will also be a region of two photon pair production of width:

$$\frac{h_{\gamma\gamma}}{R} \simeq \frac{\alpha}{2-\alpha}\frac{\varepsilon_\ell^{2-\alpha}}{a}\frac{1}{\ell_\gamma} \quad (14)$$

and Thomson depth $[4/(ab)\times\alpha/(\alpha-1)]^{1/2}$. If ε_ℓ is much larger than two and $h_{\gamma\gamma}$ is large enough for the magnetic field to fall to such a small value that Compton cooling dominates then x may be larger than one and the Thomson depth in the two photon region will be much larger than unity. It is clear that in this situation Compton scattering will destroy the line. If ε_ℓ is small (also implying that x in the single photon region cannot be large) so much of the primary luminosity is converted into pairs that an observable line may survive even if $\tau_{\gamma\gamma} > 1$. If the Thomson depth in the single photon region is greater than one, but not too large $\tau_\gamma < [4/(ab) \times \alpha/(\alpha-1)]^{1/2}$, this may increase the size of a line because Compton scattering will remove photons above $\varepsilon = 1$ reducing $\ell_\gamma(\varepsilon > 1)$. If the reduction is large enough the two photon region will be rendered optically thin.

THE SPECTRUM

There are obviously several cases to be considered and so we will consider them individually.

1) $$\ell < \frac{bR}{2h_\gamma} \quad ; \quad \ell \ll \frac{\alpha}{2-\alpha} \frac{\varepsilon_m^{2-\alpha}}{a} \quad .$$

In this case the whole region is optically thin to Compton scattering and so providing ε_ℓ is small enough for a sufficient fraction of the original luminosity to be converted into pairs we should observe a large annihilation line. The continuum spectrum will be sensitive to the angle at which we observe the emission region since ε_ℓ is determined by B_+ in the optically thin case. The spectrum will however have a sharp cutoff at an energy ε_ℓ. If we observe at a very small angle to B then the continuum may be dominated by Comptonized soft photons and curvature radiation. If the Comptonized component dominates then the spectrum will be much softer since cooling is rapid.[6] Perpendicular to B magnetic bremsstrahlung will dominate and the spectrum will be hard and have the minimum value of ε_ℓ.

2) $$\ell < \frac{bR}{2h_\gamma} \quad ; \quad \ell_\gamma(\varepsilon > 1) > \frac{\alpha}{2-\alpha} \frac{\varepsilon_\ell^{2-\alpha}}{a} \quad .$$

In this case the two photon region is optically thick to Thomson scattering. The observed spectrum will depend on ε_ℓ and the magnitude of B in the two photon region. If $\varepsilon_\ell \simeq 2$ or if the magnetic field is small enough so that $\varepsilon_\ell^2 B < 1$ then pair creation will be inefficient and the Thomson depth of the two photon region will take its minimum value $\tau \sim [4/(ab) \times \alpha/(\alpha-1)]^{1/2}$. In this case an annihilation line may be observed simply because of the very high efficiency of the single photon pair production. The line cannot contain more than $\exp[-(4/ab)\alpha/(\alpha-1)]$ of the photons. The continuum will be determined by conditions in the two photon region, which in turn will be determined by heating and cooling due to Comptonization of the continuum from the single photon region and cooling due to magnetic bremsstrahlung. These spectra would have cutoffs at $\varepsilon < 1$ due to Comptonization and a hard low energy continuum due to magnetic bremsstrahlung.

3) $$\ell > \frac{bR}{2h_\gamma} .$$

If ℓ is large enough then there will be no two photon region but the spectrum will be identical to the optically thick two photon case. For an intermediate value of ℓ such that $\tau_\gamma < [4/(ab) \times \alpha/(\alpha-1)]^{1/2}$ then the two photon region may be optically thin even if $\ell > \alpha/2-\alpha\ \varepsilon_m^{2-\alpha}/a$ due to Compton scattering removing photons above $\varepsilon = 1$. We note that in this case the line may be moderately strong and get weaker as ℓ decreases due to the single photon region becoming optically thin and so enabling the two photon region to become thick.

This work was supported by National Science Foundation grant No. AST80-19960.

REFERENCES

1. A. A. Zdziarski, Accreting Neutron Stars, eds. W. Brinkmann and J. Trumper (Max-Planck-Institut fur Physik und Astrophysik, Garching, 1982).
2. R. Ramaty and P. Meszaros, Ap.J. 250, 384 (1981).
3. R. Ramaty, R. E. Lingenfelter and R. W. Bussard, Ap. Space Sci. 75, 193 (1981).
4. S. Stepney, (These Proceedings).
5. T. Erber, Rev. Mod. Phys. 38, 626 (1966).
6. P. W. Guilbert, R. R. Ross and A. C. Fabian, M.N.R.A.S. 199, 763 (1982).

ELECTRON-POSITRON PAIR ANNIHILATION AND CREATION IN SUPERSTRONG MAGNETIC FIELDS

G. Wunner, H. Herold, H. Ruder
Lehrstuhl für Theoretische Astrophysik
Universität Tübingen
D-7400 Tübingen, West Germany

ABSTRACT

The paper reviews the results obtained so far in recalculating the processes of one-photon and two-photon pair annihilation and creation in strong magnetic fields ($B \sim 10^{11} - 10^{13}$ G, as are characteristic of neutron stars), with special emphasis being laid on annihilation.

INTRODUCTION

The self-consistent description of plasmas in the vicinity of neutron stars containing electron-positron pairs requires the accurate knowledge of 1γ and 2γ pair annihilation and creation rates in the presence of the intense magnetic fields ($B \sim 10^{11} - 10^{13}$ G) associated with these compact cosmic objects. We will consider, in order, the different processes, and, in particular, shall establish the transformation law between the cross sections of 2γ annihilation and creation which holds in strong magnetic fields.

ONE-PHOTON PAIR ANNIHILATION

In what follows, both the electron (z-momentum p_-, energy E_-) and the positron (p_+, E_+) are assumed to be initially in the lowest Landau orbital; this assumption is justified by the short lifetimes of Landau-excited states with respect to cyclotron transitions in strong magnetic fields, $\tau_{cycl} \cong 10^{-19}$ s $\times (B/B_{cr})^2$, where $B_{cr} = m_e^2/e \cong 4.41 \times 10^{13}$ G. No restrictions, however, are placed on the longitudinal momenta.

The cross sections are most conveniently expressed in the center-of-momentum (CMom) frame, which is reached by Lorentz transforming with velocity $\beta \equiv v = (p_+ + p_-)/(E_+ + E_-)$ along the direction of B. By the laws of conservation of energy and z-momentum, in the one-quantum process the photon is emitted exactly perpendicular to the field in the CMom frame ($\theta' = \pi/2$), and, written in terms of the energy E'_o ($=E'_+ =E'_- \geqslant m_e$) of either particle in the CMom frame, the 1γ annihilation cross section reads[1,2]

$$\sigma_{1\gamma}^{CMom}(E'_o) = \frac{\alpha}{2} \lambda_C^2 \frac{m_e^2}{(E'^2_o - m_e^2)^{1/2} E'_o} \frac{B_{cr}}{B} \exp(-2(E'_o/m_e)^2 B/B_{cr}) \quad , \quad (1)$$

where α is the electromagnetic coupling constant, and λ_C denotes the Compton wave length of the electron. In the lab frame, the emission occurs at $\theta=\arccos\beta$, and the cross section is given by

$$\sigma_{1\gamma}(E_+,E_-) = \sigma_{1\gamma}^{CMom}(E_o'=(E_++E_-)|\sin\theta|/2) . \qquad (2)$$

An inspection of (1) shows that $\sigma_{1\gamma}^{CMom}$ goes through a maximum at $B/B_{cr}=2(E_o'/m_e)^2$, the maximum value being proportional to $(m_e/E_o')^3$. For smaller values of B, $\sigma_{1\gamma}^{CMom}$ rapidly tends to zero as $\exp(-B_{cr}/B)$, which is characteristic of the momentum-nonconserving nature of the process. For larger values of B, the cross section slowly decreases ($\sim B_{cr}/B$), which originates in the fact that the degeneracy of the Landau levels increases linearly with B. Photon spectra emitted by a gas of electron-positron pairs can be obtained by folding (2) with the corresponding momentum distributions[2].

ONE-PHOTON PAIR CREATION

For a detailed account of pair production by single photons in strong magnetic fields the reader is referred to Daugherty[3], who studies, in particular, the properties of pair production near threshold, and compares with previous asymptotic approaches[4] that disregarded the discrete level structure of the Landau spectrum.

TWO-PHOTON PAIR ANNIHILATION

In the CMom frame, the triply differential cross section for annihilation of e^+,e^- (in the ground state Landau level, with momenta $\pm p_o$ and relativistic energy E_o) into 2 photons $\vec{k}_1'$, $\vec{k}_2'$ (with $k_1'+k_2'=2E_o$ and $k_{1z}'+k_{2z}'=0$) is given by

$$\frac{d^3\sigma_{2\gamma}^{CMom}}{dk_1' d\Omega_1' d\phi_{12}'} = \frac{\alpha^2 k_1'}{4p_o E_o} \frac{\exp(-(\vec{k}_{1\perp}'+\vec{k}_{2\perp}')^2/2eB)}{\pi \cdot eB} |M|^2 , \qquad (3)$$

$$M = m_e \Big\{ (2p_o-k_{1z}')e_{1z}e_{2z} F(1,Q_1+1,k_{1+}'k_{2-}'/2eB)/2eBQ_1$$
$$+(e_{1+}e_{2-}k_{1z}'+e_{1+}e_{2z}k_{2-}'-e_{1z}e_{2-}k_{1+}') F(1,Q_1+2,k_{1+}'k_{2-}'/2eB) \qquad (4)$$
$$/(2eBQ_1+1) + \text{ terms with subscripts interchanged } 1\leftrightarrow 2 \Big\} ,$$

where $Q_i=[m_e^2+(p_o-k_{iz}')^2-(E_o-k_i')^2]/2eB$, and $\vec{e}_i$ is the vector of polarization of photon $i(i=1,2)$. Note that, as compared to earlier treatments[1,2], the sum over the intermediate states has been carried out analytically, giving rise to the occurrence, in (4), of the numerically easily tractable hypergeometric function $F \equiv {}_1F_1$, and, in (3), of an exponential term which in the limit $B \to 0$ behaves as $\delta(\vec{k}_{1\perp}'+\vec{k}_{2\perp}')$, and thus manifestly contains the deviations from 3d momentum conservation.

Going back to the lab frame ($\beta=(p_++p_-)/(E_++E_-)$), the wave vectors transform according to $k'=k\gamma(1-\beta\cos\theta)$, $\cos\theta'=(\cos\theta-\beta)/(1-\beta\cos\theta)$, the cross section is obtained as

$$\frac{d^3\sigma_{2\gamma}}{dk_1 d\Omega_1 d\phi_{12}} = \frac{d^3\sigma_{2\gamma}^{CMom}}{dk_1' d\Omega_1' d\phi_{12}'} \cdot \frac{1}{\gamma(1-\beta\cos\theta)} \tag{5}$$

and at a given polar angle θ the maximum photon energy is

$$k_{max}(p_+,p_-,\theta) = (E_++E_-)(1\pm\beta)/(1\pm\cos\theta) \qquad (\cos\theta \gtrless \beta). \tag{6}$$

A graphical representation of (6) is shown in Fig. 1, together with the corresponding diagram for B=0. It is recognized that for emission parallel to the field the maximum energy possible remains the same as without field (viz. $(E_++E_-)(1\pm\beta)/2$), while in strong fields there exists, for any β, a polar angle $\theta=\arccos\beta$ where the transfer of the total energy of the e^+e^- pair to one of the photons becomes kinematically allowed. In particular, for relativistic β, the angular range in which practically the total energy is transferred to one photon is considerably broadened as compared to the field-free case.

Integrating (3) over the relative azimuth ϕ'_{12} of the photons yields the doubly differential cross section $d^2\sigma_{2\gamma}^{CMom}/dk_1' d\Omega_1'$, which is depicted, as an example, for $B=B_{cr}$ and non-relativistic momenta of e^+,e^-, in Fig. 2. It is evident that the steep break in the cross

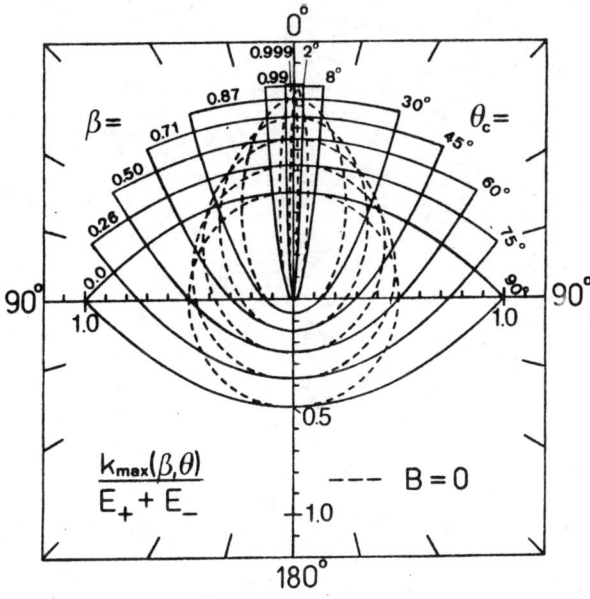

Fig. 1. Maximum energy, observable in the lab frame, of a 2γ annihilation photon as a function of the polar angle θ for several values of $\beta=(p_++p_-)/(E_++E_-)$, $0 \leq \beta < 1$. The polar axis is defined by the direction of the magnetic field. Note that the total energy of the e^+e^- pair can be transferred to one single photon at $\theta_c = \arccos\beta$. (Dashed curves: corresponding diagrams in the field-free case for motion along the z-axis).

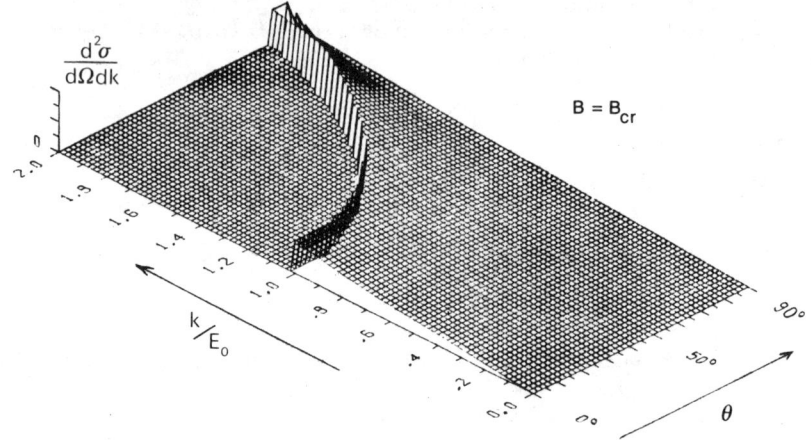

Fig. 2. Doubly-differential cross section $d^2\sigma/d(k/E_0)d\Omega$ of 2γ annihilation (in the center-of-momentum frame, i.e. $E_+ = E_- = E_0$) for $B = B_{cr}$ and non-relativistic momenta of e^+, e^- ($E_0 \simeq m_e$).

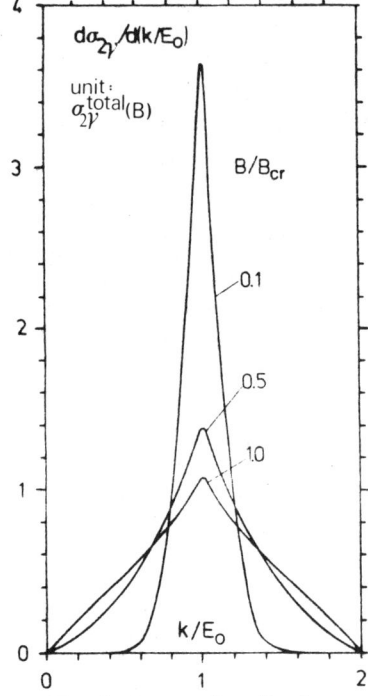

Fig. 3. Energy distribution $d\sigma_{2\gamma}/d(k/E_0)$ of the annihilation quanta (in the CMom frame) for several values of B/B_{cr} (for $E_0 \simeq m_e$).

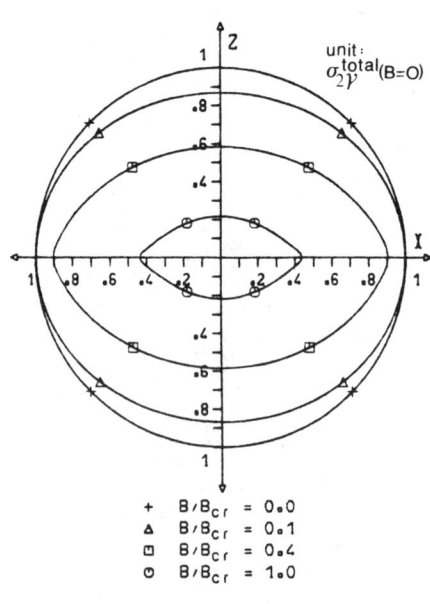

Fig. 4. Angular distribution $d\sigma_{2\gamma}/d\Omega$ of the annihilation quanta (in the CMom frame) for several values of B/B_{cr} (for $E_0 \simeq m_e$)

section along the curve $k'(\theta')/E_0 = 2/(1+|\cos\theta'|)$ in the (k',θ') plane implied by (6) will be smoothed out once momentum distributions of e^+, e^- are taken into account. Integrating $d^2\sigma_{2\gamma}^{CMom}/dk_1'd\Omega_1'$ over Ω_1' or k_1' leads to the line profiles and angular distributions of the annihilation quanta, shown in Fig. 3 and Fig. 4, respectively, for various values of B/B_{cr} and nonrelativistic momenta of e^+, e^-. It is seen that as B/B_{cr} increases the annihilation line experiences a sizeable broadening, the actual observation of which, if not blurred by other broadening effects, could provide an independent check on the magnetic field strength. Figure 4 demonstrates that with increasing B/B_{cr} emission perpendicular to the field becomes more and more favorable, and thus 2γ annihilation more and more mimics the emission pattern found for 1γ annihilation. Also, it can be seen from Fig. 4 that the absolute value of the total cross section decreases, and, in fact, it turns out[1,2] that for $B \gtrsim 0.25\ B_{cr} \cong 10^{13}$ G annihilation into 1γ becomes more probable than into 2γ (in the CMom frame, and for nonrelativistic momenta).

TWO-PHOTON PAIR CREATION

The cross section for producing e^+, e^- pairs in the zero Landau level by two-photon collisions in a strong magnetic field is related to the cross section (3) by

$$\sigma^{CMom}(\gamma\gamma \xrightarrow{B} e^+e^-) = \frac{d^3\sigma^{CMom}(e^+e^- \xrightarrow{B} \gamma\gamma)}{d(k_1'/m_e)d\Omega_1'd\phi_{12}'} \cdot \frac{4\pi(B/B_{cr})^2}{(k_1'/m_e)^2(k_2'/m_e)}, \qquad (7)$$

and thus exhibits a more complex structure than the corresponding field-free transformation law, $\sigma(\gamma\gamma) = \sigma(e^+e^-) \cdot 2p_\pm^2/E_\pm k$. Obviously this is a consequence of the different dimensionalities of the densities of the final states (1d for e^+e^- in $\gamma\gamma \xrightarrow{B} e^+e^-$, 3d for photons in $e^+e^- \xrightarrow{B} \gamma\gamma$). To the best of our knowledge, no concrete calculations have been performed so far using the exact cross section (7). In particular, it would be interesting to see in which way estimates of the relative importance of 1γ and 2γ pair creation rates in strong magnetic fields (obtained neglecting the discreteness of the Landau levels)[5] are modified by using the exact cross sections.

REFERENCES

1. G. Wunner, Phys. Rev. Lett. **42**, 79 (1979).
2. J.K. Daugherty and R. W. Bussard, Astrophys. J. **238**, 296 (1980)
3. J.K. Daugherty, this volume (1983)
4. W.-Y. Tsai and T. Erber, Phys, Rev. **D10**, 492 (1974)
5. A.K. Harding and M.L. Burns, this volume (1983)

COMPARISON OF PHOTON-PHOTON AND PHOTON-MAGNETIC FIELD PAIR PRODUCTION RATES

M. L. Burns* and A. K. Harding
NASA/Goddard Space Flight Center, Greenbelt, MD. 20771

ABSTRACT

Neutron stars have been proposed as the site of gamma-ray burst activity and the copious supply of MeV photons admits the possibility of electron-positron pair production. If the neutron star magnetic field is sufficiently intense ($> 10^{12}$ G), both photon-photon (2γ) and photon-magnetic field (1γ) pair production should be important mechanisms. Rates for the two processes have been calculated using a Maxwellian distribution for the photons. The ratio of 1γ to 2γ pair production rates has been obtained as a function of photon temperature and magnetic field strength.

INTRODUCTION

Observations of the spectra of gamma-ray bursts indicate the presence of significant numbers of high energy photons in the MeV range. Some spectra have features at energies between 350 and 450 keV[1], which have been interpreted as red-shifted annihilation lines. The amount of the shift (~10%) is the gravitational redshift expected from the potential well of a neutron star. In addition, absorption features have been observed in many of the spectra in the region 20 - 60 keV, which if interpreted as cyclotron absorption, indicate magnetic fields of order 10^{12} G. The evidence seems to suggest that the emitting regions of these sources are near the surfaces of strongly magnetized neutron stars.

If this is the case, then one-photon as well as two-photon processes might be expected to contribute to the production of pairs in gamma-ray burst sources. One-photon pair production, a first order process which is forbidden in free space, is allowed in the presence of a magnetic field. If the photons have a thermal distribution, then significant pair production rates will occur when $(kT/mc^2)(B/B_{cr}) \gtrsim 0.1$.

ONE-PHOTON PAIR PRODUCTION RATE

We first calculate the rate of 1γ pair production in a hot photon gas where the photons have a Maxwellian distribution. The rate for a single photon with energy E propagating at an angle θ to a constant, homogeneous magnetic field of strength B is[2]:

---

*Also University of Maryland, College Park, MD 20742.

$$r_{\gamma B} \sim \frac{1}{2} \frac{c\alpha}{\lambdabar_c} \frac{B\sin\theta}{B_{cr}} T(\chi), \qquad (1)$$

$$T(\chi) = 4.74 \chi^{-1/3} Ai^2 (\chi^{-2/3}) \qquad (1a)$$

$$\sim 0.46 \exp[-4/3\chi], \quad \chi \ll 1 \qquad (1b)$$

$$\sim 0.60 \chi^{-1/3}, \quad \chi \gg 1 \qquad (1c)$$

where $\chi \equiv (\frac{E}{2mc^2})(\frac{B}{B_{cr}}) \sin\theta$, $B_{cr} = 4.414 \times 10^{13}$ G, and Ai is the Airy function.

These are the asymptotic expressions in the limit where the quantum numbers of the magnetic field pair states are large. We will discuss the region of validity of this expression below. The pair production rate for a distribution of photons, $\Phi(E,T)$, will be

$$R_{\gamma B}(B,T) = 2\pi \int_0^\pi \sin\theta d\theta \int_{E_{min}}^\infty dE \, \Phi(E,T) \, r_{\gamma B}(E,B,\theta) \; s^{-1} \qquad (2)$$

where $E_{min} = 2mc^2/\sin\theta$ is the threshold photon energy. We take a Maxwellian distribution for the photons, normalized to constant, unit photon number:

$$\Phi(E,T) = \frac{E^2}{8\pi(kT)^3} \exp(-E/kT) \qquad (3)$$

If we make the further approximations, $\chi \ll 1$ and $E \gg 2mc^2$ then an analytic expression can be obtained for the pair production rate. The E integration of Eqn (2) can be performed by the method of steepest descents. The integrand has a saddle point at $E_o = mc^2 [(8/3)(kT/mc^2)(B_{cr}/B \sin\theta)]^{1/2}$, about which the major contribution to the integral is located. The θ integral can then be performed by noting that the integrand peaks very sharply around $\sin\theta = 1$. The result for the photon distribution of Eqn (3) is

$$R_{\gamma B}(B,T) \simeq 6.3 \times 10^{18} \frac{mc^2}{kT} \exp\left[-2 \left(\frac{8}{3} \frac{mc^2}{kT} \frac{B_{cr}}{B}\right)^{1/2}\right] s^{-1} \qquad (4)$$

where the above approximations translate into the regions of validity for this expression:

$$\left(\frac{kT}{mc^2} \frac{B}{B_{cr}}\right)^{1/2} \ll 1, \qquad \frac{kT}{mc^2} > \frac{3}{2} \frac{B}{B_{cr}}. \qquad (5)$$

It is evident that Eqn (4) is not valid for magnetic fields approaching the critical value or at low temperatures, where most of the photons have energies near threshold.

To obtain the behavior of the one-photon pair production rate in the full region of interest, we have numerically integrated Eqn. (2), taking into account the photon energy threshold and the full asymptotic rate [Eqns. (1), (1a)]. Figure 1 shows this rate per photon as a function of kT and B. Due to the exponential behavior of Eqn. (1b) at low photon energies, the calculated rate is a very sensitive function of T and B at low temperatures. At temperatures below $kT = mc^2$, slight variations in B can change the rate by many orders of magnitude.

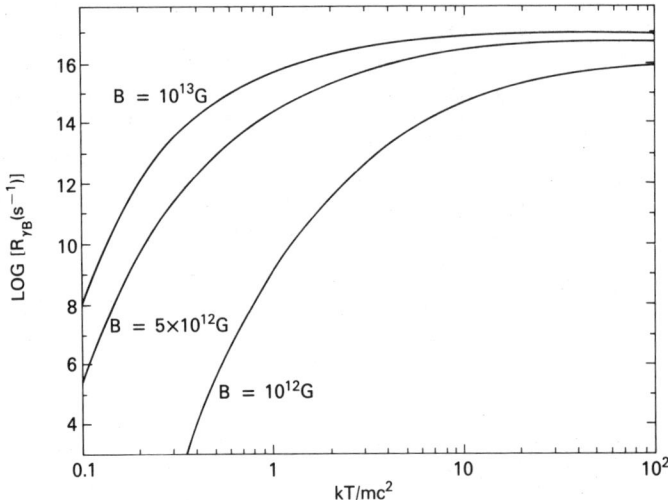

Fig. 1. 1γ pair production rate per photon versus kT/mc^2 for different magnetic field strengths.

RATIO OF ONE-PHOTON TO TWO-PHOTON PAIR PRODUCTION

The pair production rate via the two-photon process can be calculated for the same photon gas with a Maxwellian distribution of energies in order to directly compare the one-photon to the two-photon rate. In the case of photon-photon pair production, we must integrate the cross-section and the photon distribution over both photon energies, E and E', and the angle, θ, between their propagation vectors (see eg. Ref. 3):

$$\frac{R_{\gamma\gamma}(T)}{n_\gamma} = 4\pi^2 c \int_0^\pi d\theta \, \sin\theta \, (1-\cos\theta) \int_0^\infty dE \int_{E_{min}(E,\theta)}^\infty dE'$$

$$\Phi(E,T) \, \Phi(E',T) \, \sigma(E,E',\theta) \quad s^{-1} cm^3 \qquad (6)$$

where $E_{min}(E,\theta) = 2m^2c^4/E(1-\cos\theta)$

The cross-section is[4]

$$\sigma(E, E', \theta) = \frac{3}{2} \sigma_T \, \sigma(\tau) \qquad (7)$$

where

$$\sigma(\tau) = \frac{1}{\tau^3} \left\{ (\tau^2 + 4\tau - 8) \ln \left[\frac{\sqrt{\tau} + \sqrt{\tau-4}}{\sqrt{\tau} - \sqrt{\tau-4}} \right] - (\tau + 4) \sqrt{\tau(\tau-4)} \right\}$$

$$\tau(E, E', \theta) = \frac{2EE'}{m^2 c^4} (1 - \cos\theta),$$

and the photon distribution is given by Eqn (3). Numerical integration of Eqn (6) then enables us to evaluate the relative importance of the one- and two-photon processes. Figure 2 shows the ratio of the one-photon to the two-photon pair production rates as a function of kT. The vertical scale plots, on the left hand side, the photon density (which is kept constant with T) for which the two rates are equal and on the right hand side, the actual ratio of the rates at a fixed density of 10^{25} cm^{-3}. Also plotted is the blackbody photon density n_{BB} (dashed line) which is the maximum photon density achievable at a given temperature. At temperatures below $kT = mc^2$, which are the temperatures of interest for gamma-ray burst sources, a change in the magnetic field of one order of magnitude corresponds to many orders of magnitude in the ratio of the two rates. From Figure 2, one can obtain, for a given photon density and magnetic field strength, the temperature at which either process dominates. For example, for a photon density $n = 10^{25}$ cm^{-3} and $B = 10^{12}$ G, the one-photon process dominates over the two-photon process at $kT > mc^2$. For $B > 4 \times 10^{12}$ G, one-photon pair production is the dominant process at all temperatures.

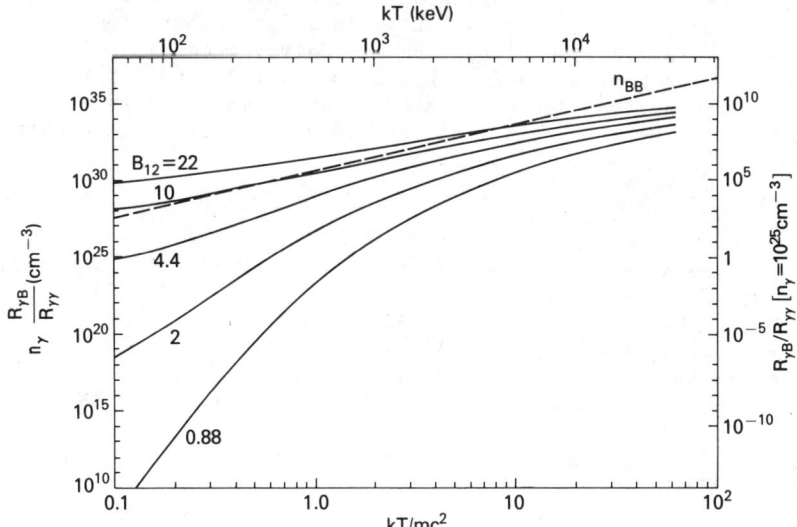

Fig. 2. Ratio of 1γ to 2γ pair production rates for Maxwellian photon distributions at temperature kT. Curves are labeled with values of the magnetic field in units of 10^{12} G.

DISCUSSION

In a magnetic field, electron and positron energies perpendicular to the field are quantized, with the energy separation between the levels increasing with field strength. When the photon energies are not large compared to the spacing between these Landau states, the pair production rates are significantly influenced by quantum effects. In the preceeding calculations, we have neglected the effect of the discrete e+e- states on the pair production rates. In the case of the one-photon rate, quantum effects are important when the quantity $2(E/2mc^2)^2/(B/B_{cr})$ is small[5] (ie. for low temperatures and high field strengths), decreasing the rate below the value given by the asymptotic expression [Eqn (1b)]. The two-photon pair production rate has not been calculated in a magnetic field, so we have used the free-space rate, where the electron and positron states are assumed to be plane waves. The effect of the magnetic field might be estimated by examining the behavior of the inverse process, two-photon annihilation, which has been calculated in a strong magnetic field.[6] There is no significant deviation from the free-space rate until the field approaches 10^{13} G, and at this point, the one-photon annihilation rate begins to dominate. Therefore, from the preceding argument, our calculation of the ratio of the two rates is probably most accurate below $B \sim 10^{13}$ G and at higher temperatures.

Photon distributions other than Maxwellian may be more realistic, since gamma-ray burst source emitting regions are not likely to be in equilibrium. The calculation presented here is meant only to give an idea of the relative behavior of the one- and two-photon processes. The actual rate of pair production in the source region would be an equilibrium solution including other processes affecting the distribution of photons and pairs, such as annihilation, synchrotron radiation and absorption, and Compton scattering. Comparisons such as this, however, may aid in the construction of the more self-consistent models by indicating which processes are important.

REFERENCES

1. E. P. Mazets, S. V. Golenetskii, R. L. Aptekar', Yu. A. Guryan, V. N. Ilinskii, Astrophys. Sp. Sci., 80, 119 (1981).
2. T. Erber, Rev. Mod. Phys., 38, 626 (1966).
3. T. A. Weaver, Phys. Rev. A, 13, 1563 (1976).
4. V. B. Berestetskii, E. M. Lifshitz, L. P. Pitaevskii, Relativistic Quantum Theory, (Pergamon, New York, 1971), p. 307.
5. J. K. Daugherty and A. K. Harding, Astrophys. J., in press (1983).
6. J. K. Daugherty and R. W. Bussard, Astrophys. J., 238, 296 (1980).

CONDITIONS FOR STIMULATED - ANNIHILATION IN A DEGENERATE $e^- - e^+$ FLUID AT THE SURFACE OF PULSARS

C. M. Varma
Bell Laboratories, Murray Hill, NJ 07974

ABSTRACT

Electrons and Positrons in the large magnetic fields at the surface of Pulsars condense into rod shaped droplets. The conditions for stimulated annihilation in such droplets are examined. The motivation for this investigation are the observation of Leventhal et al. of very narrow γ-ray lines emanating from the Crab nebula.

INTRODUCTION

Several years ago, Leventhal, McCallum and Watts (LMW)[1] detected a line feature at 400 keV emanating from the Crab nebula. This energy has the correct magnitude of red shift to be due to two-photon electron-positron annihilations at the surface of the pulsar in the Crab. 10^{-3} photons/cm^2/sec are detected, which if they are emitted isotropically, imply that the rate of annihilations at the pulsar is at least 10^{41}/sec. If the annihilations take place in the polar cap region alone, and the photons travel along the magnetic field lines, this number is reduced to about 10^{39}/sec. This is close to estimates of the number, about 10^{38}/sec, of electrons that must be injected into the Crab nebula to maintain the continuum synchrotron emission. A flux of about 10^{38}/sec in either radial direction in the polar cap region of both electrons and positrons may be consistent with current pulsar models as long as the electron-positron stream is very nearly charge neutral. The most surprising feature of LMW's observations is the width of the line which is limited by the resolution of the detector to be less than 3 keV. One expects the spectrum over which the annihilation photons are distributed to be determined by the transverse momentum of the electrons and positrons in the magnetic field or by the Doppler shift. For a field of 10^{12} Gauss, the former is about 50 keV. From the estimated surface temperature of 0.5 keV, the Doppler width is also a similar number. LMW's observations on the Crab have been confirmed by others.[2] Lot of excitement has also been caused by observation of narrow $e^- - e^+$ annihilation lines from the galactic center.

Earlier, I proposed[3] that a possible mechanism to have narrow lines from $e^- - e^+$ annihilation is for the annihilation to be stimulated. This leads to the amplification of photon density near a particular energy. Stimulated annihilation processes require population inversion, i.e., an effective negative temperature to specify the $e^- - e^+$ density. An extreme condition to achieve this is that the electron and positron fluids be nearly degenerate, i.e., the density be such that the Fermi energy is much higher than the temperature. I relied on some calculations of Ruderman[4] on the properties of matter in superstrong magnetic fields to suggest that electrons and positrons in large magnetic fields and low temperatures condense to a novel phase of matter. In this phase $e^- - e^+$ arrange themselves in rod-like structures along magnetic field lines to optimize the Coulomb interactions. They move freely along the rods and are confined in the cyclotron orbits transverse to the field. I estimated that in this phase of matter population inversion is indeed achieved in the steady state in which spontaneous annihilation is balanced by the incoming flux of e^-, e^+.

Ramaty, Mckinley and Jones[5] (RMJ) have recently investigated the conditions for stimulated annihilation in a dense three-dimensional $e^- - e^+$ plasma in zero magnetic fields. They find that for stimulated emission to win over Compton scattering a very high density (corresponding to a chemical potential ~ 1 MeV) of $e^- - e^+$ pairs is required. This has the further unfortunate effect that stimulated emission is shifted considerably to about 450 keV even without gravitational redshift.

The conditions on momentum conservation in $e^- - e^+$ annihilation in a large magnetic field are, however, quite different from that in a zero field. I will show below that in a magnetic field and in the condensed phase, conditions for stimulated emission are much less stringent, and stimulated emission can occur essentially at 511 keV (not counting the red-shift). For completeness, I will first discuss the condition for steady state density not taking into account the stimulated processes.

STEADY STATE

Following Ruderman's[4] calculations of the properties of matter in superstrong magnetic fields, Chui[6] has variationally estimated the binding energy of an electron-hole droplet in large magnetic fields in a semiconductor. We can equally well use this estimate for an electron-positron fluid. The variational estimate of the binding energy is

$$E_b = -13.4 \, (a_0/\rho_c)^{2/3} \, e^2/a_0 , \qquad (1)$$

where a_0 is the Bohr radius and ρ_c the cyclotron radius,

$$\rho_c = (\hbar c/eB)^{1/2} . \qquad (2)$$

The radial extent of the electron or positron wave function transverse to the field is

$$R_0 \cong 1.1 \, a_0 \, (\rho_c/a_0)^{2/3} , \qquad (3)$$

the linear density of the drop along the magnetic field is

$$\ell_0^{-1} \cong 1.72 \, a_0^{-1} (a_0/\rho_c)^{2/3} , \qquad (4)$$

and the Fermi energy of the one-dimensional fluid is

$$E_f = (\hbar^2/2m) \, \ell_0^{-2} . \qquad (5)$$

For $B = 10^{12}$ Gauss, one gets $\rho_c \cong 4.9 \times 10^{-2} \, a_0$, $R_0 \cong 0.15 \, a_0$, $\ell_0 \cong 0.075 \, a_0$, $E_b \cong -2.75$ keV, and $E_f \cong 3.9$ keV. Since the temperature at the surface of the neutron star in the Crab is estimated to be less than 0.5 keV, we conclude that if a steady state with the above parameters can be attained, the postulated phase will indeed be realized and conditions for population inversion are well satisfied for it.

The electrons and positrons arrange themselves[4] in parallel rods among which there is a weak van der Waal's attraction. In a droplet the rods are spaced at several times $2R_0$; say equal to αR_0. In each rod, the spontaneous annihiation rate for $e^- - e^+$ pairs is approximately given by

$$\tau_s^{-1} \cong \pi r_0^2 \rho c , \qquad (6)$$

where r_0 is the classical electron radius and ρ the density. The total number of annihilations per second per unit colume is $\pi r_0^2 \rho^2 v$. For the parameters listed above one estimates this to be $10^{40}/\sec/\text{cm}^3$.

Suppose there is a flux density of $e^- - e^+$ of $\Phi \, \text{cm}^{-2} \, \sec^{-1}$ in a given region at the surface of a pulsar. One can roughly estimate the steady state radius of the droplets, R by equating the annihilations with the introduction of pairs into the droplets assuming the latter occur uniformly over the surface of the droplet. If $R \gg R_0$, one may conclude that a steady state is possible, otherwise it is not. I estimate $R \cong 10^{-40} \, \Phi \alpha^2$ cms. For the worst case, $\alpha = 1$, one requires $\Phi \gg \Phi_c \cong 10^{31} \text{cm}^{-2} \sec^{-1}$ to get $R \gg R_0$.

In one of the more successful models for radiation mechanisms from pulsars,[7,8] electron-positron discharges play a crucial role. A flux density of only $10^{24} \text{cm}^{-2} \sec^{-1}$ is allowed in these models from discharges at the "near gap".[8] However, an "outer gap" has been proposed[8,9] which generates photons of high enough energy that pair production on magnetic field lines ensues. A flux $\Phi \gg \Phi_c$ at the surface of the neutron star in the Crab in a region of the polar cap outside the projection of the "near gap" is quite consistent with this model.[10] Thus, the existence of the $e^- - e^+$ droplets depends on this model. Such a flux also accounts for the total number of $e^- - e^+$ annihilations observed by LMW.

The length ℓ of the droplet is determined by the height to which low energy electrons and positrons are fed to the droplet, and/or by tidal effects at the surface. The length is also self-limited if nonlinear stimulated emission occurs as discussed below.

STIMULATED EMISSION

The stimulated emission process I have proposed is different from the conventional one with which earthly lasers operate. This is because we require two photons to stimulate an annihilation. Lasers based on two-photon processes have also been proposed.[11]

For $e^- - e^+$ at rest and in zero field, the sum of the momentum of the annihilation photons is zero, the two photons have equal energy and each of them is isotropically distributed. For e^-, e^+ in a large magnetic field, there is an uncertainty in momentum in the transverse direction of $\sim \hbar/\rho_c$. Correspondingly the sum of the transverse momentum of the two outgoing photons is distributed over $\sim \hbar/\rho_c$. While the sum of their energy $E_1 + E_2 = 2mc^2(g-2)$, E_1 and E_2 are individually distributed over $c\hbar/\rho_c = \omega_1$ (g is the free electron gyromagnetic ratio).

If the $e^- - e^+$ fluid were confined in a volume V, the rate equation for the number of photons in the mode i per unit volume is given by

$$\frac{dn_i}{dt} = \frac{2}{\tau_s} \rho V^2 \sum_j f_{ij} \left(n_i + \frac{1}{V}\right) \left(n_j + \frac{1}{V}\right) \delta_{\Delta\omega} (E_i + E_j - 2mc^2) - n_i/\tau_{\ell i} \quad (7)$$

where ρ is the inverted density of pairs, $\tau_{\ell i}^{-1}$ is the loss rate in the i-th mode, which includes the Compton scattering rate, and $\delta_{\Delta\omega}$ is a Lorenzion of spontaneous linewidth $\Delta\omega$. f_{ij} is the fraction of the two-photon emission into the i-th and j-th modes. f_{ij} is normalized by

$$\sum_{i,j} f_{ij} \delta_{\Delta\omega} (E_i + E_j - 2mc^2) = 1 . \quad (8)$$

The rate equation for ρ is

$$\frac{d\rho}{dt} = -\frac{\rho V^2}{\tau_s} \sum_{i,j} f_{ij} \left(n_i + \frac{1}{V}\right) \left(n_j + \frac{1}{V}\right) \delta_{\Delta\omega} (E_i + E_j - 2mc^2) + \phi , \quad (9)$$

where ϕ is the "pump rate", i.e., the flux of pairs into the rods per unit volume.

We will assume that for photons *not* emitted in the small rod solid angle, $\Omega_d \cong (\rho_o \ell)^2$, the stimulated processes are unimportant. For the total density of such photons, we have

$$\frac{dn}{dt} = \frac{\rho}{\tau_s} - \frac{n}{\tau'_\ell}, \tag{10}$$

where τ'^{-1}_ℓ is their loss rate. This gives the steady state value

$$n_s \cong \tau'_\ell \frac{\rho_s}{\tau_s}, \tag{11}$$

where ρ_s is the steady state value for ρ given by ℓ_o^{-1} and R_o of Eqs. (4) and (3). The loss time τ'_ℓ will be determined by the time to escape $\approx R_o/c$.

Let ν be the total density of photons travelling in the rod solid angle. These are the only ones important for stimulated annihilation. The rate equation for ν is obtained from (7) by integrating over all modes with propagation vectors in the rod solid angle. We will continue to assume, considering the low translational energy of pairs and $\hbar\omega_c/mc^2 \ll 1$, (omegs$_c$ is the cyclotron frequency) that the annihilation photon momentum is given by the incoming momentum. Geometrical optics is valid for our situation and the propagation vectors are uniformly distributed. We get

$$\frac{d\nu}{dt} = \frac{2\rho}{\tau_s} V^2 \left[\nu^2 + \nu\left(n + \frac{1}{V}\right) + \frac{\Omega_d}{V}\left(n + \frac{1}{V}\right)\right] - \nu/\tau_\ell \tag{12}$$

We have to solve (12) and (9), appropriately expressed in terms of n and ν, simultaneously. For small times, we assume that ρ is at its steady state value ρ_s, and find conditions for linear and nonlinear amplification of ν through (12). The ν^2 term in (12) becomes important for $\nu > \nu_c$,

$$\nu_c = \left(\frac{\tau_s}{\tau_\ell}\right) \cdot \frac{1}{2\rho_s V^2}. \tag{13}$$

One can get to a photon density ν_c either if the linear amplification rate

$$\frac{2\rho_s V^2 \left(n_s + \frac{1}{V}\right)}{\tau_s} > \frac{1}{\tau_\ell} \tag{14}$$

or if the steady state value of ν through the ν independent term,

$$2\rho\Omega_d V\left(n_s + \frac{1}{V}\right)(\tau_\ell/\tau_s) > \nu_c . \tag{15}$$

The time τ_ℓ is the smaller of the compton scattering time which is of the same order as τ_s, or ℓ/c. Since $\tau_s \cong 10^{-13}$ sec, for $\ell \ll 10^{-3}$cms the latter determines τ_ℓ. From (14) we find that the linear amplification occurs for $\ell \gtrsim 10^{-5}$cms. The inequality (15) is satisfied only for $\ell \gtrsim 10^{-2}$cm. Therefore, the linear amplification process is the more important. During the linear amplification the linewidth will be ω_T (gain per pass)$^{-\frac{1}{2}}$, which for large enough ℓ can be very small indeed. Suppose we get to a stage $\nu/\nu_c = \gamma > 1$. From (12) we can conclude that a giant pulse of photons will be stimulated in a characteristic time

$$t_{n.\ell.} = \tau_\ell \ \ell n \left(\frac{\gamma}{\gamma-1}\right) .$$

We can no longer take ρ to be at its steady state value, because of the nonlinear depletion. In a time few times $t_{n.\ell.}$, ν will decay at a rate τ_ℓ^{-1}. For low enough ν, ρ will start building up and the process will repeat -- linear amplification followed by nonlinear amplification and then decay, etc.

To get the complete time-dependent behavior, one has to solve the simultaneous nonlinear equations (9), (10) and (12), but we hope we have shown the possibility of stimulated emission with reasonable parameters.

The linewidth in the nonlinear emission regime at any time t is given approximately by

$$\omega_T /(\gamma')^{\frac{1}{2}} .$$

where γ' is ν/ν_c, a time t_ℓ earlier.

The basic equation of Ramaty, Mckinley and Jones, eqs. (7) and (15) of Ref. (5) are equivalent to eq. (7) above. As already discussed, they perform their calculation for a dense three-dimensional plasma in zero field. Their result is that the Compton loss term, the last term in eq. (7) wins out over the stimulated annihilation term unless ρ is very large. At the critical value of ρ, the degeneracy temperature of $e^- - e^+$ is $0(1\ \text{MeV})$ and the maximum gain is at 430 keV, where therefore the stimulated annihilation would occur. The density required and the energy of emitted x-rays is such that the process becomes uninteresting as far as the explaining of observations is concerned.

I have performed my calculations, assuming a large magnetic field, characteristic of the surface of pulsars and such that the rod-like $e^- - e^+$ fluid state is realized. In this situation my conclusion is that a density of $e^- - e^+$

low enough so that the chemical potential is only a few keV is enough to obtain stimulated annihilation. At these densities the annihilation photons are only near 511 keV.

I wish to thank M. Leventhal and R. Ramaty for encouragement and E. I. Blount for several discussions.

REFERENCES

1. M. Leventhal, C. J. MacCallum and A. C. Watts, Astrophys J. *216*, 491 (1977).
2. M. Yoshimori, H. Watanabe, K. Okudaira, Y. Hirasima and H. Marakami, Aust. J. Phys. *32*, 375 (1979); C. A. Ayre, P. N. Bhat, Y. Q. Ha, R. M. Myers and M. G. Thompson (1983) preprint. Null results have been obtained by J. C. Ling, W. A. Mahoney, J. B. Willet and A. S. Jacobson, Nature *270*, 36 (1977).
3. C. M. Varma, Nature *267*, 686 (1977).
4. M. Ruderman, Phys. Rev. Lett. 27, 1306 (1971).
5. R. Ramaty, J. M. McKinley and F. C. Jones, Astrophys. J. *256*, 238 (1982).
6. S.-T. Chui, Phys. Rev. *9*, 3458 (1974).
7. P. A. Sturrock, Ap. J. *164*, 529 (1971).
8. M. A. Ruderman and P. G. Sutherland, Ap. J. *196*, 51 (1975).
9. Andrew Cheng, M. Ruderman and P. Sutherland, Ap. J. *203*, 209 (1976).
10. Andrew Cheng, Thesis, Columbia University (1976) unpublished and private communication.
11. P. P. Sorokin and N. Braslaw, IBM J. Res. Develop. *8*, 177 (1964).

CRITERIA FOR GRASAR ACTION IN ASTROPHYSICAL SOURCES

J. M. McKinley* and R. Ramaty
Laboratory for High Energy Astrophysics
NASA/Goddard Space Flight Center
Greenbelt, Maryland 20771

ABSTRACT

The theory of gamma-ray amplification through stimulated annihilation radiation (grasar) was developed by Ramaty, McKinley and Jones[1] (hereafter RMJ). For gamma-ray bursts similar to the March 5, 1979 burst, an observed annihilation line of width <0.03 MeV would imply a grasar source. The minimum pair density needed for the onset of grasar action is $\sim 10^{30}$ cm^{-3} and the peak of the grasar line, without a gravitational redshift, is at < 0.5 MeV.

INTRODUCTION

An emission line at $\sim$0.43 MeV, generally believed to be spontaneous, optically thin and gravitationally redshifted e^+-e^- annihilation, has been observed[2,3] from several gamma-ray bursts. The measured photon fluxes and the likely distances and sizes of the burst sources suggest[4], however, that for at least some bursts the source regions are optically thick. Compton scattering and γ-γ pair production are the principal mechanisms that would remove photons from an emission line in a gamma-ray burst source.

In a detailed calculation of the emissivities and the absorption coefficients for two-photon pair production and annihilation and the accompanying Compton and inverse Compton scattering, RMJ[1] showed that an emission line at $\sim$0.43 MeV could be produced in an optically thick source without a gravitational redshift by amplified annihilation radiation. In the present paper we consider the observational signatures that would imply the existence of such a grasar source, the minimum e^+-e^- pair density needed for the onset of grasar action and the reasons that the central energy of the two-photon grasar line is at < 0.5 MeV.

OBSERVATIONAL REQUIREMENTS FOR A GRASAR SOURCE

The need for a grasar source can be seen from a relationship between the width of an observed annihilation line, ΔE, the fluence in the line, F, the duration of the emission of the line photons, Δt, the source distance, d, and the source projected area, A. We derive this relationship by comparing the minimum pair density required to produce the line by nonamplified radiation (i.e. radiation with total absorption coefficient K_T>0) with the maximum pair density allowed by broadening due to pair degeneracy[1] and the

*Also at Physics Department, University of Maryland, College Park

observed upper limit on the line width. We first calculate the minimum pair density.

The equation of radiative transfer, $dI/d\ell = j - K_T I$, requires that for nonamplified annihilation radiation, the ratio of the spontaneous annihilation emissivity to the absorption coefficient, j/K_T, be larger than the radiation intensity I. Otherwise $dI/d\ell < 0$ and no observable radiation is produced. RMJ[1] carried out detailed calculation of j and K_T for pair plasmas in a bath of ambient photons. Here we estimate the annihilation emissivity, j, by a simple expression which is consistent with their calculations,

$$j \simeq 2\pi r_o^2 c \, n_\pm^2/(4\pi \Delta E), \qquad (1)$$

where $n_\pm$ is the pair density (we assume equal positron and electron densities) and $r_o = 2.82 \times 10^{-13}$ cm. To calculate the minimum pair density, we can ignore the ambient photons. Thus $K_T \simeq K_C$, where

$$K_C \simeq 2 n_\pm \pi r_o^2 \qquad (2)$$

is the Compton absorption coefficient. The intensity is given by

$$I = \frac{d^2 F}{A \Delta t \Delta E} \; . \qquad (3)$$

By combining Eqs. (1), (2) and (3) and using the condition for observable nonamplified radiation ($j/K_T > I$) we obtain

$$n_\pm > \frac{4\pi d^2 F}{A \Delta t c} \simeq (4 \times 10^{23} \text{ cm}^{-3}) \, (\frac{d}{1 \text{kpc}})^2 (\frac{F}{1 \text{ ph cm}^{-2}})(\frac{A}{1 \text{km}^2})^{-1}(\frac{\Delta t}{1 \text{sec}})^{-1} . \quad (5)$$

We next consider the upper limit on $n_\pm$. For nonamplified annihilation, degeneracy broadening[1] sets a lower limit on the width of the annihilation line,

$$\Delta E > 0.8 p_F c = 0.8 (3\pi^2)^{1/3} \hbar c \, n_\pm^{1/3}, \qquad (6)$$

where p_F is the Fermi momentum of the pairs. An observed line width or upper limit on the width, therefore, sets an upper limit on $n_\pm$,

$$n_\pm < 8.5 \times 10^{27} \text{ cm}^{-3} \, (\Delta E/0.1 \text{MeV})^3. \qquad (7)$$

By combining Eqs. (5) and (7) we obtain a necessary condition for the production of an observed line by nonamplified annihilation,

$$(\frac{A}{1 \text{km}^2}) (\frac{\Delta t}{1 \text{sec}}) > 4.7 \times 10^{-5} (\frac{F}{1 \text{ ph cm}^{-2}}) (\frac{d}{1 \text{kpc}})^2 (\frac{\Delta E}{0.1 \text{MeV}})^{-3}. \quad (8)$$

The most intense astrophysical annihilation line observed so far is that seen[2] in the spectrum of the 1979, March 5 burst whose source direction coincides[5] with that of the supernova remnant N49 in the Large Magellanic Cloud (LMC). For this burst[2] $F \sim 7$ photons cm^{-2}, $\Delta E < 0.13$ MeV, and if the line-formation time equals the duration of the impulsive phase, $\Delta t \sim 0.15$ sec. Furthermore, if the source of the burst is indeed in the LMC, $d \simeq 55$ kpc. Then from Eq. (8), nonamplified annihilation requires that $A > 3$ km^2. This condition can be satisfied on a neutron star surface since its projected area is ~ 300 km^2.

For the March 5, 1979 burst, however, both Δt and ΔE could have been significantly smaller than the values given above. In particular, the line formation time could be less than the total impulsive phase duration, since the annihilation time of the pairs is extremely short[6] ($\sim 10^{-15}$ sec). If future observations should indicate short durations and narrow lines, a strong case would exist for grasar sources in astrophysics.

THRESHOLD DENSITY FOR GRASAR ACTION

Spontaneous pair annihilation into two photons and the inverse process of two-photon pair production are necessarily accompanied by stimulated annihilation (e.g. Ref. 1). The absorption coefficient due to stimulated annihilation, K_{SA}, is always negative, but the generation of amplified annihilation radiation (i.e. grasar action) requires that the total absorption coefficient be negative for at least some photon energies,

$$0 > K_T = K_C + K_{pp} + K_{SA} . \tag{9}$$

Here K_{pp} is contributed by pair production. K_{SA} is related[1] to the spontaneous emissivity of annihilation photons, j,

$$-K_{SA} = \frac{4\pi^3 (\hbar c)^3}{cE^2} j, \tag{10}$$

where E is the photon energy. This expression, like its analogue for any other radiative process, can also be found directly from the Einstein A and B coefficients.

RMJ showed[1] that for grasar action to occur the pair density $n_\pm$ must exceed a threshold value. To provide an estimate of this threshold, we consider the case of a cold pair plasma (kT << Fermi energy) with no ambient photons present. Such a degenerate system has the lowest threshold density, since $K_{pp}=0$ and K_C is reduced by the degeneracy. In this case we can use Eqs. (1), (2), (6) and (10) and $E \simeq mc^2$ to obtain

$$-\frac{K_{SA}}{K_C} \simeq 1.25 \frac{\pi^{4/3}}{3^{1/3}} \left(\frac{\hbar}{mc}\right)^2 n_\pm^{2/3} f^{-1} \simeq 5.9 \times 10^{-21} \left(\frac{n_\pm}{1 \text{cm}^{-3}}\right)^{2/3} f^{-1}, \tag{11}$$

where f is the fraction of the degenerate positrons and electrons which can contribute to Compton scattering at the photon energy of interest. The threshold for grasar action occurs when $-K_{SA} \simeq K_C$, i.e. at a pair density

$$n_\pm \simeq 2.2 \times 10^{30} \text{ cm}^{-3} \text{ f}^{3/2}. \qquad (12)$$

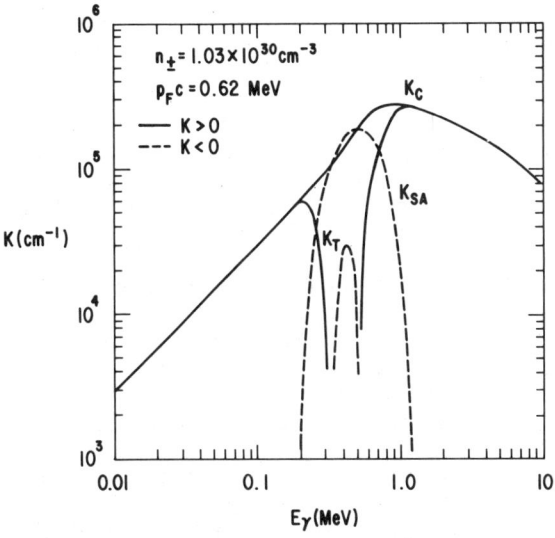

Figure 1. Absorption coefficients vs. photon energy in a degenerate positron-electron plasma.

Accurate calculation of K_{SA} and K_C for a degenerate positron-electron plasma have been carried out by RMJ[1]. In Fig. 1 we show their results for a pair density just above threshold. As can be seen, below the Fermi energy K_C indeed decreases rapidly with decreasing photon energy. By comparing the threshold density of $\sim 10^{30}$ cm^{-3}, obtained from the numerical calculations, with that given in Eq. (12), we see that $f \simeq 0.6$, i.e. approximately 40% of the degenerate positrons and electrons cannot contribute to Compton scattering at ~ 0.5 MeV.

The threshold density for grasar action would be lower than 10^{30} cm^{-3} if the Compton scattering cross section were less than πr_o^2. Indeed, in the presence of a strong magnetic field, at photon frequencies below the cyclotron frequency, ν_c, the Compton cross section is substantially reduced. But the observation[2] of cyclotron absorption features at 40-60 keV in many gamma-ray bursts suggests that for e^+-e^- annihilation photons $E \gtrsim 10 \, h\nu_c$. Daugherty and Ventura[7] have shown that at such photon energies, the scattering cross section for photons directed parallel to the field is very nearly πr_o^2. Altogether, it seems that the density of pairs in a grasar source must exceed 10^{30} cm^{-3}.

THE CENTRAL ENERGY OF A GRASAR LINE

From their numerical calculations RMJ[1] found that the maximum of $-K_T$ is at a photon energy less than 0.5 MeV and from this they concluded that the central energy of a grasar line should also be at such a photon energy. There are two qualitative effects that cause this redshift. The first one can be seen from Figure 1. Here, for a degenerate pair plasma of density close to the grasar threshold, the maximum of $-K_{SA}$ occurs very close to 0.5 MeV. But because of the steep slope of K_C, the maximum of $-K_T$ is shifted to ~ 0.42

MeV. This redshift, therefore, is a direct consequence of the effect of the degeneracy on Compton scattering and the high Fermi energy implied by the high threshold density.

The second effect is independent of the degeneracy. It is caused by either a high temperature ($kT \sim mc^2$) or a high density (Fermi energy $\sim mc^2$), both of which broaden and blueshift the emissivity[1]. Then from Eq. (10), division by E^2 moves the peak of $-K_{SA}$ to an energy <0.5 MeV. In the nondegenerate case, the peak of $-K_T$ essentially coincides with that of $-K_{SA}$, since K_C is not a strong function of photon energy. In the degenerate case, with the pair density much higher than the threshold density, the shape of K_C does not affect much the position of the maximum of $-K_T$.

Varma[8] proposed a grasar model based on degenerate pairs with Fermi energy $\ll mc^2$ for which the peak of $-K_{SA}$ would be very close to 0.511 MeV. However, Compton scattering was ignored in this model. If it were taken into account, then the relatively low pair density corresponding to this Fermi energy would imply that $K_{SA} + K_C > 0$ at all energies. Thus grasar action cannot occur in this model.

As pointed out in the Introduction, the observed annihilation lines in gamma-ray bursts are at ~ 0.43 MeV. If some of these lines are indeed from grasar sources, the implied gravitational redshift would be much smaller than that implied by spontaneous, nonamplified annihilation. The potential existence of grasar sources, therefore, allows gamma-ray bursts to be produced on neutron stars of smaller mass and larger radius than previously conjectured, or even objects other than neutron stars.

REFERENCES

1. R. Ramaty, J. M. McKinley and F. C. Jones, Ap. J. 256, 238 (1982).
2. E. P. Mazets, S. V. Golenetskii, R. L. Aptekar', Yu. A. Gur'yan, V. N. Il'inskii, Nature 290, 378 (1981).
3. B. J. Teegarden and T. L. Cline, Ap. J. 236, L67 (1980).
4. W. K. H. Schmidt, Nature 271, 525 (1978).
5. T. L. Cline et al., Ap. J. Lett. 255, L45 (1982).
6. R. Ramaty, R. E. Lingenfelter and R. W. Bussard, Astrophys. Space Sci. 75, 193 (1981).
7. J. K. Daugherty and J. Ventura, Phys. Rev. D 18, 1053 (1978).
8. C. M. Varma, Nature 267, 686 (1977).

PAIR PRODUCTION AND NON-THERMAL RADIO STARS

W. Thomas Vestrand
National Radio Astronomy Observatory, Charlottesville, VA 22901

ABSTRACT

The generation of relativistic electrons and positrons in thermal material surrounding a γ-ray source is examined. This *in situ* generation of particles can resolve many of the problems associated with the quiescent emission from non-thermal radio stars. The most luminous radio star, Cyg X-3, is used as a paradigm throughout.

INTRODUCTION

Approximately 20 radio sources have been classified as non-thermal radio stars. Virtually all of these radio stars are known to be in short period binary systems (P < 25 days)[1]. The radio emission is thought to be synchrotron or gyro-synchrotron radiation generated by relativistic electrons. The source of these electrons, however, is unknown.

Perhaps the most extreme constraints on the origin of the radiating electrons are posed by the most luminous radio star: Cyg X-3. Its quiescent radio emission has a luminosity $L_{OR} \sim 10^{31}$ erg s^{-1} and a spectrum that is slightly inverted with a peak at ~10 GHz. Using VLBI techniques at 8.4 GHz, Geldzahler et al.[2] have found that the scale size of the quiescent source is $\sim 10^{14}$ cm and the brightness temperature is $\sim 2 \times 10^9$ K. Two models for generation of the quiescent emission have been proposed: (1) synchrotron emission from relativistic electrons embedded in a stellar wind[3]; (2) radiation from a compact source that is scattered by a turbulent ISM (cf. Backer[4]). Evidence that a binary system ($a \sim 10^{11}$ cm) is enshrouded by material with electron scattering depth $\tau_{es} \sim 1$ is available both from X-ray observations[5] and radio outburst observations[6]. This circumstellar material causes problems for both models[7]. The shortcoming of the first model is that electrons generated within the binary system lose all their energy before reaching $R \sim 10^{13}$ cm, i.e. long before they reach the distances from which the radio emission is observed. The shortcoming of the second model is that the plasma frequency of the enshrouding material is so high that the observed 1 GHz radiation must originate at $R \sim 10^{13}$ cm. The question remains therefore: what process generates the radio emitting particles in the enshrouding material?

PRODUCTION OF ELECTRONS AND POSITRONS

The answer to the question posed in the introduction is apparent, I think, when it is realized that Cyg X-3 is not only enshrouded with gas, but is also the most luminous gamma-ray point source, $L_\gamma \sim 10^{37}$ erg s^{-1}, known in the galaxy[8,9]. As these gamma-rays traverse the circumstellar material they will generate energetic electrons and positrons by three processes: (1) pair production by photon-photon collisions; (2) pair production by photon-particle collisions and (3) Compton scattering. These processes act essentially as <u>in situ</u> accelerators and can easily generate enough electron-positron pairs to power the quiescent radio emission.

Within and near the Cyg X-3 binary system, where the X-ray and lower energy photon fields are nearly isotropic, photon-photon collisions ($\gamma + \gamma \rightarrow e^+ + e^-$) are the dominant mechanism of pair production. Unfortunately, a number of absorption and suppression mechanisms preclude observation of radio emission from these inner pairs[7]. In outer regions, from which radio frequency emission can escape, the reaction probability markedly decreases. Unlike its reverse reaction, two-photon electron-positron annihilation, pair production by photon-photon collisions has an energy threshold imposed by the requirement that the total photon energy in the center-of-momentum system must be larger than $2\ m_e c^2$. As a consequence, at large radii, where the photons are freely streaming, the smaller average interaction angle requires a given γ-ray to strike a higher energy target photon to meet the threshold condition. This, combined with the decreasing number of target photons, causes the pair production rate to drop rapidly with radius. If, for example, the target photons have a distribution of the form $n(E)dE = K^{-p}\ dE$, one can show that the reaction probability will decrease with radius as $R^{-2(p+2)}$. Consequently, photon-photon collisions are only likely to be an important source of radio emitting pairs for γ-ray emitters that are not enshrouded with circumstellar material.

The most important source of radio emitting pairs in an enshrouded system like Cyg X-3 is the interaction of γ-rays with charged thermal particles: $\gamma + CP \rightarrow e^+ + e^- + CP$. In order for the reaction to proceed, the photon must have an energy in the rest frame of the target exceeding $2\ m_e c^2$ ($4\ m_e c^2$ for an electron target). The total cross-section for the process is $\sim \alpha\ \sigma_T$ (where σ_T is the Thomson cross-section and $\alpha = 1/137$). Unlike photon-photon collisions, the cross-section does not depend on the incidence angle of the γ-ray. As a consequence, the reaction probability for a γ-ray of given energy is just proportional to the number density of target particles. In a constant velocity wind, for example, the pair production probability decreases as R^{-2}. Comparing this with the radial dependence for photon-photon interactions, it is clear that photon-particle interactions are a more important source of energetic pairs at large radii.

Compton scattering ($\gamma + e \rightarrow \gamma' + e'$) is another important source of energetic electrons. When a γ-ray is Compton scattered by a thermal electron, a large fraction of the photon's energy is often transferred to the electron--essentially accelerating the electron. The total cross-section for this process decreases as $\sim \sigma_T (m_e c^2 / E_\gamma)$; consequently, it is only important for medium energy γ-rays. Like photon-particle pair production, for a given γ-ray energy the reaction probability per unit length is proportional to the matter density. Compton scattering will therefore be a significant source of energetic electrons out to relatively large radii.

The source function of electrons and positrons with energies between E_e and $E_e + dE_e$ at radius R is given by

$$Q(E_e;R) = \int_{E_e}^{\infty} n_\gamma(E_\gamma,R) \, n_i(R) \left\{ 2Z^2 \sigma_p(E_e; E_\gamma) + Z\sigma_c(E_e;E_\gamma) \right\} c \, dE_\gamma \quad (1)$$

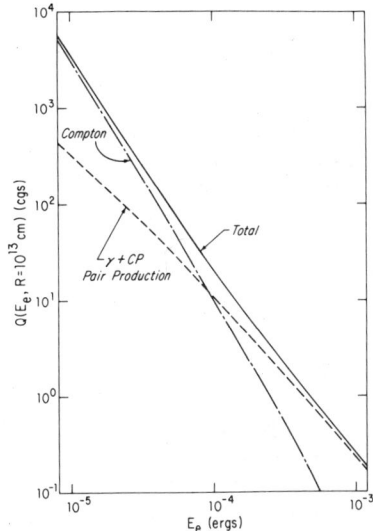

where $n_\gamma(E_\gamma,R)$ is the number density of γ-rays at R, $n_i(R)$ is the number density of ion targets, $\sigma_p(E_e;E_\gamma)$ and $\sigma_c(E_e;E_\gamma)$ are the differential cross-sections for charge-field pair production and Compton scattering given in Heitler[10]. Figure 1 shows the source function obtained for Cyg X-3 at $R = 10^{13}$ cm when a photon distribution consistent with the 10^2 MeV observations of Lamb et al.[8], and a solar abundance gas with density $n_i = 3 \times 10^9$ cm^{-3} are used. Notice that pair production is the dominant source of particles with energies $E_e \geqslant 100$ MeV.

RADIO EMISSION

Both X-ray and radio observations suggest that gas outflows from the binary system in Cyg X-3 and enshrouds it [driven perhaps by the young pulsar strongly suggested by γ-ray observations[11]]. In a steady wind, the generated pairs are convected outward at a speed V so the concentration of electrons at R with energies between E and E + dE is given by

$$N(E,R) = N(E_0(E,R,R_0);R_0)\left(\frac{R_0}{R}\right)^2 \frac{\partial E_0}{\partial E} + \int_{R_0}^{R} Q(E_i(E,R,R_i);R_i)\left(\frac{R_i}{R}\right)^2 \frac{\partial E_i}{\partial E} \frac{dR_i}{V}$$

where the function $E_i(E,R,R_i)$ gives the initial energy of an electron that is injected at R_i and has an energy E when it reaches R. The first term on the right-hand side of equation (2) corresponds to relativistic electrons initially embedded in the wind and consequently can be dropped when modeling the radio emission from Cyg X-3. The second term corresponds to electrons and positrons that are generated *in situ*.

The population of relativistic electrons and positrons will generate, in a vacuum, synchrotron radiation with emissivity

$$\varepsilon(\nu,R) = 1.86 \times 10^{-23} \int B_\perp(R) \, F(\nu/\nu_c) \, N(E_e,R) \, dE_e \quad (3)$$

where $\nu_c = 6.27 \times 10^{18} \, B_\perp \, E^2$ in cgs (Gaussian) units and F(x) is tabulated in Westfold[12]. The flux density at frequency ν is

$$S_\nu = \frac{4\pi}{D^2} \int_{R_{min}(\nu)}^{\infty} R^2 \varepsilon'(\nu, R) dR \qquad (4)$$

where D is the source distance, $\varepsilon'(\nu,R)$ is the emissivity corrected for Razin-Tsytovitch suppression[13], and $R_{min}(\nu)$ is the minimum radius from which radiation with frequency ν can escape to a distant observer. The most important absorption mechanism for the conditions likely in the Cyg X-3 system is free-free absorption[7]. In this case, the radius from which the optical depth to $R = \infty$ is unity is given by

$$R_{min}(\nu) \approx 5.3 \times 10^{13} \, \nu_{GHz}^{-2/3} \, T_7^{1/2} \, (n_0/3 \times 10^{13} \, cm^{-3})^{2/3} \, cm \qquad (5)$$

where the gas temperature is $T = T_7 \cdot 10^7$ K and the gas density profile is given by $n_i(R) = n_0 (R/10^{11} \, cm)^{-2}$.

To calculate the synchrotron spectrum, equations (2) and (4) were numerically integrated for the concentration of electrons and positrons generated in the circumstellar material. Figure 2

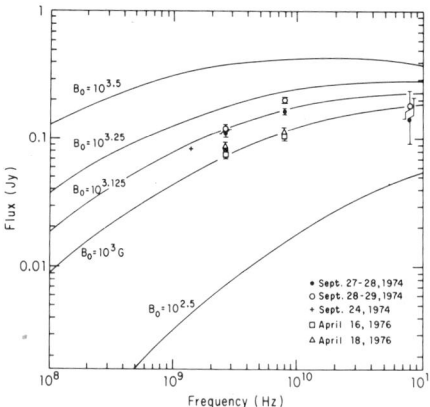

shows the results obtained using the phase averaged γ-ray flux, $\tau_{es}=1$ (i.e., $n_0=3\times10^{13}$ cm^{-3}), and magnetic field profiles of the form $B_\perp(R) = B_{\perp 0} (R/10^{11} \, cm)^{-1}$. Unfortunately, since simultaneous measurements of the medium-energy γ-ray and the radio flux are not yet available, one is forced to compare the calculated spectra, which should represent the radio flux during March 1973, with radio observations for another epoch. The radio data[14] for September 1974 are plotted for comparison in Figure 2 because they constitute the most extensive spectral coverage of the source in its quiescent state. One sees that the calculated spectra for $B_{\perp 0} \approx 10^3$ G fit the data fairly well. In particular, these predicted spectra fit the data much better above 10 GHz than previous models. Comparison of the predicted spectra for March 1973 with radio data for different epochs, should be made with some caution because at radio frequencies the averaged daily fluxes can vary by factors of ~2-3 from month-to-month even when the source is in its quiescent state. I emphasize, however, that such behavior is expected within the context of this model since the failure of the COS-B satellite to detect Cyg X-3[15] suggests that the γ-ray flux varies by at least factors of ~2.

CONCLUSIONS AND DISCUSSION

I have argued that the relativistic electrons responsible for the quiescent radio emission from Cyg X-3 are generated in circumstellar material. Specifically, it was shown that the observed flux of γ-rays can generate enough electron-positron pairs, as they traverse the amount of enshrouding material suggested by models for the X-ray light curve, to power the quiescent radio emission. When the radio frequency opacity of the enshrouding material was included, the predicted spectrum was found to match the observed spectrum quite well. In fact, the model presented here not only answers the question of where and how the radiating electrons originate, but also produces a better fit to the radio observations above 10 GHz than ad hoc models that assume a power law electron energy distribution.

A prediction of the model presented here, which can in principle be tested, is that the system should be a source of positron annihilation radiation with a flux of $\sim 10^{-8}$ photon cm^{-2} s^{-1}. In practice, however, a flux this low will probably not be detectable in the near future. A prediction that will be easier to test is that the quiescent radio flux should be causally connected with the $\sim 10^2$ MeV γ-ray flux. The next most luminous radio stars are Cir X-1, SS 433 and GT 0236. Each exhibits non-thermal quiescent radio emission and is associated with an X-ray binary system. Furthermore, there is some evidence for circumstellar material in each system. Gamma-rays, however, have only been detected from GT 0236. The luminosity of the γ-ray source identified with GT 0236, $L_\gamma (\sim 10^2$ MeV) $\approx 10^{35}$ erg s^{-1}, could easily power the system's quiescent radio emission. If the quiescent radio emission from Cir X-1 and SS 433 is powered by γ-rays, these sources should be detectable with the EGRET experiment on the GRO satellite.

REFERENCES

1. P. C. Gregory et al., A. J., **84**, 1030 (1979).
2. B. J. Geldzahler et al. A. J., **84**, 186 (1979).
3. E. R. Seaquist and P. C. Gregory, Ap. Letters, **18**, 65 (1977).
4. D. C. Backer, Ap. J., **222**, L9 (1978).
5. P. Heitz, P. Joss and S. Rappaport, Ap. J., **224**, 614 (1978).
6. P. C. Gregory and E. R. Seaquist, Ap. J., **194**, 715 (1974).
7. W. T. Vestrand, Ap. J., in press (1983).
8. R. C. Lamb, Ap. J., **212**, L63 (1977).
9. R. C. Lamb et al., Nature, **296**, 543 (1982).
10. W. Heitler, The Quantum Theory of Radiation (Oxford Press, London, 1954).
11. W. T. Vestrand and D. Eichler, Ap. J., **261**, 251 (1982).
12. K. C. Westfold, Ap. J., **130**, 241 (1959).
13. V. N. Tsytovitch, Ann. Rev. Astron. Astrophys., **11**, 363 (1973).
14. K. O. Mason et al., Ap. J., **207**, 78 (1976).
15. K. Bennett et al., A. A., **59**, 273 (1977).

RESONANT FREE-FREE EMISSION FROM ELECTRONS IN MAGNETIC POLAR REGIONS OF ACCRETING NEUTRON STARS

R. Lieu
Department of Physics, University of Calgary,
Calgary, Alberta, Canada T2N 1N4

ABSTRACT

Near the magnetic polar cap of an accreting neutron star, emission of photons at cyclotron resonance is accompanied with the transition of electrons between Landau levels as a result of free-free collisions with ions. In quantum electrodynamics, the resonance condition is satisfied whenever the energy denominators of the two possible time orderings in electron free-free emission become singular. In the absence of motion along the magnetic field, both singularities occur near the cyclotron frequency. If there is a definite longitudinal momentum due to bulk motion of the accreting plasma, the singularities for forward and backward scattering can occur at different frequencies. Moreover, the frequencies are dependent on the pitch angle of photon emission. For a given line-of-sight direction, this is a line splitting effect, the observability of which depends on the degree of comptonization and the plasma temperature.

The discovery of cyclotron line features in the binary X-ray pulsar Her X-1 (Trumper et al 1977) has led to much interest in the quantum mechanics of electrons in strong magnetic fields. In the magnetic 'hot spots' of a neutron star, the magnetic field can be as high as $H \sim 5 \times 10^{12}$ gauss, and electrons only have energy to occupy the first few Landau levels. Emission and absorption of radiation take place at integral multiples of the cyclotron frequency $\omega_c = eH/(mc)$, and are accompanied with transition of the electrons between the discrete energy levels. It has been demonstrated (Nagel 1980) that such transitions are predominantly caused by electron-ion collisions (i.e. 'free-free' or bremsstrahlung process in a strong magnetic field).

Although details of the principal plasma instabilities near the surface of the neutron star remain poorly understood, it is extremely plausible that the kinetic energy of the electrons (derived from the neutron star gravitational field and from collisions with ions) is present in the form of transverse as well as longitudinal motion (w.r.t. the magnetic field). The former takes place in quantized orbitals, i.e. the Landau levels. The latter, however, has a continuous energy spectrum which __must__ be taken into account when determining the character of the resonances. Thus, the fundamental resonance frequency can be shifted, or even 'split', from the cyclotron frequency ω_c. Fortunately, the subtle effects of electron and ionic recoil are automatically taken care of by the formalism of quantum electrodynamics, as what follows shall demonstrate.

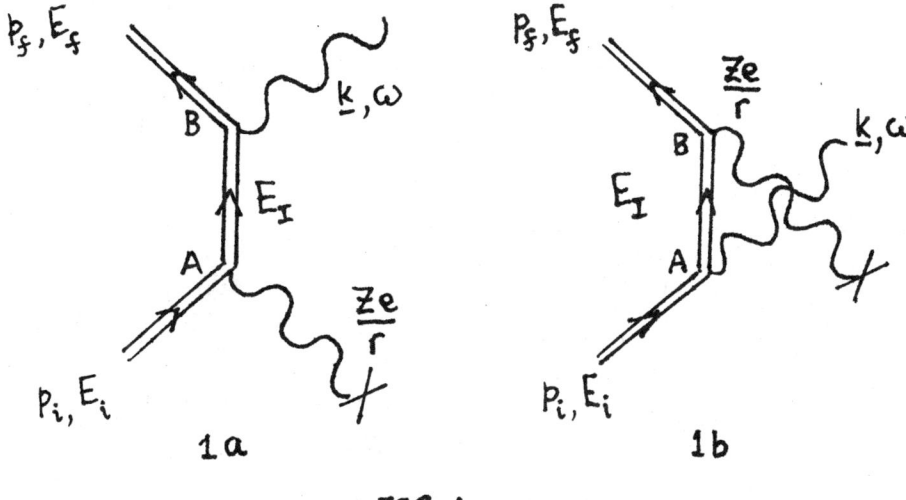

FIG 1

Calculations of the electron bremsstrahlung cross section in a strong magnetic field - the probability of transition between two Landau levels induced by collision with an ion - involve the two basic Feynman diagrams (Fig. 1). For resonant emission, the initial state of the electron may be at the first Landau level, of energy

$$E_i = (p^2 + m^2 + 2\varepsilon^2)^{1/2} \quad (1)$$

where p is the (initial) momentum of the electron along the magnetic field (z-direction), $\varepsilon^2 = (H/H_{cr})m^2$, m is the mass of the electron, and units $\hbar = c = 1$ are used for simplicity.

The final electron is then at the zeroth Landau level (ground state) of energy

$$E_f = (p'^2 + m^2)^{1/2} \quad (2)$$

where p' is the final momentum of the electron. During the emission process there is overall energy conservation between the electron and the photon, i.e.

$$(p^2 + m^2 + 2\varepsilon^2)^{1/2} = (p'^2 + m^2)^{1/2} + \omega \quad (3)$$

The energy of the 'intermediate state' (E_I) depends, however, on what method of calculation one wishes to adopt. In the 'time-ordering' approach, it is assumed that (i) the electron always remains on mass-shell, and (ii) momentum is conserved at all interaction vertices. If we apply these rules to vertex B of Fig. 1a, and vertex A of Fig. 1b (it actually does not matter which vertex one uses, because of overall energy conservation), we find that

$$E_I = \begin{cases} [(p' + \omega\cos\theta)^2 + m^2 + 2j\epsilon^2]^{1/2} & \text{(Fig. 1a)} \\ [(p - \omega\cos\theta)^2 + m^2 + 2j\epsilon^2]^{1/2} & \text{(Fig. 1b)} \end{cases} \quad (4)$$

where θ is the angle between the photon wavevector $\underline{k}$ and the z-direction. Note that j can assume values $j = 0,1,2,...$ and is a free parameter which renders the energy of the intermediate state <u>uncertain</u>. For a transition between the 1st and zeroth Landau levels, however, the probability that E_I can assume values with high j is very small.

It is well-known that the time-ordering is achieved by a θ-function which, when integrated over <u>forward</u> directions in time, yields respectively the energy denominators

and
$$\alpha_j = |E_I - E_f - \omega|^{-1}$$
$$\alpha'_j = |E_I - E_i + \omega|^{-1} \quad (5)$$

for the two diagrams Fig. 1a and Fig. 1b. The reader should check that these give rise to resonances whenever the energy of the real and virtual states are equal.

For electrons of energy ~ 10 keV and magnetic field strength $\sim 10^{12}$ gauss we may use the non-relativistic and weak field limit, viz. $p^2 \ll m^2$, $\epsilon^2 \ll m^2$, without introducing too much error. We can then make use of the approximation $(1 + x)^{1/2} \approx 1 + \frac{1}{2}x$ and the equations (1),(2),(4),(5) to show that

$$\alpha_j = 2m[2\omega(m - p'\cos\theta) - \omega^2\cos^2\theta - 2j\epsilon^2]^{-1} \quad (6a)$$
$$\alpha'_j = 2m[2(1 - j)\epsilon^2 - \omega^2\cos^2\theta - 2\omega(m - p\cos\theta)]^{-1} \quad (6b)$$

Magnetic free-free resonances occur at frequencies during which either α'_j or α_j are singular. Near the fundamental cyclotron frequency ω_c, the singularities occur at α_1 and α'_0, when ω satisfies

$$\omega^2\cos^2\theta - 2(m - p'\cos\theta)\omega + 2\epsilon^2 = 0 \quad (7a)$$

or
$$\omega^2\cos^2\theta + 2(m - p\cos\theta)\omega - 2\epsilon^2 = 0 \quad (7b)$$

For a given incident momentum p this gives in general two resonance frequencies. The positive root of (7b) gives directly one of them. The second one, from (7a), is more difficult to locate, because the final electron momentum p' is intricately related to p and ω via the energy conservation (3). In fact, in the present limit p^2, $\varepsilon^2 \ll m^2$ we can rewrite (3) as

$$p'^2 = p^2 + 2\varepsilon^2 - 2m\omega \tag{8}$$

Eliminating p' from (7a) and (8), we obtain a quartic in ω:
$\cos^4\theta\omega^4 + 4m\cos^2\theta\omega^3 + 4(m^2 - p^2\cos^2\theta - \varepsilon^2\cos^2\theta)\omega^2 - 8m\varepsilon^2\omega + 4\varepsilon^4 = 0$
The remarkable thing is that it can be factorized as follows:
$[\omega^2\cos^2\theta + 2(m - p\cos\theta)\omega - 2\varepsilon^2][\omega^2\cos^2\theta + 2(m + p\cos\theta)\omega - 2\varepsilon^2] = 0$
The expression inside the left parantheses is no different from (7b). The one on the right, viz.

$$\omega^2\cos^2\theta + 2(m + p\cos\theta)\omega - 2\varepsilon^2 = 0 \tag{9}$$

has a positive root which gives the second resonance frequency, $\omega = \omega_{R2}$.

In the absence of longitudinal motion (p=0) it is evident that $\omega_{R1} = \omega_{R2} = \omega_c$ (the discrepancy due to the term $\omega^2\cos^2\theta$ is not too large within our limit of interest). In real situations, however, p may be finite by virtue of (i) bulk motion of the accreted plasma along the magnetic field; and (ii) temperature.

We would first consider the effects of (i). The introduction of a definite momentum p w.r.t. the observer's rest frame leads not only to a shift in line frequency, but also a 'line splitting' effect, viz. $\omega_{R1} \neq \omega_{R2} \neq \omega_c$. To examine this further, we note that the energy conservation relation (8) gives, for definite values of p and ω, two solutions of p' corresponding to forward and backward scattering. At $\omega = \omega_{R1}$, the common resonance frequency of α_1 and α_o', one uses equations (7) through (9) to show that

$$p' = \pm(p - \omega\cos\theta)$$

and that α_1 resonates for both solutions, whereas α_o' resonates only for the case of forward scattering: $p' = p - \omega\cos\theta$ (during which the electron-ion momentum transfer is zero). In any case it is clear that the bremsstrahlung cross-section will resonate at $\omega = \omega_{R1}$ for both modes of scattering, since α_1 remains singular always.

At $\omega = \omega_{R2}$, however, only α_o' can be singular, and this occurs only for the case of backward scattering. Once again, making use of (7) to (9), one obtains at $\omega = \omega_{R2}$

$$p' = -(p + \omega\cos\theta)$$

i.e. the ion has clearly played a major role in the conservation of momentum.

We now take a specific example p = -140 keV/c (i.e. an electron with kinetic energy $\sim$10 keV due to bulk motion down the field line) and $H = 0.1\, H_{cr} = 4.4 \times 10^{12}$ gauss. We calculate the resonances ω_{R1}, ω_{R2} together with the corresponding values of momentum transfer between the electron and the ion, $\delta p = p - p' - \omega\cos\theta$, for various angles of photon propagation. The results are listed in table 1.

θ	ω_{R1}	δp_1 (2 values)		ω_{R2}	δp_2
0	38.85	0.0	−357.7	64.65	−254.2
π/6	40.24	0.0	−349.7	63.07	−260.2
π/3	44.42	0.0	−324.4	58.15	−273.1

Table 1 Resonant frequencies of free-free transitions in a strong magnetic field are given, together with the corresponding electron-ion momentum transfer δp, for various angles θ of photon propagation. Here ω is given in units of keV, and δp in keV/c. The cyclotron frequency is ω_c = 51 keV.

It is clear that the $\omega = \omega_{R1}$ resonance occurs in both forward and backward scattering, whereas the $\omega = \omega_{R2}$ resonance occurs in backward scattering only.

Two points become immediately obvious: (i) δp can only assume large values at $\omega=\omega_{R2}$, and increases in magnitude as θ, the pitch angle of photon propagation, approaches π/2; (ii) the values of ω_{R1} and ω_{R2} themselves vary appreciably with θ.

Thus, for a given line-of sight direction, the observer sees a 'line splitting' effect. The strengths of the lines depend on the number of scatterings with momentum transfer δp satisfying the resonance condition. As θ→π/2, an unrealistically large value of δp is required for any lines to be observable at $\omega=\omega_{R2}$.

As mentioned earlier, no detailed model of the plasma MHD situation in the neutron star polar cap exist. The overall picture, however, must be such that in any steady state, the energy loss by electron radiation is replenished by the energy release as a result of proton accretion. It is therefore plausible to assume that the protons have a slightly higher temperature than the electrons, and this allows a constant flow of energy and momenta from the former to the latter. In particular there may exist a mean and *finite* momentum transfer <δp> per e-p collision. This in turn will determine a mean pitch angle <θ> for resonant emission, if we assume a well-defined bulk momentum p in the plasma.

Moreover, photons emitted at different angles have different resonant frequencies. If compton scattering is important, then resonant photons observed from a given line-of-sight direction may have originated from a range of values of θ. This will significantly broaden the lines. A detailed treatment is beyond the scope of the present paper.

The effect of plasma temperature is again broadening of lines. For the line splitting effect to be observable, the broadening must be small compared with the difference in frequency $\omega_{R2} - \omega_{R1}$. For a given temperature there is a momentum distribution about the mean (bulk) momentum, of width $\Delta p \sim (2mkT)^{1/2}$. If the electron temperature is high enough that the kinetic energy of random motion approaches that of bulk motion, one tail of the Maxwellian distribution approaches p = 0, which implies that the tails of the resonances ω_{R1}, ω_{R2} will touch at the cyclotron frequency $\omega=\omega_c$. This may be taken as a criterion for the limit of line resolvability. If the electron temperature goes much higher, the lines will be significantly broadened, and are no longer distinguishable from each other.

More detailed theoretical and experimental work is necessary to determine the observability of such a 'line splitting' effect.

I am grateful to Professor T.W.B. Kibble of Imperial College London for his invaluable advice and insight in quantum electrodynamics. The University of Calgary is acknowledged for the award of a research fellowship to work with Professor D. Venkatesan (DV). This work is also supported by NSERC (Ottawa) grant A-1565 awarded to DV.

REFERENCES

Nagel, W., 1980, Astrophys. J., 236, 904.

Trümper, J., Pietsch, W., Reppin, C., Sacco, B., Kendziorra, E., Staubert, R., 1977, Ann. N.Y. Acad. Sci., 302, 538.

THE PRODUCTION OF SPINLESS HADRON PAIRS VIA VIRTUAL PHOTON EXCHANGE IN UNIFORM MAGNETIC FIELDS

D. White
Department of Physics, Roosevelt University
430 S. Michigan Ave., Chicago, Illinois 60605

ABSTRACT

The method involved in bridging the gap between production of lepton pairs (as per the Källén virtual photon formalism) and production of spinless hadron pairs is outlined. Some results associated with transitions to the ground state are displayed.

I. INTRODUCTION

Inasmuch as copious lepton pair production rates[1] via virtual synchrotron radiation (VSR) (although small compared to ordinary synchrotron radiation rates) are present for magnetic field strengths $H \gtrsim 0.04 \, H_o$, where $H_o \simeq 4 \times 10^{13}$ gauss (the quantum mechanical critical field strength), and inasmuch as $H \gtrsim 0.10 \, H_o$ is representative of pulsar conditions, an investigation of the characteristics of hadronic emission via VSR in intense magnetic fields has particular interest for pulsar physics research. The above-mentioned investigation bears special interest for astrophysics in light of the fact that neutrino emission frequently accompanies hadronic decay.

Starting from the framework as developed, for example, by Källén[2] regarding production of lepton pairs via virtual photon exchange in electromagnetic fields, we will outline below the steps taken in reaching formulas descriptive of hadronic pair production via virtual photon exchange in intense uniform magnetic fields. We shall find, as in the case of lepton pair production (including $\mu^+\mu^-$ and $\tau^+\tau^-$)[3], relative to ordinary synchrotron emission rates the hadronic pair production rates grow logarithmically with the energy of the electron initially undergoing VSR.

II. CONNECTIONS BETWEEN THE FORMAL THEORY OF e^+e^- PAIR PRODUCTION (KÄLLÉN FORMALISM) AND THE PRODUCTION OF HEAVY LEPTONS

In ordinary synchrotron radiation (SR) in which real photons of four-wave vector k are emitted by virtue of the electron-field interaction characterized by the four-current J_R, the formal amplitude descriptive of the rate of photon production is essentially the following[4]:

$$W_\gamma = -\left(\frac{1}{2\pi}\right)^3 \int d^4k \, |J_R|^2 \, \delta(k^2), \qquad (2.1)$$

where the factor $\delta(k^2)$ is a "form factor" of sorts for the photon in that it restricts the real photon to the light cone.

In e^+e^- production the amplitude descriptive of the rate of pair production in which the current, J_V, is associated with the mediation of a virtual (massive) photon is[5]

$$W_p = -\left(\frac{\alpha}{3\pi}\right)\left(\frac{1}{2\pi}\right)^3 \int d^4k \, |J_V|^2 \, (1/k^2) P(k,m), \qquad (2.2)$$

where α is the fine structure constant $\cong (1/137)$ and $P(k,m)$ is a form factor for the virtual photon analogous to $\delta(k^2)$ in Eq. (2.1) in that $P(k,m) = [1-(2m^2/k^2)][1+(4m^2/k^2)]^{1/2}$ restricts k^2 to be $-4m^2$, where $m \equiv$ electron mass.

In calculating an e^+e^- production rate associated with an electron moving transversely to a uniform magnetic field one must incorporate into J_V the initial and final state electron wave functions along with the virtual photon mediating function (e^{ikr}) with k restricted as mentioned and perform the integral represented by Eq. (2.2). Such calculations have been performed[6] for a large number of cases in which the final electron state is the ground state, represented by principal quantum number $n_f = 0$. Results gained from these calculations show pair production rates, $\lambda_{no}^{(e^+e^-)}$, where n = initial state of the electron, to grow logarithmically with initial electron energy relative to the corresponding SR rate, λ_{no}, as shown below:

$$\lambda_{no}^{(e^+e^-)} \cong (\alpha/3\pi) \ln(E_n/2mc^2) \lambda_{no}, \qquad (2.3)$$

where E_n = initial electron energy $\gg mc^2$. Furthermore, because of the nature of SR matrix elements in general and P(k,m) in particular it is probably true that for any $n \to n_f$ transition taking place via VSR in which $E_n \gg 2mc^2$,

$$\lambda_{nn_f} \cong (\alpha/3\pi)\ln(E_n/2mc^2)\,\lambda_{nn_f}. \quad (2.4)$$

The onset of significant VSR rates in ground state termination is $H \sim 0.04\,H_o$, and it is probably true that the onset of significant VSR rates generally occurs at such field strengths.

Because of lepton universality in Q.E.D., the rates for muon or tauon pair production may be gained by straightforward substitution of muon mass (m_μ) or tauon mass (m_τ) for the electron mass in Eq. (2.2). Heavy lepton pair production calculations for $n \to 0$ transitions have also been performed[7] with results completely analogous to Eq. (2.3), viz.,

$$\lambda_{no}^{(\mu^+\mu^-)} \cong (\alpha/3\pi)\ln(E_n/2m_\mu c^2)\,\lambda_{no}$$
$$\text{for } E_n \gg 2m_\mu c^2, \quad (2.5)$$

and
$$\lambda_{no}^{(\tau^+\tau^-)} \cong (\alpha/3\pi)\ln(E_n/2m_\tau c^2)\,\lambda_{no}$$
$$\text{for } E_n \gg 2m_\tau c^2. \quad (2.6)$$

III. BRIDGING THE GAP FROM HEAVY LEPTONS TO SPINLESS HADRONS

When considering the production of hadron pairs, we must take into account two additional facets peculiar to $\pi\pi$ and KK systems (besides mass substitution in Eq. (2.2)), viz., spin-one resonances (at the ρ-mass and ϕ-mass) and the fact that the π's and K's are spinless. A uniform magnetic field is unique in that it provides for a system in which only energy and the z-components ($\vec{H} \parallel$ z-axis) of angular and linear momentum (ℓ_z and p_z) need be conserved. The ℓ_z conservation criterion means that background and resonant contributions to the $\pi\pi$ or KK pair production amplitude can be easily worked in to the Q.E.D. framework in which a virtual photon is involved: The background (which for $\pi^o\pi^o$ production only is the entire contribution) is provided by coupling the electron-magnetic field vertex to the hadron pair vertex via a virtual photon in the $m_\ell = 0$ helicity state. Mathematically, the above corresponds to considering only the non-spin-flip matrix elements in Eq. (2.2). The resonant contribution for the amplitude, present in $\pi^+\pi^-$, $K^0 K^0$, and $K^+ K^-$ production, is included by multiplying the proper Lorentz amplitude function associated with either the ρ^0 (for $\pi^+\pi^-$) or the ϕ^0 (for $K^0 K^0$ and $K^+ K^-$). Interference terms involving spin-flip and non-spin-flip matrix elements are handled appropriately.[8]

In the asymptotic limit of $E_n \to \infty$ and $n \to \infty$ we find:

$$\lambda_{no}^{(\pi^0\pi^0)} \cong (1/2)(\alpha/3\pi)\ln(E_n/2m_{\pi^0}c^2) \tag{3.1}$$

and $\lambda_{no}^{(\pi^+\pi^-)} \cong (1/2)(\alpha/3\pi)[\ln(E_n/2m_{\pi^+}c^2)+(3/5)]. \tag{3.2}$

Assuming similar behavior for $K^0 K^0$ and $K^+ K^-$ production we see that the asymptotic results are in complete agreement with the QCD prediction that for pair production in general in the asymptotically high energy regime that the rate of production via virtual photons of all hadron pairs below the charm threshold approximately equals twice the rate of muon pair production at the same energy. In VSR the factor (1/2) stems from the fact that only half the matrix elements (non-spin-flip) effectively contribute relative to muon production for any given type of hadronic production. Whether any relationship between straightforward QED calculations in the non-asymptotic limits (high energy and high n) and gluon correction formulae (which become important in non-asymptotic limits) in QCD remains to be seen.

REFERENCES

1. D. White, Phys. Rev. D24, 526 (1981).
2. G. Källén, *Quantum Electrodynamics* (Springer, N.Y., 1972), pp. 107-112.
3. D. White, Phys. Rev. D26, 2924 (1982).
4. See Ref. 1, section IB.
5. See Ref. 1, section IB.
6. Ref. 1.
7. Ref. 3.
8. See D. White, Phys. Rev. D26, 3009 (1982).

AIP Conference Proceedings

		L.C. Number	ISBN
No.1	Feedback and Dynamic Control of Plasmas	70-141596	0-88318-100-2
No.2	Particles and Fields - 1971 (Rochester)	71-184662	0-88318-101-0
No.3	Thermal Expansion - 1971 (Corning)	72-76970	0-88318-102-9
No.4	Superconductivity in d-and f-Band Metals (Rochester, 1971)	74-18879	0-88318-103-7
No.5	Magnetism and Magnetic Materials - 1971 (2 parts) (Chicago)	59-2468	0-88318-104-5
No.6	Particle Physics (Irvine, 1971)	72-81239	0-88318-105-3
No.7	Exploring the History of Nuclear Physics	72-81883	0-88318-106-1
No.8	Experimental Meson Spectroscopy - 1972	72-88226	0-88318-107-X
No.9	Cyclotrons - 1972 (Vancouver)	72-92798	0-88318-108-8
No.10	Magnetism and Magnetic Materials - 1972	72-623469	0-88318-109-6
No.11	Transport Phenomena - 1973 (Brown University Conference)	73-80682	0-88318-110-X
No.12	Experiments on High Energy Particle Collisions - 1973 (Vanderbilt Conference)	73-81705	0-88318-111-8
No.13	π-π Scattering - 1973 (Tallahassee Conference)	73-81704	0-88318-112-6
No.14	Particles and Fields - 1973 (APS/DPF Berkeley)	73-91923	0-88318-113-4
No.15	High Energy Collisions - 1973 (Stony Brook)	73-92324	0-88318-114-2
No.16	Causality and Physical Theories (Wayne State University, 1973)	73-93420	0-88318-115-0
No.17	Thermal Expansion - 1973 (lake of the Ozarks)	73-94415	0-88318-116-9
No.18	Magnetism and Magnetic Materials - 1973 (2 parts) (Boston)	59-2468	0-88318-117-7
No.19	Physics and the Energy Problem - 1974 (APS Chicago)	73-94416	0-88318-118-5
No.20	Tetrahedrally Bonded Amorphous Semiconductors (Yorktown Heights, 1974)	74-80145	0-88318-119-3
No.21	Experimental Meson Spectroscopy - 1974 (Boston)	74-82628	0-88318-120-7
No.22	Neutrinos - 1974 (Philadelphia)	74-82413	0-88318-121-5
No.23	Particles and Fields - 1974 (APS/DPF Williamsburg)	74-27575	0-88318-122-3
No.24	Magnetism and Magnetic Materials - 1974 (20th Annual Conference, San Francisco)	75-2647	0-88318-123-1
No.25	Efficient Use of Energy (The APS Studies on the Technical Aspects of the More Efficient Use of Energy)	75-18227	0-88318-124-X

No.	Title		
No. 26	High-Energy Physics and Nuclear Structure - 1975 (Santa Fe and Los Alamos)	75-26411	0-88318-125-8
No. 27	Topics in Statistical Mechanics and Biophysics: A Memorial to Julius L. Jackson (Wayne State University, 1975)	75-36309	0-88318-126-6
No. 28	Physics and Our World: A Symposium in Honor of Victor F. Weisskopf (M.I.T., 1974)	76-7207	0-88318-127-4
No. 29	Magnetism and Magnetic Materials - 1975 (21st Annual Conference, Philadelphia)	76-10931	0-88318-128-2
No. 30	Particle Searches and Discoveries - 1976 (Vanderbilt Conference)	76-19949	0-88318-129-0
No. 31	Structure and Excitations of Amorphous Solids (Williamsburg, VA., 1976)	76-22279	0-88318-130-4
No. 32	Materials Technology - 1976 (APS New York Meeting)	76-27967	0-88318-131-2
No. 33	Meson-Nuclear Physics - 1976 (Carnegie-Mellon Conference)	76-26811	0-88318-132-0
No. 34	Magnetism and Magnetic Materials - 1976 (Joint MMM-Intermag Conference, Pittsburgh)	76-47106	0-88318-133-9
No. 35	High Energy Physics with Polarized Beams and Targets (Argonne, 1976)	76-50181	0-88318-134-7
No. 36	Momentum Wave Functions - 1976 (Indiana University)	77-82145	0-88318-135-5
No. 37	Weak Interaction Physics - 1977 (Indiana University)	77-83344	0-88318-136-3
No. 38	Workshop on New Directions in Mossbauer Spectroscopy (Argonne, 1977)	77-90635	0-88318-137-1
No. 39	Physics Careers, Employment and Education (Penn State, 1977)	77-94053	0-88318-138-X
No. 40	Electrical Transport and Optical Properties of Inhomogeneous Media (Ohio State University, 1977)	78-54319	0-88318-139-8
No. 41	Nucleon-Nucleon Interactions - 1977 (Vancouver)	78-54249	0-88318-140-1
No. 42	Higher Energy Polarized Proton Beams (Ann Arbor, 1977)	78-55682	0-88318-141-X
No. 43	Particles and Fields - 1977 (APS/DPF, Argonne)	78-55683	0-88318-142-8
No. 44	Future Trends in Superconductive Electronics (Charlottesville, 1978)	77-9240	0-88318-143-6
No. 45	New Results in High Energy Physics - 1978 (Vanderbilt Conference)	78-67196	0-88318-144-4
No. 46	Topics in Nonlinear Dynamics (La Jolla Institute)	78-057870	0-88318-145-2
No. 47	Clustering Aspects of Nuclear Structure and Nuclear Reactions (Winnepeg, 1978)	78-64942	0-88318-146-0
No. 48	Current Trends in the Theory of Fields (Tallahassee, 1978)	78-72948	0-88318-147-9
No. 49	Cosmic Rays and Particle Physics - 1978 (Bartol Conference)	79-50489	0-88318-148-7

No.	Title		
No. 50	Laser-Solid Interactions and Laser Processing - 1978 (Boston)	79-51564	0-88318-149-5
No. 51	High Energy Physics with Polarized Beams and Polarized Targets (Argonne, 1978)	79-64565	0-88318-150-9
No. 52	Long-Distance Neutrino Detection - 1978 (C.L. Cowan Memorial Symposium)	79-52078	0-88318-151-7
No. 53	Modulated Structures - 1979 (Kailua Kona, Hawaii)	79-53846	0-88318-152-5
No. 54	Meson-Nuclear Physics - 1979 (Houston)	79-53978	0-88318-153-3
No. 55	Quantum Chromodynamics (La Jolla, 1978)	79-54969	0-88318-154-1
No. 56	Particle Acceleration Mechanisms in Astrophysics (La Jolla, 1979)	79-55844	0-88318-155-X
No. 57	Nonlinear Dynamics and the Beam-Beam Interaction (Brookhaven, 1979)	79-57341	0-88318-156-8
No. 58	Inhomogeneous Superconductors - 1979 (Berkeley Springs, W.V.)	79-57620	0-88318-157-6
No. 59	Particles and Fields - 1979 (APS/DPF Montreal)	80-66631	0-88318-158-4
No. 60	History of the ZGS (Argonne, 1979)	80-67694	0-88318-159-2
No. 61	Aspects of the Kinetics and Dynamics of Surface Reactions (La Jolla Institute, 1979)	80-68004	0-88318-160-6
No. 62	High Energy e^+e^- Interactions (Vanderbilt, 1980)	80-53377	0-88318-161-4
No. 63	Supernovae Spectra (La Jolla, 1980)	80-70019	0-88318-162-2
No. 64	Laboratory EXAFS Facilities - 1980 (Univ. of Washington)	80-70579	0-88318-163-0
No. 65	Optics in Four Dimensions - 1980 (ICO, Ensenada)	80-70771	0-88318-164-9
No. 66	Physics in the Automotive Industry - 1980 (APS/AAPT Topical Conference)	80-70987	0-88318-165-7
No. 67	Experimental Meson Spectroscopy - 1980 (Sixth International Conference, Brookhaven)	80-71123	0-88318-166-5
No. 68	High Energy Physics - 1980 (XX International Conference, Madison)	81-65032	0-88318-167-3
No. 69	Polarization Phenomena in Nuclear Physics - 1980 (Fifth International Symposium, Santa Fe)	81-65107	0-88318-168-1
No. 70	Chemistry and Physics of Coal Utilization - 1980 (APS, Morgantown)	81-65106	0-88318-169-X
No. 71	Group Theory and its Applications in Physics - 1980 (Latin American School of Physics, Mexico City)	81-66132	0-88318-170-3
No. 72	Weak Interactions as a Probe of Unification (Virginia Polytechnic Institute - 1980)	81-67184	0-88318-171-1
No. 73	Tetrahedrally Bonded Amorphous Semiconductors (Carefree, Arizona, 1981)	81-67419	0-88318-172-X
No. 74	Perturbative Quantum Chromodynamics (Tallahassee, 1981)	81-70372	0-88318-173-8

No.	Title	LCCN	ISBN
No. 75	Low Energy X-ray Diagnostics-1981 (Monterey)	81-69841	0-88318-174-6
No. 76	Nonlinear Properties of Internal Waves (La Jolla Institute, 1981)	81-71062	0-88318-175-4
No. 77	Gamma Ray Transients and Related Astrophysical Phenomena (La Jolla Institute, 1981)	81-71543	0-88318-176-2
No. 78	Shock Waves in Condensed Matter - 1981 (Menlo Park)	82-70014	0-88318-177-0
No. 79	Pion Production and Absorption in Nuclei - 1981 (Indiana University Cyclotron Facility)	82-70678	0-88318-178-9
No. 80	Polarized Proton Ion Sources (Ann Arbor, 1981)	82-71025	0-88318-179-7
No. 81	Particles and Fields - 1981: Testing the Standard Model (APS/DPF, Santa Cruz)	82-71156	0-88318-180-0
No. 82	Interpretation of Climate and Photochemical Models, Ozone and Temperature Measurements (La Jolla Institute, 1981)	82-071345	0-88318-181-9
No. 83	The Galactic Center (Cal. Inst. of Tech., 1982)	82-071635	0-88318-182-7
No. 84	Physics in the Steel Industry (APS.AISI, Lehigh University, 1981)	82-072033	0-88318-183-5
No. 85	Proton-Antiproton Collider Physics - 1981 (Madison, Wisconsin)	82-072141	0-88318-184-3
No. 86	Momentum Wave Functions - 1982 (Adelaide, Australia)	82-072375	0-88318-185-1
No. 87	Physics of High Energy Particle Accelerators (Fermilab Summer School, 1981)	82-072421	0-88318-186-X
No. 88	Mathematical Methods in Hydrodynamics and Integrability in Dynamical Systems (La Jolla Institute, 1981)	82-072462	0-88318-187-8
No. 89	Neutron Scattering - 1981 (Argonne National Laboratory)	82-073094	0-88318-188-6
No. 90	Laser Techniques for Extreme Ultraviolt Spectroscopy (Boulder, 1982)	82-073205	0-88318-189-4
No. 91	Laser Acceleration of Particles (Los Alamos, 1982)	82-073361	0-88318-190-8
No. 92	The State of Particle Accelerators and High Energy Physics (Fermilab, 1981)	82-073861	0-88318-191-6
No. 93	Novel Results in Particle Physics (Vanderbilt, 1982)	82-73954	0-88318-192-4
No. 94	X-Ray and Atomic Inner-Shell Physics-1982 (International Conference, U. of Oregon)	82-74075	0-88318-193-2
No. 95	High Energy Spin Physics - 1982 (Brookhaven National Laboratory)	83-70154	0-88318-194-0
No. 96	Science Underground (Los Alamos, 1982)	83-70377	0-88318-195-9

No. 97	The Interaction Between Medium Energy Nucleons in Nuclei-1982 (Indiana University)	83-70649	0-88318-196-7
No. 98	Particles and Fields - 1982 (APS/DPF University of Maryland)	83-70807	0-88318-197-5
No. 99	Neutrino Mass and Gauge Structure of Weak Interactions (Telemark, 1982)	83-71072	0-88318-198-3
No. 100	Excimer Lasers - 1983 (OSA, Lake Tahoe, Nevada)	83-71437	0-88318-199-1
No. 101	Positron-Electron Pairs in Astrophysics (Goddard Space Flight Center, 1983)	83-71926	0-88318-200-9

RAYMOND H. FOGLER LIBRARY
DATE DUE

BOOKS ARE SUBJECT TO
RECALL AFTER TWO WEEKS

NOV 20